Classical and Modern Approaches in the Theory of Mechanisms

Classical and Modern Approaches in the Theory of Mechanisms

Nicolae Pandrea, Dinel Popa and Nicolae-Doru Stănescu
University of Piteşti, Argeş, Romania

Registered Office(s)
John Wiley & Sons, Inc., 111 River Street, Hoboken, NJ 07030, USA
John Wiley & Sons Ltd, The Atrium, Southern Gate, Chichester, West Sussex, PO19 8SQ, UK

Editorial Office
The Atrium, Southern Gate, Chichester, West Sussex, PO19 8SQ, UK

For details of our global editorial offices, customer services, and more information about Wiley products visit us at www.wiley.com.

Wiley also publishes its books in a variety of electronic formats and by print-on-demand. Some content that appears in standard print versions of this book may not be available in other formats.

Library of Congress Cataloging-in-Publication Data

Names: Pandrea, Nicolae, author. | Popa, Dinel, 1958- author. | Stănescu, Nicolae-Doru, author.
Title: Classical and modern approaches in the theory of mechanisms / Professor Nicolae Pandrea, University of Pitești, Argeș, Romania, Prof Dinel Popa, University of Pitești, Argeș, Romania, Nicolae-Doru Stănescu, University of Pitești, Argeș, Romania.
Description: Chichester, West Sussex, UK : John Wiley & Sons, Inc., [2017] | Includes bibliographical references and index.
Identifiers: LCCN 2016045899 (print) | LCCN 2016059812 (ebook) | ISBN 9781119221616 (cloth) | ISBN 9781119221722 (pdf) | ISBN 9781119221760 (epub)
Subjects: LCSH: Mechanical movements. | Machine parts.
Classification: LCC TJ181 .P35 2017 (print) | LCC TJ181 (ebook) | DDC 621.8/11—dc23
LC record available at https://lccn.loc.gov/2016045899

Hardback: 9781119221616

Cover design by Wiley
Cover image: © Thorsten Gast/EyeEm/Gettyimages

Set in 10/12pt Warnock by SPi Global, Chennai, India

Printed in Singapore by C.O.S. Printers Pte Ltd

10 9 8 7 6 5 4 3 2 1

Contents

Preface

The study of mechanisms was one of the first applied sciences; its development was closely related to humanity's socio-economic situation; as the state of mankind improved, there was a great diversification of the problems dealt with using mechanisms.

At first, only very simple mechanisms were examined. Later, mechanisms were categorized and the general characteristics (kinematics, dynamics, synthesis and so on) of each class were determined. The absence of modern calculation tools limited the study of mechanisms to simple problems for which there were solutions in closed form.

Because practical problems and the mechanisms involved are complicated, specialists have to use numerical methods to obtain solutions. We believe that in the coming years new algorithms will be written and new mechanisms will be developed; using them, more complex practical problems will be solved. The theory of mechanisms is closely related to the areas of mechanics, numerical methods, mathematical calculus, algebra and so on.

The book is divided into ten chapters. Chapter 1 is dedicated to the structures of mechanisms and it analyzes the components of mechanisms and the general conditions for the transmission of motion.

In Chapter 2 we examine the positions, velocities, and accelerations of the elements of planar mechanisms with bars. In the kinematic analysis of modular components, we use the classic grapho-analytic and analytic methods. We also propose two new methods: the analytic assisted method, for which we will present the calculation programs written in Turbo Pascal, and the graphical method. We will show how the latter approach can be performed using AutoCAD. The vector approach, in which the CAD software uses graphic entities, leads, in the presence of a programming language (AutoLisp), to an easy transposition of the vector relations. The examples presented in this chapter are solved using all four methods.

Chapter 3 is dedicated to the kinetostatic analysis of planar mechanisms with bars. We firstly present general aspects of the equilibrium of rotors, and then we discuss calculation methods for obtain the reactions in the linkages, with and without friction kinematic joints. As in the previous chapter, we describe the programs and functions used to obtain the numerical results.

Chapter 4 is entitled 'Dynamics of Machines'. We present the dynamic model of the machine, and we determine the differential equations of motion, the machine being driven by forces and torques. The integration of the moving equation is performed for the regime phase in the most general case. At the end of the chapter we study the adjustment of the machine's motion and the dynamics of the multi-mobile mechanisms with application to the Constantinescu dynamic convertor.

Chapter 5, entitled 'Synthesis of Planar Mechanisms with Bars', deals with aspects of the two-dimensional synthesis of planar mechanisms with bars using analytic methods, as well as up-to-date graphical methods with AutoLisp functions in AutoCAD. The applications highlight the advantages of the latter method.

Mechanisms with cams are described in Chapter 6. The analysis of the followers' displacements, velocities and accelerations, the dynamic analysis, and the expressions of the fundamental laws of the followers' displacements are all accompanied by numerical examples, which are solved either analytically or using graphical methods. Finally, the problem of the synthesis of the mechanisms with cams in the general case of a curvilinear follower is solved, alongside a presentation of some applications, with different shapes and motions of the follower.

Chapter 7 is dedicated to mechanisms with gears. First of all, we present general aspects of rolling surfaces, reciprocal wrapping surfaces and the fundamental law of toothing. We describe the geometric dimensions for cylindrical gears with right and inclined teeth, conical gears with right teeth, and gears with crossed axes. We pay great attention to the generation of solids that represent the gears, using AutoLisp functions in AutoCAD.

Spatial mechanisms are covered in Chapter 8. We study, from the kinematic point of view, the most-used mechanisms. Great attention is paid to the study of transmission, especially tripod transmission. The chapter ends with the presentation of a way to create such a mechanism with application to the spatial 7R mechanism.

In Chapter 9 we consider industrial robots. We discuss their mechanical systems, as well as general aspects of their command and control systems. The chapter ends with the presentation of a walking mechanism known as 'Tchebychev's horse'.

The last chapter, Chapter 10, is entitled 'Variators of angular velocity with bars'. We present the main kinematic schemata of such mechanisms, which are also known as *variators with impulses*. We discuss the kinematic aspects of such variators, determining the characteristic kinematic parameters and their calculation in examples, also presenting the programs that were used to obtain the numerical values.

Unlike other books, we deal with mechanisms in a most general way, rather than focusing on particular applications. We also introduce new methods to approach the kinematics and kinetostatics of mechanisms. The book also deals with the analysis and synthesis of planar or spatial mechanisms, the applications being created using analytic and graphical methods using AutoLisp functions in CAD software. The proposed algorithms for synthesis and kinematic analysis can be used in the CAD software in a machine tool, allowing users to obtain solids that represent the kinematic elements, cams, and gears, or providing them with a file for the control of mechanisms and robots.

The authors are grateful to Mrs Eng. Ariadna-Carmen Stan for her valuable help in the presentation of this book. The excellent cooperation with the team of John Wiley & Sons is gratefully acknowledged.

The book is addressed to a large audience: to all those interested in designing mechanisms, in fields such as mechanics, physics, and civil and mechanical engineering, people involved in teaching, research, or design, and students.

The book can be also used either as a standalone course for postgraduate students, or as supplemental reading for courses on mechanisms, computational mechanics, analytical mechanics, multibody mechanics, and so on. The prerequisites are courses in elementary algebra and analysis and mechanics.

Piteşti, Argeş, Romania
September, 2016

Nicolae Pandrea
Dinel Popa
Nicolae-Doru Stănescu

About the Companion Website

Don't forget to visit the companion website for this book:

www.wiley.com/go/pandmech17

There you will find valuable material designed to enhance your learning, including:

1) Links to AutoCAD and related programs
2) Animations and AutoLisp
3) Word listings

Scan this QR code to visit the companion website

1

The Structure of Mechanisms

The *structure* of a mechanism is its composition (*elements, kinematic pairs*), its classification, and the conditions in which motion is transmitted in a predeterminate way (the conditions of desmotomy). The variety and the complexity of technical systems gives a requirement for conventional representations of the elements of mechanisms and the linkages between them (kinematic pairs), and the establishment of certain rules for their representation, formation and decomposition.

1.1 Kinematic Elements

The component parts of mechanisms that have motions relative to one another are called the*kinematic elements* of the mechanisms. The elements may consist of one or more component parts. Figure 1.1a,b shows two cranks, the first consisting of one piece, and the second one consisting of more (the body, the cap, the screws and so on). The conventional representation of these elements is given in Figure 1.1c,d.

The *rank j* of an element is the number of kinematic pairs of that element. The conventional representation of the elements as a function of their rank is shown in Table 1.1.

1.2 Kinematic Pairs

The direct and permanent link between two elements is called a *kinematic pair*. The *class* of the kinematic pair represents the number of restrictions (the non-permitted motions) of an element in relative motion with respect to another one. For example, if one considers elements 1 and 2, which are supported by one another (Figure 1.2), then it can be seen that element 2 has only one restriction on its relative displacement with respect to element 1 ($v_z = 0$).

Denoting the number of restrictions by m and the number of the degrees of freedom in relative motion by l, it can be seen, in the case considered, that $m = 1, l = 5$, and the kinematic pair has a class equal to one. In general:

$$m + l = 6. \tag{1.1}$$

The classification of the kinematic pairs is according to the criteria in Table 1.2 (see also Table 1.3).

Classical and Modern Approaches in the Theory of Mechanisms, First Edition.
Nicolae Pandrea, Dinel Popa and Nicolae-Doru Stănescu.
© 2017 John Wiley & Sons Ltd. Published 2017 by John Wiley & Sons Ltd.
Companion Website: www.wiley.com/go/pandmech17

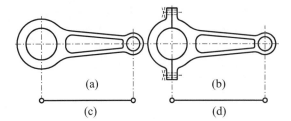

Figure 1.1 Two elements and their conventional representations: (a) single-piece crank and (c) its conventional representation; (b) multi-piece crank and (d) its conventional representation.

(a) (b)

(c) (d)

Table 1.1 Rank of elements.

Rank	Name	Representation
1	Unitary ($j = 1$)	
2	Binary ($j = 2$)	
3	Polynary ($j = 3$)	
4	Polynary ($j = 4$)	

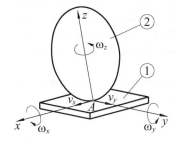

Figure 1.2 Representation of two elements.

1.3 Kinematic Chains

The system of elements jointed by kinematic links is called a *kinematic chain*. Figure 1.3a,b shows two planar kinematic chains, while Figure 1.3c shows a spatial kinematic chain. Kinematic chains are classified by taking into account the rank of the elements, the shape of the chain, and the motion of the elements. This classification, together with some examples, is presented in Table 1.4.

The *degree of freedom* of the kinematic chain is denoted L. Consider a kinematic chain having c elements and c_m kinematic links of class m. In the absence of any common restriction, the number of restrictions introduced by the kinematic links is equal to $\sum_{m=1}^{5} mc_m$.

If the elements are not jointed to one another, they will have a total number of $6e$ degrees of freedom. The degree of freedom L of the kinematic chain is therefore given by:

$$L = 6e - \sum_{m=1}^{5} mc_m. \tag{1.2}$$

The *degree of mobility* of the kinematic chain M is defined as the degree of freedom of a kinematic chain having a fixed element. It has the relation:

Table 1.2 The classification of kinematic pairs.

Nr. crt.	Classification criterion	Name	Condition	Examples
1	Geometric	Superior (Higher pair)	Contact at a point or on a curve (curve–curve; curve–surface; surface–surface; point–surface)	$S–P\ C–P$
		Inferior (Lower pair)	Contact on a surface	S, C, R, H
2	Kinematic	Plane	Motion of the two elements in same or in parallel planes	R, T
		Spatial	The motion of the two elements in different planes	S, C
3	Constructive	Closed	Must be destroyed in order to separate the elements	R
		Open	Separation of the elements without destruction	$S–P, C–P$
4	Structural	Of class 1, 2, 3, 4, or 5	See Table 1.3	
5	Number of axes along which independent motions take place	With one axis	Axis along or around which translational, rotational or helical motion takes place	R, T, H, C
		With two axes	Axes along or around which translational, rotational or helical motion takes place	
		With three axes	Axes along or around which translational, rotational or helical motion takes place	S

$$M = L - 6 \tag{1.3}$$

or, if we denote by $n = e - 1$ the number of the elements considered mobile, we get, for the degree of mobility, the expression

$$M = 6n - \sum_{m=1}^{5} mc_{m.} \tag{1.4}$$

1.4 Mobility of Mechanisms

1.4.1 Definitions

The mechanism is the kinematic chain having one fixed element (the base, the frame) and in which all the elements have predeterminate (*desmotomic*) motions. According to their functional role, the component elements of the mechanism can be:

- *driving (motor, input)*; the elements that receive the motion from outside the mechanism
- *driven (commanded, output)*; the elements whose motion depends on the motion of the driving elements.

Table 1.3 Classes of kinematic pairs.

Class	Structural schema	Notation	Technical name	Restrictions	Conventional representation
1		S–P	Sphere on a plane	$v_z = 0$	
2		C–P	Cylinder on a plane	$\omega_y = 0,\ v_z = 0$	
3		S	Spherical pair	$v_x = 0,\ v_y = 0,$ $v_z = 0$	
4		C	Cylindrical kinematic pair	$v_y = 0,\ v_z = 0,$ $\omega_y = 0,\ \omega_z = 0$	
5		R	Rotational kinematic link (Revolving pair, hinge)	$v_x = 0,\ v_y = 0,$ $v_z = 0,\ \omega_y = 0,$ $\omega_z = 0$	
		T	Translational kinematic link (prismatic pair)	$v_y = 0,\ v_z = 0,$ $\omega_x = 0,\ \omega_y = 0,$ $\omega_z = 0$	
		H	Helical kinematic link (h is the twist of the screw)	$v_x = 0,\ v_y = 0,$ $\omega_x = 0,\ \omega_y = 0,$ $v_z = p\omega_z$ $\left(p = \dfrac{h}{2\pi}\right)$	

Figure 1.3 Kinematic chains: (a) planar kinematic chain with three elements; (b) planar kinematic with six elements; (c) spatial kinematic chain with four elements.

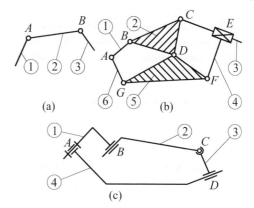

Representations of the mechanisms and the kinematic chains are called the *representation schema*. We define:

- *constructive schemata*, in which the representations of the kinematic links and elements contain some structural details
- *kinematic schemata*, in which the elements and kinematic links have conventional representations
- *structural schemata*, which give the fundamental structural mechanisms of the mechanism.

1.4.2 Mobility Degree of Mechanisms without Common Constraints

The mobility degree of a mechanism is the degree of mobility of the respective kinematic chain; that is,

$$M = 6n - \sum_{m=1}^{5} mc_m, \tag{1.5}$$

where n is the number of the mobile elements of the mechanism.

1.4.3 Mobility Degree of Mechanisms with Common Constraints

If the elements of the mechanism have f *common restrictions*, we say that the mechanism is of family f. For instance, a planar mechanism, where each element has only three degrees of freedom (two translation and one rotation), is of family $f - 3$, because all the elements have three common restrictions (two rotations about the axes contained in the plane and one translation along the axis perpendicular to the plane).

Returning to the general case, if the elements have f restrictions, then the number of the degrees of freedom is equal to $6 - f$ for each element, while a kinematic joint of class m has only $m - f$ restrictions, where $m > f$.

To determine the mobility, 6 has to be replaced with $6 - f$, m with $m - f$, and M with M_f in (1.5). This gives the Dobrovolski formula:

$$M_f = (6 - f) n - \sum_{m=f+1}^{5} (m - f) c_m, \tag{1.6}$$

that is,

$$M_0 = 6n - c_1 - 2c_2 - 3c_3 - 4c_4 - 5c_5, \text{ if } f = 0, \tag{1.7}$$

$$M_1 = 5n - c_2 - 2c_3 - 3c_4 - 4c_5, \text{ if } f = 1, \tag{1.8}$$

Table 1.4 Classification of kinematic chains.

No.	Classification criterion	Name	Condition	Examples	
				Structural schema	Kinematic schema
1	Rank of elements	Simple	If rank of elements $j \leq 2$		
		Complex	If kinematic chain has elements with $j > 2$		
2	Shape of kinematic chain	Open			
		Closed			
3	Motion of elements	Planar	If motion is planar		
		Spatial	If motion is spatial		

$$M_2 = 4n - c_3 - 2c_4 - 3c_5, \text{ if } f = 2, \tag{1.9}$$

$$M_3 = 3n - c_4 - 2c_5, \text{ if } f = 3, \tag{1.10}$$

$$M_4 = 2n - c_5, \text{ if } f = 4. \tag{1.11}$$

As already described, planar mechanisms have family $f = 3$. Their degree of mobility is given by relation (1.10).

1.4.4 Mobility of a Mechanism Written with the Aid of the Number of Loops

To determine the mobility of a mechanism, we take into account the fact that a mechanism with j loops is obtained from a mechanism with $j - 1$ loops, to which we add an open kinematic chain with n_j elements and $c_j = n_j + 1$ kinematic pairs.

In Figure 1.4 we have shown how a mechanism with three loops can be obtained by adding an open kinematic chain with three elements and four kinematic pairs to a mechanism with two loops. For the first loop, taking into account that it has a fixed element, the expression $c_1 = n_1 + 1$ is obtained. This then results in

$$c_1 = n_1 + 1, \ c_2 = n_2 + 1, \ ..., \ c_N = n_N + 1, \tag{1.12}$$

where N is the number of loops.

Denoting by c the total number of kinematic links and by n the number of elements, summing (1.12) gives:

$$c = n + N. \tag{1.13}$$

Developing expression (1.6) gives:

$$\begin{aligned} M_f &= 6n - \sum_{m=f+1}^{5} mc_m - f\left(n - \sum_{m=f+1}^{5} c_m\right) \\ &= 6n - \sum_{m=f+1}^{5} mc_m - f(n - c). \end{aligned} \tag{1.14}$$

Taking into account (1.13) gives:

$$M_f = 6n - \sum_{m=f+1}^{5} mc_m + fN. \tag{1.15}$$

1.4.5 Families of Mechanisms

We defined the family f of a mechanism as the number of the restrictions (*constraints*) common to all elements. If a mechanism is formed with loops of different families, then, conventionally,

Figure 1.4 Obtaining a mechanism with three loops from a mechanism with two loops.

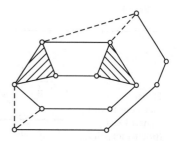

Table 1.5 The determination of the families of the kinematic chains.

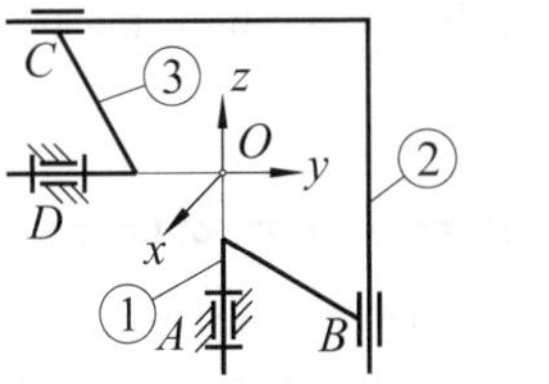

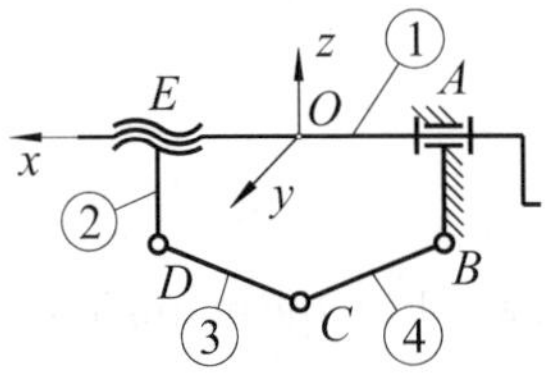

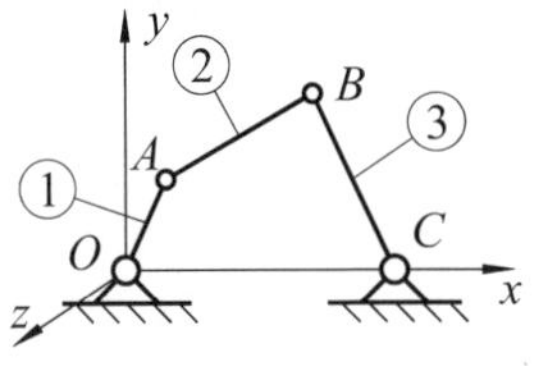

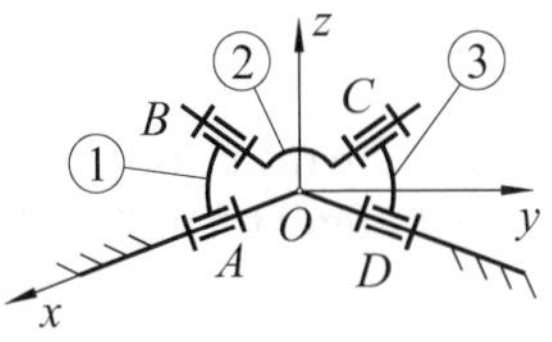

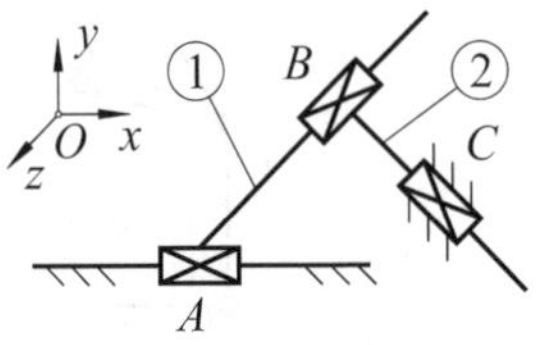

Examples of mechanisms	Analysis of the common restrictions	Family

Petersburg screw

Element	Possible displacements					
	ω_x	ω_y	ω_z	v_x	v_y	v_z
1				*		
2				*	*	*
3		*				

Common restrictions $\omega_x = 0$ — Family **1**

RHRR

Element	Possible displacements					
	ω_x	ω_y	ω_z	v_x	v_y	v_z
1	*					
2				*		
3	*			*		*
4	*			*		*

Common restrictions $\omega_z = 0$, $v_y = 0$ — Family **2**

Planar mechanisms situated in the *Oxy* plane

Element	Possible displacements					
	ω_x	ω_y	ω_z	v_x	v_y	v_z
1			*			
2			*	*	*	
3			*			

Common restrictions $\omega_x = 0$, $\omega_y = 0$, $v_z = 0$ — Family **3**

Spherical mechanisms (with axes concurrent at same point)

Element	Possible displacements					
	ω_x	ω_y	ω_z	v_x	v_y	v_z
1	*					
2	*	*	*			
3	*	*	*			

Common restrictions $v_x = 0$, $v_y = 0$, $v_z = 0$ — Family **3**

Planar mechanism with translational (prismatic) pairs

Element	Possible displacements					
	ω_x	ω_y	ω_z	v_x	v_y	v_z
1				*		
2				*	*	

Common restrictions $\omega_x = 0$, $\omega_y = 0$, $\omega_z = 0$, $v_z = 0$ — Family **4**

we define the *apparent family* by the relation

$$f = \frac{\sum_{i=1}^{N} f_i}{N}, \tag{1.16}$$

where f_i marks the family of the mechanism i.

For a series of mechanisms, the family may be determined based on the analysis of the common constraints for all the elements in regard to a dextrosum reference system. Table 1.5 shows how the family can be established for five types of simple mechanism by analyzing the common restrictions.

1.4.6 Actuation of Mechanisms

To assure desmotomy, it is necessary that the number of the acted (motor, driving) elements is equal to the degree of mobility M of the mechanism. If the driving elements are defined, the mechanism is called a *driving (motor) mechanism*. The driving elements, in most cases, are linked to a base or frame, but there are also driving elements that are not linked to a base or frame. Figure 1.5 shows different types of driving mechanism and how they are actuated. In Figure 1.5a,b, the driving elements, denoted by 1, are adjacent to the base, while in Figure 1.5c,d the driving elements, denoted by 2, are not adjacent to the frame.

1.4.7 Passive Elements

There are situations in which, due to the geometric particularities of certain mechanisms, some elements can be eliminated, or others added, but the real mobility of the mechanism remains unchanged. So, for instance, in the example in Figure 1.6, in which $OC = AB$, $OA = BC$, and $AD = BE$, the loops $ODEC$, $OABC$ are parallelograms and the element denoted by 4 can be eliminated without changing the mobility of the mechanism. Such an element is called a *passive element*. For the calculation of the mobility, the passive elements must be eliminated.

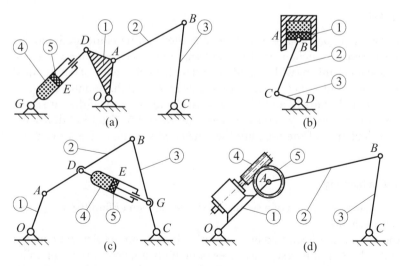

Figure 1.5 Actuation of mechanisms: in (a) and (b) the driving element is adjacent to the base; in (b) and (d) the driving element is not adjacent to the base.

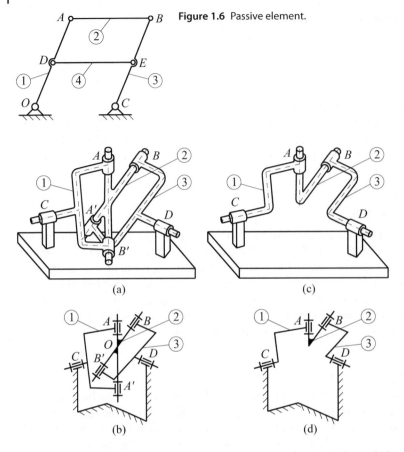

Figure 1.6 Passive element.

Figure 1.7 Passive kinematic links: (a) constructive schema of Cardan linkage; (b) kinematic schema of spherical mechanism; (c) kinematic schema of Cardan linkage; (d) kinematic schema of spherical mechanism without passive linkages.

1.4.8 Passive Kinematic Pairs

In some cases, certain kinematic *pairs* have a passive role: they can be eliminated and the mobility of the mechanism remains unchanged. These links are called *passive kinematic pairs*. Figure 1.7 shows a spherical mechanism (with a Cardan link) of three elements, of family equal to three. The axes being concurrent at O (Figure 1.7b), the elements can have only rotational motions. The rotational kinematic pairs A' and B' are passive kinematic pairs. If they are eliminated, the structural variant in Figure 1.7c and the kinematic variant in Figure 1.7d, respectively, are obtained. For calculation of the mobility the kinematic schema in Figure 1.7d has to be used.

1.4.9 Redundant Degree of Mobility

Figure 1.8a shows the constructive schema of a spatial RSSR mechanism of family $f = 0$, while Figure 1.8b is its kinematic schema.

In this mechanism, the motion is transmitted in a determined way from the element denoted by 1 to the element denoted by 3. In this case, the degree of mobility is given by the expression $M = 6n - 5c_5 - 3c_3$; since $c_3 = 2$ (the pairs at the points O and C) and $c_3 = 2$ (the pairs at the points A and B), this gives $M = 2$.

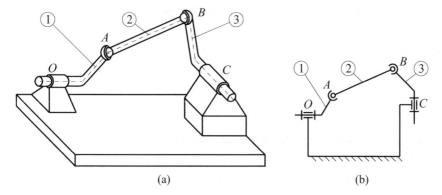

Figure 1.8 Redundant degree of mobility: (a) constructive schema of the RSSR mechanism and (b) kinematic schema of the same mechanism.

Figure 1.9 Multiple kinematic pair (at point *B*).

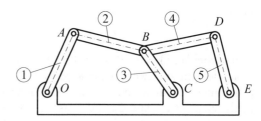

One of the two degrees of mobility is redundant; that is, the rotation of crank 2 around axis *AB*. In the calculation of the real degree of mobility, any redundant degrees of mobility have to be eliminated.

1.4.10 Multiple Kinematic Pairs

Consider the planar mechanism in Figure 1.9. For the calculation of the mobility of this mechanism, one has to take into account that at point *B* there are two fifth-class kinematic pairs. Generally, the number of kinematic pairs is one less than the number of elements concurrent at the same point.

A pair such as that presented in Figure 1.9 is called a *multiple kinematic pair*, and it has to be taken into account when calculating the mobility. In the present case, the number of elements of the mechanism is $n = 5$ and the number of class 5 kinematic pairs is c_5, so the mobility $M = 3 \times 5 - 2 \times 7 = 1$.

1.5 Fundamental Kinematic Chains

By definition, *general kinematic chains* contain kinematic pairs of different classes, while *fundamental kinematic chains* contain only kinematic pairs of the fifth class; that is, R, T, H. For the equivalence of a general kinematic chain with a fundamental kinematic chain the theorem of the equivalence of the kinematic pair developed by Gruebler and Harisberger must be used. This states that a kinematic pair of class k in a kinematic chain of any family can be replaced with a simple open kinematic chain consisting of $(5 - k)$ elements jointed to one another by $(6 - k)$ fundamental kinematic pairs (of class 5).

To demonstrate this theorem, we equate the degree of mobility of the kinematic chain consisting of zero elements and one kinematic pair of class k with the degree of mobility of the

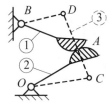

Figure 1.10 The equivalence of a class 4 kinematic pair.

replacing kinematic chain consisting of n elements and c kinematic pairs of class 5. Applying the Dobrovolski formula gives:

$$M_f = (6-f) \cdot 0 - (k-f) \cdot 1 = (6-f)n - (5-f)c.$$

Knowing that for a kinematic chain with one loop $c = n + 1$, this gives:

$$-k + f = 6n + f(c-n) - 5(n+1)$$

or

$$n = 5 - k, \; c = 6 - k. \tag{1.17}$$

In particular, a kinematic pair of the fourth class is equivalent to a kinematic chain consisting of an element and two kinematic pairs of the fifth class. Such a situation is presented by the planar mechanism in Figure 1.10, in which the fourth-class kinematic pair at point A is replaced by the element CD and the fifth-class kinematic pairs at the points C and D, finally obtaining the mechanism $OCDB$; from the structural point of view, this mechanism is equivalent to the mechanism OAB. If the kinematic equivalence is also required, then the points D and C are situated at the curvature centres of the contact curves.

For the fundamental kinematic chains, the degree of mobility is given by

$$M_f = 6n - 5c + fN;$$

replacing $c = n + N$ gives:

$$M_f = n - N(5-f). \tag{1.18}$$

In the case of a spatial fundamental chain with one loop and with the degree of mobility $M_f = 1$, $n = 6 - f$ and the fundamental kinematic chains in Table 1.6 are obtained.

The planar fundamental kinematic chains ($f = 3$) have the degree of mobility given by:

$$M = n - 2N. \tag{1.19}$$

Table 1.6 Fundamental kinematic chains ($M_f = 1, N = 1$).

Family	Number of elements	Number of fundamental kinematic pairs	Examples
0	6	7	7R (Figure 1.11)
1	5	6	6R (the mechanisms of Franke – Figure 1.12 Bricard – Figure 1.13
2	4	5	5R (the Golberg mechanism – Figure 1.14)
3	3	4	4R (the Bennett mechanism – Figure 1.15)

Figure 1.11 The 7R mechanism.

Figure 1.12 The Franke mechanism.

Figure 1.13 The Bricard mechanism.

Figure 1.14 The Goldberg mechanism.

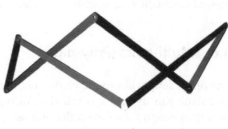

Figure 1.15 The Bennett mechanism.

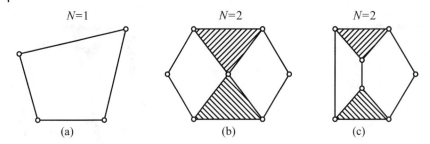

Figure 1.16 Planar fundamental kinematic chains ($M = 1$): (a) fundamental four-bar kinematic chain; (b) fundamental Watt kinematic chain; (c) fundamental Stephenson kinematic chain.

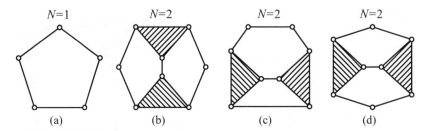

Figure 1.17 Planar fundamental kinematic chains ($M = 2$): (a) fundamental kinematic chain with five elements; (b), (c) and (d) fundamental kinematic chains with seven elements.

Planar fundamental mechanisms with mobility $M = 1$ have the numerical solutions:

$$n = 2N + 1, \ e = n + 1 = 2N + 2. \ c = 3N + 1. \tag{1.20}$$

This gives for $N = 1$ the fundamental four-bar (quadrilateral) kinematic chain (Figure 1.16a; $c = 4$, $e = 4$) and for $N = 2$ the fundamental kinematic chains of Watt (Figure 1.16b) and Stephenson (Figure 1.16c).

For $M = 2$:

$$n = 2N + 2, \ e = 2n + 3, \ c = 3N + 2, \tag{1.21}$$

resulting in a fundamental kinematic chain with five elements (Figure 1.17a) and three fundamental kinematic chains with seven elements (Figure 1.17b–d).

1.6 Multi-pairs (Poly-pairs)

The open, simple kinematic chain, that is equivalent, from the point of view of mobility, to a kinematic pair of class k is called a *multi-pair* or a *poly-pair* of class k. Multi-pairs are used in system design for constructive and technological reasons, because, on the one hand, not all independent motions can be realized by direct (simple or complex) contacts, and, on the other hand, because of the tendency to replace superior kinematic pairs with kinematic chains formed from inferior kinematic pairs (lower pairs) of increased mobility. Since $N = 1$, by equating the mobilities $M_f = -\left(k - f\right) - 6n - mc_m + f$ or

$$- k = 6n - 5c_5 - 4c_4 - 3c_3 - 2c_2 - c_1. \tag{1.22}$$

Based on this relation, in Table 1.7 we give a few examples of simple superior kinematic pairs replaced by multi-pairs. Important categories of multi-pairs are *couplings* and especially *homokinetic couplings*.

Table 1.7 The equivalence between the simple kinematic links and the multi-links.

Class k	Simple link	n	c_5	c_4	c_3	c_2	Constructive schema	Kinematic schema
					Equivalent multi-link			
1		1	1			1		
		$-1 = 6 \cdot 1 - 5 \cdot 1 - 2 \cdot 1$						
2		1	1		1			
		$-2 = 6 \cdot 1 - 5 \cdot 1 - 3 \cdot 1$						
3		2	3					
		$-3 = 6 \cdot 2 - 5 \cdot 3$						
4		1	2					
		$-4 = 6 \cdot 1 - 5 \cdot 2$						

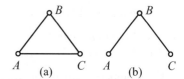

Figure 1.18 Obtaining a dyad from a fundamental kinematic chain with three elements: (a) fundamental kinematic chain with mobility; (b) the dyad.

1.7 Modular Groups

The modular groups are obtained from the fundamental kinematic chains with mobility M by eliminating an element. For $M = 0$ this gives the *passive* (kinematic) modular groups, while for $M = 1$ it gives the *driving* (motor) modular groups. For planar mechanisms, the numerical relations used to obtain the modular groups are: $n = M + 2N$ and $c = M + 3N$.

The important passive modular groups are:

1) *Dyad* ($N = 1$, $n = 2$, $c = 3$; Figure 1.18b), which is obtained from the fundamental zero-mobile kinematic chain with three elements (see Figure 1.18a). The dyad has five aspects: RRR, RTR, RRT, TRT, RTT (Figure 1.19).
2) *Triad* ($N = 2$, $c = 6$; Figure 1.20b), which is obtained from the zero-mobile kinematic chain (Figure 1.20a) by eliminating an element of rank $j = 3$.

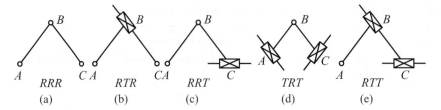

Figure 1.19 Aspects of the dyad: (a) the RRR dyad; (b) the RTR dyad; (c) the RRT dyad; (d) the TRT dyad; (e) the RTT dyad.

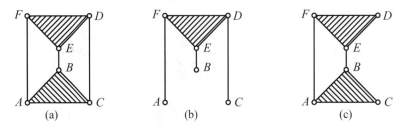

Figure 1.20 Obtaining a triad and tetrad from a fundamental zero-mobile chain with five elements: (a) fundamental kinematic chain with five elements and mobility; (b) triad; (c) tetrad.

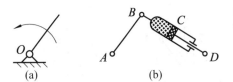

Figure 1.21 Driving modular groups ($M = 1$): (a) driving element; (b) driving dyad.

3) *Tetrad* ($N = 2$, $c = 6$; Figure 1.20c), which is obtained from the same kinematic chain (Figure 1.20a) by eliminating an element of rank $j = 2$.

The important driving (motor) modular groups are:

1) the *driving element*: $M = 1$, $N = 0$, $n = 1$, $c = 1$; see Figure 1.21a
2) the *driving dyad*: $M = 1$, $N = 1$, $n = 3$, $c = 4$; see Figure 1.21b.

The kinematic pairs of the modular groups can be

- *exterior*: A, B, C (Figure 1.20b); D, C (Figure 1.20c)
- *interior*: D, E, F (Figure 1.20b); A, B, E, F (Figure 1.20c)

and

- *active*: C (Figure 1.21b)
- *passive*: A, B, C, D, E, F (Figure 1.20a,b)

respectively.

1.8 Formation and Decomposition of Planar Mechanisms

The general principle for the formation of planar mechanisms is the successive connection of the modular groups; the mobility degree of the resulting mechanism is equal to the sum of the mobility degrees of the component mechanisms. The following rules have to be observed:

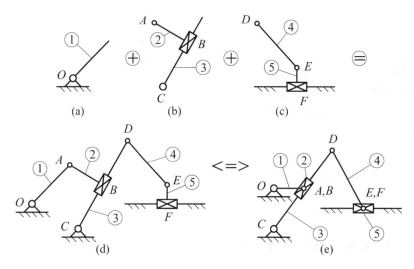

Figure 1.22 Formation of a shaper mechanism: (a) driving element; (b) RTR dyad; (c) RTR dyad; (d) and (e) the shaper mechanism.

- the exterior pairs of the passive groups are not all connected with the same element
- at least one driving modular group is linked with all its exterior pairs to the frame.

Figure 1.22d shows the mode of formation of the mechanism of a shaper with mobility $M = 1$ from a driving element (Figure 1.22a) with $M_a = 1$ and two dyads (Figure 1.22b,c) with $M_b = M_c = 0$. Figure 1.22e shows the same mechanism in another variant in which the lengths AB and EF are equal to zero.

The decomposition of a mechanism into modular groups, which is necessary in reaching geometric, kinematic, and kinetostatic solutions, involves the following steps:

1) Create the kinematic schema of the mechanism.
2) Create the structural schema of the fundamental mechanism.
3) Separate the driving elements linked to the frame.
4) Identify the other modular groups.

An example of decomposition is shown in Figures 1.23 and 1.24. These show the constructive schema of an intake mechanism (Figure 1.23a), then the kinematic schema (Figure 1.23b), the structural schema (Figure 1.24a), and the modular groups (Figure 1.24b).

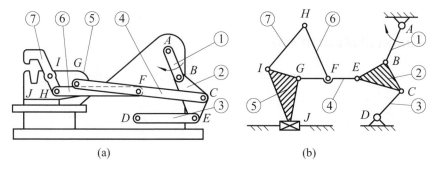

Figure 1.23 Intake mechanism schemata: (a) structural; (b) kinematic.

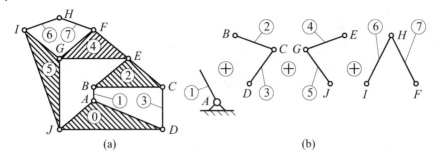

Figure 1.24 The decomposition of an intake mechanism into modular groups: (a) structural schema; (b) modular groups.

1.9 Multi-poles and Multi-polar Schemata

Recalling elementary notions of the theory of systems, modular groups are subsystems, called *multi-poles*, which transmit, in a unique determined way, the information of motion (positions, velocities, accelerations), the mechanism being formed by interconnected multi-poles. The multi-poles are referred to as *active* if $M \geq 1$, and *passive* if $M = 0$.

A multi-pole can have

- exterior input poles, through which it receives information about the motion from the exterior
- exterior output poles, through which it transmits in exterior the information of motion
- interior poles, through which information is transmitted and manipulated inside the multi-pole.

The exterior input poles can be active (driving pairs) or passive. Table 1.8 shows the correspondence of representation between modular groups and multi-poles.

The *multi-polar schema* is the representation of the component multi-poles of the mechanism and the interconnections between them. Figure 1.25 shows the multi-polar schema of the intake mechanism drawn in Figure 1.23.

The structural relation is the sum of the component multi-poles, in their connection order.

$$Z(0) + ME(1) + D(2, 3) + D(4, 5) + D(6, 7).$$

The cover sense of the schema is from left to right for the kinematic analysis, and from right to left for the kinetostatic analysis.

1.10 Classification of Mechanisms

The classification of mechanisms according to different criteria is given in Table 1.9.

Table 1.8 The representation of modular groups.

Modular group		Multi-pole		
Name	**Representation**	**Name**	**Representation**	**Observations**
Base/frame (fixed element)		Zero-pole		• Z, zero-pole • 0, component element • A, B, C, output poles (it has no input poles)
Driving element		Driving uni-pole		• ME, driving element • 1, component element • A, active input pole
Dyad		Bi-pole		• D, dyad • (2, 3), component elements • B, interior pole • A, C, exterior input poles • D, E, output poles
Triad		Tri-pole		• TR - triad • (2, 3, 4, 5), component elements • B, C, F, interior poles • A, D, E, input poles • G, H, output poles
Tetrad		Multi-pole tetrad		• TE - Tetrad • (2, 3, 4, 5), component elements • B, C, D, E, interior poles • A, F, input poles • G, H, output poles

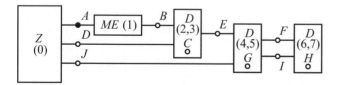

Figure 1.25 The multi-polar schema of the intake mechanism in Figure 1.23.

Table 1.9 Classification of mechanisms.

Nr. crt.	Criteria of classification	Name of the mechanisms	Conditions
1	Motion of the elements	Planar	The motion of the elements is in the same plane or in parallel planes
		Spatial	The motion of the elements is spatial
2	Family	Of family 1, 2, 3, 4	The family of the mechanism is $f = 1, 2, 3,$
3	Number of independent loops	Mono-loop	$N = 1$
		Poly-loop	$N \geq 2$
4	Mobility	Mono-mobile	$M = 1$
		Multi-mobile	$M \geq 2$
5	Destination	Function-generating mechanisms	The input angle depends on the output angle by a required function
		Path-generating mechanisms	A point of an element describes a required trajectory (straight line, circle etc)
		Adjustable mechanisms	They adjust certain parameters of some technological processes

2

Kinematic Analysis of Planar Mechanisms with Bars

In this chapter we will study the positions, velocities, and accelerations of the elements of planar mechanisms with bars. The kinematic parameters (positions, velocities, accelerations) are necessary for the calculation of the forces that act upon the elements, while the forces are necessary in the strength of materials calculation (checkout and sizing).

2.1 General Aspects

In the kinematic calculation we will use two methods: the grapho-analytic method and the analytic method. Graphical methods were designed in the 1970s and used until the use of computers became widespread. They facilitated the development of the analytic methods, the graphical ones being retained only for the approximate determination of positions, which is necessary for initiating numerical analytic calculations. The use of the personal computers enabled the assisted graphical approaches, which use computer-aided design (CAD) products. These allow geometric constructions to be created and the coordinates of any desired point to be determined.

From these ideas it may be concluded that graphical methods, with their characteristic simplicity, nevertheless become applicable and also 'analytic' through the use of CAD products. For these reasons we will present graphical methods starting from this premise. We will discuss the planar mechanisms that are the most used in practice; from the structural point of view, the majority can be created using dyads.

We will perform the kinematic analysis of the five types of dyad (RRR, RRT, RTR, TRT, RTT). We will also study the 6R triad, and in the final section we will present examples based on programs that use algorithms derived from graphical methods. At the end of the chapter we perform a kinematic analysis for the most used design: planar mechanisms with bars.

2.2 Kinematic Relations

2.2.1 Plane-parallel Motion

Distribution of velocities

The distribution of velocities for an element in plane-parallel motion is given by the *Euler relation*,

$$\mathbf{v}_B = \mathbf{v}_A + \boldsymbol{\omega} \times \mathbf{A}B, \tag{2.1}$$

Classical and Modern Approaches in the Theory of Mechanisms, First Edition.
Nicolae Pandrea, Dinel Popa and Nicolae-Doru Stănescu.
© 2017 John Wiley & Sons Ltd. Published 2017 by John Wiley & Sons Ltd.
Companion Website: www.wiley.com/go/pandmech17

where A and B are two arbitrary points, while $\boldsymbol{\omega}$ is the angular velocity vector, which is perpendicular to the plane of motion. Using the notation

$$\mathbf{v}_{BA} = \boldsymbol{\omega} \times \mathbf{AB}, \tag{2.2}$$

the following formula is obtained, known from the kinematic analysis of the mechanisms

$$\mathbf{v}_B = \mathbf{v}_A + \mathbf{v}_{BA} \tag{2.3}$$

The velocity has the following properties:

- it is perpendicular to $\mathbf{AB}$
- it rotates segment AB around point A in the sense given by the angular velocity $\boldsymbol{\omega}$ (Figure 2.1)
- it satisfies the scalar relation

$$v_{BA} = \omega \cdot AB. \tag{2.4}$$

Rotational motion is a particular case of plane-parallel motion. Noting that the centre of rotation is point A, then $\mathbf{v}_A = \mathbf{0}$, and, from (2.1) and (2.4):

$$\mathbf{v}_B = \boldsymbol{\omega} \times \mathbf{AB}, \; v_B = \omega \cdot AB. \tag{2.5}$$

Translational motion is another particular case for which $\boldsymbol{\omega} = \mathbf{0}$; from (2.1) the equality of the velocities for all points results.

Distribution of accelerations

For the accelerations recall the *Rivals' relations*, which, in the case of plane-parallel motion, becomes

$$\mathbf{a}_B = \mathbf{a}_A - \omega^2 \mathbf{AB} + \boldsymbol{\varepsilon} \times \mathbf{AB}, \tag{2.6}$$

where $\mathbf{a}_A$ and $\mathbf{a}_B$ denote the accelerations of the points A and B, respectively, while $\boldsymbol{\varepsilon}$ denotes the angular acceleration. Using the notation

$$\mathbf{a}_{BA}^{\nu} = -\omega^2 \mathbf{AB}, \tag{2.7}$$

$$\mathbf{a}_{BA}^{\tau} = \boldsymbol{\varepsilon} \times \mathbf{AB}, \tag{2.8}$$

gives:

$$\mathbf{a}_B = \mathbf{a}_A + \mathbf{a}_{BA}^{\nu} + \mathbf{a}_{BA}^{\tau}. \tag{2.9}$$

The normal acceleration $\mathbf{a}_{BA}^{\nu}$ has the following properties:

- it has the sense from point B to point A (Figure 2.2)
- it satisfies the scalar relation

$$a_{BA}^{\nu} = AB \cdot \omega^2, \tag{2.10}$$

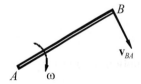

Figure 2.1 The velocity $\mathbf{v}_{BA}$.

Figure 2.2 The accelerations $\mathbf{a}_{BA}^{\nu}$ and $\mathbf{a}_{BA}^{\tau}$.

while the tangential acceleration has the following properties:

- it is perpendicular to the segment AB
- it rotates the segment AB around point A, in the sense given by the angular acceleration ε (Figure 2.2)
- it satisfies the scalar relation

$$a_{BA}^\tau = AB \cdot \varepsilon. \tag{2.11}$$

In the particular case of rotational motion around point A, we have $\mathbf{a}_A = \mathbf{0}$, $\mathbf{a}_B^\nu = \mathbf{a}_{BA}^\nu$, and $\mathbf{a}_B^\tau = \mathbf{a}_{BA}^\tau$, while from (2.6)–(2.11) we obtain

$$\mathbf{a}_B^\nu = \mathbf{a}_{BA,}^\nu \tag{2.12}$$

$$\mathbf{a}_B^\tau = \mathbf{a}_{BA,}^\tau \tag{2.13}$$

$$a_B^\nu = AB \cdot \omega^2, \; a_B^\tau = AB \cdot \varepsilon. \tag{2.14}$$

For translational motion ($\boldsymbol{\omega} = \mathbf{0}$, $\boldsymbol{\varepsilon} = \mathbf{0}$), the known property of all the points of the element having the same acceleration results.

2.2.2 Relative Motion

Distribution of velocities

Relative motion presents some particularities in the case of the existence of translational (prismatic) kinematic pairs (Figure 2.3), and of superior class joints (higher pairs); the former are dealt with later on. For this reason, the formula for compounding the velocities in the case of relative motion takes the form

$$\mathbf{v}_a = \mathbf{v}_t + \mathbf{v}_{r,} \tag{2.15}$$

where $\mathbf{v}_a$ is the absolute velocity, $\mathbf{v}_t$ is the transport velocity (frame velocity), and $\mathbf{v}_r$ is the relative velocity.

Element 2 in Figure 2.3 has relative motion with respect to element 1 and, if we study the motion of point A_2 (the notation meaning point A that belongs to element 2), then we may use the notation $\mathbf{v}_a = \mathbf{v}_{A_2}$, $\mathbf{v}_t = \mathbf{v}_{A_1}$ (the velocity of point A that belongs to element 1), $\mathbf{v}_r = \mathbf{v}_{A_2 A_1}$, and (2.15) becomes

$$\mathbf{v}_{A_2} = \mathbf{v}_{A_1} + \mathbf{v}_{A_2 A_1,} \tag{2.16}$$

where the relative velocity $\mathbf{v}_{A_2 A_1}$ is situated on the straight line (Δ).

The relative motion of element 1 with respect to element 2 gives

$$\mathbf{v}_{A_1} = \mathbf{v}_{A_2} + \mathbf{v}_{A_1 A_2,} \tag{2.17}$$

where $\mathbf{v}_{A_1 A_2} = -\mathbf{v}_{A_2 A_1}.$

Distribution of accelerations

The analysis starts from the formula that compounds the accelerations in relative motion

$$\mathbf{a}_a = \mathbf{a}_t + \mathbf{a}_C + \mathbf{a}_{r,} \tag{2.18}$$

Figure 2.3 Relative motion in a translational (prismatic) kinematic pair.

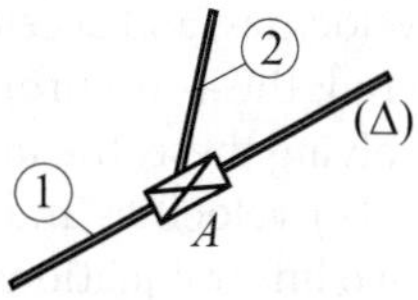

$\mathbf{a}_a$, $\mathbf{a}_t$, $\mathbf{a}_C$, and $\mathbf{a}_r$ being the absolute acceleration, the transport (frame) acceleration, the Coriolis acceleration, and the relative acceleration, respectively. The Coriolis acceleration is given by:

$$\mathbf{a}_C = 2\boldsymbol{\omega} \times \mathbf{v}_r, \tag{2.19}$$

where $\boldsymbol{\omega}$ is the angular velocity of the support straight line (Δ).

Here, too, if we use the same kind of notation:

$$\mathbf{a}_a = \mathbf{a}_{A_2}, \ \mathbf{a}_t = \mathbf{a}_{A_1}, \ \mathbf{a}_r = \mathbf{a}_{A_2 A_1}, \tag{2.20}$$

$$\mathbf{a}_C = \mathbf{a}^C_{A_2 A_1} = 2\boldsymbol{\omega} \times \mathbf{v}_{A_2 A_1}, \tag{2.21}$$

then we get

$$\mathbf{a}_{A_2} = \mathbf{a}_{A_1} + \mathbf{a}^C_{A_2 A_1} + \mathbf{a}_{A_2 A_1}. \tag{2.22}$$

The relative motion of point A_1 with respect to point A_2 gives

$$\mathbf{a}_{A_1} = \mathbf{a}_{A_2} + \mathbf{a}^C_{A_1 A_2} + \mathbf{a}_{A_1 A_2}, \tag{2.23}$$

where

$$\mathbf{a}^C_{A_1 A_2} = -\mathbf{a}^C_{A_2 A_1} = 2\boldsymbol{\omega} \times \mathbf{v}_{A_1 A_2}, \tag{2.24}$$

$$\mathbf{a}_{A_1 A_2} = -\mathbf{a}_{A_2 A_1}. \tag{2.25}$$

2.3 Methods for Kinematic Analysis

2.3.1 The Grapho-analytical Method

To perform the kinematic analysis using this method, the mechanism must be represented at a certain scale and in a given position and then the vector equations for the velocities and accelerations must be solved graphically, and at scale. The effective magnitude of an unknown is given by the product of its length, measured in millimetres in the drawing, and the representation scale.

The system of the measurement units for the real parameters is the international one – the SI system (m for length, m/s for velocities, and m/s^2 for accelerations) – while the measurement unit in the drawing is the millimetre. Because of this, the following scales are obtained:

- the scale of lengths $\left[\dfrac{m}{mm}\right]$
- the scale of velocities $\left[\dfrac{m/s}{mm}\right]$
- the scale of accelerations $\left[\dfrac{m/s^2}{mm}\right]$.

For instance, if we want an element of length $l = 0.6$ m to be represented in drawing by a length $[l]$ of 30 mm, then the scale of lengths is $k_l = \dfrac{0.6}{30} = 0.02 \left[\dfrac{m}{mm}\right]$, and if an element has, in the drawing, a length of 25 mm, then the real length of the element is $0.02 \times 25 = 0.5$ m.

2.3.2 The Method of Projections

In this method, a reference system is chosen, and this is generally the same for the mechanism, velocities, and accelerations. To determine the positions, the vector loop equations are considered; these are projected onto the axes and two non-linear scalar equations are obtained; by solving these, the unknowns of the system are obtained.

For velocities and accelerations, the position equations are also projected onto the axes, and two linear equations are obtained, from which the unknowns of the problem can be deduced.

2.3.3 The Newton–Raphson Method

We have shown that the system of the positional equations is a non-linear one. The numerical solution of such a system is possible if an approximate solution is known.

Define the system of non-linear equations as

$$F_i\left(x_1, x_2, \ldots, x_i, \ldots, x_n\right) = 0, \ i = 1, \ 2, \ \ldots, \ n, \tag{2.26}$$

and the approximate solution as $\left(x_1^0, x_2^0, \ldots, x_n^0\right)$. We want to determine the variations Δx_i, $i = 1, \ 2, \ \ldots, \ n$, so that the values $x_i + \Delta x_i$ verify the system of equations; that is,

$$F_i\left(x_1^0 + \Delta x_1, x_2^0 + \Delta x_2, \ldots, x_i^0 + \Delta x_i, \ldots, x_n^0 + \Delta x_n\right) = 0, \ i = 1, \ 2, \ \ldots, \ n. \tag{2.27}$$

If we develop this into a Taylor series and neglect the non-linear terms in Δx_i, then we obtain the linear system

$$\sum_{k=1}^{n} \Delta x_i \left.\frac{\partial F_i}{\partial x_k}\right|_{\mathbf{x}=\mathbf{x}^0} = -\left.F_i\right|_{\mathbf{x}=\mathbf{x}^0}, \ i = 1, \ 2, \ \ldots, \ n. \tag{2.28}$$

By solving this system, the variations Δx_i are obtained and, using these, the new values that approximate the solution become $x_i^0 + \Delta x_i$, $i = 1, \ 2, \ \ldots, \ n$. The iteration procedure then continues until the required calculation precision is reached:

$$\left|\Delta x_i\right| \leq v. \tag{2.29}$$

As an example, let us consider the system

$$\begin{cases} F_1\left(x_1, x_2\right) = x_1^2 + x_2 - 3 = 0, \\ F_2\left(x_1, x_2\right) = x_1 x_2^2 - 4 = 0, \end{cases} \tag{2.30}$$

and the approximate solution $x_1 = 1.1$, $x_2 = 2.1$. We obtain the system

$$\begin{cases} 2x_1 \Delta x_1 + \Delta x_2 = -F_1\left(x_1, x_2\right), \\ x_2^2 \Delta x_1 + 2x_1 x_2 \Delta x_2 = -F_2\left(x_1, x_2\right), \end{cases} \tag{2.31}$$

and, if we replace x_1 and x_2 with the corresponding values 1.1 and 2.1, respectively:

$$\begin{cases} 2.2\Delta x_1 + \Delta x_2 = -0.01, \\ 4.41\Delta x_1 + 4.62\Delta x_2 = -0.851, \end{cases}$$

that is

$$\Delta x_1 = -0.101, \ \Delta x_2 = 0.088.$$

The new initialization values are $x_1 = 0.999$, $x_2 = 2.012$, while the system in (2.31) becomes

$$\begin{cases} 1.998\Delta x_1 + \Delta x_2 = -0.01, \\ 4.048\Delta x_1 + 4.0199\Delta x_2 = -0.044, \end{cases}$$

from which:

$$\Delta x_1 = 0.000954, \ \Delta x_2 = 0.001196,$$

and the solution

$$x_1 = 0.999954, \ x_2 = 2.0004.$$

If we limit ourselves to a precision of $v = 0.0015$, then $\left|\Delta x_i\right| \leq v$ and the previous values are the system's solution. We may observe that we have obtained a good approximation of the exact solution $x_1 = 1$, $x_2 = 2$ in only two iterations.

2.3.4   Determination of Velocities and Accelerations using the Finite Differences Method

Determination of the positions in mechanisms is a function of the driving element, which in most cases has rotational motion. The positions of the elements are given either by angles or by lengths. The derivatives of these elements with respect to time represent either angular or relative velocities.

Let us denote such a positional parameter by φ, and the rotational angle of the driving element by θ. If the driving element has angular velocity ω and the angular acceleration ε,

$$\omega = \frac{d\theta}{dt}, \ \varepsilon = \frac{d^2\theta}{dt^2}, \tag{2.32}$$

then the derivatives of the parameter φ with respect to time are

$$\frac{d\theta}{dt} = \frac{d\varphi}{d\theta}\omega, \ \frac{d^2\varphi}{d\theta^2} = \frac{d^2\varphi}{d\theta^2}\omega^2 + \frac{d\varphi}{d\theta}\varepsilon. \tag{2.33}$$

Therefore, these derivatives may be expressed using the reduced angular velocity $\dfrac{d\varphi}{d\theta}$ and reduced angular acceleration $\dfrac{d^2\varphi}{d\theta^2}$.

If θ has a constant step variation $\Delta\theta$ (for instance, degree by degree), then these derivatives may be calculated using the finite differences. Because of this, we use the notation

$$\theta_i = \theta_1 + (i-1)\,\Delta\theta, \tag{2.34}$$

$$\varphi_i = \varphi\left(\theta_i\right), \tag{2.35}$$

and the development into a Taylor series is limited to the first three terms

$$\begin{aligned}\varphi\left(\theta_i - \Delta\theta\right) &= \varphi\left(\theta_i\right) - \Delta\theta\left.\frac{d\varphi}{d\theta}\right|_{\theta=\theta_i} + \frac{(\Delta\theta)^2}{2}\left.\frac{d^2\varphi}{d\theta^2}\right|_{\theta=\theta_i} \\ \varphi\left(\theta_i + \Delta\theta\right) &= \varphi\left(\theta_i\right) + \Delta\theta\left.\frac{d\varphi}{d\theta}\right|_{\theta=\theta_i} + \frac{(\Delta\theta)^2}{2}\left.\frac{d^2\varphi}{d\theta^2}\right|_{\theta=\theta_i}\end{aligned} \tag{2.36}$$

Taking into account (2.35), then from (2.36), by subtraction and addition:

$$\left.\frac{d\varphi}{d\theta}\right|_{\theta=\theta_i} = \frac{\varphi_{i+1} - \varphi_{i-1}}{2\Delta\theta}, \tag{2.37}$$

$$\left.\frac{d^2\varphi}{d\theta^2}\right|_{\theta=\theta_i} = \frac{\varphi_{i+1} - 2\varphi_i + \varphi_{i-1}}{(\Delta\theta)^2}. \tag{2.38}$$

When θ is an angle, and $\Delta\theta = 1° = \dfrac{\pi}{180}$ rad, we get

$$\left.\frac{d\varphi}{d\theta}\right|_{\theta=\theta_i} = \frac{90}{\pi}\left(\varphi_{i+1} - \varphi_{i-1}\right), \tag{2.39}$$

$$\left.\frac{d^2\varphi}{d\theta^2}\right|_{\theta=\theta_i} = \left(\frac{180}{\pi}\right)^2\left(\varphi_{i+1} - 2\varphi_i + \varphi_{i-1}\right). \tag{2.40}$$

Since the value φ_0 does not exist, (2.39) and (2.40) cannot be used for the starting values ($i = 1$). In this situation, the derivative $\left.\dfrac{d\varphi}{d\theta}\right|_{\theta=\theta_i}$ is calculated from the second relation (2.36),

$$\left.\frac{d\varphi}{d\theta}\right|_{\theta=\theta_1} = \frac{\varphi_2 - \varphi_1}{\Delta\theta}, \tag{2.41}$$

while the second-order derivative is given by

$$\frac{d^2\varphi}{d\theta^2}\bigg|_{\theta=\theta_1} = \frac{\dfrac{d\varphi}{d\theta}\bigg|_{\theta=\theta_2} - \dfrac{d\varphi}{d\theta}\bigg|_{\theta=\theta_1}}{\Delta\theta}. \tag{2.42}$$

Similarly, the derivatives at the last point can be calculated too.

2.4 Kinematic Analysis of the RRR Dyad

2.4.1 The Grapho-analytical Method

Formulation of the problem

Consider the dyad ABC in Figure 2.4a. The following are known:

- the positions of points A and C
- the lengths of segments AB and CB, equal to l_2 and l_3, respectively
- the velocities $\mathbf{V}_A$ and $\mathbf{V}_C$ of points A and C, respectively
- the accelerations $\mathbf{a}_A$ and $\mathbf{a}_C$ of points A and C, respectively.

The following values are to be determined:

- the position of point B and, eventually, the position of a point K that belongs to element 2 or 3
- the angular velocities ω_2 and ω_3, the velocity of point B and, eventually, the velocity of a point K that belongs to element 2 or 3
- the angular accelerations ε_2 and ε_3, the acceleration of point B and, eventually, the acceleration of a point K that belongs to element 2 or 3.

Determination of the positions

The grapho-analytic determination of the position for the dyad is based on elementary geometric constructions, which lead to the following sequence of operations:

- Choose the scale of the lengths k_l.
- Represent the points A and C.
- Divide the lengths l_2, l_3 by the scale k_l and determine the lengths of representation $[l_2]$, $[l_3]$ in millimetres.
- Construct two circles: one of centre A and radius $[l_2]$, and the other of centre C and radius $[l_3]$.
- Obtain the points B and B' at the intersection of these circles (usually, the problem has two solutions).
- Choose the convenient solution according to the initial position.
- Determine point K (the lengths l_{KB} and l_{KC} are known), and proceed in an analogous way to the determination of the point B (Figure 2.4b).

Determination of the velocities

For the determination of the velocities, note that point B belongs to both element 2 and element 3 and, consequently, applying (2.3), we get

$$\mathbf{v}_B = \mathbf{v}_A + \mathbf{v}_{BA} = \mathbf{v}_C + \mathbf{v}_{BC}. \tag{2.43}$$

We underline the known vectors with two lines, but use only one line for vectors for which the direction is known, but the magnitude is not. Based on these rules, (2.43) may be written as

$$\mathbf{v}_B = \underline{\underline{\mathbf{v}_A}} + \underbrace{\mathbf{v}_{BA}}_{\perp AB} = \underline{\underline{\mathbf{v}_C}} + \underbrace{\mathbf{v}_{BC}}_{\perp BC}. \tag{2.44}$$

We then represent the vector relation (2.44), each member being constructed starting from the pole of velocities, which is denoted by i.

The representations of the magnitudes $[v_A]$, $[v_C]$ of the velocities v_A, v_C expressed in millimetres, are determined by dividing the real values by the scale of velocities k_v; that is (Figure 2.4c):

$$(ia) = [v_A] = \frac{v_A}{k_v}, \quad (ic) = [v_C] = \frac{v_C}{k_v}. \tag{2.45}$$

At the intersection of the directions of the vectors underlined with only one line (Figure 2.4c), point b is found. The segments (ab), (cb), (ib) are then measured and the velocities v_{BA}, v_{BC}, v_B are calculated using:

$$v_{BA} = k_v (ab), \quad v_{BC} = k_v (bc), \quad v_B = k_v (ib). \tag{2.46}$$

The angular velocities are calculated using (2.4):

$$\omega_2 = \frac{v_{BA}}{l_2}, \quad \omega_3 = \frac{v_{Bc}}{l_3}, \tag{2.47}$$

their senses being given in Figure 2.4c.

For point K:

$$\mathbf{v}_K = \underline{\underline{\mathbf{v}_B}} + \underbrace{\mathbf{v}_{KB}}_{\perp KB} = \underline{\underline{\mathbf{v}_C}} + \underbrace{\mathbf{v}_{KC}}_{\perp KC} \tag{2.48}$$

and point K is then situated at the intersection of the perpendiculars on BK and CK constructed through the points b and c, respectively (Figure 2.4c). The magnitude of the velocity of point K is given by:

$$v_K = k_v (ik). \tag{2.49}$$

If points B, K, and C are collinear, then the theorem of similarity is needed[1], so that points b, k, and c are also collinear; in addition

$$\frac{bk}{BK} = \frac{ck}{CK}. \tag{2.50}$$

Determination of the accelerations

Given the relations

$$\mathbf{a}_B = \underline{\underline{\mathbf{a}_A}} + \underbrace{\mathbf{a}_{BA}^{\nu}}_{\substack{B \to A \\ l_2 \omega_2^2}} + \underbrace{\mathbf{a}_{BA}^{\tau}}_{\perp AB}$$

$$= \underline{\underline{\mathbf{a}_C}} + \underbrace{\mathbf{a}_{BC}^{\nu}}_{\substack{B \to C \\ l_3 \omega_3^2}} + \underbrace{\mathbf{a}_{BC}^{\tau}}_{\perp BC} \tag{2.51}$$

---

1 Also known as the Burmester and Mehmke theorem.

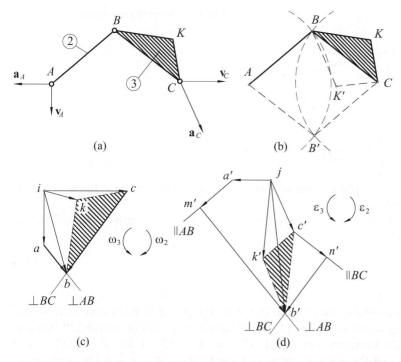

Figure 2.4 Kinematic analysis of the RRR dyad using the grapho-analytic method: (a) RRR dyad; (b) determination of positions of the points; (c) plane of velocities; (d) plane of accelerations.

the normal accelerations $a_{BA}^v = l_2\omega_2^2$, $a_{BC}^v = l_3\omega_3^2$ can be calculated. Then the scale of accelerations k_a is chosen and the parameters representing (2.51) are calculated:

$$\left(ja'\right) = \frac{a_A}{k_a}, \ \left(jc'\right) = \frac{a_C}{k_a}, \ \left(a'm'\right) = \frac{a_{BA}^v}{k_a}, \ \left(c'n'\right) = \frac{a_{BC}^v}{k_a}. \tag{2.52}$$

Each member is constructed from the pole of accelerations j. At the intersection of the directions of the vectors underlined with only one line we find point b' (Figure 2.4d). Segments $(m'b'), (n'b')$, and (jb') are measured and the accelerations a_{BA}^τ, a_{BC}^τ, and a_B are calculated using:

$$a_{BA}^\tau = k_a\left(m'b'\right), \ a_{BC}^\tau = k_a\left(n'b'\right), \ a_B = k_a\left(jb'\right). \tag{2.53}$$

The angular accelerations are calculated from:

$$\varepsilon_2 = \frac{a_{BA}^\tau}{l_2}, \ \varepsilon_3 = \frac{a_{BC}^\tau}{l_3}, \tag{2.54}$$

their senses being as shown in Figure 2.4d.

For the determination of the acceleration of point K, the following relations are used:

$$
\begin{aligned}
\mathbf{a}_K &= \underline{\underline{\mathbf{a}_A}} + \underset{\underset{l_{KB}\omega_2^2}{K \to B}}{\underline{\underline{\mathbf{a}_{KB}^v}}} + \underset{\perp KB}{\underline{\mathbf{a}_{KB}^\tau}} \\
&= \underline{\underline{\mathbf{a}_C}} + \underset{\underset{l_{KC}\omega_3^2}{K \to C}}{\underline{\underline{\mathbf{a}_{KC}^v}}} + \underset{\perp KC}{\underline{\mathbf{a}_{KC}^\tau}},
\end{aligned} \tag{2.55}
$$

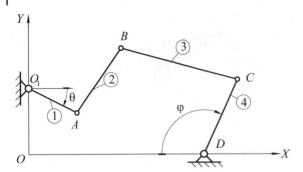

Figure 2.5 Bi-mobile mechanism.

These are represented in the plane of the accelerations; point k' is situated at the intersection of the vectors $\mathbf{a}^{\tau}_{KB}$ and $\mathbf{a}^{\tau}_{KC}$.

When the points B, K, and C are collinear, from the theorem of similarity, points b', k', and c' are also collinear and

$$\frac{b'k'}{BK} = \frac{c'k'}{CK}. \tag{2.56}$$

Numerical example

Problem: Consider the bi-mobile mechanism OO_1ABCD (Figure 2.5), relative to the reference system of axes OXY, at which the joints O_1 and D have coordinates (in metres) $(0, 0.4)$ and $(1.3, 0)$, respectively. The elements have lengths $O_1A = 0.4\,\text{m}$, $AB = 0.5\,\text{m}$, $BC = 0.67\,\text{m}$, and $CD = 0.5\,\text{m}$. Knowing that the angles θ and φ are given by $\theta = 15t$, $\varphi = \frac{\pi}{2} + 12t$, perform the grapho-analytic kinematic analysis of the ABC dyad at time $t = 0$.

Solution: Choose the scale of lengths so that a representation of 20 mm corresponds to a length of 0.4 m; that is, $k_l = \frac{0.4}{20} = 0.02 \left[\frac{\text{m}}{\text{mm}}\right]$. The angular velocities ω_1 and ω_4 are equal to $\dot{\theta}$ and $\dot{\varphi}$, respectively, so $\omega_1 = 15\,\text{rad/s}$ and $\omega_4 = 12\,\text{rad/s}$. At time $t = 0$ the angles are $\theta = 0$, $\varphi = \frac{\pi}{2}$. The scale drawing in Figure 2.6a is obtained.

The velocities of points A and C, shown in Figure 2.6a, have magnitudes $v_A = 0.4 \times 15 = 6\,\text{m/s}$ and $v_C = 0.5 \times 12 = 6\,\text{m/s}$. Choosing the scale of velocities so that at 6 [m/s] corresponds to a representation of 30 [mm], we obtain $k_v = \frac{6}{30} = 0.2 \left[\frac{\text{m/s}}{\text{mm}}\right]$.

By scale representation of the vector relation

$$\mathbf{v}_B = \underset{}{\underline{\underline{\mathbf{v}_A}}} + \underset{\perp AB}{\underline{\mathbf{v}_{BA}}} = \underset{}{\underline{\underline{\mathbf{v}_C}}} + \underset{\perp BC}{\underline{\mathbf{v}_{BC}}},$$

the polygon of velocities in Figure 2.6b is obtained, from which, by measuring:

$$[ab] = 13\,\text{mm}, \quad [bc] = 43\,\text{mm}, \quad [ib] = 40\,\text{mm}.$$

Then

$$v_{BA} = 2.6\,\text{m/s}, \; v_{BC} = 8.6\,\text{m/s}, \; v_B = 8\,\text{m/s}.$$

The angular velocities have the values

$$\omega_2 = \frac{2.6}{0.5} = 5.2\,\text{rad/s}, \; \omega_3 = \frac{8.6}{0.7} = 12.83\,\text{rad/s}$$

and the senses are as represented in Figure 2.6b.

The accelerations of points A and C, shown in Figure 2.6b, have the values

$$a_A = 0.4 \times 15^2 = 90\,\text{m/s}^2, \; a_C = 0.5 \times 12^2 = 72\,\text{m/s}^2.$$

Choosing the scale of accelerations so that at $90\left[\text{m/s}^2\right]$ corresponds a representation of $20\,[\text{mm}]$, we get $k_a = \dfrac{90}{20} = 4.5\left[\dfrac{\text{m/s}^2}{\text{mm}}\right]$. So $\left[ja'\right] = \dfrac{90}{4.5} = 20\,\text{mm}$, $\left[jc'\right] = \dfrac{72}{4.5} = 16\,\text{mm}$. For the acceleration of point B we write the vector relation

$$\begin{aligned}
\mathbf{a}_B &= \underset{=}{\mathbf{a}_A} + \underset{\substack{B \to A \\ l_2\omega_2^2}}{\mathbf{a}_{BA}^{\nu}} + \underset{\perp AB}{\mathbf{a}_{BA}^{\tau}} \\[2mm]
&= \underset{=}{\mathbf{a}_C} + \underset{\substack{B \to C \\ l_3\omega_3^2}}{\mathbf{a}_{BC}^{\nu}} + \underset{\perp BC}{\mathbf{a}_{BC}^{\tau}}
\end{aligned}$$

and calculate the values

$$\begin{aligned}
a_{BA}^{\nu} &= l_2\omega_2^2 = 0.5 \times 5.2^2 = 13.52 \text{ m/s}^2 \\
a_{BC}^{\nu} &= l_3\omega_3^2 = 0.67 \times 12.83^2 = 110.28 \text{ m/s}^2.
\end{aligned}$$

and the representation values

$$\left[a_{BA}^{\nu}\right] = \frac{13.52}{4.5} = 3.0\,\text{mm}, \quad \left[a_{BC}^{\nu}\right] = \frac{110.28}{4.5} = 24.5\,\text{mm}.$$

This results in the representation in Figure 2.6c, from which, by measuring, we deduce

$$\left[m'b'\right] = 50\,\text{mm}, \quad \left[n'b'\right] = 6\,\text{mm}, \quad \left[jb'\right] = 37\,\text{mm}.$$

or

$$a_{BA}^{\tau} = k_a\left(m'b'\right) = 225 \text{ m/s}^2, \ a_{BC}^{\tau} = k_a\left(n'b'\right) = 27 \text{ m/s}^2, \ a_B = k_a\left(jb'\right) = 166.5 \text{ m/s}^2.$$

Then the angular accelerations are:

$$\varepsilon_2 = \frac{225}{0.5} = 450 \text{ m/s}^2, \ \varepsilon_3 = \frac{27}{0.67} = 40.29 \text{ m/s}^2,$$

and their senses are as drawn in Figure 2.6c.

2.4.2 The Analytical Method

Formulation of the problem

Consider the dyad in Figure 2.7. The following are known:

- the coordinates X_A, Y_A and X_C, Y_C of points A and C, respectively
- the lengths l_2 and l_3 of segments AB and BC, respectively
- the velocities v_{Ax}, v_{Ay} and v_{Cx}, v_{Cy} of points A and C, respectively
- the accelerations a_{Ax}, a_{Ay} and a_{Cx}, a_{Cy} of points A and C, respectively.

The following values are to be determined:

- the coordinates X_B, Y_B of point B
- the velocities v_{Bx}, v_{By} of point B
- the angular velocities ω_2, ω_3
- the accelerations a_{Bx}, a_{By} of point B
- the angular accelerations ε_2, ε_3.

In addition, for a point K of segment 3, for which the distances KB and KC are known and equal to l_{KB} and l_{KC}, respectively, the coordinates X_K, Y_K, the velocities v_{Kx}, v_{Ky}, and the accelerations a_{Kx}, a_{Ky} are to be determined.

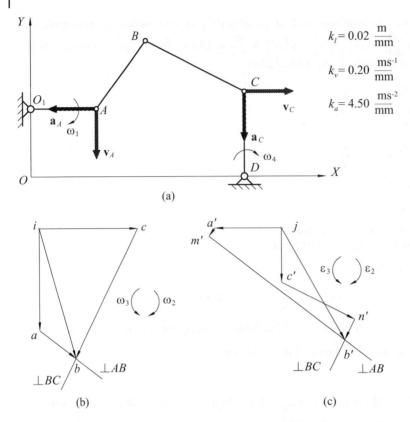

Figure 2.6 Numerical example of an RRR dyad: (a) position of the mechanism; (b) plane of velocities; (c) plane of accelerations.

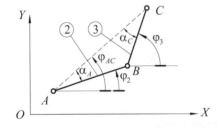

Figure 2.7 The RRR dyad.

Determination of the positions

The determination of the positions for the dyad reduces to the determination of angles φ_2 and φ_3. For this, the following algorithm is needed:

$$l_{AC} = \sqrt{\left(X_C - X_A\right)^2 + \left(Y_C - Y_A\right)^2}, \tag{2.57}$$

$$\alpha_A = \arccos\left(\frac{l_{AC}^2 + l_2^2 - l_3^2}{2l_2 l_{AC}}\right), \tag{2.58}$$

$$\varphi_{AC} = \arctan\left(\frac{Y_C - Y_A}{X_C - X_A}\right), \tag{2.59}$$

$$\varphi_2 = \varphi_{AC} - \alpha_A, \tag{2.60}$$

$$\alpha_C = \arccos\left(\frac{l_{AC}^2 + l_3^2 - l_2^2}{2l_3 l_{AC}}\right), \tag{2.61}$$

$$\varphi_3 = \varphi_{AC} - \alpha_C, \tag{2.62}$$

$$X_B = X_A + l_2 \cos \varphi_2, \quad Y_B = Y_A + l_2 \sin \varphi_2. \tag{2.63}$$

Determination of the velocities
Projecting the equalities

$$\mathbf{v}_B = \mathbf{v}_A + \boldsymbol{\omega}_2 \times \mathbf{AB} = \mathbf{v}_C + \boldsymbol{\omega}_3 \times \mathbf{CB} \tag{2.64}$$

on the axes OX and OY, the following system is obtained

$$\begin{cases} v_{Ax} - \omega_2 \left(Y_B - Y_A\right) = v_{Cx} - \omega_3 \left(Y_B - Y_C\right), \\ v_{Ay} + \omega_2 \left(X_B - X_A\right) = v_{Cy} + \omega_3 \left(X_B - X_C\right), \end{cases} \tag{2.65}$$

with angular velocities:

$$\begin{aligned} \omega_2 &= \frac{\left(v_{Cx} - v_{Ax}\right)\left(X_B - X_C\right) + \left(v_{Cy} - v_{Ay}\right)\left(Y_B - Y_C\right)}{\left(X_B - X_A\right)\left(Y_B - Y_C\right) - \left(X_B - X_C\right)\left(Y_B - Y_A\right)}, \\ \omega_3 &= \frac{\left(v_{Cx} - v_{Ax}\right)\left(X_B - X_A\right) + \left(v_{Cy} - v_{Ay}\right)\left(Y_B - Y_A\right)}{\left(X_B - X_A\right)\left(Y_B - Y_C\right) - \left(X_B - X_C\right)\left(Y_B - Y_A\right)}. \end{aligned} \tag{2.66}$$

The components of the velocity of point B are:

$$v_{Bx} = v_{Ax} - \omega_2 \left(Y_B - Y_A\right), \quad v_{By} = v_{Ay} + \omega_2 \left(X_B - X_A\right). \tag{2.67}$$

Determination of the accelerations
By projection of:

$$\mathbf{a}_B = \mathbf{a}_A + \boldsymbol{\varepsilon}_2 \times \mathbf{AB} - \omega_2^2 \mathbf{AB} = \mathbf{a}_C + \boldsymbol{\varepsilon}_3 \times \mathbf{CB} - \omega_3^2 \mathbf{CB} \tag{2.68}$$

onto the axes, the following system is obtained:

$$\begin{cases} -\varepsilon_2 \left(Y_B - Y_A\right) + \varepsilon_3 \left(Y_B - Y_C\right) = A_x, \\ \varepsilon_2 \left(X_B - X_A\right) - \varepsilon_3 \left(X_B - X_C\right) = A_y, \end{cases} \tag{2.69}$$

where

$$\begin{aligned} A_x &= a_{Cx} - a_{Ax} + \omega_2^2 \left(X_B - X_A\right) - \omega_3^2 \left(X_B - X_C\right), \\ A_y &= a_{Cy} - a_{Ay} + \omega_2^2 \left(Y_B - Y_A\right) - \omega_3^2 \left(Y_B - Y_C\right). \end{aligned} \tag{2.70}$$

This gives the angular accelerations:

$$\begin{aligned} \varepsilon_2 &= \frac{A_x \left(X_B - X_C\right) + A_y \left(Y_B - Y_C\right)}{\left(X_B - X_A\right)\left(Y_B - Y_C\right) - \left(X_B - X_C\right)\left(Y_B - Y_A\right)}, \\ \varepsilon_3 &= \frac{A_x \left(X_B - X_A\right) + A_y \left(Y_B - Y_A\right)}{\left(X_B - X_A\right)\left(Y_B - Y_C\right) - \left(X_B - X_C\right)\left(Y_B - Y_A\right)}, \end{aligned} \tag{2.71}$$

while the components a_{Bx} and a_{By} of the acceleration $\mathbf{a}_B$ read:

$$\begin{aligned} a_{Bx} &= a_{Ax} - \varepsilon_2 \left(Y_B - Y_A\right) - \omega_2^2 \left(X_B - X_A\right), \\ a_{By} &= a_{Ay} + \varepsilon_2 \left(X_B - X_A\right) - \omega_2^2 \left(Y_B - Y_A\right). \end{aligned} \tag{2.72}$$

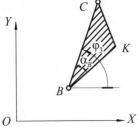

Figure 2.8 The position of a point K belonging to element BC.

Kinematic elements of a point K belonging to an element

Refer to Figure 2.8. For the positions, we have the successive relations

$$\alpha_B = \arccos\left(\frac{l_{BK}^2 + l_3^2 - l_{CK}^2}{2l_3 l_{BK}}\right), \tag{2.73}$$

$$X_K = X_B + l_{BK}\cos\left(\varphi_3 - \alpha_B\right), \quad Y_K = Y_B + l_{BK}\sin\left(\varphi_3 - \alpha_B\right). \tag{2.74}$$

The components of the velocity of point K are

$$v_{Kx} = v_{Bx} - \omega_3\left(Y_K - Y_B\right), \quad v_{Ky} = v_{By} + \omega_3\left(X_K - X_B\right), \tag{2.75}$$

while the components of the acceleration of the same point read

$$\begin{aligned}
a_{Kx} &= a_{Bx} - \varepsilon_3\left(Y_K - Y_B\right) - \omega_3^2\left(X_K - X_B\right), \\
a_{Ky} &= a_{By} + \varepsilon_3\left(X_K - X_B\right) - \omega_3^2\left(Y_K - Y_B\right).
\end{aligned} \tag{2.76}$$

Numerical example

Problem: Using the analytical method, perform the kinematic analysis for the mechanism in Figure 2.6a, considering the same conditions as in the previous example, which was solved by the grapho-analytical method. Compare the results.

Solution: From the data given in the problem, points A and C have coordinates

$$X_A = 0.4\,\text{m}, \ Y_A = 0.4\,\text{m}, \ X_C = 1.3\,\text{m}, \ Y_C = 0.5\,\text{m}.$$

From (2.57)–(2.63)

$$l_{AC} = 0.9055\,\text{m}, \ \alpha_A = 46.6981°, \ \varphi_{AC} = 6.3402°.$$

In this case, point B being situated in the left part of the segment AC, the formula $\varphi_2 = \varphi_{AC} + \alpha_A = 53.0383°$ is needed in order to calculate angle φ_2; this gives the coordinates

$$X_B = 0.4 + 0.5\cos\varphi_2 = 0.7006\,\text{m}, \ Y_B = 0.4 + 0.5\sin\varphi_2 = 0.7995\,\text{m}.$$

For the components of the velocities we have

$$v_{Ax} = 0\,\text{m/s}, \ v_{Ay} = -6\,\text{m/s}, \ v_{Cx} = 6\,\text{m/s}, \ v_{Cy} = 0\,\text{m/s},$$

while from (2.66) and (2.67)

$$\omega_2 = -5.46\,\text{rad/s}, \ \omega_3 = 12.75\,\text{rad/s},$$

$$v_{Bx} = 2.18\,\text{m/s}, \ v_{By} = -7.64\,\text{m/s}, \ v_B = 7.94\,\text{m/s}.$$

For the accelerations we have

$$a_{Ax} = -90\,\text{m/s}^2, \ a_{Ay} = 0\,\text{m/s}^2, \ a_{Cx} = 0\,\text{m/s}^2, \ a_{Cy} = -72\,\text{m/s}^2,$$

and (2.70)–(2.72) give

$$A_x = 196.40, \; A_y = -108.7777, \; \varepsilon_2 = -456.17 \text{ rad/s}^2, \; \varepsilon_3 = 47.28 \text{ rad/s}^2,$$

$$a_{Bx} = 83.28 \text{ m/s}^2, \; a_{By} = -149.03 \text{ m/s}^2, \; a_B = 170.72 \text{ m/s}^2.$$

Comparing the values obtained using the grapho-analytical method with those obtained using the analytical method, the errors are acceptable.

2.4.3 The Assisted Analytical Method

Based on the relations established for the analytical method, we have formulated a procedure *Dyad_RRR*, which can be found in the companion website of this book. The procedure is written in Turbo Pascal and has 14 input data and 14 output data. The input data are the coordinates of the kinematic pairs A and C (X_A, Y_A and X_C, Y_C, respectively), the lengths of elements 2 and 3 (l_2 and l_3, respectively), the input kinematic parameters of poles A and C (v_{Ax}, v_{Ay}, a_{Ax}, a_{Ay} and v_{Cx}, v_{Cy}, a_{Cx}, a_{Cy}, respectively). The output data are the angles of position for elements 2 and 3 (φ_2 and φ_3, respectively), the coordinates of point B (X_B and Y_B), the angular velocities of elements 2 and 3 (ω_2 and ω_3, respectively), the components of the velocity of point B (v_{Bx} and v_{By}), the angular accelerations of elements 2 and 3 ($\varepsilon 2$ and ε_3, respectively), and the components of the acceleration of point B (a_{Bx} and a_{By}).

The procedure follows the calculation algorithm described in (2.57)–(2.72). By calling the procedure with input data for the mechanism in Figure 2.6, and under the same conditions as for the example solved using the grapho-analytical and analytical methods, that is,

Dyad_RRR $(0.4, 0.4, 1.3, 0.5, 0.5, 0.67, 0, -6, 6, 0, -90, 0, 0, -72,$ *Phi2, Phi3,*
XB, YB, Omega2, Omega3, vBx, vBy, vB, Epsilon2, Epsilon3, aBx, aBy, aB$)$

the following results were obtained

$$\varphi_2 = 53.034534°, \; \varphi_3 = 153.447765°,$$

$$X_B = 0.700667 \text{ m}, \; Y_B = 0.799499 \text{ m},$$

$$\omega_2 = -5.460093 \text{ rad/s}, \; \omega_3 = 12.750284 \text{ rad/s},$$

$$v_{Bx} = 2.181302 \text{ m/s}, \; v_{By} = -7.641669 \text{ m/s}, \; v_B = 7.946897 \text{ m/s},$$

$$\varepsilon_2 = -456.129366 \text{ rad/s}^2, \; \varepsilon_3 = 47.325219 \text{ rad/s}^2,$$

$$a_{Bx} = 83.259585 \text{ m/s}^2, \; a_{By} = -149.053057 \text{ m/s}^2, \; a_B = 170.730701 \text{ m/s}^2.$$

2.4.4 The Assisted Graphical Method

Using CAD, precise geometric constructions can be simulated. The term 'CAD' has a poly-semantic character; it is an acronym for different terms in English: computer-aided/assisted drawing/design/drafting. Generally speaking, it refers to the process of using of a computer to assist in the creation, modification, and representation of a drawing or project. By this definition, CAD is much more than a sophisticated computing program for graphical representations.

The central point of a CAD-type system is the association between a human being and informatics. When speaking about such a system, three basic components have to be taken into account: the human, the computer, and the database. The CAD system represents a technique by which the human and the computer form a team. The precision of drawing and calculation possessed by the well-known CAD soft is marked. The notion of scale vanishes, due to the possibility of representing vector parameters at their natural magnitudes. The main advantages of the CAD soft when solving the problems of mechanisms are:

- the double precision of drawing and calculation (one unit in the drawing means, in real numbers, the value of 1.0000000000000000)
- the almost unlimited possibilities for increasing and decreasing (*Zoom*) the size of the entities represented (10^{16} : 1)
- the coordinates of points can be absolute Cartesian, relative Cartesian, absolute polar, or relative polar
- the realization of planar or spatial constructions, no matter how complicated
- accurate determination of:
  - points of intersection (*INTersection*)
  - centre points of segments or circular arcs (*MIDpoint*)
  - centre points of circles or circular arcs (*CENter*)
  - tangent point between a circular arcs or circles and straight lines (*TANgent*)
  - construction of perpendiculars to segment of straight lines (*PERpendicular*)
- the dimensioning of the drawings with the posting of the linear dimensions of entities at their natural magnitude or multiplied by a scale factor
- use of *script* files to transfer drawing instructions from a programming language (Pascal, C++ and so on) to AutoCAD, thus obtaining
  - bi- or tri-dimensional graphics
  - drawings of successive positions of certain elements of mechanisms when the driving element occupies different positions
  - animation of the planar or spatial mechanisms.

In light of the advantages presented above, we may conclude that graphical methods, which are special due to their simplicity, become actual and analytical as well, when CAD soft is used. As an example, using AutoCAD, we will perform a kinematic analysis of the bi-mobile mechanism in Figure 2.5.

According to the algorithm described in Section 2.4.1, at $t = 0$, the angles are $\theta = 0$ and $\varphi = \frac{\pi}{2}$; so the positions of point A (0.4, 0.4) and C (1.3, 0.5) are known. For the determination of the coordinates of point B, a convenient point is at the intersection of:

- the circle with the centre at A with radius AB
- the circle at centre C with radius CB.

To create the geometrical construction having as its result the coordinates of point B, the following steps are taken.

1) Open an AutoCAD working session and choose a window of visualization

   Zoom;−0.5, −0.2, 2.5, 1; (where ; stands for 'Enter')

2) Construct two circles using the command **Circle**. The first has its centre at point A and radius 0.5, while the second has its centre at point C and radius 0.67:

   Circle;0.4, 0.4; 0.5;Circle;1.3, 0.5; 0.67;

3) The intersection of the two circles gives two solutions for point B. Choose the convenient solution, depending on the position of the dyad, and draw segment AB using the command **Line**:

   Line;0.4, 0.4;Int;\;

4) The symbol \ is used for the selection of point B using the mouse (we choose the superior point of intersection of the two circles). Using the command **Int** of the **Osnap** modes, the intersection point can be precisely located. The user saves the bar AB using the command **Block**. In the dialog box that appears, the name of the block is completed, along

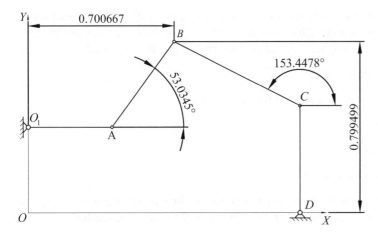

Figure 2.9 Construction of the mechanism using AutoCAD.

with the coordinate of the base point (point A), and the object (the segment AB) selected. If the command **−Block** is used, the dialog box does not appear, and the name of the block, the coordinates of the point used as base, and then the selection of the object have to be manually entered:

$$-\text{Block};\text{BarAB};0.4,0.4;\backslash;$$

5) Proceed in an analogous manner with the construction and saving of bar BC, with point C as the base point.

$$\text{Line};1.3,0.5;\text{Int};\backslash;;-\text{Block};\text{BarBC};1.3,0.5;\backslash;$$

6) Using the command **Pline**, construct the loop O_1ABCD, as in Figure 2.9, and then erase the two circles.

$$\text{Pline};0,0.4;\,W;0.002;\,0.002;\,0.4,0.4;\text{Int};\backslash;1.3,0.5;\,1.3,0;\,;\text{Erase};\backslash\backslash;$$

Draw the axes of coordinates:

$$\text{Pline};0,0;W;0.002;\,;\,1.5,0;W;0.02;\,0;L;0.07;\,;$$
$$\text{Pline};0,0;W;0.002;\,;\,0,1;W;0.02;\,0;L;0.07;\,;$$

the kinematic pairs and dimension the drawing as in Figure 2.9.

7) For the determination of the velocities use the vector relation (2.44). The angular velocities ω_1 and ω_4 are equal to $\dot\theta$ and $\dot\varphi$, respectively; hence $\omega_1 = 15$ rad/s and $\omega_4 = 12$ rad/s. At time $t = 0$, $v_A = 0.4 \cdot 15 = 6$ m/s and $v_C = 0.5 \cdot 12 = 6$ m/s. The representation of the vector polygon given by (2.44) is made in a new window of visualization:

$$\text{Zoom};0,0;12,12;$$

8) Choose as the pole of velocities the point at coordinates $i\,(5, 10)$. Starting from point i, construct the velocity of point A (**ia**). The symbol @ in front of a coordinate moves the reference frame to the last inserted point, thus point a from the polygon of velocities localizes at 0 units in the horizontal direction relative to the x-coordinate of point i and at -6 units in vertical direction relative to the y-coordinate of point i):

$$\text{Line};5,10;\,@0,-6;\,;$$

9) Construct a perpendicular to AB by inserting the block $BarAB$ at the last inserted point, using the command **Insert**. In the dialog box of the command introduce: the name of the

block, the insertion point (the end point of the velocity $\mathbf{v}_A$), the scale factor 5, equal for all three axes, and the rotation angle 270°. The sign − in front of the command **Insert** leads to the suppression of the dialog box, the data being typed using the keyboard:

−Insert;BarAB;5, 4; 5; 5; 270;

10) Proceed in an analogous way to construct the velocity $\mathbf{v}_C$:

Line;5, 10; @6, 0; ;

and the perpendicular to *BC*:

−Insert;BarBC;11, 10; 5; 5; 90;

11) To extend the two perpendiculars constructed above to their point of intersection, it is necessary to unselect the blocks *BarAB* and *BarBC* using the command **Explode**:

Explode;\\;

In this way, the extension of the two segments of straight lines is created using the command **Fillet**. At the first calling of the command, sets the fillet radius $R = 0$, and at the second calling select the two entities:

Fillet;R;0; ;Fillet;\\

12) Construct the segment *ib* with the command **Line**:

Line;5, 10;End;\;

13) The polygon drawn in Figure 2.10 is obtained, and this completes with arrows and dimensions.

14) Using the command **Cal**, perform the calculation to determine the angular velocities:

$$\omega_2 = \frac{v_{BA}}{l_2} = \frac{2.73004670}{0.5} = 5.4600093, \ \omega_3 = \frac{v_{BC}}{l_3} = \frac{8.54269005}{0.67} = 12.750284$$

and their squares

$$\omega_2^2 = 5.4600093\hat{\ }2 = 29.81170166, \ \omega_3^2 = 12.750284\hat{\ }2 = 162.56974208$$

15) We may pass to the determination of the accelerations using the vector relation (2.51). The accelerations of points *A* and *C* have the values:

$$a_A = 0.4 \cdot 15^2 = 90 \text{ m/s}^2, \ a_C = 0.5 \cdot 12^2 = 72 \text{ m/s}^2$$

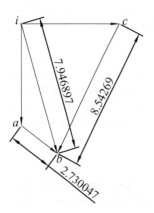

Figure 2.10 RRR dyad: the polygon of velocities.

Define a new window of visualization:

> Zoom;20, −50; 300, 160;

and successively construct (Figure 2.11):
- the acceleration of point A ($\mathbf{ja'}$) starting from point j at coordinates (150, 150)

> Line;150, 150; @ − 90, 0; ;

- the acceleration $\mathbf{a}_{BA}^{\nu}$ ($\mathbf{a'm'}$) by inserting the block *BarAB* at point a', the scale factor being $\omega_2^2 = 29.81170166$ (equal for all three axes), and the rotation angle 180°:

> −Insert;BarAB;60, 150; 29.81170166; 29.81170166; 180;

16) The perpendicular to AB is obtained by inserting the same block *BarAB* at point m', the scale factor being 100, equal for all three axes, the block being rotated with an angle of 270°:

> −Insert;BarAB;End;\100; 100; 270;

17) Returning to point j, successively construct:
- the acceleration of point C:

> Line;150, 150; @0, −72; ;

- the acceleration $\mathbf{a}_{BC}^{\nu}$ ($\mathbf{c'm'}$) by inserting the block *BarBC* at point c', the scale factor being $\omega_3^2 = 162.56974208$, equal for the all three axes, the block being rotated with an angle of 180°:

> −Insert;BarBC;150, 78; 162.56974208; 162.56974208; 180;

- the perpendicular to BC by inserting the block *BarBC* at point n', multiplied by the scale factor 100 (equal for the three axes) and rotated with an angle equal to 90°

> −Insert;BarBC;end;\100; 100; 90;

18) In order to extend the two perpendiculars constructed above to their intersection, it is necessary to un-select the blocks *BarAB* and *BarBC*,

> Explode;\\;;

and to fillet them using a radius of 0, which in fact leads to their extension to the point of intersection,

> Fillet;\\

19) The construction completes with the segment $\mathbf{jb'}$, as in Figure 2.11,

> Line;150, 150;End\;

We put the dimensions on the polygon of accelerations, and complete it with letters and arrows for the vectors.

The results obtained using this method and the assisted analytical one are identical, a fact that highlights the advantages of using the assisted design approach in solving problems of mechanisms.

We may conclude that CAD soft has a pleasant graphical interface and is similar to grapho-analytical methods. Moreover, because the solution is obtained in graphical form, and not as a listing of values as in the case of analytical methods, the approach is an attractive one.

The vector mode in which the entities drawn in AutoCAD are seen and deposited determines a greater similarity to the vector relations used in the grapho-analytical kinematic analysis of the mechanisms.

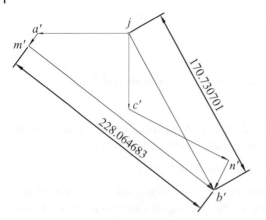

Figure 2.11 RRR dyad: the polygon of accelerations.

The graphical method presented above permits to obtain the polygons of velocities and accelerations using AutoCAD. The method is simple, but when we deal with more than one representation for one dyad, the volume of work is large and the method is no longer efficient. In this case, AutoLisp offers a solution because, being a programming language, it can implement, using procedures and functions, the calculation relations of an analytical method. We are using an analytical method in a graphical medium, but in the end we have a graphical representation of the results.

We will set out a working method based on AutoLisp. The method is intuitive, simple, and produces results close to those from a grapho-analytical method. Starting from the vector relations, vector polygons are obtained without using graphical constructions, the vector mode used being perfectly integrated in AutoCAD.

To highlight this aspect, we now present an AutoLisp function that constructs the polygon of velocities of the RRR dyad of the mechanism in Figure 2.5, and determines the velocity v_B and the angular velocities of elements 2 and 3, as in Figure 2.10. The name of the function is *Velocities* and it is set out below:

```
(Defun VELOCITIES ()
(Setq phi2 (angle ACapital BCapital)
phi3 (angle BCapital CCapital)
i (Getpoint "Choose the pole of velocities i:" )
asmall(Polar i alphav va)
rsmall (Polar asmall (+ phi2 (/ Pi 2)) 1)
csmall (Polar i betav vc)
ssmall (Polar csmall (+ phi 3 (Pi 2)) 1)
bsmall(Inters asmall rsmall csmall ssmall Nil)
vitb(Distance i bsmall)
om2(/ Distance asmall bsmall) (Distance ACapital BCapital))
om3(/ Distance csmall bsmall) (Distance BCapital CCapital)) )
```

The function is without local parameters or variables and is called from AutoCAD. The positional angles of the elements *AB* and *BC* of the dyad (φ_2 and φ_3, Figure 2.9) are first determined, using the function of multiple assignments *Setq*. Thus, the angle made by the straight line *AB* with the axis *OX* is assigned to *phi2*. The angle is determined using the function *Angle* by specifying the end points (*ACapital* and *Bcapital*) in a trigonometric sense. In an analogous way, the angle made by the straight line *BC* with the horizontal is assigned to *phi3*.

Then, there is the choice for the point (*i*), which becomes the pole of velocities, using the function *Getpoint*. Then, as in the grapho-analytical method, point *a* is obtained, starting from the pole *i* of velocities, with angle α_v and distance v_A, and using the AutoLisp function *Polar*.

With the same function an arbitrary point r that belongs to the perpendicular to AB is determined, starting (in polar coordinates) from point a with angle $\varphi_2 + \frac{\pi}{2}$ and an arbitrary distance (we have chosen the value 1). point c and then point s, which belongs to the perpendicular to BC, are then obtained in an analogous way. We start (in polar coordinates) from c with angle $\varphi_3 + \frac{\pi}{2}$ and an arbitrary distance (again, we have chosen the value 1).

Using the function *Inters* the point of intersection of the segments ar and cs, denoted by b, is determined.

The function *Inters* analyzes two straight lines, each given by two points, and returns their intersection point (if it exists). The distance between i and b is determined using the function *Distance* and it represents the magnitude of the velocity of point b.

In this way, we have analytically determined the value of the velocity of point B, using a graphical method and making no geometric construction.

The angular velocities of elements 2 and 3 (ω_2 and ω_3) are determined as the ratios between the distances ab and AB, and the distances bc and BC, respectively ($\omega_2 = \frac{ab}{AB}$, $\omega_3 = \frac{bc}{BC}$).

If a graphical representation is also required, then the function *Command* is used, from which AutoCAD functions may be called, drawing lines between the obtained points. The geometric figure is completed with arrows that indicate the senses of the vectors, using the AutoCAD command **Pline**. The completion of the previous function is given below:

```
(Command "Line" i asmall bsmall i csmall bsmall)
(Setq l(/va 12) g(* i 0.035))
(Command "Pline" asmall "W" "0" g (Polar asmall (-alphav Pi) l) "")
(Command "Pline" csmall "W" "0" g (Polar csmall (-betav Pi) l) "")
(Command "Pline" bsmall "W" "0" g (Polar bsmall (- (angle i bsmall)
  Pi) l) "")
(Command "Pline" bsmall "W" "0" g (Polar bsmall (- (angle asmall bsmall)
  Pi) l) "")
(Command "Pline" bsmall "W" "0" g (Polar bsmall (- (angle csmall bsmall)
  Pi) l) "")
```

The acceleration of point b and the angular accelerations of elements 2 and 3, can be determined in a similar way, as in Figure 2.11.

The AutoLisp function that constructs the polygon of accelerations and determines from it the acceleration $\mathbf{a}_B$ and the angular accelerations of elements 2 and 3 is as follows:

```
(Defun ACCELER ()
(Setq j (Getpoint "Choose the pole of accelerations j: ")
aprime (Polar j alphaa acca)
mprime (Polar aprime (+ phi2 Pi) (* om2 om2 (Distance ACapital
  BCapital)))
rsmall (Polar mprime (+ phi2 (/Pi 2)) 10)
cprime (Polar j betaa accc)
(Polar cprime phi3 (* om3 om3 (Distance BCapital CCapital)))
ssmall (Polar nprime (+ phi3 (/ Pi 2)) 10)
bprime (Inters mprime rsmall nprime ssmall Nil)
accb (Distance j bprime)
eps2 (/ (Distance mprime bprime) (Distance ACapital BCapital))
eps4 (/ (Distance nprime bprime) (Distance BCapital CCapital)))
(Command "Line" j aprime mprime bprime j nprime bprime "")
(Arrows aprime j)
(Arrows cprime j)
(Arrows bprime j)
(Arrows bprime mprime)
```

```
(Arrows bprime nprime)
(Arrows mprime aprime)
(Arrows nprime cprime)
```

As in the previous case, the function has no local parameters or variables. The function requires a point j, which will be used as the pole of accelerations. From this point, point a' is obtained, starting in polar coordinates with angle α_A made by the acceleration $\mathbf{a}_A$ with the horizontal, and with the distance a_A (the magnitude of the acceleration). Next, point m' (Figure 2.11) is determined, starting in polar coordinates from point a' with angle $\varphi_2 + \pi$ and distance $\omega_2^2 AB$ (`(* om2 om2 (Distance ACapital BCapital))`). Then, this point is used to construct a perpendicular to AB starting, in polar coordinates, from point m' with angle $\varphi_2 + \frac{\pi}{2}$, and with an arbitrary distance (`(Polar mprime (+ phi2 (/Pi 2)) 10)`).

According to the vector relation (2.51), we construct, from point j, the acceleration $\mathbf{a}_C$ (the segment jc') of point C (`(Polar j betaa accc)`) and then the normal acceleration $\mathbf{a}_{BC}^v$, starting, in polar coordinates, from point c' previously determined, with angle φ_3 and distance $\omega_3^2 BC$ (segment $c'n'$ in Figure 2.11).

For the construction of the perpendicular to BC we start, in polar coordinates, from point n', with angle $\varphi_3 + \frac{\pi}{2}$ and an arbitrary distance 10. We thus obtain the point sprime.

Using the function *Inters*, we get point b' by intersecting segments $m'r'$ and $n's'$. The distance between points j and b' represents the magnitude of the acceleration of point B (`(accb (Distance j bprime))`).

The angular accelerations of elements 2 and 3 (ε_2 and ε_3) are determined as the ratios between distances $m'b'$ and AB, and distances $b'n'$ and BC, respectively ($\varepsilon_2 = \frac{m'b'}{AB}, \varepsilon_3 = \frac{b'n'}{BC}$).

Using function *Command*, the AutoCAD command **Line** is called to construct the loop $ja'm'b'jc'n'b'$. Then, the function called *Arrows* can be used to complete the polygon of accelerations with arrows.

```
(Defun Arrows (head tail)
(Command "Pline" head "W" "0" g(Polar head (angle head tail) 1) "") )
```

Because the arrows have always the same shape, we agreed this function to avoid the repetition of the same expression, but with other points as head and tail of the arrow. In this way we simplify the writing. The same function may be used in the previous function *Velocities*.

Further on we will present an AutoLisp function which performs the kinematic analysis of the RRR dyad. This function is saved as *DyadRRR.lsp*.

```
(Defun C:RRR ()
(Setq xO1 0.0 yO1 0.4 xD 1.3 yD 0.0 lAB 0.5 lBC 0.67 xA 0.4 yA 0.4
  xC 1.3 yC 0.5 ismall
(List 5 10) vA 6.0 angvA 270.0 vC 6.0 angvC 0.0 jsmall(List 100 50) accA
90.0 angaA 180.0 accC 72.0 angaC 270.0)
;Calculation of the positions
(Setq a1 xA b1 yA R1 lAB a2 xC b2 yC R2 lBC)
(Int_2C)
(Setq xB x1 yB y1)
;Position of the mechanism
(Command "Erase" "All" "" "Osnap" "Off" "Ortho" "Off")
(Command "Zoom" "-0.5,-0.2" "2.5,1")
(Command "-Layer" "N" "Positions" "C" "7" "Positions" "S"
  "Positions" "" "")
```

```
(Command "Pline" (List xO1 yO1) "W" "0.002" "" (List xA yA) BCapital
(List xC yC)
  (List xD yD) "")
(Command "Pline" "0,0" "W" "0.002" "" "1.5,0" "W" "0.02" "0" "L"
  "0.07" "")
(Command "Pline" "0,0" "W" "0.002" "" "0,1" "W" "0.02" "0" "L"
  "0.07" "") ;Polygon of velocities
(Setq alfav(/ (* angvA PI) 180) betav(/ (* angvC PI) 180))
(Command "Zoom" "0,0" "12,12")
(Command "-Layer" "N" "Velocities" "C" "1" "Velocities" "S"
  "Velocities" "" "")
(Velocities)
;Polygon of accelerations
(Setq alfaa(/ (* angaA PI) 180) betaa(/ (* angaC PI) 180))
(Command "Zoom" "20,-50" "300,160")
(Command "-Layer" "N" "Accelerations" "C" "5" "Accelerations" "S"
  "Accelerations" "" "")
(Accelerations)
)
```

The function is called *Dyad_RRR.lsp* and can be called from AutoCAD. After loading (*Load "Dyad_RRR.lsp"*) it is called simply *RRR*. The function begins with the assignment of values (*Setq*) for the poles of the RRR dyad. From what we have already presented, several problems result and must be solved in AutoLisp.

A first problem is the determination of the intersection of two straight lines. This can be directly solved using the AutoLisp function *Inters*, which analyzes the two straight lines and returns their intersection point, if it exists (Nil). The two straight lines are defined by two points for each.

The second problem is the determination of the intersection points of two circles. Denoting the centres of the two circles by $O_1(a_1, b_1)$ and $O_2(a_2, b_2)$, respectively, and the radii by R_1 and R_2, respectively, with the notation in Figure 2.12, we obtain the vector relation

$$\mathbf{OO_1} + \mathbf{O_1P} = \mathbf{OO_2} + \mathbf{O_2P}, \tag{2.77}$$

which, projected onto the axes of the system OXY, leads to the linear system

$$\begin{cases} a_1 + R_1 \cos \varphi_1 = a_2 + R_2 \cos \varphi_2, \\ b_1 + R_1 \sin \varphi_1 = b_2 + R_2 \sin \varphi_2. \end{cases} \tag{2.78}$$

From the system (2.78) the solutions for angle φ_1 are obtained as:

$$\varphi_{1_{1,2}} = 2 \arctan\left(\frac{-B \pm \sqrt{A^2 + B^2 - C^2}}{C - A} \right), \tag{2.79}$$

Figure 2.12 The intersection of two circles.

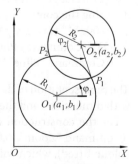

where

$$A = 2R_1 \left(a_1 - a_2 \right), \ B = 2R_1 \left(b_1 - b_2 \right),$$
$$C = \left(a_1 - a_2 \right)^2 + \left(b_1 - b_2 \right)^2 + R_1^2 - R_2^2. \tag{2.80}$$

Thus, if they exist, the intersection points $P_1 \left(x_1, y_1 \right)$ and $P_2 \left(x_2, y_2 \right)$ are obtained, where

$$x_1 = a_1 + R_1 \cos \varphi_{1_1}, \ y_1 = a_1 + R_1 \sin \varphi_{1_1}, \tag{2.81}$$
$$x_2 = a_1 + R_1 \cos \varphi_{1_2}, \ y_2 = a_1 + R_1 \sin \varphi_{1_2}. \tag{2.82}$$

The existence of the intersection points is given by the sign of the expression under the radical in (2.79); that is,

- if $A^2 + B^2 - C^2 > 0$, then there are two points of intersection
- if $A^2 + B^2 - C^2 = 0$, then there is only one point of intersection (the circles are tangential)
- if $A^2 + B^2 - C^2 < 0$, then there is no point of intersection.

To obtain the intersection points in AutoCAD, without geometrical constructions, based on (2.78)–(2.82), we wrote the AutoLisp function *Int_2C*, which is shown in Table 2.1. This returns the intersection points $P_1 \left(x_1, y_1 \right)$ and $P_2 \left(x_2, y_2 \right)$ between two circles and assigns to variable *error* the value 0 when there are two points of intersection, and the value 1 if the circles do not intersect. For better understanding of the logic of the function, we have inserted some comments in the right-hand column of the table. The function may be easily called and, further on, it will be used in the function *RRR* (*Defun C:RRR*).

The AutoLisp functions can be directly edited in AutoCAD using the Visual Lisp Editor (Tools ≫ AutoLisp ≫ Visual Lisp editor). As a rule, the AutoLisp functions are situated in the folder 'Support' of AutoCAD and they have the extension .lsp. Therefore, we have to indicate the path to the folder in which is the file is located. The functions may be typed using any editing program that can save in ASCII format (Save As... ≫ simple text (*.txt) ≫ MS-DOS ≫ name_of_the_function.lsp); we prefer the Visual Lisp Editor because of its editing facilities.

Before the function is called, the following assignments are made: $a_1 = x_A$, $b_1 = y_A$, $a_2 = x_C, b_2 = y_C, R_2 = BC$ ((Setq a1 xA b1 yA a2 xC b2 yC R2 lBC)). After it has been called (*Int_2C*) it returns the intersection points $P_1(x_1, y_1)$ and $P_2(x_2, y_2)$; that is, the two solutions for point *B*. The convenient solution is the first one ((Setq xB x1 yB y1)).

The mechanism's loop can be readily drawn.

1) Firstly, erase all previous entities (if they exist) and inhibit the AutoCAD modes **Osnap** and **Ortho** ((Command "Erase" "All" "" "Osnap" "Off" ' Ortho" "Off")). The mechanism will be positioned in a window defined with the command **Zoom** ((Command "Zoom" "−0.5, −0.2" "2.5,1")) and in a layer named *Positions*.
2) Create the layer, and assign the number 7 for the colour, and then set it as the current layer ((Command "−Layer" "N" "Positions" "C" "7" "Positions" "S" "Positions" "" "")).
3) Using the command **Pline**, draw the loop O_1ABC ((Command "Pline" (List xO1 yO1) "W" "0.002" "" (List xA yA) BCapital (List xC yC) (List xD yD) "")) and the axes of coordinates *OX* ((Command "Pline" "0,0" "W" "0.002" "" "1.5,0" "W" "0.02" "0" "L" "0.07" "")) and *OY* ((Command "Pline" "0,0" "W" "0.002" "" "0,1" "W" "0.02" "0" "L" "0.07" "")).

The representation of the mechanism, as in Figure 2.9, will be displayed. The figure is completed with dimensions and the conventional representations of the kinematic pairs.

For the construction of the polygon of velocities a window of visualization is established ((Command "Zoom" "0,0" "12,12")) and also a new layer called *Velocities*, with the number of colour 1 (red), which will become the current layer ((Command "−Layer" "N" "Velocities" "C"

Table 2.1 The function *Int_2C*.

Program	Comments
`(Defun Int_2C ()`	Definition of the function
`(Setq A (* 2 R1 (- a1 a2))`	point *A* relation (2.80)
`B (* 2 R1 (- b1 b2))`	point *B* relation (2.80)
`C (+ (* (- a1 a2) (- a1 a2)) (* (- b1 b2) (- b1 b2))` `(* R1 R1) (* R2 R2 -1))`	point *C* relation (2.80)
`aa (+ (* A A) (* B B) (* C C -1)))`	Discriminant Δ
`(If (>= aa 0)`	If *aa* ≥ 0
`(PROGN`	then instructions
`(Setq error 0)`	
`(Setq bb (/ (+ (* -1 B) (Sqrt aa)) (- C A)))`	
`(If (> bb 0)`	If *bb* > 0
`(Setq phi11 (* 2 (atan bb)))`	then
`(Setq phi11 (+ (* 2 Pi) (* 2 (atan bb))))`	else
`)`	End If *bb* > 0
`(Setq bb (/ (- (* -1 B) (Sqrt aa)) (- C A)))`	
`(If (> bb 0)`	If *bb* > 0
`(Setq phi12 (* 2 (atan bb)))`	then
`(Setq phi12 (+ (* 2 Pi) (* 2 (atan bb))))`	else
`)`	End If *bb* > 0
`)`	End PROGN
`(Setq error 1)`	else from *aa* ≥ 0
`)`	End If *aa* ≥ 0
`(If (= error 0)`	
`(PROGN`	then instructions
`(Setq x1 (+ a1 (* R1 (cos phi11)))`	x_1 relation (2.81)
`y1 (+ b1 (* R1 (sin phi11)))`	y_1 relation (2.81)
`x2 (+ a1 (* R1 (cos phi12)))`	x_2 relation (2.82)
`y2 (+ b1 (* R1 (sin phi12)))`	y_2 relation (2.82)
`P1 (List x1 y1)`	
`P2 (List x2 y2))`	
`)`	End from PROGN
`(Print "The circles do not intersect")`	else from error=0
`)`	End IF error=0
`)`	End function

"1" "Velocities" "S" "Velocities" "" "")). Calling the function *Velocities* defined above ((Velocities)), the representation given in Figure 2.10 will be displayed, the polygon being drawn with red lines.

The polygon of accelerations is constructed in a similar way. A new window of visualization ((Command "Zoom" "20,−50" "300,160")) is defined, and also a new layer, which becomes the current one ((Command "−Layer" "N" "Accelerations" "C" "5" "Accelerations" "S"

"Accelerations" "" "")). Calling the function *Accelerations* ((Accelerations)) presented above, the polygon in Figure 2.11 will be displayed, drawn in blue.

For the dimensioning of the drawings, the layers that are not of interest are hidden, leaving three layers: the position of the mechanism and the polygons of velocities and accelerations. By modifying the input data, other representations of the mechanism can be obtained without any other intervention. The input data can also be the result of queries and in this case there will be no intervention required in the body of the function.

2.5 Kinematic Analysis of the RRT Dyad

2.5.1 The Grapho-analytical Method

Formulation of the problem

Consider the dyad ABC in Figure 2.13a, for which the following are known:

- the positions of point A, the straight line Δ, and one of its points D, lengths l_2 and l_3 of segments AB and BC, respectively, and angle α
- the velocities $\mathbf{v}_A$ and $\mathbf{v}_D$ of points A and D, respectively, and the angular velocity ω_4 of the straight line Δ
- the accelerations $\mathbf{a}_A$ and $\mathbf{a}_D$ of the points A and D, respectively, and the angular velocity ε_4 of the straight line Δ.

The following values are to be determined:

- the positions of points B and C
- the angular velocity ω_2, the velocity of point B, and the relative velocity of slider C.

Determination of the positions

The scale k_l of the lengths is chosen. Since the distance d_3 from point B to the straight line Δ is known:

$$d_3 = l_3 \sin \alpha, \tag{2.83}$$

the following working procedure for the graphical construction results (Figure 2.13b):

1) Construct the straight line Δ', parallel to the straight line Δ and situated at the distance $[d_3]$ relative to it (two solutions).
2) With the centre at point A, construct the circle arc of radius $[l_2]$ that intersects the straight line Δ' at points B and B' (two solutions), and choose the solution that corresponds to the initial conditions (Figure 2.13b).
3) With the centre at point B, construct the circle arc of radius $[l_3]$ that intersects the straight line Δ at points C and C'; choose the solution that corresponds to the initial conditions (Figure 2.13b).

Determination of the velocities

The scale of velocities k_v is chosen and the velocity $\mathbf{v}_{BA}$ (Figure 2.13c) of point B, rigidly linked to the straight line Δ is determined. The following relation is used:

$$\mathbf{v}_{B_4} = \underset{=}{\mathbf{v}_D} + \underset{\underset{\perp BD}{=}}{\mathbf{v}_{B_4D}}, \tag{2.84}$$

where

$$v_{B_4D} = \omega_4 DB = \omega_4 k_l [DB]. \tag{2.85}$$

Since points B_2 and B_3 coincide, $\mathbf{v}_{B_2} = \mathbf{v}_{B_3}$.

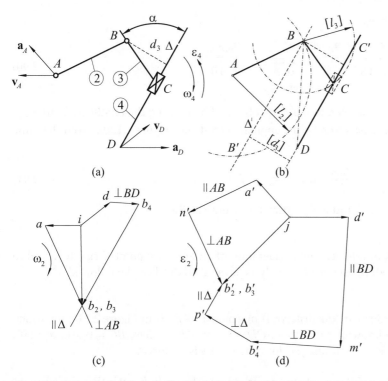

Figure 2.13 The grapho-analytical kinematic analysis of the RRT dyad: (a) RRT dyad; (b) position of the RRT dyad; (c) plane of velocities; (d) plane of accelerations.

Writing the Euler relation for point B_2 and the relation of the composition of velocities in the case of the relative motion for point B_3, we have

$$\mathbf{v}_{B_2} = \mathbf{v}_A + \mathbf{v}_{B_2 A}, \ \mathbf{v}_{B_3} = \mathbf{v}_{B_4} + \mathbf{v}_{B_3 B_4}, \tag{2.86}$$

resulting in the equality

$$\underline{\underline{\mathbf{v}_A}} + \underline{\underline{\mathbf{v}_{B_2 A}}} = \underline{\underline{\mathbf{v}_{B_4}}} + \underline{\underline{\mathbf{v}_{B_3 B_4}}}, \tag{2.87}$$
$$\perp AB \qquad\qquad \parallel \Delta$$

which has the representation shown in Figure 2.13c.

Finally, this results in:

$$\mathbf{v}_{B_2 A} = k_v \left(\mathbf{ab}_3 \right), \ \mathbf{v}_{B_3} = \mathbf{v}_{B_2} = k_v \left(\mathrm{i} b_2 \right),$$
$$\mathbf{v}_{B_3 B_4} = k_v \left(\mathbf{b}_4 \mathbf{b}_3 \right), \ \omega_2 = \frac{v_{B_2 A}}{l_2}. \tag{2.88}$$

The sense of the angular velocity ω_2 is represented in Figure 2.13c.

Determination of the accelerations

The scale k_a of the accelerations is chosen and the acceleration of point B_4 is determined using (Figure 2.13d):

$$\mathbf{a}_{B_4} = \underline{\underline{\mathbf{a}_D}} + \mathbf{a}^{\nu}_{B_4 D} + \mathbf{a}^{\tau}_{B_4 D}$$
$$B \to D \qquad \perp BD \tag{2.89}$$
$$\omega_4^2 BD \qquad \varepsilon_4 BD$$

From the equality $\mathbf{a}_{B_2} = \mathbf{a}_{B_3}$:

$$\underbrace{\mathbf{a}_A}_{} + \underbrace{\mathbf{a}^{\nu}_{B_2A}}_{\substack{B \to A \\ l_3\omega_2^2}} + \underbrace{\mathbf{a}^{\tau}_{B_2A}}_{\perp AB} = \underbrace{\mathbf{a}_{B_4}}_{} + \underbrace{\mathbf{a}^{C}_{B_3B4}}_{\substack{\perp \Delta \\ 2\omega_4 v_{B_3B_4}}} + \underbrace{\mathbf{a}_{B_3B_4}}_{\parallel AB}. \tag{2.90}$$

The sense of the Coriolis acceleration $\mathbf{a}^{C}_{B_3B_4}$ is obtained by rotating the relative velocity $\mathbf{v}_{B_3B_4}$ by an angle of $90°$ in the sense given by the angular velocity ω_4 (Figure 2.13d). From the same figure:

$$\mathbf{a}^{\tau}_{B_2A} = k_a\left(\mathbf{n'}\mathbf{b'}_3\right), \; \varepsilon_2 = \frac{\mathbf{a}^{\tau}_{B_2A}}{l_2}, \; \mathbf{a}_{B_3B_4} = k_a\left(\mathbf{p'}\mathbf{b'}_3\right). \tag{2.91}$$

The sense of the angular acceleration ε_2 is drawn in Figure 2.13d.

Observation

To determine the position, velocity, and acceleration of an arbitrary point K rigidly linked to element 2 or 3, the procedure used for the analysis of the RRR dyad can be followed.

Numerical example

Problem: Consider the four-bar (quadrilateral) mechanism $ABCD$ in Figure 2.14a, for which the following values are known: $OA = 0.4$ m, $OD = 0.8$ m, $AB = 0.5$ m, $BC = 0.6$ m, $\alpha = 60°$, $\varphi = 10t + \frac{\pi}{2} - 10$ rad. A kinematic analysis for time $t = 1$ s is required.

Solution: For the position of the mechanism, we choose the scale k_l such that the segment OA has a representation of 20 mm; this gives $k_l = 0.02 \left[\frac{m}{mm}\right]$. In this case, the distance $d_3 = BC \sin\alpha$ has the value 0.5196 m, so $[d_3] = 25.98$ mm. In addition, $[OA] = 20$ mm, $[OB] = 40$ mm, $[AB] = 25$ mm. Since at $t = 1$ s, $\varphi = \frac{\pi}{2}$ rad, the scale representation in Figure 2.13b is obtained, from which we get $[DB] = 48$ mm; that is, $DB = 0.96$ m.

The angular velocity ω_4 of the straight line Δ has the value $\omega_4 = \dot{\varphi} = 10$ rad/s. Since the straight line Δ has a rotational motion, the velocity of point B_4 has the value $v_{B_4} = BD\omega_4 = 9.6$ rad/s and it is perpendicular to BD.

Taking into account that the velocity of point A is zero, (2.87) becomes

$$\underbrace{\mathbf{v}_{B_3}}_{\perp AB} = \underbrace{\mathbf{v}_{B_4}}_{} + \underbrace{\mathbf{v}_{B_3B_4}}_{\parallel \Delta}.$$

Choosing the scale of velocities such that 10 m/s corresponds to 20 mm, we obtain the values $k_v = 0.5 \left[\frac{m/s}{mm}\right]$ and $[v_{B_4}] = 19.2$ mm, and the representation in Figure 2.14c. From this, $[ib_3] = 19$ mm and $[b_4b_3] = 21$ mm and:

$$v_B = v_{B_2} = v_{B_3} = 9.5 \text{ m/s}, \; \omega_2 = 19 \text{ rad/s}, \; v_{B_3B_4} = 10.5 \text{ m/s}.$$

The sense of the angular velocity ω_2 is shown in Figure 2.14b. Since element 4 has an uniform rotational motion, the acceleration of point B_4 has magnitude $a_{B_4} = \omega_4^2 BD = 96$ m/s^2 and the sense from point B to point D.

In the conditions of this example, (2.90) becomes:

$$\underbrace{\mathbf{a}^{\nu}_{B_2A}}_{B \to A} + \underbrace{\mathbf{a}^{\tau}_{B_2A}}_{\perp AB} = \underbrace{\mathbf{a}_{B_4}}_{B \to D} + \underbrace{\mathbf{a}^{C}_{B_3B4}}_{\perp \Delta} + \underbrace{\mathbf{a}_{B_3B_4}}_{\parallel \Delta},$$

where $a^{\nu}_{B_2A} = AB\omega_2^2 = 180.5$ m/s^2, $a_{B_4} = 96$ m/s^2, and $a^{C}_{B_3B_4} = 2\omega_4 v_{B_3B_4} = 210$ m/s^2.

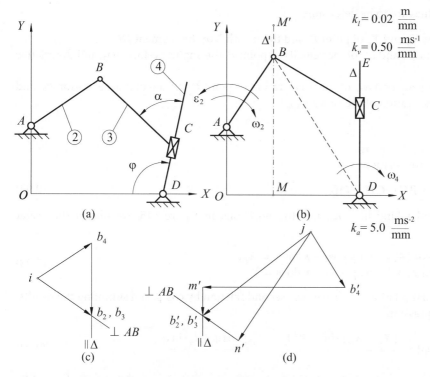

$$k_l = 0.02 \frac{\text{m}}{\text{mm}}$$

$$k_v = 0.50 \frac{\text{ms}^{-1}}{\text{mm}}$$

$$k_a = 5.0 \frac{\text{ms}^{-2}}{\text{mm}}$$

Figure 2.14 Four-bar (quadrilateral) mechanism with RRT dyad: (a) four-bar mechanism; (b) position of the mechanism; (c) plane of velocities; (d) plane of accelerations.

Choosing the scale of accelerations such that 100 m/s^2 corresponds to a representation of 20 mm, we obtain $k_a = 5 \left[\frac{\text{m/s}^2}{\text{mm}} \right]$, and:

$$\left[a_{B_2A} \right] = 36.1 \text{ mm}, \quad \left[a_{B_4} \right] = 19.2 \text{ mm}, \quad \left[a_{B_3B_4}^C \right] = 42 \text{ mm}.$$

The representation in Figure 2.14d is obtained, from which $\left[m'b_3' \right] = 7 \text{ mm}$, $\left[n'b_3' \right] = 13 \text{ mm}$, $\left[jb_3' \right] = 39 \text{ mm}$; hence

$$a_{B_2A}^\tau = k_a \left(n'b_3' \right) = 65 \text{ m/s}^2, \quad \varepsilon_2 = \frac{a_{B_2A}}{AB} = 130 \text{ rad/s}^2,$$
$$a_{B_3B_4} = k_a \left(m'b_3' \right) = 35 \text{ m/s}^2, \quad a_B = a_{B_2} = a_{B_3} = k_a \left(jb_3' \right) = 195 \text{ m/s}^2.$$

2.5.2 The Analytical Method

Formulation of the problem

Consider the dyad ABC in Figure 2.15, for which the following values are known:

- the coordinates X_A, Y_A and X_D, Y_D of points A and D, respectively; the angle φ made by the straight line Δ with the axis OX; the lengths l_2 and l_3 of segments AB and BC, respectively; the angle α between the segment BC and the straight line Δ;
- the components v_{Ax}, v_{Ay} and v_{Dx}, v_{Dy} of the velocities of points A and D, respectively; the angular velocity ω_4 of the straight line Δ;
- the components a_{Ax}, a_{Ay} and a_{Dx}, a_{Dy} of the accelerations of points A and D, respectively; the angular acceleration ε_4 of the straight line Δ.

The following values are to be determined:

- the coordinates X_B and Y_B of point B, and the length λ of the segment DC
- the components v_{Bx} and v_{By} of the velocity of point B, the angular velocity ω_2, and the relative velocity $v_{B_3B_4}$
- the components a_{Bx} and a_{By} of the acceleration of point B, the angular acceleration ε_2, and the relative acceleration of the slider $a_{B_3B_4}$.

Determination of the positions

Projecting the vector relation

$$\mathbf{OA} + \mathbf{AB} + \mathbf{BC} + \mathbf{CD} = \mathbf{OD} \tag{2.92}$$

onto the axes and taking into account the notations in Figure 2.15, we obtain the scalar equations

$$\begin{aligned}
X_A + l_2 \cos\varphi_2 + l_3 \cos(\varphi - \alpha) - \lambda\cos\varphi = X_D, \\
Y_A + l_2 \sin\varphi_2 + l_3 \sin(\varphi - \alpha) - \lambda\sin\varphi = Y_D.
\end{aligned} \tag{2.93}$$

Multiplying the first relation by $-\sin\varphi$, the second relation by $\cos\varphi$, and summing the results, we obtain the expression

$$\sin(\varphi_2 - \varphi) = \frac{(X_A - X_D)\sin\varphi - (Y_A - Y_D)\cos\varphi + l_3 \sin\alpha}{l_2}. \tag{2.94}$$

from which angle φ_2 can be deduced. Similarly, if the first relation in (2.93) is multiplied by $\cos\varphi$ and the second one by $\sin\varphi$, then parameter λ can be obtained:

$$\lambda = (X_A - X_D)\cos\varphi + (Y_A - Y_D)\sin\varphi + l_2 \cos(\varphi_2 - \varphi) + l_3 \cos\alpha. \tag{2.95}$$

The coordinates of point B are given by:

$$X_B = X_A + l_2 \cos\varphi_2, \quad Y_B = Y_A + l_2 \sin\varphi_2. \tag{2.96}$$

Determination of the velocities

From the vector relation

$$\mathbf{v}_{B_4} = \mathbf{v}_D + \boldsymbol{\omega}_4 \times \mathbf{DB}, \tag{2.97}$$

the components of the velocity of point B_4 can be obtained:

$$v_{B_4x} = v_{Dx} - \omega_4(Y_B - Y_D), \quad v_{B_4y} = v_{Dy} + \omega_4(X_B - X_D) \tag{2.98}$$

Now writing the vector relation

$$\mathbf{v}_A + \boldsymbol{\omega}_2 \times \mathbf{AB} = \mathbf{v}_{B_4} + \mathbf{v}_{B_3B_4} \tag{2.99}$$

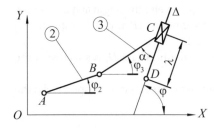

Figure 2.15 The RRT dyad.

and projecting it onto the axes, we obtain the scalar system

$$
\begin{aligned}
v_{Ax} - \omega_2 \left(Y_B - Y_A \right) &= v_{B_4x} + v_{B_3B_4} \cos \varphi, \\
v_{Ay} + \omega_2 \left(X_B - X_A \right) &= v_{B_4y} + v_{B_3B_4} \sin \varphi.
\end{aligned}
\tag{2.100}
$$

Then, by eliminating the parameter $v_{B_3B_4}$ we get:

$$
\omega_2 = \frac{- \left(v_{B_4x} - v_{Ax} \right) \sin \varphi + \left(v_{B_4y} - v_{Ay} \right) \cos \varphi}{\left(X_B - X_A \right) \cos \varphi + \left(Y_B - Y_A \right) \sin \varphi}.
\tag{2.101}
$$

Multiplying the first relation in (2.100) by $\cos \varphi$ and the second one by $\sin \varphi$, the relative velocity can be deduced:

$$
\begin{aligned}
v_{B_3B_4} &= \left[v_{Ax} - v_{B_4x} - \omega_2 \left(Y_B - Y_A \right) \right] \cos \varphi \\
&+ \left[v_{Ay} - v_{B_4y} + \omega_2 \left(X_B - X_A \right) \right] \sin \varphi.
\end{aligned}
\tag{2.102}
$$

The components of the velocity of point B are given by:

$$
v_{Bx} = v_{Ax} - \omega_2 \left(Y_B - Y_A \right), \quad v_{By} = v_{Ay} + \omega_2 \left(X_B - X_A \right).
\tag{2.103}
$$

Determination of the accelerations

From the vector relation

$$
\mathbf{a}_{B_4} = \mathbf{a}_D + \boldsymbol{\varepsilon}_4 \times \mathbf{DB} - \omega_4^2 \mathbf{DB},
\tag{2.104}
$$

the components of the acceleration of point B_4 can be deduced

$$
\begin{aligned}
a_{B_4x} &= a_{Dx} - \varepsilon_4 \left(Y_B - Y_D \right) - \omega_4^2 \left(X_B - X_D \right), \\
a_{B_4y} &= a_{Dy} + \varepsilon_4 \left(X_B - X_D \right) - \omega_4^2 \left(Y_B - Y_D \right).
\end{aligned}
\tag{2.105}
$$

The vector relation

$$
\mathbf{a}_A + \boldsymbol{\varepsilon}_2 \times \mathbf{AB} - \omega_2^2 \mathbf{AB} = \mathbf{a}_{B_4} + \mathbf{a}_{B_3B_4}^C + \mathbf{a}_{B_3B_4}.
\tag{2.106}
$$

can be projected onto the axes OX and OY, and using:

$$
\begin{aligned}
A_x &= a_{B_4x} - a_{Ax} + \omega_2^2 \left(X_B - X_A \right) - 2\omega_4 v_{B_3B_4} \sin \varphi, \\
A_y &= a_{B_4y} - a_{Ay} + \omega_2^2 \left(Y_B - Y_A \right) + 2\omega_4 v_{B_3B_4} \cos \varphi.
\end{aligned}
\tag{2.107}
$$

we obtain the system

$$
\begin{aligned}
-\varepsilon_2 \left(Y_B - Y_A \right) - a_{B_3B_4} \cos \varphi &= A_x, \\
\varepsilon_2 \left(X_B - X_A \right) - a_{B_3B_4} \sin \varphi &= A_y,
\end{aligned}
\tag{2.108}
$$

from which the unknowns can be determined:

$$
\varepsilon_2 = \frac{-A_x \sin \varphi + A_y \cos \varphi}{\left(Y_B - Y_A \right) \sin \varphi + \left(X_B - X_A \right) \cos \varphi},
\tag{2.109}
$$

$$
a_{B_3B_4} = - \left[A_x + \varepsilon_2 \left(Y_B - Y_A \right) \right] \cos \varphi - \left[A_y - \varepsilon_2 \left(X_B - X_A \right) \right] \sin \varphi.
\tag{2.110}
$$

The components of the acceleration of point B are given by:

$$
\begin{aligned}
a_{Bx} &= a_{Ax} - \varepsilon_2 \left(Y_B - Y_A \right) - \omega_2^2 \left(X_B - X_A \right), \\
a_{By} &= a_{Ay} + \varepsilon_2 \left(X_B - X_A \right) - \omega_2^2 \left(Y_B - Y_A \right).
\end{aligned}
\tag{2.111}
$$

Numerical example

Problem: Using the analytical method, perform the kinematic analysis of the mechanism in Figure 2.14a, considering the same conditions as in the numerical example used to demonstrate the grapho-analytical method. Compare the results.

Solution: From the problem, points A and B have the coordinates

$$X_A = 0, Y_A = 0.4\,\text{m}, X_D = 0.8\,\text{m}, Y_D = 0.$$

We also have

$$\varphi = \frac{\pi}{2}, \ \alpha = \frac{2\pi}{3}, \ l_2 = 0.5\,\text{m}, \ l_3 = 0.6\,\text{m}.$$

and from t (2.83) we obtain

$$\sin\left(\varphi_2 - \frac{\pi}{2}\right) = -0.560759 \text{ or } \varphi_2 = 55.891°.$$

From Equation 2.95, $\lambda = 0.514\,\text{m}$, while (2.96) gives the coordinates $X_B = 0.2804\,\text{m}$ and $Y_B = 0.8140\,\text{m}$. Taking into account the sense of angle φ, we obtain $\omega_4 = -10\,\text{rad/s}$; since $v_{Dx} = v_{Dy} = 0$, from (2.98) $v_{B_4 x} = 8.12\,\text{m/s}$ and $v_{B_4 y} = 5.196\,\text{m/s}$. From (2.101), $\omega_2 = -19.66\,\text{rad/s}$, and from (2.102) $v_{B_3 B_4} = -10.709\,\text{m/s}$. The magnitude of the relative velocity is in good agreement with that deduced using the grapho-analytical method, while the – sign shows that it is orientated from point C to point D. Equation 2.103 gives the components of the velocity of point B: $v_{Bx} = 8.139\,\text{m/s}$ and $v_{By} = -5.513\,\text{m/s}$; hence, $v_B = 9.83\,\text{m/s}$.

Passing now to the accelerations, we have $a_{Ax} = a_{Ay} = a_{Dx} = a_{Dy} = 0, \varepsilon_4 = 0$, and from (2.105) we obtain $a_{B_4 x} = 51.96\,\text{m/s}^2$ and $a_{B_4 y} = -81.40\,\text{m/s}^2$. Equation (2.107) gives the parameters $A_x = -53.841$ and $A_y = 78.617$, while (2.109)–(2.111) give:

$$\varepsilon_2 = 130.05\,\text{rad/s}^2, \ a_{B_3 B_4} = -42.151\,\text{m/s}^2,$$

$$a_{Bx} = -162.219\,\text{m/s}^2, \ a_{By} = -123.55\,\text{m/s}^2, \ a_B = 204.3\,\text{m/s}^2.$$

Comparing the results obtained using the analytical method with those obtained using the grapho-analytical method, we may state that the errors are acceptable, the largest being that obtained for the relative acceleration, which is equal to 17%. To reduce the errors, a smaller scale can be used.

2.5.3 The Assisted Analytical Method

Based on the relations established for the analytical method and following the calculation algorithm given in (2.94)–(2.111), we have created the procedure *Dyad_RRT*, the content of which is given in the companion website of this book.

The procedure has 18 input data and 18 output data. The input data are:

- the coordinates of the kinematic pairs A and D (X_A, Y_A, X_D, Y_D)
- the lengths of elements 2 and 3 (l_2 and l_3, respectively)
- the input angle φ and the structural angle α
- the kinematic values for the input poles A and D ($v_{Ax}, v_{Ay}, v_{Dx}, v_{Dy}, \omega_4, a_{Ax}, a_{Ay}, a_{Dx}, a_{Dy}, \varepsilon_4$).

The output data are:

- the position angle φ_2 for element 2 and the variable length λ
- the coordinates of point B (X_B, Y_B)
- the angular velocity ω_2
- the relative velocity $v_{B_3 B_4}$
- the components of the velocity of point B (v_{Bx}, v_{By}), the components of the acceleration $\mathbf{a}_{B_4}$ ($a_{B_4 x}, a_{B_4 y}$)
- the angular acceleration ε_2
- the relative acceleration $a_{B_3 B_4}$
- the components of the acceleration of point B (a_{Bx}, a_{By}).

By calling the procedure with the same input data as in the example solved by the grapho-analytical and the analytical method, namely:

Dyad_RRT $(0, 0.4, 0.8, 0, 0.5, 0.6, Pi/2, 2 * Pi/3, 0, 0, 0, 0, -10, 0, 0, 0, 0, 0, Phi2,$
$Lambda, XB, YB, vB4x, vB4y, Omega2, vB3B4, vBx, vBy, vB, aB4x, aB4y,$
$Epsilon2, aB3B4, aBx, aBy, aB)$,

the following results are obtained for the assisted analytical method:

$\varphi_2 = 55.890968, \ \lambda = 0.513986,$

$X_B = 0.280385, \ Y_B = 0.813986,$

$v_{B_4 x} = 8.139860, \ v_{B_4 y} = 5.196152,$

$\omega_2 = -19.662163, \ v_{B_3 B_4} = -10.709123,$

$v_{Bx} = 8.139860, \ v_{By} = -5.512971, \ v_B = 9,831082,$

$a_{B_4 x} = 51.961524, \ a_{B_4 y} = -81.398598,$

$\varepsilon_2 = 130,014089, \ a_{B_3 B_4} = -42.194683,$

$a_{Bx} = -162,220941, \ a_{By} = -123.593281, \ a_B = 203.938551.$

2.5.4 The Assisted Graphical Method

Recalling the observations and notations from Section 2.4.4, we will use an AutoCAD session for a kinematic analysis of the *R-RRT* mechanism in Figure 2.14a, using numerical data from the example in Section 2.5.1.

At time $t = 1$ s, the coordinates of points A $(0, 0.4)$ and D $(0.8, 0)$, the lengths of the elements $(AB = 0.5, BC = 0.6)$, and angle $\varphi = \frac{\pi}{2}$ are all known. A window of visualization is chosen:

Zoom;$-0.5, -0.2; 2.5, 1;$

The horizontal straight line Δ' that passes through point E is constructed:

(Setq M(List $(-0.8 \ (* \ 0.6 \ (Sin \ (/ \ Pi \ 3)))) \ 0))$;Line;!M;@1<90;;

For a precise construction, it is necessary to use the AutoLisp function *Setq* to assign the coordinates of point M (x_M, y_M), $x_M = x_D - AB \sin \alpha$, $y_M = 0$ or $M \left(0.8 - 0.6 \sin \frac{\pi}{3}, 0 \right)$. For the introduction of the point the function *!* is used; this evaluates the AutoLisp expressions. If a line starts with the exclamation mark, the AutoCAD compiler knows that it is to evaluate an AutoLisp expression.

A circle with centre at point A and radius $AB = 0.5$ is constructed:

Circle;0,0.4;0.5;

and the upper point of intersection is chosen for the construction of the straight line AB:

Line;0,0.4;Int;\;

Segment AB is saved with the command **Block**, using point A as reference:

−Block;BarAB;0,0.4;\;

Straight line Δ is created at point D:

Line;0.8,0;@1<90;;

as well as a circle with its centre at point B, previously determined, and of radius $BC = 0.6$

Circle;Int;\0.6;

Segment BC is constructed, point C being the lower point of intersection of the previously constructed circle and straight line Δ

Line;Int;\Int;\;

Segment BD, is constructed and saved, point B being the point of reference

Line;Int;\0.8,0;;–Block;DistBD;Int;\;

as well as straight line Δ, having point D as reference point

–Block;Delta;0.8,0;\;

The loop of the mechanism is drawn as a line with non-zero thickness

Pline;0.8,0;W;0.004;0.004;@1<90;;PLine;0,0.4;Int;\int;\;

and the axes of coordinates:

Pline;0,0;W;0.002;;1.1,0;W;0.02;0;L;0.07;;
Pline;0,0;W;0.002;;0,1;W;0.02;0;L;0.07;;

The two circles and the straight line Δ' are erased:

Erase;\\\;

the kinematic pairs are drawn, and the notations and dimensions are added, as in Figure 2.16.

The polygon of velocities is obtained using the vector relation (2.87), in which we take into account that the velocity of point A is zero. Knowing that at $t = 0$, $\omega_4 = \dot{\varphi} = 10$ rad/s, the graphical construction in Figure 2.17 is obtained. A new window of visualization is chosen

Zoom;0,0;12,12;

The pole of velocities $i\,(1, 6)$ is selected, from which the velocity $\mathbf{v}_{B4}$ is drawn ($\mathbf{v}_{B4}$ is perpendicular to BD and it has magnitude equal to $\omega_4 BD$) by inserting BD, amplified by ω_4, and rotated by an angle of $90°$ (ib_4)

–Insert;DistBD;1,6;10;10;90;

To determine point b_4, block $DistBD$ is unselected

Explode;\;

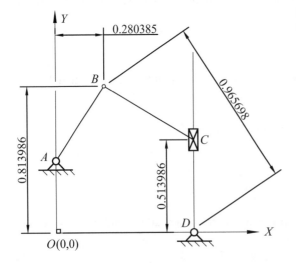

Figure 2.16 The construction of the mechanism's position using AutoCAD.

From point b_4 a parallel line to Δ is constructed by inserting the block *Delta*, amplified by 10 and rotated by an angle of 180°

−Insert;Delta;End\;10;10;180;

Now we return to point i to construct a perpendicular to AB by inserting the block *BarAB* at point i, multiplied by 10 and rotated by an angle of −90°

−Insert;BarAB;1, 6; 10; 10; −90;

The last two inserted blocks are unselected and extended to their intersection:

Explode;\\;Fillet;R;0;;Fillet;\\

The vector polygon in Figure 2.17 results, which is completed with arrows and dimensions. Because we need the relative velocity $\mathbf{v}_{B_3B_4}$ for the construction of the polygon of accelerations, (used for the calculation of the Coriolis acceleration), segment b_4b_3 is saved:

−Block;vB3B4;Int;\;Oops;

using point b_4 as reference.

The polygon of accelerations is constructed using the vector relations (2.89)–(2.90), resulting in the construction in Figure 2.18. A new window of visualization is chosen:

Zoom;20, −50; 300, 160;

and starting from pole j (200, 180), the following are constructed:

- the acceleration $\mathbf{a}_{B_4}$ ($\mathbf{jb}'_4$), by inserting BD, amplified by 100 and rotated through an angle of 0° (that is, no rotation),

  −Insert;DistBD;200,180;100;100;0;

  before unselecting the inserted block

  Explode;\;

- $\mathbf{a}^C_{B_3B_4}$ ($\mathbf{b}'_4\mathbf{m}$), by inserting the relative velocity $\mathbf{v}_{B_3B_4}$ at point b'_4, amplified by $2\omega_4 = 20$ and rotated through an angle of 270°

  −Insert;vB3B4;End;\;20;20;270;

  before unselecting the inserted block

  Explode;\;

and inserting the straight line Δ at point m' and then unselecting the block

  −Insert;Delta;End;\;100;100;180;Explode;\;

Figure 2.17 The polygon of velocities for the RRT dyad.

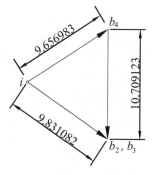

- $\mathbf{a}_{BA}^{\nu}$ (ja'), by inserting the segment AB at point j, multiplied by $\omega_2^2 = 386.600654$ and rotated through an angle of $180°$

    −Insert;BaraAB;200,180;386.600654;386.600654;180;

before unselecting the inserted block

    Explode;\;

inserting the segment AB at point n', rotated by $90°$

    −Insert;BarAB;End;\;100;100;90;

and then unselecting the block

    Explode;\;

The segments inserted at points m' and n' are extended until they intersect

    Fillet;R;0;Fillet;\\

The acceleration of point B is drawn by inserting a line between points j and b_2'

    Line;200,180;Int;\;

resulting in the vector polygon in Figure 2.18, which is completed with arrows and dimensions.

As in the case of the RRR dyad we want an AutoLisp function, which, no matter what the position of the dyad, can determined the position of the RRT dyad and the linear and angular velocities and accelerations. This will give us, with a single function call, and in separate layers, the representations in Figures 2.16–2.18.

For the position of the RRT dyad, point B (Figure 2.16) is obtained as the intersection between a straight line and a circle. Because there is no AutoLisp function that returns the intersection points between a circle and a straight line, we have to write one.

For the circle with radius R (Figure 2.19) and centre $C(a, b)$, and the straight line defined by points $M(X_M, Y_M)$ and $N(X_N, Y_N)$, from the vector relation

$$\mathbf{OC} + \mathbf{CP} = \mathbf{OM} + \mathbf{MP}, \tag{2.112}$$

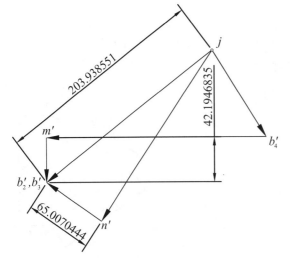

Figure 2.18 The polygon of accelerations for the RRT dyad.

Figure 2.19 The intersection between a circle and a straight line.

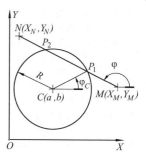

the following system is obtained:

$$\begin{cases} a + R\cos\varphi_C = X_M + MP\cos\varphi, \\ b + R\sin\varphi_C = Y_M + MP\sin\varphi. \end{cases} \tag{2.113}$$

Denoting

$$\varphi = \arctan\left(\frac{Y_N - Y_M}{X_N - X_M}\right),$$
$$A = (X_M - a)\cos\varphi + (Y_M - b)\sin\varphi, \tag{2.114}$$
$$B = (X_M - a)^2 + (Y_M - b)^2 - R^2,$$

we obtain

$$MP_{1,2} = -A \pm \sqrt{A^2 - B}. \tag{2.115}$$

The two points of intersection, P_1 and P_2, have coordinates

$$X_{P_1} = X_M + MP_1\cos\varphi, \ Y_{P_1} = Y_M + MP_1\sin\varphi, \tag{2.116}$$

$$X_{P_2} = X_M + MP_2\cos\varphi, \ Y_{P_2} = Y_M + MP_2\sin\varphi. \tag{2.117}$$

The existence of the two points of intersection is indicated by the plus/minus signsign of the expression situated under the radical in (2.115).

The AutoLisp function that incorporates (2.114)–(2.117) is called *Int_CL* (see Table 2.2). It has no local parameters and variables. As for function *Int_2C* for determining the intersection of two circles, we have added some comments to give readers a better understanding.

The function returns the points of intersection P_1 and P_2 between a circle and a straight line, and it assigns to the variable *error* the value 0 for two points of intersection, and the value 1 if the circle and the straight line do not intersect.

This function is called from the function *Dyad_RRT*, which has the following content:

```
(Defun C:RRT ()
(Setq xA 0.0 yA 0.4 xD 0.8 yD 0.0 xE 0.8 yE 1.0 AB 0.5 BC 0.6 alpha 60.0
om4 10 ismall (List 1.0 6.0) jsmall (List 200.0 180.0))
;Calculation of the positions
(Setq dsmall (- xD (* BC (Sin (/ (* alpha Pi) 180))))) xM dsmall yM 0.0
xN dsmall yN 1.0 a xA b yA R AB)
(Int_CL)
(Setq xB xP1 yB yP1)
(Setq xM xD yM yD xN xE yN yE a xB b yB R BC)
(Int_CL)
(Setq xC xP2 yC yP2)
;Positions of the mechanism
(Command "Erase" "All" "" "Osnap" "Off" "Ortho" "Off")
(Command "Zoom" "-0.5,-0.2" "2.5,1")
```

```
(Command "-Layer" "N" "Positions" "C" "7" "Positions" "S"
  "Positions" "" "")
(Command "PLine" (List xA yA) "W" "0.002" "" (List xB yB) (List xC
  yC) "")
(Command "PLine" (List xD yD) (List xE yE) "")
(Command "Pline" "0,0" "W" "0.002" "" "1.1,0" "W" "0.02" "0" "L"
  "0.07" "")
(Command "Pline" "0,0" "W" "0.002" "" "0,1" "W" "0.02" "0" "L"
  "0.07" "")
;Calculation of velocities
(Setq phiAB(Angle (List xA yA) (List xB yB))
phiDelta(Angle (List xD yD) (List xE yE))
phiDB(Angle (List xD yD) (List xB yB))
lBD(Distance (List xB yB) (List xD yD))
vB4(* lBD om4)
bsmall4(Polar ismall (- phiDB (/ pi 2)) vb4)
bsmall2(Polar ismall (+ phiAB (* 3 (/ Pi 2))) 1.0)
bsmall3(Polar bsmall4 phiDelta 1.0)
bsmall23(Inters ismall bsmall2 bsmall4 bsmall3 Nil)
om2(/ (Distance ismall bsmall23) AB))
;Polygon of velocities
(Command "Zoom" "0,0" "12,12")
(Command "-Layer" "N" "Velocities" "C" "1" "Velocities" "S"
  "Velocities" "" "")
(Command "Line" ismall bsmall23 bsmall4 "C")
(Setq l(/ vb4 20) g(* l 0.353))
(Arrows bsmall4 ismall)
(Arrows bsmall23 ismall)
(Arrows bsmall23 bsmall4)
;Calculation of accelerations
(Setq aB4(* lBD om4 om4)
aB2An(* AB om2 om2)
aCB3B4(* 2.0 om4 (Distance bsmall4 bsmall23))
nprime(Polar jsmall (+ phiAB Pi) aB2An)
bprime2(Polar nprime (+ phiAB (/ Pi 2)) 10.0)
bprime4(Polar jsmall (+ phiDB Pi) aB4)
mprime(Polar bprime4 (+ phiDelta (/ Pi 2)) aCB3B4)
bprime3(Polar mprime (+ phiDelta Pi) 10.0)
bprime23(Inters nprime bpriem2 mprime bprime3 Nil)
eps2(/ (Distance nprime bprime23) AB))
;Polygon of accelerations
(Command "Zoom" "20,-50" "300,160")
(Command "-Layer" "N" "Accelerations" "C" "5" "Accelerations" "S"
"Accelerations" "" "")
(Command "Line" jsmall bprime4 mprime bprime23 nprime "C" "Zoom" "E")
(Setq l(/ ab4 20) g(* l 0.353))
(Arrows bprime4 jsmall)
(Arrows nprime jsmall)
(Arrows mprime bprime4)
(Arrows bprime23 mprime)
(Arrows bprime23 nprime) )
```

Table 2.2 The function *Int_CL*.

Program	Comments
`(Defun Int_CL ()`	
`(Setq y (- yN yM) x (- xN xM))`	
`(If (/= x 0)`	If x<>0
`(PROGN`	Start I
`(If (> x 0)`	If x>0
`(PROGN`	Start II
`(If (>= y 0)`	If y>=0
`(Setq phi (Atan y x))`	Then
`(Setq phi (+ (* 2 Pi) (Atan y x)))`	Else
`)`	End If y>=0
`)`	End II
`(Setq phi (+ Pi (Atan y x)))`	Else x>0
`)`	End If x>0
`)`	End I
`(Setq phi (/ Pi 2))`	Else x<>0
`)`	End x<>0
`(Setq aa (+ (* (- xM a) (Cos phi)) (* (- yM b) (Sin phi)))`	
`bb (+ (* (- xM a) (- xM a)) (* (- yM b) (- yM b)) (* R R -1))`	
`delta (- (* aa aa) bb) error 1)`	
`(If (>= delta 0)`	
`(PROGN`	
`(Setq error 0`	
`MP1 (+ (* aa -1) (Sqrt delta))`	
`MP2 (- (* aa -1) (Sqrt delta)))`	
`)`	
`(Print "The circle and the straight line do not intersect")`	
`)`	
`(Setq xP1 (+ xM (* MP1 (Cos phi)))`	
`yP1 (+ yM (* MP1 (Sin phi)))`	
`xP2 (+ xM (* MP2 (Cos phi)))`	
`yP2 (+ yM (* MP2 (Sin phi)))`	
`P1 (List xP1 yP1)`	
`P2 (List xP2 yP2))`	
`)`	

The function is called from AutoCAD, after it has been loaded (*Load "Dyad_RRT.lsp"*), with the aid of the command *RRT*. The most important steps in the function are described next.

1) After the assignments of values to:

- the coordinates of the points (Figure 2.14) A, D, E (point E being situated on the straight line Δ at a distance of 1 relative to point D)
- the dimensions of the elements AB and BC
- angle α,

values are also assigned to the angular velocity ω_4 and to the coordinates of the poles i (the pole of velocities) and j (the pole of accelerations).

2) For the position of the dyad of the mechanism, the distance d_3 and the coordinates of points M and M' that belong to the straight line Δ' (Figure 2.14) are calculated.

3) To determine the intersection points between the circle centred on A with radius AB and the straight line Δ', function *Int_CL* is called. Preliminarily, the correspondence between the points used in the function *Int_CL* and the points from the function *RRT* is used. Point P_1 is retained in the function *Int_CL* as the intersection point and this becomes point B.

4) *Int_CL* is called again to determine point C, the intersection between the circle centred on B with radius BC and the straight line Δ defined by the coordinates of its points D and E. The convenient solution – point P_2 – is retained in the function *Int_CL*; this becomes point C.

5) The representation of the mechanism is passed to a visualization window and is placed in its own layer, which becomes the current layer. The loop ABC and the straight line Δ are drawn with a line of non-zero thickness, the latter using the coordinates of its points D and E. Finally, the axes of coordinates are drawn, following the procedure described for the AutoLisp function *Dyad_RRR* in Section 2.4.4.

6) Later, the values of the parameters necessary to determine the velocities are calculated:
- angle φ_{AB} between the straight line AB and the axis OX
- angle φ_Δ between the straight line Δ and the axis OX
- angle φ_{BD} between the straight line BD and the axis OX
- the velocity v_{B_4}.

Points b_4, b_2, and b_3 are determined using the function *Polar*, and point b_{23} is found by taking into account the vector relation (2.87) and using the function *Inters*. The angular velocity ω_2 is given by the ratio between the distances ib_3 and AB.

7) The polygon of velocities is constructed in a new window and in a new layer called *Velocities*. This is done by drawing the loop ib_3b_4i, and then by completing the polygon with the arrows of the vectors, using the function *Arrows*, as defined in Section 2.4.4.

8) For the calculation of the accelerations, the accelerations: a_{B_4}, $a^v_{B_2}$, $a^C_{B_3B_4}$ and then points n', b'_2, b'_4, m', b'_3, and b'_{23} of the polygon of accelerations are determined, as is the angular acceleration ε_2, using (2.91). The polygon of accelerations is placed in a new layer, by drawing the loop $jb'_4m'b'_{23}n'j$ and completing the polygon with arrows (using the function *Arrows*).

2.6 Kinematic Analysis of the RTR Dyad

2.6.1 The Grapho-analytical Method

Formulation of the problem

Consider the dyad $AA'BC$ in Figure 2.20a, for which the following are known:

- the positions of points A and C, the lengths l_2 and l_3 of the segments AA' and BC, respectively, and the angles α and β
- the velocities $\mathbf{v}_A$ and $\mathbf{v}_C$ of points A and C, respectively
- the accelerations $\mathbf{a}_A$ and $\mathbf{a}_C$ of points A and C, respectively.

The following values are to be determined:

- the positions of points A' and B

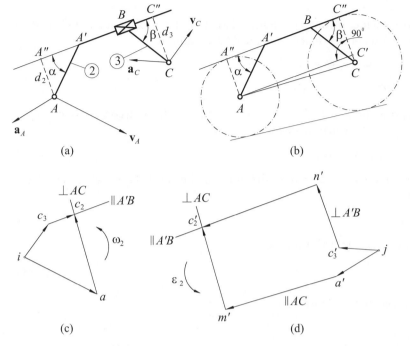

Figure 2.20 The kinematic analysis of the RTR dyad: (a) the RRT dyad; (b) position of the dyad; (c) plane of velocities; (d) plane of accelerations.

- the angular velocity $\omega_2 = \omega_3$ and the relative velocity of the slider
- the angular acceleration $\varepsilon_2 = \varepsilon_3$, and the relative acceleration of the slider.

Determination of the positions

We will denote by d_2 and d_3 the distances from points A and B, respectively, to the straight line $A'B$. These distances are given by:

$$d_2 = l_2 \sin \alpha, \quad d_3 = l_3 \sin \beta. \tag{2.118}$$

For the construction of the dyad the following procedure is used:

1) Using relations (2.118), calculate the distances d_2 and d_3.
2) Choose the scale of lengths k_l.
3) At point C', construct the right-angled triangle ACC' (Figure 2.20b), in which $CC' = d_3 - d_2$.
4) From point A construct the perpendicular to AC' and choose point A'' so that $AA'' = d_2$.
5) Extend the segment CC' and construct point C'' so that $CC'' = d_3$.
6) Unite points A'' and C''.
7) Construct the circle arcs with centres at points A and C, and having radii equal to l_2 and l_3, respectively; these circle arcs intersect the straight line $A''C''$ at points A' and B.

   As the reader can see in Figure 2.20b, the problem reduces to the construction of the tangents to the circles with the centres at points A and C, having radii equal to d_2 and d_3, respectively; this means that the problem has more than one solution. A solution has to be chosen based on the initial conditions.

Determination of the velocities

Determination of the velocities involves the Euler relation for the velocity of point C_2, and the relation for the composition of the velocities in the relative motion

$$\mathbf{v}_{C_2} = \underline{\underline{\mathbf{v}_A}} + \underbrace{\mathbf{v}_{C_2 A}}_{\perp AC} = \underline{\underline{\mathbf{v}_{C_3}}} + \underbrace{\mathbf{v}_{C_2 C_3}}_{\parallel A'B}, \tag{2.119}$$

where $\mathbf{v}_{C_3} = \mathbf{v}_C$. Representing at scale the relation (2.119), the polygon of velocities in Figure 2.20c is obtained, from which we deduce

$$\mathbf{v}_{C_2 A} = k_v \left(\mathbf{ac}_2 \right), \ \mathbf{v}_{C_2 C_3} = k_v \left(\mathbf{c}_2 \mathbf{c}_3 \right). \tag{2.120}$$

The relative velocity of the displacement of the slider on the bar $A'B$ is equal to the velocity $\mathbf{v}_{C_3 C_2}$, which is given by:

$$\mathbf{v}_{C_3 C_2} = -\mathbf{v}_{C_2 C_3}. \tag{2.121}$$

The angular velocity ω_2 is obtained from:

$$\omega_2 = \frac{v_{C_2 A}}{AC}, \tag{2.122}$$

and has the sense represented in Figure 2.20c.

Determination of the accelerations

The accelerations are determined from the vector equation

$$\begin{aligned}
\mathbf{a}_{C_2} &= \underline{\underline{\mathbf{a}_A}} + \underbrace{\mathbf{a}_{C_2 A}^\nu}_{C \to A} + \underbrace{\mathbf{a}_{C_2 A}^\tau}_{\perp AC} \\
&= \underline{\underline{\mathbf{a}_{C_3}}} + \underbrace{\mathbf{a}_{C_2 C_3}^C}_{\perp A'B} + \underbrace{\mathbf{a}_{C_2 C_3}}_{\parallel A'B}
\end{aligned} \tag{2.123}$$

where

$$\mathbf{a}_{C_3} = \mathbf{a}_C, \ a_{C_2 A} = \omega_2^2 CA, \ a_{C_2 C_3}^C = 2\omega_2 v_{C_2 C_3}. \tag{2.124}$$

The Coriolis acceleration $\mathbf{a}_{C_2 C_3}^C$ is perpendicular to $A'C$ and its sense is obtained by rotating the relative velocity $\mathbf{v}_{C_2 C_3}$ through an angle of $90°$ in the sense of the angular velocity ω_2.

By a scale representation of (2.123), the polygon in Figure 2.20d is obtained, from which we deduce:

$$a_{C_2 A}^\tau = k_a \left(m'c_2' \right), \ a_{C_2 C_3} = k_a \left(n'c_2' \right). \tag{2.125}$$

The angular acceleration ε_2 has the sense represented in Figure 2.20d, while its magnitude is given by:

$$\varepsilon_2 = \frac{a_{C_2 A}^\tau}{AC}. \tag{2.126}$$

Numerical example

Problem: Consider the mechanism $O_1 AA'BCD$ in Figure 2.21a, for which the following are known: $OO_1 = 0.6\,\mathrm{m}$, $OD = 1.2\,\mathrm{m}$, $O_1A = 0.4\,\mathrm{m}$, $AA' = 0.2\,\mathrm{m}$, $AA' \perp A'B$, $BC = \dfrac{0.8}{\sqrt{3}}\,\mathrm{m}$, $CD = 0.5\,\mathrm{m}$, $\beta = 60°$, $\theta = 20t$, and $\varphi = 10t + \dfrac{\pi}{2}$.

We want to obtain the kinematic analysis at time $t = 0$, using the grapho-analytical method.

Solution: At time $t = 0, \theta = 0$ and $\varphi = \frac{\pi}{2}$. From the conditions given in the problem, $d_2 = 0.2$ m, $d_3 = 0.4$ m; if the scale of the lengths $k_l = 0.02 \left[\frac{m}{mm} \right]$, then the representation in Figure 2.21b is obtained, from which $[AC] = 40.5$ mm; that is, $AC = 0.81$ m.

From the expressions for the angles θ and φ, the angular velocities $\omega_1 = \dot{\theta}$ and $\omega_4 = \dot{\varphi}$ have constant values $\omega_1 = 20$ rad/s and $\omega_4 = 10$ rad/s. We also obtain $v_A = O_1 A \omega_1 = 8$ m/s and $v_C = DC\omega_4 = 5$ m/s.

Choosing the scale of velocities such that velocity v_A is represented by a segment of 40 mm, we obtain $k_v = 0.2 \left[\frac{m/s}{mm} \right]$. The relation (2.119), represented at scale, leads to the polygon in Figure 2.21c, from which we deduce that $(ac_2) = 45$ mm and $(c_2c_3) = 31$ mm. This leads to $v_{C_2A} = 9$ m/s, $v_{C_2C_3} = 6.2$ m/s, and $\omega_2 = \frac{v_{C_2A}}{CA} = 11.11$ m/s. The angular velocity ω_2 has the sense represented in Figure 2.21c.

The accelerations of points A and C have the values

$$a_A = O_1 A \omega_1^2 = 160 \text{ m/s}^2, \quad a_C = DC\omega_4^2 = 50 \text{ m/s}^2.$$

while $a_{C_2A} = \omega_2^2 AC = 99.98$ m/s^2 and $a_{C_2C_3}^C = 2\omega_2 v_{C_2C_3} = 137.76$ m/s^2.

Choosing the scale of accelerations so that the acceleration of point A is represented by a segment of 40 mm, $k_a = 4 \left[\frac{m/s^2}{mm} \right]$. We deduce that

$$[a_C] = 12.5 \text{ mm}, \quad [a_{C_2A}] = 25 \text{ mm}, \quad [a_{C_2C_3}] = 34.4 \text{ mm}.$$

and representing the vector equation (2.123), the polygon in Figure 2.21d is obtained. From here we obtain the segments $(m'c_2') = 11$ mm and $(n'c_2') = 58$ mm, so that:

$$a_{C_2A}^\tau = 4 \cdot 11 = 44 \text{ m/s}^2, \quad a_{C_2C_3} = 4 \cdot 58 = 232 \text{ m/s}^2, \quad \varepsilon_2 = \frac{a_{C_2A}}{CA} = 54.32 \text{ rad/s}^2.$$

The sense of the angular acceleration ε_2 is as represented in Figure 2.21d.

2.6.2 The Analytical Method

Formulation of the problem

Consider the RTR dyad in Figure 2.22, for which the following known:

- the coordinates X_A, Y_A and X_C, Y_C of points A and C, respectively, the lengths l_2 and l_3 of the segments AA' and BC, respectively, and the angles α and β
- the components v_{Ax}, v_{Ay} and v_{Cx}, v_{Cy} of the velocities of points A and C, respectively
- the components a_{Ax}, a_{Ay} and a_{Cx}, a_{Cy} of the accelerations of points A and C, respectively.

The following values are to be determined:

- angle φ_2 between the straight line $A'B$ and the axis OX, and the distance $\lambda = A'B$
- the angular velocity ω_2 and the relative velocity of the slider in regard to the straight line $A'B$
- the angular acceleration ε_2 and the relative acceleration of the slider relative to the straight line $A'B$.

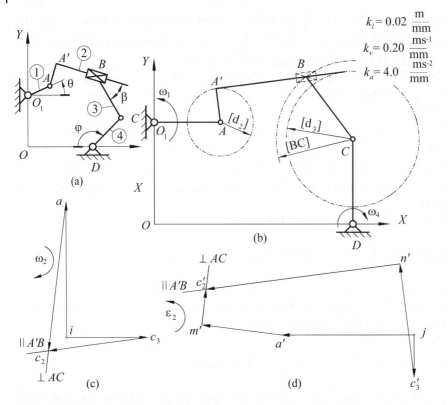

$$k_l = 0.02 \ \frac{m}{mm}$$
$$k_v = 0.20 \ \frac{ms^{-1}}{mm}$$
$$k_a = 4.0 \ \frac{ms^{-2}}{mm}$$

Figure 2.21 Mechanism with an RTR dyad: (a) mechanism; (b) position of the mechanism; (c) plane of velocities; (d) plane of accelerations.

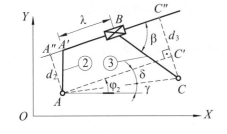

Figure 2.22 The RTR dyad.

Determination of the positions

The parameters λ and φ_2 are deduced using the algorithm:

$$AC = \sqrt{\left(X_C - X_A\right)^2 + \left(Y_C - Y_A\right)^2}, \tag{2.127}$$

$$d_2 = l_2 \sin \alpha, \ d_3 = l_3 \sin \beta, \tag{2.128}$$

$$AC' = \sqrt{AC^2 - \left(d_3 - d_2\right)^2}, \tag{2.129}$$

$$\lambda = \sqrt{AC^2 - \left(d_3 - d_2\right)^2} - l_2 \cos \alpha - l_3 \cos \beta, \tag{2.130}$$

$$\gamma = \arctan \left(\frac{Y_C - Y_A}{X_C - X_A} \right), \tag{2.131}$$

$$\delta = \arcsin\left(\frac{d_3 - d_2}{AC}\right), \tag{2.132}$$

$$\varphi_2 = \gamma + \delta. \tag{2.133}$$

Determination of the velocities

Projecting (2.119) onto the axes, written in the form

$$\mathbf{v}_A + \boldsymbol{\omega}_2 \times \mathbf{AC} = \mathbf{v}_C + \mathbf{v}_{C_2 C_3}, \tag{2.134}$$

leads to the scalar equations

$$\begin{aligned}
v_{Ax} - \omega_2\left(Y_C - Y_A\right) &= v_{Cx} + v_{C_2 C_3}\cos\varphi_2, \\
v_{Ay} + \omega_2\left(X_C - X_A\right) &= v_{Cy} + v_{C_2 C_3}\sin\varphi_2,
\end{aligned} \tag{2.135}$$

from which are deduced the unknowns

$$\omega_2 = \frac{-\left(V_{Cx} - V_{Ax}\right)\sin\varphi_2 + \left(v_{Cy} - v_{Ay}\right)\cos\varphi_2}{\left(Y_C - Y_A\right)\sin\varphi_2 + \left(X_C - X_A\right)\cos\varphi_2}, \tag{2.136}$$

$$v_{C_2 C_3} = \left[v_{Ax} - v_{Cx} - \omega_2\left(Y_C - Y_A\right)\right]\cos\varphi_2 + \left[v_{Ay} - v_{Cy} + \omega_2\left(X_C - X_A\right)\right]\sin\varphi_2. \tag{2.137}$$

Determination of the accelerations

Projecting (2.123) onto the axes, written as

$$\mathbf{a}_A + \boldsymbol{\varepsilon}_2 \times \mathbf{AC} - \omega_2^2 \mathbf{AC} = \mathbf{a}_C + 2\boldsymbol{\omega}_2 \times \mathbf{v}_{C_2 C_3} + \mathbf{a}_{C_2 C_3}, \tag{2.138}$$

leads to the system of equations

$$\begin{aligned}
a_{Ax} - \varepsilon_2\left(Y_C - Y_A\right) - \omega_2^2\left(X_C - X_A\right) &= a_{Cx} - 2\omega_2 v_{C_2 C_3}\sin\varphi_2 + a_{C_2 C_3}\cos\varphi_2, \\
a_{Ay} + \varepsilon_2\left(X_C - X_A\right) - \omega_2^2\left(Y_C - Y_A\right) &= a_{Cy} + 2\omega_2 v_{C_2 C_3}\cos\varphi_2 + a_{C_2 C_3}\sin\varphi_2.
\end{aligned} \tag{2.139}$$

If we denote

$$\begin{aligned}
A_x &= a_{Cx} - a_{Ax} + \omega_2^2\left(X_C - X_A\right) - 2\omega_2 v_{C_2 C_3}\sin\varphi_2, \\
A_y &= a_{Cy} - a_{Ay} + \omega_2^2\left(Y_C - Y_A\right) + 2\omega_2 v_{C_2 C_3}\cos\varphi_2,
\end{aligned} \tag{2.140}$$

then we obtain the unknowns

$$\varepsilon_2 = \frac{-A_x \sin\varphi_2 + A_y \cos\varphi_2}{\left(Y_C - Y_A\right)\sin\varphi_2 + \left(X_C - X_A\right)\cos\varphi_2}, \tag{2.141}$$

$$a_{C_2 C_3} = -\left[A_x + \varepsilon_2\left(Y_C - Y_A\right)\right]\cos\varphi_2 - \left[A_y - \varepsilon_2\left(X_C - X_A\right)\right]\sin\varphi_2. \tag{2.142}$$

Numerical example

Problem: Using the analytical method, perform a kinematic analysis of the mechanism in Figure 2.21a, considering the same conditions as in the example solved using the grapho-analytical method; compare the results.

Solution: From the problem: $\alpha = 90°$, $l_2 = 0.2\,\text{m}$, $l_3 = \dfrac{0.8}{\sqrt{3}}\,\text{m}$, $d_2 = 0.2\,\text{m}$, $d_3 = 0.4\,\text{m}$, $X_A = 0.4\,\text{m}$, $X_C = 1.2\,\text{m}$, $Y_C = 0.5\,\text{m}$. Using (2.127) and (2.133), we obtain

$$AC = 0.8062\,\text{m}, \quad AC' = 0.7810\,\text{m}, \quad \lambda = 0.55\,\text{m},$$
$$\gamma = -7.125°, \quad \delta = 14.364°, \quad \varphi_2 = 7.239°.$$

Taking into account the senses of the angles θ and φ, $\omega_1 = 20$ rad/s and $\omega_4 = -10$ rad/s, and from Figure 2.21b we get $v_{Ax} = 0$, $v_{Ay} = 8$ m/s, $v_{Cx} = 5$ m/s, and $v_{Cy} = 0$. In these conditions, from (2.136) and (2.137) $\omega_2 = -10.96$ rad/s and $v_{C_2 C_3} = -6.144$ m/s.

For the accelerations we have

$$a_{Ax} = -160 \text{ m/s}^2, \ a_{Ay} = 0, \ a_{Cy} = -50 \text{ m/s}^2,$$

while from (2.140) $A_x = 239.128$ and $A_y = 71.586$. With these results, from (2.141) and (2.142) is obtained:

$$\varepsilon_2 = 52.34 \text{ rad/s}^2, \ a_{C_2 C_3} = -238.4 \text{ m/s}^2.$$

Comparing the results obtained using the analytical method with those obtained with the grapho-analytical method, there is a good correspondence, the errors being in the interval 0–6%.

2.6.3 The Assisted Analytical Method

Based on the relations established for the analytical method and following the calculation algorithm in (2.127)–(2.142), we wrote the procedure *Dyad_RTR*, the content of which is given in the companion website of this book.

The procedure has 16 input data and 6 output data. The input data are:

- the coordinates of the kinematic pairs at points A and C (X_A, Y_A, X_C, Y_C)
- the lengths of elements 2 and 3 (l_2, l_3)
- the structural angles α, β
- the kinematic parameters of the input poles A and C (v_{Ax}, v_{Ay}, v_{Cx}, v_{Cy}, a_{Ax}, a_{Ay}, a_{Cx}, a_{Cy}).

The output data are:

- the position angle φ_2 of element 2
- the variable length λ
- the angular velocity ω_2
- the relative velocity $v_{C_2 C_3}$
- the angular acceleration ε_2
- the relative acceleration $a_{C_2 C_3}$.

By calling the procedure with the same input data as in the example solved using the grapho-analytical and analytical methods, namely

$$\textbf{Dyad\_RTR}\,(0.4, 0.6, 1.2, 0.5, 0.2, 0.8/\text{sqrt}(3), Pi/2, Pi/3, 0, 8, 5, 0, -160, 0, 0, -50,$$
$$Phi2, Lambda, Omega2, vC2C3, Epsilon2, aC2C3),$$

the following results are obtained:

$$\gamma = -7,125016, \ \delta = 14.363303,$$

$$\varphi_2 = 7.238286, \ \lambda = 0.550085,$$

$$\omega_2 = -10.967929, \ v_{C_2 C_3} = -6.145770,$$

$$\varepsilon_2 = 52.485813, \ a_{C_2 C_3} = -235.881714.$$

2.6.4 The Assisted Graphical Method

At time $t = 0$ the following are known:

- the coordinates of points O_1 (0, 0.6), A (0.4, 0.6) and D (1.2, 0)
- the structural angle $\beta = 60°$

- the lengths of the elements ($O_1A = 0.4$, $AA' = 0.2$, $DC = 0.5$, $BC = \frac{0.8}{\sqrt{3}}$)
- angle $\varphi_2 = \frac{\pi}{2}$.

Distances $d_2 = 0.2$ and $d_3 = BC \sin 60° = 0.4$ are calculated.

For the construction of the mechanism's position using AutoCAD, start by choosing a window of visualization:

> Zoom;−0.5, −0.2; 2.5, 1;

and construct two circles, the first centred on point A and with radius 0.2, the second centred on point C and having radius 0.4,

> Circle;0.4,0.6;0.2; Circle;1.2,0.5;0.4;

For the construction of the common tangent to the two circles, use the facility offered by the command **Line** to directly draw the straight-line tangent, selecting the circles one by one (the selection is made at their upper parts)

> Line;Tan;\Tan;\;

This gives points $A'' \equiv A'$ and C'' in Figure 2.20. Then construct the straight lines AA' and CC''

> Line;0.4,0.6;Int;\;Line;1.2,0.5;Int;\;

and determine point B as the intersection between the circle centred at point C with radius BC, and the straight line $A'C''$

> (Setq BC(/ 0.8 (Sqrt 3)));Circle;1.2,0.5;!BC;

Draw the segments CB and AC

> Line;1.2,0.5;Int;\; Line;0.4,0.6;1.2,0.5;

Save as blocks the bar $\Delta = A'C''$ having the end point A' as reference point and the distance AC having the end point A as reference point

> −Block;Delta;End;\;−Block;Dist_AC;0.4,0.6;\;

Draw the loop $O_1AA'C''$ as the contour of the mechanism and the loop DCB using line with thickness

> Pline;0,0.6;W;0.004;0.004;0.4,0.6;End;\End;\;
>
> Pline;1.2,0;1.2,0.5;End;\;

and the axes of coordinates

> Pline;0,0;W;0.002;;1.5,0;W;0.02;0;L;0.07;;
>
> Pline;0,0;W;0.002;;0,1;W;0.02;0;L;0.07;;

Erase the circles and the segment CC''

> Erase;\\\\;

Draw the kinematic pairs, make the annotations and add the dimensions, as in Figure 2.23.

The polygon is obtained using the vector relation (2.119), in which at $t = 0$, $\omega_2 = \dot{\varphi} = 10$ rad/s. Determine the velocities of points A and C ($v_A = 8$ m/s, $v_C = 5$ m/s). This leads to the construction in Figure 2.24. Choose a new window of visualization

> Zoom;0,0;12,12;

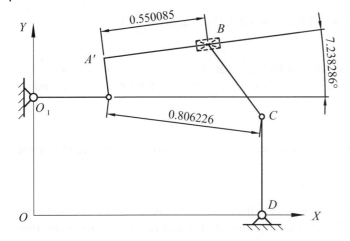

Figure 2.23 The RTR dyad: analysis of the positions.

and the pole of velocities $i\,(3,3)$, from which the velocity $\mathbf{v}_A$ is drawn with the vertical segment **ia**. Then at point a insert the distance AC, amplified by 10 and rotated by 270°

    Line;3,3;@8<90;;–Insert;Dist_AC;End;\;10;10;270;

Return to pole i and draw the velocity $\mathbf{v}_{C_3}$, by the horizontal segment $\mathbf{ic}_3$, then at point c_3 insert the straight line Δ, amplified by 10 and rotated through an angle of 270°

    Line;3,3;@5<0;;–Insert;Delta;End;\;10;10;180;

Unselect the two introduced blocks

    Explode;\\;

and extend them to their intersection at point c_2

    Fillet;R;0;;Fillet;\\

resulting in the vector polygon in Figure 2.24, which is completed with arrows and dimensions.

Since for the construction of the polygon of accelerations we need $\mathbf{v}_{C_2C_3}$ for the calculation of the Coriolis acceleration, the latter is saved as block, having point c_3 as reference point

    –Block;vC2C3;Int;\\;Oops;

To get to the polygon of accelerations in Figure 2.25, characterized by the vector relation (2.123), calculate the values of the expressions $2\omega_2 = 2\dfrac{v_{C_2A}}{AC}$ and $\omega_2^2 = \left(\dfrac{v_{C_2A}}{AC}\right)^2$. The accelerations of points A and C are known: $a_A = 160$ m/s^2 and $a_C = 50$ m/s^2. The construction is created in a window

    Zoom;20,–50;300,160;

from pole $j\,(100,50)$, from which is successively constructed:

- $\mathbf{a}_{C_3}$ (the segment $\mathbf{jc}_3'$), starting from point j

    Line;100,50;@50<270;;

- $\mathbf{a}_{C_2C_3}$, by inserting at point c_3' the relative velocity $\mathbf{v}_{C_2C_3}$, multiplied by $2\omega_2$ and rotated through an angle of 270°

    –Insert;vC2C3;End;\21.935858;21.935858;270;

Figure 2.24 The RTR dyad: the polygon of velocities.

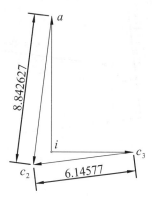

after which the inserted block is unselected

> Explode;\;

the straight line Δ is inserted at point n', multiplied by 100 and rotated through an angle of 180°

> −Insert;Delta;End;\100;100;180

and the inserted block is unselected

> Explode;\;

- $\mathbf{a}_A$ ($\mathbf{j}a'$), starting from point j

> Line;100,50;@160<180;;

- $\mathbf{a}^{\nu}_{C_2A}$, by inserting the straight line AC at point a', multiplied by ω^2_2 and rotated with an angle of 180°

> −Insert;Dist_AC;End;\120.2954665;120.2954665;180;

after which the inserted block is unselected

> Explode;\;

the straight line AC is inserted at point m', rotated through an angle of 90°

> −Insert_Dist_AC;End;\100;100;90;

and the inserted block is unselected

> Explode;\;

The inserted entities at points m' and n' are extended to their intersection point c'_2

> Fillet;R;0;;Fillet;\\

The points are labelled and the polygon is completed with arrows and dimensions, as in Figure 2.25.

To write an AutoLisp function for kinematic analysis of the RTR dyad, no matter what the position of the dyad, and with no manual intervention to select points and so on, as was the case for the direct construction described earlier in this section, it is necessary to identify the optimum method for the determination of the tangent points of a straight line with two circles.

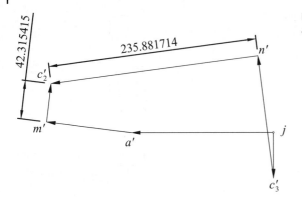

Figure 2.25 The RTR dyad: the polygon of accelerations.

In AutoCAD, using the command **Line** the segment $A''C''$ in Figure 2.20 can be drawn, specifying only the circles; depending on the points chosen, one of the four solutions (two common exterior and two common interior tangents) of the problem will be obtained.

The selection of an entity is performed using the mouse and specifying a point that belongs to it, or by using a window of selection. In AutoCAD, the windows of selection are of two types: 'window' type, in which only entities that are completely in the window are selected, and 'crossing' type, in which objects that intersect the window are selected. The selection is made in the dialog boxes corresponding to the commands, using 'W' for window type or 'C' for crossing type.

Crossing-type windows may be used in our case, the windows of selection being sufficiently small to permit the selection of the parts of the circles situated in the second quadrant. If the dimensions of the dyad are changed, or if it significantly changes its position, then, to be efficient, the selections must be precisely calculated from the corners of the window of selection.

For this reason, we have chosen the classical geometrical construction for the determination of the common exterior tangent of two circles of radii R_1 and R_2 and arbitrary centres O_1 and O_2, respectively; the construction is shown in Figure 2.26.

The geometrical construction consists of:

- the construction of the circle of radius $R_2 - R_1$ ($R_2 \geq R_1$), centred at point O_2;
- the determination of the intersection points A, B between the circle centred at point O_1 ($O_1E = EO_2$) and with radius $\frac{O_1O_2}{2}$, and the circle centred at O_2 of radius $R_2 - R_1$;

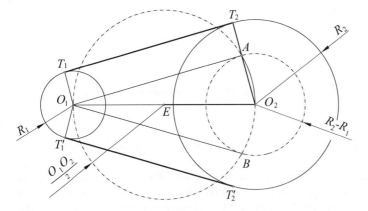

Figure 2.26 The construction of the common tangent of two circles.

- constructing the perpendiculars to the tangent O_1A at points O_1 and A; these intersect the circles with the centres at points O_1 and O_2 at points T_1 and T_2, respectively.

Similarly, the perpendiculars to the tangent O_1B are constructed at points O_1 and B, and the intersection points T_1' and T_2', respectively, can be found.

The common tangent construction is preferred because the intersection of two circles is solved by the AutoLisp function *Int_2C*, while the intersection between a straight line and a circle is solved by the AutoLisp function *Int_CL*.

The AutoLisp function for kinematic analysis of the RTR dyad is named *Dyad_RTR.lsp* and it is called in AutoCAD using the command *RTR*.

```
(Defun C:RTR ()
(Setq xO1 0.0 yO1 0.6 xD 1.2 yD 0.0 lAAp 0.2 lBC (/ 0.8 (Sqrt 3.0)) lCD
0.5 alfa 90.0 beta 60.0 xA 0.4 yA 0.6 xC 1.2 yC 0.5 ismall(List 3 3)
vA 8.0 angvA 90.0 vC 5.0 angvC 0.0 jsmall(list 100 50) accA 160.0 angaA
180.0 accC 50.0 angaC 270.0)
;Calculation of positions
(Setq lAC (Distance (List xA yA) (List xC yC)) xE(/ (+ xA xC) 2) yE(/ (+
yA yC) 2) alphaRad(* alpha (/ Pi 180)) betaRad(* beta (/ Pi 180)) d2(*
lAAp (Sin AlphaRad)) d3(* lBC (Sin BetaRad)) a1 xE b1 yE R1 (/ lAC 2) a2
xC b2 yC R2 (- d3 d2))
(Int_2C)
(Setq xM xC yM yC xN x1 yN y1 a xC b yC R d3)
(Int_CL)
(Setq xCsec xP2 yCsec yP2)
(Setq phi(Angle (List xC yC) (List xCsec yCsec)))
(Setq xM xA yM yA xN (+ xA (Cos phi)) yN (+ yA (Sin phi)) a xA b yA R
  d2)
(Int_CL)
(Setq xAsec xP2 yAsec yP2)
(Setq xM xAsec yM yAsec xN xCsec yN yCsec a xC b yC R lBC)
(Int_CL)
(Setq xB xP2 yB yP2)
;Position of the mechanism
(Command "Erase" "All" "" "Osnap" "Off" "Ortho" "Off")
(Command "Zoom" "-0.5,-0.2" "2.5,1")
(Command "-Layer" "N" "Positions" "C" "7" "Positions" "S"
  "Positions" "" "")
(Command "Pline" (List xO1 yO1) "W" "0.002" "" (List xA yA) (List xAsec
yAsec) (List xB yB) (List xCsec yCsec) "L" "0.25" "" "Pline" (List xD
yD) (List xC yC) (List xB yB) "")
(Command "Pline" "0,0" "W" "0.002" "" "1.7,0" "W" "0.02" "0" "L"
  "0.07" "")
(Command "Pline" "0,0" "W" "0.002" "" "0,1.1" "W" "0.02" "0" "L"
  "0.07" "")
;Calculation of velocities
(Setq phiAC(Angle (List xA yA) (List xC yC)) phiApB(Angle (List xAsec
yAsec) (List xB yB)) asmall(Polar ismall (* angva (/ Pi 180)) va)
rsmall(Polar asmall (- phiAC (/ Pi 2)) 1) csmall3(Polar ismall (* angvC
(/ Pi 180)) vC) ssmall(Polar csmall3 phiApB 1) csmall2(Inters asmall
rsmall csmall3 ssmall Nil) om2(/ (Distance asmall csmall2) lAC))
;Polygon of velocities
(Command "Zoom" "0,0" "12,12")
```

```
(Command "-Layer" "N" "Velocities" "C" "1" "Velocities" "S"
   "Velocities" "" "" "Line" ismall asmall csmall2 csmall3 "C")
(Setq l(/ vA 20) g(* l 0.353))
(Arrows asmall ismall)
(Arrows csmall2 asmall)
(Arrows csmall2 csmall3)
(Arrows csmall3 ismall)
;Calculation of accelerations
(Setq aC2An(* om2 om2 lAC) acC2C3(* 2 om2 (Distance csmall3 csmall2))
aprime(Polar jsmall (* angaa (/ Pi 180)) accA) mprime(Polar aprime
(- phiAC Pi) aC2An) rprime(Polar mprime (- phiAC (/ Pi 2)) 10)
cprime3(Polar jsmall (* angaC (/ Pi 180)) accC) nprime(Polar cprime3 (+
phiApB (/ Pi 2)) acC2C3) sprime(Polar nprime phiApB 10) cprime2(Inters
mprime rprime nprime sprime Nil) eps2(/ (Distance mprime cprime2) lAC))
;Polygon of accelerations
(Command "Zoom" "20,-50" "300,160")
(Command "-Layer" "N" "Accelerations" "C" "5" "Accelerations" "S"
   "Accelerations" "" "")
(Command "Line" jsmall aprime mprime cprime2 nprime cprime3 "C" "Zoom"
   "E")
(Setq l(/ accC 20) g(/ accC 50))
(Arrows aprime jsmall)
(Arrows mprime aprime)
(Arrows cprime2 mprime)
(Arrows cprime3 jsmall)
(Arrows nprime cprime3)
(Arrows cprime2 nprime)
)
```

The dimensions of the dyad are those given in Section 2.6.1 and they are declared at the beginning of the function. Similar to the other dyads, we preferred direct assignment of the values of the dimensions and of the poles of the dyad in the body of the function. Values may also come from queries, with the values introduced from the keyboard or read from a data file. The only difference would be a more complicated way of introducing of the values and verifying their correctness, but the structure of the function would remain unchanged.

The calculation of the positions starts with the determination of the distance AC (l_{AC}) and the coordinates of point E (x_E, y_E), situated in the middle of the segment.

In the AutoLisp functions the angles are measured in radians and, for this reason, if they are assigned values in degrees, then these must be converted to radians. Angles α and β have input values in degrees (*alphaRad*, *betaRad*).

Using (2.118), the values of the distances d_2, d_3 are calculated and then the correspondence of the values is used for the function *Int_2C* ($a_1 = x_E$, $b_1 = y_E$, $R_1 = \frac{l_{AC}}{2}$, $a_2 = x_C$, $b_2 = y_C$, $R_2 = d_3 - d_2$).

One thus starts with the determination of the points of the common tangent to the circles with the centres at points A and C (Figure 2.20), of radii l_2 and l_3, respectively.

After calling function *Int_2C*, the intersection points in Figure 2.26 are obtained. Only point P_1 (x_1, y_1) is retained. A new correspondence is made between the values of the input data for the function *Int_CL* for the determination of point C'' ($x_M = x_C$, $y_M = y_C$, $x_N = x_1$, $y_N = y_1$, $a = x_C$, $b = y_C$, $R = d_3$). The convenient solution (point P_2) is chosen.

For the determination of point A'' angle φ between the horizontal and the straight line CC'' is calculated. Then the intersection point between a straight line of angle φ, which passes through point A, and the circle of radius d_2 is determined by calling function *Int_CL*. The convenient solution $(x_{A''} = x_{P_2}, y_{A''} = y_{P_2})$ is retained.

For the determination of point B, function *Int_CL* is used again, and again point P_2 $(x_B = x_{P_2}, y_B = y_{P_2})$ is retained.

The representation of the mechanism in AutoCAD uses a dedicated layer, as in the earlier examples, drawing the loops $O_1AA'BC''$ and DCB, and the axes of coordinates with thin lines (see Figure 2.23).

For the determination of the velocities, angle φ_{AC} between the straight line AC and the horizontal, and angle $\varphi_{A'B}$ between the straight line $A'B$ and the horizontal are calculated. Using the function *Polar* (Figure 2.24) points a, a point r that belongs to the perpendicular to AC, c_3, and a point s that belongs to the parallel to $A'B$ are determined. According to (2.119), at the intersection of the straight lines defined by these points, is point c_2. The angular velocity ω_2 is given by the ratio between the distances ac_2 and AC.

The polygon of velocities is constructed in a new window and a new layer, called *Velocities*, by drawing the loop ia_2c_3i and then, using the function *Arrows*, by completing the polygon with the arrows corresponding to the velocities.

For the calculation of the accelerations, the values: $a_{C_2A}^v$ and $a_{C_2C_3}^C$, the points of the polygon of accelerations: a', m', r', c_3', n', s' and, finally, c_2' are determined. The angular acceleration ε_2 is found from (2.126).

The polygon of accelerations is represented in a new layer, by drawing the loop $ja'm'c_2'n'c_3'j$ and by completing this polygon with arrows (function *Arrows*).

2.7 Kinematic Analysis of the TRT Dyad

2.7.1 The Grapho-analytical Method

Formulation of the problem

Consider the TRT dyad ($DABCE$) in Figure 2.27a, for which the following are known:

- the positions of the straight lines Δ_1 and Δ_4, the positions of points D and E that belong to these straight lines, the lengths l_2 and l_3 of the segments AB and BC, respectively, and the angles α and β;
- the velocities $\mathbf{v}_D$ and $\mathbf{v}_E$ of points D and E, respectively, and the angular velocities ω_1 and ω_4 of the straight lines Δ_1 and Δ_4, respectively;
- the velocities $\mathbf{a}_D$ and $\mathbf{a}_E$ of points D and E, respectively, and the angular accelerations ε_1 and ε_4 of the straight lines Δ_1 and Δ_4, respectively.

The following values are to be determined:

- the position of point B and the lengths of the segments DA and CE
- the velocity of point B and the relative velocities of the sliders relative to the straight lines Δ_1 and Δ_4
- the acceleration of point B and the relative accelerations of the sliders relative to the straight lines Δ_1 and Δ_4.

Determination of the positions

The distances d_2 and d_3 from point B to the axes Δ_1 and Δ_4, respectively, are given by:

$$d_2 = l_2 \sin \alpha, \ d_3 = l_3 \sin \beta. \tag{2.143}$$

Point B is therefore situated at the intersection of:

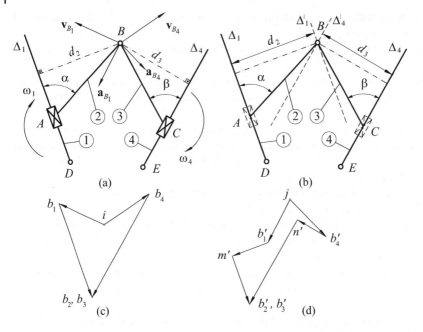

Figure 2.27 The grapho-analytical analysis of the TRT dyad:(a) TRT dyad; (b) position of the dyad; (c) plane of velocities; (d) plane of accelerations.

- the straight line Δ_1', parallel to the straight line Δ_1, and situated at a distance d_2 from it
- the straight line Δ_4', parallel to the straight line Δ_4 and situated at a distance d_3 from it (Figure 2.27b).

Two straight lines can now be constructed at point B, inclined with angles α and β with respect to the straight lines Δ_1 and Δ_4, respectively. Points A and C are obtained. Obviously, the lengths of the segments DA and CE are given by:

$$DA = k_l\,[DA], \quad CE = k_l\,[CE], \tag{2.144}$$

where k_l is the scale of the lengths.

Determination of the velocities
First, the velocities $\mathbf{v}_{B_1}$ and $\mathbf{v}_{B_4}$ of points B_1 and B_4 rigidly linked to the straight lines Δ_1 and Δ_4, respectively are determined:

$$\mathbf{v}_{B_1} = \underline{\underline{\mathbf{v}_D}} + \underbrace{\mathbf{v}_{B_1 D}}_{\perp BD}, \quad \mathbf{v}_{B_4} = \underline{\underline{\mathbf{v}_E}} + \underbrace{\mathbf{v}_{B_4 E}}_{\perp BE} \tag{2.145}$$

where

$$\mathbf{v}_{B_1 D} = \boldsymbol{\omega}_1 \times \mathbf{DB}, \quad v_{B_1 D} = \omega_1 DB, \quad \mathbf{v}_{B_4 E} = \boldsymbol{\omega}_4 \times \mathbf{EB}, \quad v_{B_4 E} = \omega_4 EB. \tag{2.146}$$

From the relative motions of points B_2 and B_3 (on directions parallel to the straight lines Δ_1 and Δ_4, respectively) we have:

$$\mathbf{v}_{B_2} = \mathbf{v}_{B_1} + \mathbf{v}_{B_1 B_2}, \quad \mathbf{v}_{B_3} = \mathbf{v}_{B_4} + \mathbf{v}_{B_3 B_4}, \tag{2.147}$$

and taking into account that points B_2 and B_3 coincide, we get

$$\mathbf{v}_{B_2} = \mathbf{v}_{B_3} = \underline{\underline{\mathbf{v}_{B_1}}} + \underbrace{\mathbf{v}_{B_2 B_1}}_{\parallel\,\Delta_1} = \underline{\underline{\mathbf{v}_{B_4}}} + \underbrace{\mathbf{v}_{B_3 B_4}}_{\parallel\,\Delta_4}. \tag{2.148}$$

If the velocities $\mathbf{v}_{B_1}$ and $\mathbf{v}_{B_4}$ have the representations in Figure 2.27a, then the polygon of velocities that corresponds to the vector equation (2.148) is as drawn in Figure 2.27c; this gives:

$$\mathbf{v}_B = k_v \left(\mathbf{i}\mathbf{b}_2 \right), \ \mathbf{v}_{B_1 B_2} = k_v \left(\mathbf{b}_1 \mathbf{b}_2 \right), \ \mathbf{v}_{B_3 B_4} = k_v \left(\mathbf{b}_3 \mathbf{b}_4 \right), \tag{2.149}$$

where k_v is the scale of velocities.

Determination of the accelerations
In this case too, the accelerations $\mathbf{a}_{B_1}$ and $\mathbf{a}_{B_4}$ of points B_1 and B_4, respectively are determined, using:

$$\begin{aligned}
\mathbf{a}_{B_1} &= \underbrace{\mathbf{a}_D}_{} + \underbrace{\mathbf{a}^\nu_{B_1 D}}_{B \to D} + \underbrace{\mathbf{a}^\tau_{B_1 D}}_{\perp BD}, \\
\mathbf{a}_{B_4} &= \underbrace{\mathbf{a}_E}_{} + \underbrace{\mathbf{a}^\nu_{B_4 E}}_{B \to E} + \underbrace{\mathbf{a}^\tau_{B_4 E}}_{\perp BE},
\end{aligned} \tag{2.150}$$

where

$$a^\nu_{B_1 D} = \omega_1^2 BD, \ a^\tau_{B_1 D} = \varepsilon_1 BD, \ a^\nu_{B_4 E} = \omega_4^2 BE, \ a^\tau_{B_4 E} = \varepsilon_4 BE. \tag{2.151}$$

Then, from the relative motions of points B_2 and B_3, we have

$$\begin{aligned}
\mathbf{a}_{B_2} = \mathbf{a}_{B_3} &= \underbrace{\mathbf{a}_{B_1}}_{} + \underbrace{\mathbf{a}^C_{B_2 B_1}}_{\perp \Delta_1} + \underbrace{\mathbf{a}_{B_2 B_1}}_{\parallel \Delta_1} \\
&= \underbrace{\mathbf{a}_{B_4}}_{} + \underbrace{\mathbf{a}^C_{B_3 B_4}}_{\perp \Delta_4} + \underbrace{\mathbf{a}_{B_3 B_4}}_{\parallel \Delta_4}
\end{aligned} \tag{2.152}$$

where

$$\begin{aligned}
\mathbf{a}^C_{B_2 B_1} &= 2\boldsymbol{\omega}_1 \times \mathbf{v}_{B_2 B_1}, \ a_{B_2 B_1} = 2\omega_1 v_{B_2 B_1}, \\
\mathbf{a}^C_{B_3 B_4} &= 2\boldsymbol{\omega}_4 \times \mathbf{v}_{B_3 B_4}, \ a_{B_3 B_4} = 2\omega_4 v_{B_3 B_4}.
\end{aligned} \tag{2.153}$$

Taking into account the senses of the angular velocities ω_1 and ω_4 in Figure 2.27a, and the senses of the velocities in Figure 2.24c, the senses for the Coriolis accelerations are as represented in Figure 2.24d.

If the accelerations $\mathbf{a}_{B_1}$ and $\mathbf{a}_{B_4}$ have the representations given in Figure 2.27a, then the polygon of accelerations that corresponds to (2.152) is as drawn in Figure 2.27d; this gives:

$$\mathbf{a}_B = k_a \left(\mathbf{i}\mathbf{b}'_2 \right), \ \mathbf{a}_{B_2 B_1} = k_a \left(\mathbf{m}'\mathbf{b}'_2 \right), \ \mathbf{a}_{B_3 B_4} = k_a \left(\mathbf{n}'\mathbf{b}'_4 \right). \tag{2.154}$$

Numerical example
Problem: Consider the mechanism *ECBA* in Figure 2.28a, in which the following are known: $OE = 0.6\,\text{m}$, $AB = \dfrac{0.8}{\sqrt{3}}\,\text{m}$, $BC = \dfrac{1.2}{\sqrt{3}}\,\text{m}$, $\alpha = \beta = \dfrac{\pi}{3}$, and $\varphi = 20t$. Using the grapho-analytical method, perform the kinematic analysis of the mechanism at time $t = \dfrac{\pi}{60}\,\text{s}$.

Solution: At time $t = \dfrac{\pi}{60}\,\text{s}$, $\varphi = \dfrac{\pi}{3}$. In the conditions of the problem, $d_2 = 0.4\,\text{m}$ and $d_3 = 0.6\,\text{m}$. Choose the scale of the lengths $k_l = 0.02 \left[\dfrac{\text{m}}{\text{mm}} \right]$ which gives the representation of the *ABC* dyad in Figure 2.28b from is deduced $[OA] = 30\,\text{mm}$, $[EC] = 14\,\text{mm}$, and $[BE] = 43\,\text{mm}$; that is, $OA = 0.6\,\text{m}$, $EC = 0.28\,\text{m}$, and $BE = 0.86\,\text{m}$. The angular velocity of the straight line Δ_4 is $\omega_4 = 20\,\text{rad/s}$.

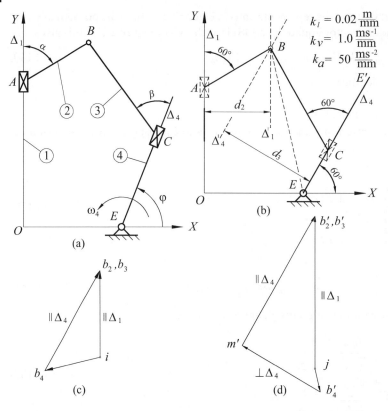

Figure 2.28 Numerical example of mechanism for the TRT dyad: (a) mechanism with TRT dyad; (b) position of the mechanism; (c) plane of velocities; (d) plane of accelerations

The straight line Δ_1 being fixed, $\mathbf{v}_{B_1} = \mathbf{0}$, and the velocity of point B_4 has magnitude $v_{B_4} = BE\omega_4$ and is perpendicular to BE.

If the scale of the velocities is chosen so that v_{B_4} is represented by a segment of 17.2 mm, then $k_v = 1 \left[\dfrac{\text{m/s}}{\text{mm}} \right]$ and from (2.148), the polygon in Figure 2.28c is obtained, from which we deduce that $ib_2 = 25$ mm and $b_4 b_3 = 34$ mm; that is

$$v_{B_2 B_1} = 25 \text{ m/s}, \ v_{B_3 B_4} = 34 \text{ m/s}.$$

The acceleration of point B_1 is zero, and the acceleration of point B_4 has magnitude $a_{B_4} = BE\omega_4^2 = 344$ m/s^2 and the sense from B to E.

The Coriolis acceleration $\mathbf{a}_{B_3 B_4}^C$ is perpendicular to the straight line Δ_4 and has magnitude $a_{B_3 B_4}^C = 2\omega_4 v_{B_3 B_4} = 1360$ m/s^2. Choosing for the Coriolis acceleration a representation of 27.2 mm, $k_a = 50 \left[\dfrac{\text{ms}^{-2}}{\text{mm}} \right]$; Equation (2.150) written in the form

$$\mathbf{a}_{B_2} = \underbrace{\mathbf{a}_{B_2 B_1}}_{\parallel \ \Delta_1} = \underbrace{\mathbf{a}_{B_4}}_{\perp \ \Delta_4} + \underbrace{\mathbf{a}_{B_3 B_4}^C}_{\perp \ \Delta_4} + \underbrace{\mathbf{a}_{B_3 B_4}}_{\parallel \ \Delta_4},$$

has the representation shown in Figure 2.28d, from which we deduce

$$a_{B_2} = 2350 \text{ m/s}^2, \ a_{B_2 B_3} = 2050 \text{ m/s}^2.$$

2.7.2 The Analytical Method

Formulation of the problem

Consider the TRT dyad in Figure 2.29, for which the following are known:

- the coordinates X_D, Y_D and X_E, Y_E of points D and E, respectively, angles φ_1, φ_4 between the straight lines Δ_1 and Δ_4, respectively, and the axis OX, the lengths l_2 and l_3 of segments BA and BC, respectively, and the angles α and β;
- the components v_{Dx}, v_{Dy} and v_{Ex}, v_{Ey} of the velocities of points D and E, respectively, and the angular velocities ω_1 and ω_4;
- the components a_{Dx}, a_{Dy} and a_{Ex}, a_{Ey} of the accelerations of points D and E, respectively, and the angular accelerations ε_1 and ε_4.

The following values are to be determined:

- the coordinates X_B, Y_B of point B, and the lengths λ_1, λ_4 of the segments DA and EC, respectively
- the components v_{Bx} and v_{By} of the velocity of point B, and the relative velocities of the sliders relative to the straight lines Δ_1 and Δ_4
- the components a_{Bx} and a_{By} of the acceleration of point B, and the relative accelerations of the sliders relative to the straight lines Δ_1 and Δ_4.

Determination of the positions

Projecting the vector relation

$$\mathbf{OD} + \mathbf{DA} + \mathbf{AB} + \mathbf{BC} + \mathbf{CE} = \mathbf{OE} \tag{2.155}$$

onto the axes and taking into consideration the considered notations (Figure 2.29), the following scalar relations are obtained

$$\begin{cases} \lambda_1 \cos \varphi_1 - \lambda_4 \cos \varphi_4 = X_E - X_D - l_2 \cos (\alpha - \varphi_1) + l_3 \cos (\beta + \varphi_4), \\ \lambda_1 \sin \varphi_1 - \lambda_4 \sin \varphi_4 = Y_E - Y_D + l_2 \sin (\alpha - \varphi_1) + l_3 \cos (\beta + \varphi_4), \end{cases} \tag{2.156}$$

from which are determined the unknowns

$$\lambda_1 = \frac{(X_E - X_D) \sin \varphi_4 - (Y_E - Y_D) \cos \varphi_4 - l_2 \sin (\alpha + \varphi_4 - \varphi_1) - l_3 \sin \beta}{\sin (\varphi_4 - \varphi_1)}, \tag{2.157}$$

$$\lambda_4 = \frac{(X_E - X_D) \sin \varphi_1 - (Y_E - Y_D) \cos \varphi_1 - l_3 \sin (\beta + \varphi_4 - \varphi_1) - l_2 \sin \alpha}{\sin (\varphi_4 - \varphi_1)}. \tag{2.158}$$

The coordinates of point B are obtained by projecting the vector relation $\mathbf{OB} = \mathbf{OD} + \mathbf{DA} + \mathbf{AB}$, giving

$$X_B = X_D + \lambda_1 \cos \varphi_1 + l_2 \cos (\alpha - \varphi_1), \quad Y_B = Y_D + \lambda_1 \sin \varphi_1 - l_2 \sin (\alpha - \varphi_1). \tag{2.159}$$

Figure 2.29 The TRT dyad.

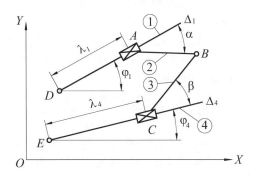

Determination of the velocities

From the vector relations

$$\mathbf{v}_{B_1} = \mathbf{v}_D + \boldsymbol{\omega}_1 \times \mathbf{DB}, \; \mathbf{v}_{B_4} = \mathbf{v}_E + \boldsymbol{\omega}_4 \times \mathbf{EB}, \tag{2.160}$$

are obtained the results

$$v_{B_1x} = v_{Dx} - \omega_1 \left(Y_B - Y_D \right), \; v_{B_1y} = v_{Dy} + \omega_1 \left(X_B - X_D \right), \tag{2.161}$$

$$v_{B_4x} = v_{Ex} - \omega_4 \left(Y_B - Y_E \right), \; v_{B_4y} = v_{Ey} + \omega_4 \left(X_B - X_E \right), \tag{2.162}$$

and from the vector relations

$$\mathbf{v}_{B_2} = \mathbf{v}_{B_3} = \mathbf{v}_{B_1} + \mathbf{v}_{B_2B_1} = \mathbf{v}_{B_4} + \mathbf{v}_{B_3B_4}, \tag{2.163}$$

are obtained the system

$$\begin{cases} v_{B_1x} + v_{B_2B_1} \cos \varphi_1 = v_{B_4x} + v_{B_3B_4} \cos \varphi_4, \\ v_{B_1y} + v_{B_2B_1} \sin \varphi_1 = v_{B_4y} + v_{B_3B_4} \sin \varphi_4, \end{cases} \tag{2.164}$$

from which results the relative velocities

$$v_{B_2B_1} = \frac{\left(v_{B_4x} - v_{B_1x} \right) \sin \varphi_4 - \left(v_{B_4y} - v_{B_1y} \right) \cos \varphi_4}{\sin \left(\varphi_4 - \varphi_1 \right)},$$
$$v_{B_3B_4} = \frac{\left(v_{B_4x} - v_{B_1x} \right) \sin \varphi_1 - \left(v_{B_4y} - v_{B_1y} \right) \cos \varphi_1}{\sin \left(\varphi_4 - \varphi_1 \right)}. \tag{2.165}$$

The components of the velocity of point B read

$$v_{Bx} = v_{B_1x} + v_{B_2B_1} \cos \varphi_1, \; v_{By} = v_{B_1y} + v_{B_2B_1} \sin \varphi_1. \tag{2.166}$$

Determination of the accelerations

From the vector relations

$$\mathbf{a}_{B_1} = \mathbf{a}_D + \boldsymbol{\varepsilon}_1 \times \mathbf{DB} - \omega_1^2 \mathbf{DB}, \; \mathbf{a}_{B_4} = \mathbf{a}_E + \boldsymbol{\varepsilon}_4 \times \mathbf{EB} - \omega_4^2 \mathbf{EB}, \tag{2.167}$$

are obtained:

$$a_{B_1x} = a_{Dx} - \varepsilon_1 \left(Y_B - Y_D \right) - \omega_1^2 \left(X_B - X_D \right),$$
$$a_{B_1y} = a_{Dy} + \varepsilon_1 \left(X_B - X_D \right) - \omega_1^2 \left(Y_B - Y_D \right), \tag{2.168}$$

$$a_{B_4x} = a_{Ex} - \varepsilon_4 \left(Y_B - Y_E \right) - \omega_4^2 \left(X_B - X_E \right),$$
$$a_{B_4y} = a_{Ey} + \varepsilon_4 \left(X_B - X_E \right) - \omega_4^2 \left(Y_B - Y_E \right), \tag{2.169}$$

and from the vector expressions

$$\mathbf{a}_{B_2} = \mathbf{a}_{B_3} = \mathbf{a}_{B_1} + 2\boldsymbol{\omega}_1 \times \mathbf{v}_{B_2B_1} + \mathbf{a}_{B_2B_1} = \mathbf{a}_{B_4} + 2\boldsymbol{\omega}_4 \times \mathbf{v}_{B_3B_4} + \mathbf{a}_{B_3B_4} \tag{2.170}$$

are obtained the system

$$\begin{cases} a_{B_2B_1} \cos \varphi_1 - a_{B_3B_4} \cos \varphi_4 = A_x, \\ a_{B_2B_1} \sin \varphi_1 - a_{B_3B_4} \sin \varphi_4 = A_y, \end{cases} \tag{2.171}$$

where

$$A_x = a_{B_4x} - a_{B_1x} + 2\omega_1 v_{B_2B_1} \sin \varphi_1 - 2\omega_4 v_{B_3B_4} \sin \varphi_4,$$
$$A_y = a_{B_4y} - a_{B_1y} + 2\omega_1 v_{B_2B_1} \cos \varphi_1 + 2\omega_4 v_{B_3B_4} \cos \varphi_4. \tag{2.172}$$

Finally this gives

$$a_{B_2B_1} = \frac{A_x \sin \varphi_4 - A_y \cos \varphi_4}{\sin \left(\varphi_4 - \varphi_1 \right)}, \; a_{B_3B_4} = \frac{A_x \sin \varphi_1 - A_y \cos \varphi_1}{\sin \left(\varphi_4 - \varphi_1 \right)}, \tag{2.173}$$

$$a_{Bx} = a_{B_1x} - 2\omega_1 v_{B_2B_3} \sin \varphi_1 + a_{B_2B_1} \cos \varphi_1,$$
$$a_{By} = a_{B_1y} - 2\omega_1 v_{B_2B_3} \cos \varphi_1 + a_{B_2B_1} \sin \varphi_1. \tag{2.174}$$

Numerical examples

Problem: Using the analytical method, perform the kinematic analysis of the mechanism in Figure 2.28a. Consider the same conditions as in the example solved using the grapho-analytical method.

Solution: From the problem:

$$X_D = Y_D = 0, \ X_E = 0.6 \, \text{m}, \ Y_E = 0, \ l_2 = \frac{0.8}{\sqrt{3}} \, \text{m}, \ l_1 = \frac{1.2}{\sqrt{3}} \, \text{m},$$

$$\varphi_1 = \frac{\pi}{2}, \ \varphi_4 = \frac{\pi}{3}, \alpha = \beta = \frac{\pi}{3}$$

and, substituting in (2.157):

$$\lambda_1 = 0.623 \, \text{m}. \ \lambda_4 = 0.293 \, \text{m}, \ X_B = 0.4 \, \text{m}, \ Y_B = 0.855 \, \text{m}.$$

For the velocities:

$$v_{Ex} = v_{Ey} = v_{Dx} = v_{Dy} = 0, \ \omega_1 = 0, \omega_4 = 20 \, \text{rad/s},$$

and, substituting in (2.161), (2.162), (2.165), and (2.166), we obtain

$$v_{B_1x} = v_{B_1y} = 0, \ v_{B_4x} = -17.1 \, \text{m/s}, \ v_{B_4y} = -4 \, \text{m/s},$$

$$v_{B_2B_1} = 25.62 \, \text{m/s}, \ v_{B_3B_4} = 34.2 \, \text{m/s}, \ v_{Bx} = 12.81 \, \text{m/s}, \ v_{By} = 22.18 \, \text{m/s}.$$

For the accelerations, we have $a_{Ex} = a_{Ey} = a_{Dx} = a_{Dy} = 0$, $\varepsilon_1 = \varepsilon_4 = 0$ and, substituting in (2.168), (2.169), and (2.172)–(2.174), we successively deduce

$$a_{B_1x} = a_{B_1y} = 0, \ a_{B_4x} = 80 \, \text{m/s}^2, \ a_{B_4y} = -342 \, \text{m/s}^2,$$

$$A_x = -1104.72, \ A_y342, \ a_{B_2B_1} = 2209.44 \, \text{m/s}^2, \ a_{B_3B_4} = 2255.43 \, \text{m/s}^2.$$

Comparing the results with those obtained using the grapho-analytical method, there is an acceptable error, the highest value of 10% being obtained for $a_{B_3B_4}$.

2.7.3 The Assisted Analytical Method

Based on the relations established for the analytical method and following the algorithm given by (2.157)–(2.174), we wrote the procedure *Dyad_TRT*, the content of which is given in the companion website of this book. The procedure has 22 input data and 20 output data. The procedure's input data are:

- the coordinates of the kinematic pairs D and E (X_D, Y_D and X_E, Y_E)
- the position angles φ_1, φ_4
- the lengths of elements 2 and 3 (l_2, l_3)
- the structural angles α and β
- the kinematic parameters of the input poles D and E ($v_{Dx}, v_{Dy}, v_{Ex}, v_{Ey}, \omega_1, \omega_4, a_{Dx}, a_{Dy}, a_{Ex}, a_{Ey}, \varepsilon_1, \varepsilon_4$).

The procedure's output data are:

- the variable lengths λ_2 and λ_4
- the coordinates of point B (X_B, Y_B)
- the components of the relative velocities and accelerations ($v_{B_1x}, v_{B_1y}, v_{B_4x}, v_{B_4y}, v_{B_2B_1}, v_{B_3B_4}, v_{Bx}, v_{By}, a_{B_1x}, a_{B_1y}, a_{B_4x}, a_{B_4y}, a_{B_2B_1}, a_{B_3B_4}, a_{Bx}, a_{By}$).

Calling the procedure with the same input parameters as in the example solved by the grapho-analytical and analytical methods, namely

$\quad$ **Dyad_TRT** $(0, 0, 0.6, 0, Pi/2, Pi/3, 0.8/\text{sqrt}(3), 1.2/\text{sqrt}(3), Pi/3, Pi/3, 0, 0, 0, 0, 0, 20,$
$\quad 0, 0, 0, 0, 0, 0, Lambda1, Lambda4, XB, YB, vB1x, vB1y, vB4x, vB4y, vB2B1, vB3B4,$
$\quad\quad vBx, vBy, aB1x, aB1y, aB4x, aB4y, aB2B1, aB3B4, aBx, aBy)$

the following results are obtained

$$\lambda_1 = 0.622650, \ \lambda_4 = 0.292820,$$
$$X_B = 0.400000, \ Y_B = 0.853590,$$
$$v_{B_1x} = 0.000000, \ v_{B_1y} = 0.000000,$$
$$v_{B_4x} = -17.071797, \ v_{B_4y} = -4.000000,$$
$$v_{B_2B_1} = 25.569219, \ v_{B_3B_4} = 34.143594,$$
$$v_{Bx} = 0.000000, \ v_{By} = 25.569219,$$
$$a_{B_1x} = 0.000000, \ a_{B_1y} = 0.000000,$$
$$a_{B_4x} = 80.000000, \ a_{B_4y} = -341.435935,$$
$$a_{B_2B_1} = 2251.487483, \ a_{B_3B_4} = 2205.537551,$$
$$a_{Bx} = 0.000000, \ a_{By} = 2251.487483.$$

2.7.4 The Assisted Graphical Method

At time $t = \frac{\pi}{60}$ s, the coordinates of point $E\,(0.6, 0)$, the structural angles $\alpha = \beta = 60°$, the lengths of the elements $(AB = \dfrac{0.8}{\sqrt{3}}, \ BC = \dfrac{1.2}{\sqrt{3}})$, and angle $\varphi = \dfrac{\pi}{3}$ are all known.

1) In a window of visualization:

$\quad$ Zoom;−0.5, −0.2; 2.5, 1;

2) Construct
   - the straight line Δ_4, starting from point E with an angle of $60°$ $(EE' = 0.8$)

$\quad$ PLine;0.6,0;W;0.005;0.005;@0.8<60;;

   - the straight line Δ_4' parallel to the straight line Δ_4 and situated at a distance $\dfrac{d_3}{\sin 60°}$

$\quad$ (Setq Ep(List (- 0.6 (/ 0.6 (sin (/ Pi 3)))) 0.0));Line;!Ep;@1<60;;

3) Draw the vertical straight line Δ_1', situated at the distance $d_2 = 0.4$ relative to the axis OY

$\quad$ Line;0.4,0;@1<90;;

4) At the intersection of the straight lines Δ_1' and Δ_4' is point B. With the centre at point B construct a circle of radius AB, intersecting axis OY at point A

$\quad$ (Setq lAB(/ 0.8 (Sqrt 3)));Circle;Int;\!lAB;Line;0,0;0,1;;

5) Draw the line AB

$\quad$ PLine;Int;\Int;\;

6) With the centre at point B construct a circle of radius BC

$\quad$ (Setq lBC(/ 1.3 (Sqrt 3)));Circle;Int;\!lBC;

Figure 2.30 The TRT dyad: analysis of the positions.

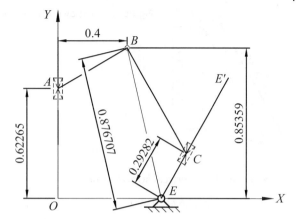

which intersects the straight line Δ'_4 at point C. Draw the segment BC

> PLine;Int;\Int;\;

7) Construct and save the straight line EB

> Line;0.6,0;Int;\;- Block;Dist_EB;0.6,0;\;

8) Put the straight line Δ'_4 into memory, erase the circles and the straight line Δ'_1

> −Block;Delta4;!Ep;\;Erase;\\\;

then draw the axes of coordinates, kinematic pairs and draws the dimensions, as in Figure 2.30:

> Pline;0,0;W;0.002;;1.2,0;W;0.02;0;L;0.07;;Pline;0,0;W;0.002;;0,1;W;0.02;0;L;0.07;;

The polygon of velocities is obtained using the vector relation (2.148), knowing that at $t = \dfrac{\pi}{60}$ s the angular velocity is $\omega_4 = \dot\varphi = 20$ rad/s.

1) In a new window of visualization

> Zoom;−50, 0; 35, 40;

2) Start from the pole of velocities $i\,(1, 6)$ and taking into account the notation in Figure 2.28c, successively construct
   - the velocity $\mathbf{v}_{B_4}$ ($\mathbf{ib}_4$), by inserting the distance EB, multiplied by ω_4 and rotated through 90°

> −Insert;Dist_EB;1,6;20;20;90;

then unselecting the inserted block

> Explode;\;

   - at point b_4, the parallel to Δ_4, by inserting the block *Delta4*, multiplied by 20 and rotated by 0°

> −Insert;Delta4;End;\;20;20;0;

then unselecting the inserted block

> Explode;\;

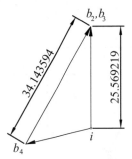

Figure 2.31 TRT dyad: the polygon of velocities.

- at point i, a vertical straight line

    Line;1,6;@20<90;;

  then intersecting the last two straight lines

    Fillet;R;0;;Fillet;\\

3) This gives point b_2 and the polygon in Figure 2.31, which is completed with arrows and dimensions. Save $\mathbf{v}_{B_3B_4}$ having point b_4 as base point

    −Block;vB3B4;END;\;Oops;

The polygon of accelerations can be obtained by taking into account the vector relation (2.152) and the notations in Figure 2.28d.

1) Choose a window of visualization

    Zoom;−3000, 50; 3000, 3000;

2) From pole j (500, 500) construct
   - $\mathbf{a}_{B_4}$ ($\mathbf{j}b'_4$), by inserting the distance EB at the pole of accelerations j, multiplied by $\omega_2^2 = 400$ and rotated by an angle of 180°

    −Insert;Dist_EB;500,500;400;400;180;

  then unselecting the inserted block

    Explode;\;

   - $\mathbf{a}_{B_3B_4}^C$ ($\mathbf{b}'_4\mathbf{m}$), by inserting $\mathbf{v}_{B_3B_4}$ at point b'_4 multiplied by $2\omega_4$ and rotated through an angle of 90°

    −Insert;vB3B4;End\;40;40;90;

  then unselecting the inserted block

    Explode;\;

   - the parallel to Δ_4, by inserting the block *Delta4* at point m' previously obtained

    −Insert;Delta4;End;\;1000;1000;0;

  then unselecting the inserted block

    Explode;\;

   - a parallel to the axis Oy at pole j

    Line;500,500;@2000<90;

Figure 2.32 TRT - the polygon of accelerations.

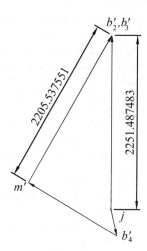

3) The last two constructed entities are filleted, completing the polygon with arrows, letters and the dimensions, as in Figure 2.32.

Fillet;\\

The AutoLisp function for kinematic analysis of the TRT dyad in Figure 2.28, based on the relations established in the Section 2.7.1, no matter the positions of the kinematic pairs and the dimensions of the mechanism, is as follows:

```
(Defun C:TRT ()
(Setq lAB(/ 0.8 (Sqrt 3)) lBC (/ 1.2 (Sqrt 3)) xE 0.6 yE 0.0 lEEp
0.8 alpha 60.0 beta 60.0 phiG (/ Pi 3) om4 20 ismall(List 1.0 6.0)
jsmall(List 500.0 500.0))
;Calculation of positions
(Setq alphaG(* alfa (/ Pi 180)) betaG(* alpha (/ Pi 180)) d2(* lAB (Sin
alphaG)) d3(* lBC (Sin betaG)) xD2 d2 yD2 0.0 xD2p d2 yD2p 1.0
xD3(- xE (/ d3 (Sin fiG))) yD3 0 xD3p(+ xD3 (Cos phiG)) yD3p(Sin phiG)
BCapital(Inters (List xD2 yD2) (List xD2p yD2p) (List xD3 yD3) (List
xD3p yD3p) Nil) xB(Princ(car BCapital)) yB(Princ(car(cdr BCapital))))
(Setq xBp(+ xB (Cos (- (/ Pi 2) alphaG))) yBp(+ yB (Sin (- (/ Pi 2)
alphaG))) xM xB yM yB xN xBp yN yBp a xB b yB R lAB)
(Int_CL)
(Setq xA xP2 yA yP2)
(Setq xM xE yM yE xN (+ xE (Cos phiG)) yN (Sin phiG) a xB b yB R lBC)
(Int_CL)
(Setq xC xP2 yC yP2)
;Position of mechanism
(Command "Erase" "All" "" "Osnap" "Off" "Ortho" "Off")
(Command "Zoom" "-0.5,-0.2" "2.5,1")
(Command "-Layer" "N" "Positions" "C" "7" "Positions" "S"
  "Positions" "" "")
(Command "Pline" (List xA yA) "W" "0.005" "" (List xB yB) (List xC
  yC) "")
(Command "Pline" (List xE yE) (List xC yC) "L" "0.3" "")
(Command "Pline" "0,0" "W" "0.002" "" "1.1,0" "W" "0.02" "0" "L"
  "0.07" "")
(Command "Pline" "0,0" "W" "0.002" "" "0,1" "W" "0.02" "0" "L"
  "0.07" "")
```

```
;Calculation of velocities
(Setq vB4(* om4 (Distance (List xB yB) (List xE yE))) phiBE (Angle
(List xB yB) (List xE yE)) bsmall4(Polar ismall (- phiBE (/ Pi 2))
vB4) rsmall(Polar bsmall4 phiG 1) ssmall(Polar ismall (/ Pi 2) 1)
bsmall23(Inters bsmall4 rsmall ismall ssmall Nil))
;Polygon of velocities
(Command "Zoom" "-50,0" "35,40")
(Command "-Layer" "N" "Velocities" "C" "1" "Velocities" "S"
  "Velocities" "" "")
(Command "Line" ismall bsmall4 bsmall23 "C")
(Setq l(/ vb4 7) g(* l 0.353))
(Arrows bsmall4 ismall)
(Arrows bsmall23 bsmall4)
(Arrows bsmall23 ismall)
;Calculation of accelerations
(Setq aB4(* om4 om4 (Distance (List xB yB) (List xE yE))) acB3B4(*
2.0 om4 (Distance bsmall4 bsmall23)) bprime(Polar jsmall phiBE aB4)
mprime(Polar bprime (+ phiG (/ Pi 2)) acB3B4) rprime(Polar mprime phiG
10.0) sprime(Polar jsmall (/ Pi 2) 10.0) bprime23(Inters mprime rprime
jsmall ssmall Nil))
;Polygon pf accelerations
(Command "Zoom" "-3000,-50" "3000,3000")
(Command "-Layer" "N" "Accelerations" "C" "5" "Accelerations" "S"
  "Accelerations" "" "")
(Command "Line" jsmall bprime mprime bprime23 "C" "Zoom" "E")
(Setq l(/ ab4 4) g(* l 0.353))
(Arrows bprime jsmall)
(Arrows mprime bprime)
(Arrows bprime23 mprime)
(Arrows bprime23 jsmall)
)
```

The function is saved as *Dyad_TRT.lsp* and, after it is loaded ((Load "Dyad_RRT .lsp")), it can be called with the command *TRT*. The dimensions of the dyad are those given in the Section 2.7.1, but other values may be used.

The calculation of the positions starts with the transformation of the angles α and β from degrees to radians. The following are calculated:

- the distances d_2, d_3
- the coordinates of two points that belong to the straight line Δ_1' $(D_2\left(x_{D_2}, y_{D_2}\right), D_2'\left(x_{D_2}', y_{D_2}'\right))$
- the coordinates of two points that belong to the straight line Δ_4' $(D_3\left(x_{D_3}, y_{D_3}\right), D_3'\left(x_{D_3}', y_{D_3}'\right))$.

Using the function *Inters*, point *B* – the intersection of the straight lines D_2D_2' and D_3D_3' – is determined. Function *INT_CL* is called to determine point *A* and then point *C*. The convenient solutions are retained and the loop of the mechanism is drawn using a line with thickness, in a layer called *Positions*. The construction is completed with the arrows of the axes, giving the construction in Figure 2.30. This figure is completed with dimensions, kinematic pairs and notations.

For the calculation of the velocities, calculate v_{B_4} as the product of the angular velocity ω_4 and the distance *BE*. Using the function *Polar* determine (Figure 2.31) point b_4, a point *r* that belongs to the straight line Δ_4, and a point *s* that belongs to the straight line Δ_1. At the intersection of the straight lines determined by points b_4, *r*, *i*, and *s* is point b_{23}. The representation of the polygon of velocities is constructed in the layer *Velocities*, by drawing the loop

$ib_4b_{23}i$. The polygon is completed with arrows (function *Arrows*), obtaining the representation in Figure 2.31; at the end the dimensions are added.

For the calculation of the accelerations the accelerations a_{B_4} and $a^C_{B_3B_4}$ are determined using (2.153), and then points b'_4, m', r' (a point that belongs to the straight line Δ_4), and s' (a point that belongs to the straight line Δ_1), of the polygon of accelerations. Using the function *Inters*, point b'_{23} is determined as the intersection of the straight lines defined by points m', r' and j, s', respectively. The polygon of accelerations is represented in the layer *Accelerations*, by drawing the loop $jb'_4m'b'_{23}j$. The polygon completes with arrows (with the function *Arrows*) resulting the representation in Figure 2.32. After the dimension and notations for the points, the AutoCAD working session is closed and the output file is saved (with a name).

2.8   Kinematic Analysis of the RTT Dyad

2.8.1   The Grapho-analytical Method

Formulation of the problem.
Consider the RTT dyad ($ABCD$) in Figure 2.33a, for which the following are known:

- the position of point A, the position of the straight line Δ_4, the position of a point D belonging to the straight line Δ_4, the angles α and β, and the length l_2 of the segment AB;
- the velocities $\mathbf{v}_A$ and $\mathbf{v}_D$ of points A and D, respectively, and the angular velocity ω_4 of the straight line Δ_4;
- the accelerations $\mathbf{a}_A$ and $\mathbf{a}_D$ of points A and D, respectively, and the angular acceleration ε_4 of the straight line Δ_4.

The following values are to be determined:

- the positions of points B and C, and the lengths of the segments $BC = \lambda_3$ and $CD = \lambda_4$
- the relative velocities of the sliders B and C relative to the straight lines Δ_3 and Δ_4, respectively
- the relative accelerations of the sliders B and C relative to the straight lines Δ_3 and Δ_4, respectively.

Determination of the positions
The construction of the dyad starts with the observation that the straight line Δ_3 is a tangent to the circle centred at point A, and of radius $d_2 = l_2 \sin \alpha$. This results in the following succession of graphical constructions (Figure 2.33b):

1) Construct the circle centred at point A and of radius $d_2 = l_2 \sin \alpha$.
2) Draw the circle centred at point A and radius d_2
3) Construct the straight line Δ_3 tangent to the circle and making an angle β with the straight line Δ_4.

Thus points A' and C are obtained, and then a straight line passing through point A and making an angle α with the straight line Δ_3 can be drawn. In this way point B is determined (Figure 2.33b).

Determination of the velocities
First determine the velocity $\mathbf{v}_{A_4}$ of point A_4 (Figure 2.33c), which is rigidly linked to the straight line Δ_4

$$\mathbf{v}_{A_4} = \underline{\underline{\mathbf{v}_D}} + \underset{\perp AD}{\underline{\underline{\mathbf{v}_{A_4D}}}},$$

$$(2.175)$$

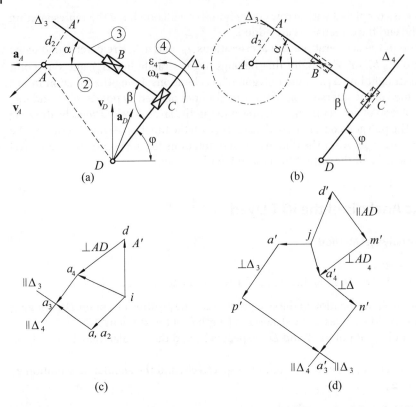

Figure 2.33 The grapho-analytical analysis of the RTT dyad: (a) RTT dyad; (b) position of the dyad; (c) plane of velocities; (d) plane of accelerations.

where

$$v_{A_4D} = \omega_4 AD. \tag{2.176}$$

The relative velocities are determined by the representation (Figure 2.33c) of the vector relations

$$\mathbf{v}_{A_3} = \underline{\underline{\mathbf{v}_{A_2}}} + \underline{\mathbf{v}_{A_3A_2}}_{\parallel \Delta_3} = \underline{\underline{\mathbf{v}_{A_4}}} + \underline{\mathbf{v}_{A_3A_4}}_{\parallel \Delta_4}. \tag{2.177}$$

The relative velocities $\mathbf{v}_{A_3A_2}$, $\mathbf{v}_{A_3A_4}$ of the displacements of the sliders B and C along the axes Δ_3 and Δ_4, respectively, are determined from the scale k_v of the velocities, using:

$$\mathbf{v}_{A_3A_2} = k_v\left(\mathbf{a}_3\mathbf{a}_2\right), \; \mathbf{v}_{A_3A_4} = k_v\left(\mathbf{a}_3\mathbf{a}_4\right). \tag{2.178}$$

Determination of the accelerations
The acceleration of point A_4 is determined using:

$$\mathbf{a}_{A_4} = \underline{\underline{\mathbf{a}_D}} + \underline{\mathbf{a}^\nu_{A_4D}}_{A \to D} + \underline{\mathbf{a}^\tau_{A_4D}}_{\perp AD}, \tag{2.179}$$

where

$$a^\nu_{A_4D} = \omega_4^2 AD, \; a^\tau_{A_4D} = \varepsilon_4 AD. \tag{2.180}$$

The relative accelerations $\mathbf{a}_{A_3A_2}$ and $\mathbf{a}_{A_3A_4}$ are deduced by the representation of the vector relations

$$
\begin{aligned}
\mathbf{a}_{A_3} &= \underline{\underline{\mathbf{a}_{A_2}}} + \underline{\mathbf{a}^C_{A_3A_2}} + \underline{\mathbf{a}_{A_3A_2}}, \\
&\qquad\quad \perp \Delta_3 \qquad \parallel \Delta_3 \\
&= \underline{\underline{\mathbf{a}_{A_4}}} + \underline{\mathbf{a}^C_{A_3A_4}} + \underline{\mathbf{a}_{A_3A_4}}, \\
&\qquad\quad \perp \Delta_3 \qquad \parallel \Delta_3
\end{aligned}
\tag{2.181}
$$

where the Coriolis accelerations are perpendicular to the straight lines Δ_3 and Δ_4, having the magnitudes

$$
a^C_{A_3A_2} = 2\omega_4 v_{A_3A_2}, \; a^C_{A_3A_4} = 2\omega_1 v_{A_3A_4}.
\tag{2.182}
$$

The relative accelerations (Figure 2.33d) are deduced from the scale k_a of the accelerations, using:

$$
\mathbf{a}_{A_3A_2} = k_a \left(\mathbf{p'a'}_3\right), \; \mathbf{a}_{A_3A_4} = k_a \left(\mathbf{n'a'}_3\right).
\tag{2.183}
$$

Numerical example
Problem: Consider the mechanism $ABCD$ in Figure 2.34a, in which the following are known: $OA = 0.6\,\mathrm{m}$, $OD = 0.8\,\mathrm{m}$, $l_2 = AB = 0.3\sqrt{3}\,\mathrm{m}$, $\alpha = \frac{\pi}{4}$, $\beta = \frac{\pi}{3}$, and $\varphi = 20t$. Using the grapho-analytical method, perform the kinematic analysis of the mechanism at time $t = \frac{\pi}{80}$ s.

Solution: At time $t = \frac{\pi}{80}$ s we have $\varphi = \frac{\pi}{4}$; since $d_2 = 0.3\,\mathrm{m}$ and choosing the scale $k_l = 0.02\left[\frac{\mathrm{m}}{\mathrm{mm}}\right]$, the representation in Figure 2.34b is obtained, from which $[DC] = 40\,\mathrm{mm}$ and $[BC] = 50\,\mathrm{mm}$. This gives $DC = 0.8\,\mathrm{m}$ and $BC = 1\,\mathrm{m}$.

The angular velocity $\omega_4 = \dot\varphi$ has magnitude $\omega_4 = 20\,\mathrm{rad/s}$; in these conditions the velocity of point A_4 has magnitude $v_{A_4} = \omega_4 AD = 20\,\mathrm{m/s}$ and it is perpendicular to AD (Figure 2.34b).

Choosing the scale of the velocities $k_v = 0.5\left[\frac{\mathrm{m/s}}{\mathrm{mm}}\right]$, taking into account that $\mathbf{v}_{A_2} = \mathbf{0}$, and representing the vector equations (2.177), the triangle in Figure 2.34c is obtained, from which $[ia_3] = 7\,\mathrm{mm}$ and $[a_4a_3] = 43\,\mathrm{mm}$; that is, $v_{A_3A_2} = 3.5\,\mathrm{m/s}$ and $v_{A_3A_4} = 21.5\,\mathrm{m/s}$.

The acceleration of point A_4 is equal to the acceleration $a^v_{A_4D} = \omega_4^2 DA = 400\,\mathrm{m/s^2}$. Choosing the scale of the accelerations $k_a = 20\left[\frac{\mathrm{m/s^2}}{\mathrm{mm}}\right]$ and taking into account the equalities $\mathbf{a}_{A_2} = \mathbf{0}$ and $a^C_{A_3A_2} = 140\,\mathrm{m/s^2}$, for the vector equations (2.181) the representation in Figure 2.34d is obtained, from which we deduce that $[n'a'_3] = 23\,\mathrm{mm}$ and $[m'a'_3] = 8\,\mathrm{mm}$; that is

$$
a_{A_3A_2} = 460\,\mathrm{m/s^2}, \; a_{A_3A_4} = 160\,\mathrm{m/s^2}.
$$

2.8.2 The Analytical Method

Formulation of the problem
Consider the RTT dyad in Figure 2.35, for which the following are known:

- the coordinates X_A, Y_A and X_D, Y_D of points A and D, respectively, angle φ_4, the length l_2 of the segment AB, and the angles α and β;
- the components v_{Ax}, v_{Ay} and v_{Dx}, v_{Dy} of the velocities of points A and D, respectively, and the angular velocity ω_4;
- the components a_{Ax}, a_{Ay} and a_{Dx}, a_{Dy} of the accelerations of points A and D, respectively, and the angular acceleration ε_4.

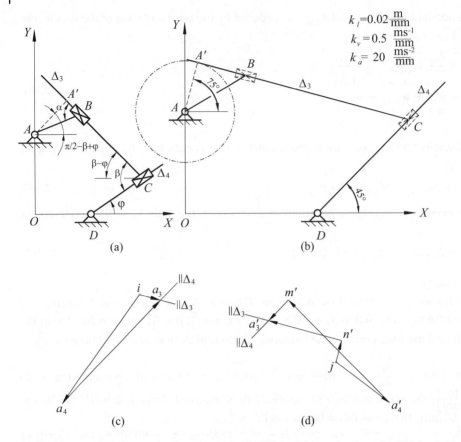

$$k_l = 0.02 \frac{m}{mm}$$
$$k_v = 0.5 \frac{ms^{-1}}{mm}$$
$$k_a = 20 \frac{ms^{-2}}{mm}$$

Figure 2.34 Numerical example of mechanism with RTT dyad: (a) mechanism with RTT dyad; (b) position of the mechanism; (c) plane of velocities; (d) plane of accelerations..

Figure 2.35 The RTT dyad.

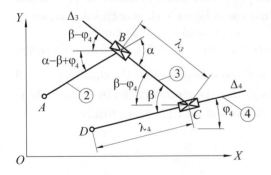

The following values are to be determined:

- the lengths λ_3 and λ_4 of the segments BC and CD, respectively
- the relative velocities of displacements of the sliders relative to the axes Δ_3 and Δ_4
- the relative accelerations of displacements of the sliders relative to the axes Δ_3 and Δ_4.

Determination of the lengths

Projecting the vector relation

$$\mathbf{OA} + \mathbf{AB} + \mathbf{BC} + \mathbf{CD} = \mathbf{OD} \tag{2.184}$$

onto the axes, we obtain the scalar relations

$$
\lambda_3 \cos \lambda \left(\beta - \varphi_4 \right) - \lambda_4 \cos \varphi_4 = X_D - X_A - l_2 \cos \left(\alpha - \beta + \varphi_4 \right),
$$
$$
-\lambda_3 \sin \left(\beta - \varphi_4 \right) - \lambda_4 \cos \left(\beta - \varphi_4 \right) = Y_D - Y_A - l_2 \sin \left(\alpha - \beta + \varphi_4 \right), \tag{2.185}
$$

from which

$$
\lambda_3 = \frac{\left(X_D - X_A \right) \sin \varphi_4 - \left(Y_D - Y_A \right) \cos \varphi_4 + l_2 \sin \left(\alpha - \beta \right)}{\sin \beta}, \tag{2.186}
$$

$$
\lambda_4 = \frac{\left(X_D - X_A \right) \sin \left(\varphi_4 - \beta \right) - \left(Y_D - Y_A \right) \cos \left(\varphi_4 - \beta \right) + l_2 \sin \alpha}{\sin \beta}. \tag{2.187}
$$

Determination of the velocities
The components of the velocity of point A_4 are obtained from the vector relation

$$
\mathbf{v}_{A_4} = \mathbf{v}_D + \boldsymbol{\omega}_4 \times \mathbf{DA}, \tag{2.188}
$$

and they have the expressions

$$
v_{A_4 x} = v_{Dx} - \omega_4 \left(Y_A - Y_D \right), \; v_{A_4 y} = v_{Dy} + \omega_4 \left(X_A - X_D \right). \tag{2.189}
$$

Projecting the vector relation (2.188) onto the axes gives the scalar relations

$$
v_{Ax} + v_{A_3 A_2} \cos \left(\pi - \beta + \varphi_4 \right) = v_{A_4 x} + v_{A_3 A_2} \cos \varphi_4,
$$
$$
v_{Ay} + v_{A_3 A_2} \sin \left(\pi - \beta + \varphi_4 \right) = v_{A_4 y} + v_{A_3 A_2} \sin \varphi_4,
$$

from which the relative velocities are obtained:

$$
v_{A_3 A_2} = \frac{-\left(v_{A_4 x} - v_{Ax} \right) \sin \varphi_4 + \left(v_{A_4 y} - v_{Ay} \right) \cos \varphi_4}{\sin \beta},
$$
$$
v_{A_3 A_4} = \frac{-\left(v_{A_4 x} - v_{Ax} \right) \sin \left(\beta - \varphi_4 \right) - \left(v_{A_4 y} - v_{Ay} \right) \cos \left(\beta - \varphi_4 \right)}{\sin \beta}. \tag{2.190}
$$

Determination of the accelerations
The components of the acceleration of point A_4 are obtained from the vector relation

$$
\mathbf{a}_{A_4} = \mathbf{a}_D + \boldsymbol{\varepsilon}_4 \times \mathbf{DA} - \omega_4^2 \mathbf{DA}, \tag{2.191}
$$

and they read as

$$
a_{A_4 x} = a_{Dx} - \varepsilon_4 \left(Y_A - Y_D \right) - \omega_4^2 \left(X_A - X_D \right),
$$
$$
a_{A_4 y} = a_{Dy} + \varepsilon_4 \left(X_A - X_D \right) - \omega_4^2 \left(Y_A - Y_D \right). \tag{2.192}
$$

The relative accelerations are determined from the system obtained with the vector relations (2.169). If, in addition, we use:

$$
A_x = a_{A_4 x} - a_{Ax} + 2\omega_4 v_{A_3 A_2} \sin \left(\beta - \varphi_4 \right) - 2\omega_4 v_{A_3 A_4} \sin \varphi_4,
$$
$$
A_y = a_{A_4 y} - a_{Ay} + 2\omega_4 v_{A_3 A_2} \cos \left(\beta - \varphi_4 \right) + 2\omega_4 v_{A_3 A_4} \cos \varphi_4. \tag{2.193}
$$

then, ultimately, we obtain the expressions for the relative accelerations

$$
a_{A_3 A_2} = \frac{-A_x \sin \varphi_4 + A_y \cos \varphi_4}{\sin \beta}, \; a_{A_3 A_4} = \frac{-A_x \sin \left(\beta - \varphi_4 \right) - A_y \cos \left(\beta - \varphi_4 \right)}{\sin \beta}. \tag{2.194}
$$

Numerical example
Problem: Using the analytical method, perform a kinematic analysis of the mechanism in Figure 2.35a, considering the same conditions as in the example solved by the grapho-analytical method.

Solution: From the figure and the hypothesis the analytical data of the problem are as follows:
$X_A = 0$, $Y_A = 0.6$ m, $X_D = 0.8$ m, $Y_D = 0$, $l_2 = 0.3\sqrt{3}$ m, $\alpha = \frac{\pi}{4}$, $\beta = \frac{\pi}{3}$, $\varphi_4 = \frac{\pi}{4}$, $v_{Ax} = v_{Ay} = 0$,
$v_{Dx} = v_{Dy} = 0$, $\omega_4 = 20$ rad/s, $\varepsilon_4 = 0$, $a_{Ax} = a_{Ay} = a_{Dx} = a_{Dy} = 0$. From these, and from (2.186),
(2.187), (2.189), (2.190), and (2.192)–(2.194), we obtain

$$\lambda_3 = 1.016 \text{ m}, \quad \lambda_4 = 0.776 \text{ m}, \quad v_{A_3A_2} = -3.26 \text{ m/s}, \quad v_{A_3A_4} = 21.43 \text{ m/s},$$
$$a_{A_3A_2} = 457.28 \text{ m/s}^2, \quad a_{A_3A_4} = -172.27 \text{ m/s}^2.$$

The results obtained using the analytical method and those obtained using the grapho-analytical have a good correspondence, the errors of measurement leading to errors up to 5%.

2.8.3 The Assisted Analytical Method

Based on the relations established for the analytical method and following the calculation algorithm described by (2.186)–(2.194), we wrote the procedure *Dyad_RTT*, the content of which is given in the companion website of this book. The procedure has 18 input data and 6 output data.

The input data of the procedure are:

- the coordinates of the kinematic pairs A and D (X_A, Y_A and X_D, Y_D, respectively)
- the length of element 2 (l_2)
- the structural angles α and β
- the kinematic parameters of the input poles A and D (v_{Ax}, v_{Ay}, v_{Dx}, v_{Dy}, ω_4, a_{Ax}, a_{Ay}, a_{Dx}, a_{Dy}, ε_4).

The output data of the procedure are:

- the variable lengths λ_3 and λ_4
- the relative velocities and accelerations ($v_{A_3A_2}$, $v_{A_3A_4}$, $a_{A_3A_2}$, $a_{A_3A_4}$).

Calling the procedure with the same input data as used in the example solved by the grapho-analytical and analytical methods, namely

Dyad_RTT $(0, 0.6, 0.8, 0, 0.3 * \text{sqrt}(2), Pi/4, Pi/3, Pi/4, 0, 0, 0, 0, 20, 0, 0, 0, 0, 0,$
$Lambda3, Lambda4, vA3A2, vA3A4, aA3A2, aA3A4)$,

the following results are obtained

$$\lambda_3 = 1.016300, \quad v_{A_3A_2} = -3.265986, \quad a_{A_3A_2} = 457.238085$$
$$\lambda_4 = 0.776536, \quad v_{A_3A_4} = 21.431983, \quad a_{A_3A_4} = -172.050500$$

2.8.4 The Assisted Graphical Method

At time $t = \frac{\pi}{80}$ s, the coordinates of point $D\,(0.8, 0)$, the structural angles $\alpha = \frac{\pi}{4}$ and $\beta = \frac{\pi}{3}$, the lengths $AB = 0.3\sqrt{2}$, and angle $\varphi = \frac{\pi}{4}$ are all known. The graphical construction in AutoCAD of the mechanism that contains the RTT dyad in Figure 2.34 is as follows:

1) Start by choosing a window of visualization

Zoom;−0.5, −0.2; 2.5, 1;

2) In this, starting from point A, construct the segment AA' having length $d_2 = 0.3$ and making an angle $\varphi_{AA'} = \frac{\pi}{2} - \beta + \varphi = 90 - 60 + 45 = 75°$ with the horizontal

(Setq Ap(Polar (List 0 0.6) (* 75 (/ Pi 180)) 0.3));Line;0,0.6;!Ap;;

3) From point A' construct the straight line Δ_3 with a segment of length 1.0, which makes an angle $\varphi_{A'A''} = 2\pi - \beta + \varphi = 345°$ with the horizontal

        Line;End;\@1.5<345;;

4) At point D draw the straight line Δ_4

        Line;0.8,0;@1<45;;

5) At the intersection of the straight lines Δ_3 and Δ_4 is point B. Cuts the inconvenient part of the straight line Δ_3, the straight line Δ_4 being the cutting edge

        Trim;\;\;

6) Draw the straight line AB that makes angle $\varphi_{AB} = \alpha - \beta + \varphi = 30°$ with the horizontal. For a correct construction, point B belongs to the straight line Δ_3, too. For this reason, we use AutoLisp in order to determine point B

        (Setq B(Polar (List 0 0.6) (/ Pi 6)(* 0.3 (Sqrt 2))));Pline;0,0.6;W;0.004;0.004;!B;

7) Construct the segment AD and save it

        Line;0,0.6;0.8,0;;–Block;Dist_AD;0,0.6;\;

8) Save straight line Δ_3 having point A' as reference point

        –Block;Delta3;!Ap;\;Oops;

9) The same thing is done for the straight line Δ_4 having point D as reference point

        –Block;Delta4;0.8,0;\;Oops;

The command **Oops** is used to bring back the blocks into the current drawings.

10) Draw, using a line with non-zero thickness, the straight lines $A'C$ and Δ_4

        Pline;!Ap;End;\;;Pline;0.8,0;@1<45;;

11) Erase the segment AA' and construct the axes of coordinates

        Pline;0,0;W;0.002;;1.5,0;W;0.02;0;L;0.07;;Pline;0,0;W;0.002;;0,1;W;0.02;0;L;0.07;;

the kinematic pairs, and the dimensions, as in Figure 2.36.

The polygon of velocities is obtained from the vector relation (2.177), knowing that at time $t = \frac{\pi}{80}$ s, the angular velocity is $\omega = \dot{\varphi} = 20$ rad/s.

Figure 2.36 The RTT dyad: analysis of the positions.

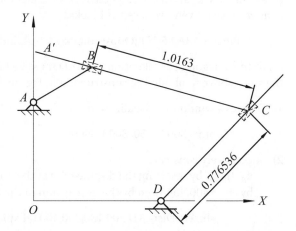

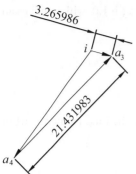

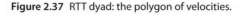

Figure 2.37 RTT dyad: the polygon of velocities.

1) The construction in Figure 2.37 is created in the same drawing, but in a new window of visualization.

    Zoom;$-20, -5; 25, 20$;

2) Starting from pole $i\,(5, 15)$, construct the velocity $\mathbf{v}_{A_4}$ (the vector $\mathbf{i}a_4$) by inserting the distance AD multiplied by ω_4 and rotated through an angle of $-90°$. To localize point a_4, the block is unselected after the insertion

    $-$Insert;Dist_AD;5, 15; 20; 20; -90;Explode;\;

3) At point a_4, construct a parallel to Δ_4 by inserting the block *Delta4*, amplified by 20, and rotated through an angle of $0°$; then this block is unselected

    $-$Insert;Delta4;End;\;20;20;0;Explode;\;

4) At point i, construct a parallel to Δ_3 by inserting the block *Delta3*, multiplied by 20, and rotated through an angle of $0°$; then this block is unselected

    $-$Insert;Delta3;5,15;20;20;0;Explode;\;

5) Extend the last entities to their intersection

    Fillet;R;0;;Fillet;\\

This gives the vector polygon in Figure 2.37, which is completed with arrows and dimensions. For the calculation of the Coriolis accelerations we need the relative velocities $v_{A_3A_2}$ and $v_{A_3A_4}$ (the segments ia_3 and a_4a_3, respectively). The velocities $v_{A_3A_2}$ and $v_{A_3A_4}$ having as starting points i and a_4, respectively are saved as blocks:

    $-$Block;vA3A2;5,15;\;Oops;$-$Block;vA3A4;End;\;Oops;

For the polygon of accelerations, (2.181) must be taken into account and the process follows the steps above to obtain the construction in Figure 2.38.

1) Defines a window of visualization

    Zoom;$-3000, -50; 3000, 3000$;

2) Successively construct
   - $\mathbf{a}_{B_4}$ ($\mathbf{ja'}_4$), by inserting the distance AD at the pole of accelerations $j\,(100, 200)$, multiplied by $\omega_4^2 = 400$, followed by the deselection of the block

    $-$Insert;Dist_AD;200,180;400;400;0;Explode;\;

Figure 2.38 RTT dyad: the polygon of accelerations.

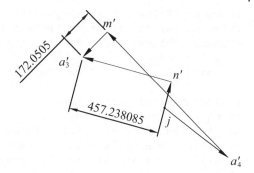

- $\mathbf{a}^{C}_{A_3A_4}$ ($\mathbf{a}'_4\mathbf{m}'$), by inserting $v_{A_3A_4}$ at point a'_4, multiplied by $2\omega_4$, and rotated through an angle of 90°, then unselecting the inserted block

    −Insert;vA3A4;End;\40;40;90;Explode;\;

- a parallel to Δ'_4 passing through point m'

    −Insert;Delta4;End;\400;400;180;Explode;\;

- $\mathbf{a}^{C}_{A_3A_2}$ ($\mathbf{j}n'$), by inserting $v_{A_3A_2}$ at point j, multiplied by $2\omega_4$ and rotated through an angle of 90°, then unselecting the inserted block

    −Insert;vA3A2;200,180;40;40;90;Explode;\;

- a parallel to Δ_3 passing through point n'

    −Insert;Delta3;End;\400;400;180;Explode;\;

3) Fillet the straight lines that start from points m' and n'

    Fillet;\\

mark the points and add the dimensions, as in Figure 2.38.

As for the previous four dyads, we will set out the AutoLisp function for kinematic analysis of the RTT dyad. The function is called *RTT* and it is saved as *Dyad_RTT.lsp*.

```
(Defun C:RTT ()
(Setq xA 0.0 yA 0.6 xD 0.8 yD 0 alphaG (/ Pi 4) betaG (/ Pi 3) phiG
(/ Pi 4) lAB(* 0.3 (Sqrt 2.0)) om4 20 ismall(List 5.0 15.0) jsmall(List
100.0 200.0))
;Calculation of positions
(Setq d2(* lAB (Sin alphaG)) phi2(+ (/ Pi 2) (* -1 betaG) phiG) phi3(-
betaG fiG) Aprime(Polar (List xA yA) phi2 d2) Cprime(Polar Aprime (-
Pi fi3) 0.1) Dprime(Polar (List xD yD) phiG 0.1) CCapital(Inters Cprime
Aprime (List xD yD) Dprime Nil) xAp(Princ(car Aprime)) yAp(Princ(car(cdr
Aprime))) xCp(princ(car Cprime)) yCp(Princ(car(cdr Cprime)))))
(Setq xM xAp yM yAp xN xCp yN yCp a xA b yA R lAB)
(Int_CL)
(Setq xB xP1 yB yP1)
;Position of the mechanism
(Command "Erase" "All" "" "Osnap" "Off" "Ortho" "Off")
(Command "Zoom" "-0.5,-0.2" "2.5,1")
(Command "-Layer" "N" "Position" "C" "7" "Position" "S"
  "Position" "" "")
```

```
(Command "Pline" (List xA yA) "W" "0.005" "" (List xB yB) "" "Pline"
(List xD yD) "L" "1.1" "" "Pline" CCapital (List xB yB) "L" "0.3" "")
(Command "Pline" "0,0" "W" "0.002" "" "1.7,0" "W" "0.02" "0" "L"
  "0.07" "")
(Command "Pline" "0,0" "W" "0.002" "" "0,0.9" "W" "0.02" "0" "L"
  "0.07" "")
;Calculation of velocities
(Setq vA4(* om4 (Distance (List xA yA) (List xD yD))) phiAD(Angle (List
xA yA) (List xD yD)) asmall4(Polar ismall (- phiAD (/ Pi 2)) vA4)
rsmall(Polar asmall4 phiG 0.1) ssmall(Polar ismall (* -1 phi3) 0.1)
asmall3(Inters asmall4 rsmall ismall ssmall Nil))
;Polygon of velocities
(Command "Zoom" "-20,-5" "25,20")
(Command "-Layer" "N" "Velocities" "C" "1" "Velocities" "S"
  "Velocities" "" "")
(Command "Line" ismall asmall4 asmall3 "C")
(Setq l(/ vA4 20) g(* l 0.353))
(Arrows asmall4 ismall)
(Arrows asmall3 asmall4)
(Arrows asmall3 ismall)
;Calculation of accelerations
(Setq aA4(* om4 vA4) acA3A4(* 2.0 om4 (Distance asmall3 asmall4))
acA3A2(* 2.0 om4 (Distance ismall asmall3)) aprime4(Polar jsmall phiAD
aA4) mprime (Polar aprime4 (+ phiG (/ Pi 2)) acA3A4) rprime (Polar
mprime phiG 10) nprime (Polar jsmall (- (/ Pi 2) phi3) acA3A2)
sprime(Polar nprime (* -1 phi3) 10) aprime3(Inters mprime rprime nprime
sprime Nil))
;Polygon of accelerations
(Command "Zoom" "-3000,-50" "3000,3000")
(Command "-Layer" "N" "Accelerations" "C" "5" "Accelerations" "S"
  "Accelerations" "" "")
(Command "Line" jsmall aprime4 mprime aprime3 nprime "C" "Zoom" "E")
(Setq l(/ aa4 15) g(* l 0.353))
(Arrows aprime4 jsmall)
(Arrows mprime aprime4)
(Arrows aprime3 mprime)
(Arrows nprime jsmall)
(Arrows aprime3 nprime)
)
```

At the beginning of the function the dimensions of the mechanism in Section 2.8.1 and the coordinates of the poles *i* and *j* are assigned. The calculations of the positions start with the determination of the distance d_2 and the angles φ_2 and φ_3 made by the straight lines *AB* and *BC*, respectively, with the horizontal. Points *A'*, *C'* (a point on the straight line Δ_3), and *D'* (a point on the straight line Δ_4) are determined using the function *Polar*.

point *C* is determined using the function *Inters*, and then, using the function *Princ*, the coordinates of points $A'\left(x_{A'}, y_{A'}\right)$ and $C'\left(x_{C'}, y_{C'}\right)$ are retained. point *B* is determined using the function *INT_CL* as the intersection between the circle centred at point *A* and the straight line *A'C'*. This results in the values: $x_M = x_A, y_M = y_A, x_N = x_C, y_N = y_C, a = x_A, b = y_A, R = AB$.

One may also determine point *B*, knowing angle φ_{AB} made by the straight line *AB* with the horizontal, using the expression

```
(Setq BCapital(Polar (List 0 0.6) (/ Pi 6)(* 0.3 (Sqrt 2))))
```

followed by the retaining of the coordinates

```
(Setq (xB(Princ(car BCapital)) yB(Princ(car(cdr BCapital)))))
```

For the representation of the mechanism in AutoCAD, a new layer called *Positions* is defined, and this is set as the current layer. Using the command **Pline**, and setting a line thickness of 0.005, the straight lines AB, DD' (D' being a point on Δ_4, at distance $DD' = 1.1$), and CA' are drawn. After adding the axes of coordinates, the representation in Figure 2.36 is obtained, and the dimensions are drawn. Then the points are annotated and the kinematic pairs are drawn.

The calculations for the determination of the velocities involve the determination of the velocity v_{A_1} and angle φ_{AD} made by the straight line AD with the horizontal. The points a_4, r (a point that on the straight line Δ_4), and s (a point on the straight line Δ_3) are determined using the function *Polar*. At the intersection of the straight lines defined by points a_4, r, i, and s, and using the function *Inters*, point a_3 is determined. The representation of the polygon is made in a new layer (*Velocities*) by drawing the loop ia_4a_3i and the arrows of the vectors. The dimensions are drawn, as in Figure 2.37. For the calculation of the accelerations, first the magnitudes of the accelerations $\mathbf{a}_{A_4}$, $\mathbf{a}_{A_3A_4}^C$, $\mathbf{a}_{A_3A_2}^C$, then points a_4', m', r', n', and s', and finally point a_3' are determined from the polygon of accelerations. The polygon of accelerations is represented in a new layer (*Accelerations*) by drawing the loop $ja_4'm'a_3'n'j$; this polygon is completed with arrows (the function *Arrows*). After the dimensions have been drawn and the points annotated, the representation in Figure 2.38 is obtained.

2.9 Kinematic Analysis of the 6R Triad

2.9.1 Formulation of the Problem

Consider the 6R triad ($A_2A_3A_4A_5A_6A_7$) in Figure 2.39 for which the following are known:

- the positions of points A_2, A_3, $A_{4,}$ and the lengths l_2, l_3, l_4, l_5, l_6, l_7 of the segments A_2A_5, A_3A_6, A_4A_7, A_5A_6, A_6A_7, A_7A_5, respectively
- the velocities and accelerations of points A_2, A_3, and A_4.

The following values are to be determined:

- the positions of points A_5, A_6, and A_7
- the angular velocities ω_2, ω_3, ω_4, and ω_5 and the velocities of points A_5, A_6, and A_7
- the angular accelerations ε_2, ε_3, ε_4, and ε_5 and the accelerations of points A_5, A_6, and A_7.

Figure 2.39 The 6R triad.

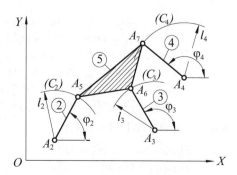

2.9.2 Determination of the Positions

The graphical solution for the position of the triad is in essence the pattern method. Cut from a piece of cardboard a triangle (pattern) equal (at scale k_l) to the triangle $A_5A_6A_7$. Draw the circular arcs (C_2), (C_3), and (C_4), centred at points A_2, A_3, and A_4, respectively, and having radii (at scale k_l) of l_2, l_3, and l_4, respectively. Move the pattern with points A_5 and A_6, on the circles (C_2) and (C_3), respectively, until point A_7 is situated on the circle (C_4). The position of the triad is thus determined by a graphical method.

As shown in Figure 2.39, the position of the triad is known if angles φ_2, φ_3, and φ_4 are known. These angles can be determined numerically (at any required precision) using the Newton–Raphson method. In this case, the values found using the graphical method may be considered as initial approximate values.

To obtain the non-linear systems, denote by X_i, Y_i, $i = 2, 3, 4$, the known coordinates of points A_2, A_3, and A_4, and then write the coordinates of points A_{i+3}, $i = 2, 3, 4$, as

$$X_{i+3} = X_i + l_i \cos \varphi_i, \quad Y_{i+3} = Y_i + l_i \sin \varphi_i. \tag{2.195}$$

The system of non-linear equations is obtained from the expression for the lengths l_5, l_6, and l_7 of the segments A_5A_6, A_6A_7, and A_7A_5, respectively. If the following notations is used:

$$\begin{aligned}
B_{23} &= X_2 - X_3 + l_2 \cos \varphi_2 - l_3 \cos \varphi_3, & C_{23} &= Y_2 - Y_3 + l_2 \sin \varphi_2 - l_3 \sin \varphi_3, \\
B_{34} &= X_3 - X_4 + l_3 \cos \varphi_3 - l_4 \cos \varphi_4, & C_{34} &= Y_3 - Y_4 + l_3 \sin \varphi_3 - l_4 \sin \varphi_4, \\
B_{42} &= X_4 - X_2 + l_4 \cos \varphi_4 - l_2 \cos \varphi_2, & C_{42} &= Y_4 - Y_2 + l_4 \sin \varphi_4 - l_2 \sin \varphi_2,
\end{aligned} \tag{2.196}$$

$$\begin{aligned}
F_1 (\varphi_2, \varphi_3, \varphi_4) &= B_{23}^2 + C_{23}^2 - l_5^2, \\
F_2 (\varphi_2, \varphi_3, \varphi_4) &= B_{34}^2 + C_{34}^2 - l_6^2, \\
F_3 (\varphi_2, \varphi_3, \varphi_4) &= B_{42}^2 + C_{42}^2 - l_7^2,
\end{aligned} \tag{2.197}$$

then the system reads

$$F_i (\varphi_2, \varphi_3, \varphi_4) = 0, \quad i = 1, 2, 3. \tag{2.198}$$

Considering the initial values φ_i^0, $i = 2, 3, 4$, developing into a Taylor series and retaining only the linear terms, the following system is obtained:

$$\sum_{k=2}^{4} \Delta\varphi_k \left. \frac{\partial F_i}{\partial \varphi_k} \right|_{\varphi_k = \varphi_k^0} = -F_i^0, \quad i = 1, 2, 3 \tag{2.199}$$

where

$$F_i^0 = F_i (\varphi_2^0, \varphi_3^0, \varphi_4^0), \quad i = 1, 2, 3. \tag{2.200}$$

Denoting by D_{ik} the partial derivatives $\left. \dfrac{\partial F_i}{\partial \varphi_k} \right|_{\varphi_k = \varphi_k^0}$, we have the following equalities

$$\begin{aligned}
D_{11} &= 2l_2 \left(-B_{23} \sin \varphi_2^0 + C_{23} \cos \varphi_2^0 \right), \\
D_{12} &= 2l_3 \left(B_{23} \sin \varphi_3^0 - C_{23} \cos \varphi_3^0 \right), \quad D_{13} = 0, \\
D_{21} &= 0, \quad D_{22} = 2l_3 \left(-B_{34} \sin \varphi_3^0 + C_{34} \cos \varphi_3^0 \right), \\
D_{23} &= 2l_4 \left(B_{34} \sin \varphi_4^0 - C_{34} \cos \varphi_4^0 \right) \\
D_{31} &= 2l_2 \left(B_{42} \sin \varphi_2^0 - C_{42} \cos \varphi_2^0 \right), \quad D_{32} = 0, \\
D_{33} &= 2l_4 \left(-B_{42} \sin \varphi_4^0 + C_{42} \cos \varphi_{42}^0 \right),
\end{aligned} \tag{2.201}$$

and the system (2.199) reads

$$\begin{aligned}
D_{11}\Delta\varphi_2 + D_{12}\Delta\varphi_3 &= -F_1^0, \\
D_{22}\Delta\varphi_3 + D_{23}\Delta\varphi_4 &= -F_2^0, \\
D_{31}\Delta\varphi_2 + D_{33}\Delta\varphi_4 &= -F_3^0.
\end{aligned} \tag{2.202}$$

If we denote

$$D = D_{11}D_{22}D_{33} + D_{12}D_{23}D_{31}, \tag{2.203}$$

then we get the variations

$$\Delta\varphi_2 = \frac{1}{D}\left(-F_1^0 D_{22}D_{33} + F_2^0 D_{12}D_{33} - F_3^0 D_{12}D_{23}\right),$$

$$\Delta\varphi_3 = \frac{1}{D}\left(-F_1^0 D_{23}D_{31} + F_2^0 D_{11}D_{33} + F_3^0 D_{11}D_{23}\right), \tag{2.204}$$

$$\Delta\varphi_4 = \frac{1}{D}\left(F_1^0 D_{22}D_{31} + F_2^0 D_{12}D_{31} - F_3^0 D_{11}D_{22}\right).$$

The new initial values are $\varphi_i^0 + \Delta\varphi_i$, $i = 2, 3, 4$.

The iterations continue until the required precision is reached; that is, until $|\Delta\varphi_i| < \nu$, $i = 2, 3, 4$, (ν being the imposed deviation).

The calculations for the given initial values can be obtained either in an elementary way, or using a calculation program, successively calling (2.196), (2.197), and (2.201)–(2.204). The angle φ_5, which defines the position of the triangle $A_5 A_6 A_7$ (Figure 2.39), is determined from the scalar relations

$$l_2 \cos\varphi_2 + l_5 \cos\varphi_5 - l_3 \cos\varphi_3 = X_3 - X_2,$$
$$l_2 \sin\varphi_2 + l_5 \sin\varphi_5 - l_3 \sin\varphi_3 = Y_3 - Y_2. \tag{2.205}$$

This gives the equations from which are deduced the unique value of angle φ_5

$$\cos\varphi_5 = \frac{X_3 - X_2 - l_2 \cos\varphi_2 + l_3 \cos\varphi_3}{l_3},$$

$$\sin\varphi_5 = \frac{Y_3 - Y_2 - l_2 \sin\varphi_2 + l_3 \sin\varphi_3}{l_3}. \tag{2.206}$$

If the triad is a component of a mechanism with only one degree of mobility, and if the parameter that defines the position of the driving element is θ, then the angular parameters φ_i, $i = 2, 3, 4, 5$, are functions depending on the parameter θ.

2.9.3 Determination of the Velocities and Accelerations

If the variational step of angle θ is $\Delta\theta$ and if we use the notation $\dot\theta = \omega$, $\ddot\theta = \varepsilon$, then the reduced angular velocities $\dfrac{d\varphi_i}{d\theta}$ are determined using the finite differences. The angular velocities are given by:

$$\omega_i = \omega\frac{d\varphi_i}{d\theta}, \ i = 2, 3, 4, 5. \tag{2.207}$$

The linear velocities of points A_5, A_6, and A_7 are determined using the Euler relation:

$$\mathbf{v}_{i+3} = \mathbf{v}_i + \boldsymbol{\omega}_i \times \mathbf{A}_i\mathbf{A}_{i+3}, \ i = 2, 3, 4. \tag{2.208}$$

Similarly, the angular accelerations are determined from relations of the form

$$\varepsilon_i = \omega^2\frac{d^2\varphi_i}{d\theta^2} + \varepsilon\frac{d\varphi_i}{d\theta}, \tag{2.209}$$

where the derivative $\dfrac{d^2\varphi_i}{d\theta^2}$ is determined using the finite differences. The linear accelerations of points A_5, A_6, and A_7 are calculated with:

$$\mathbf{a}_{i+3} = \mathbf{a}_i + \boldsymbol{\varepsilon}_i \times \mathbf{A}_i\mathbf{A}_{i+3} - \omega_i^2\mathbf{A}_i\mathbf{A}_{i+3}, \ i = 2, 3, 4. \tag{2.210}$$

2.9.4 The Assisted Analytical Method

Based on the analytical relations previously established, we wrote the procedure *Triad_6R*, which has six input data and three output data.

The input parameters are the positions of the kinematic pairs at points A_2, A_3, and A_4 (X_2, Y_2, X_3, Y_3, and X_4, Y_4, respectively), while the output parameters are the positional angles φ_2, φ_3, and φ_4. The content of this procedure is given in the companion website of this book.

Numerical example

Problem: Perform the kinematic analysis for the mechanism in Figure 2.40, in which $l_1 = A_1A_2 = 0.15\,\text{m}$, $l_2 = A_2A_5 = 0.5\,\text{m}$, $l_3 = A_3A_6 = 0.6\,\text{m}$, $l_4 = A_4A_7 = 0.84\,\text{m}$, $l_5 = A_5A_6 = 0.5\,\text{m}$, $l_6 = A_6A_7 = 0.5\,\text{m}$, $l_7 = A_5A_7 = 0.92\,\text{m}$, $X_1 = 0\,\text{m}$, $Y_2 = 0.3\,\text{m}$, $X_3 = 0.9\,\text{m}$, $Y_3 = 0.2\,\text{m}$, $X_4 = 1.5\,\text{m}$, $Y_4 = 0.4\,\text{m}$, and $\theta = 20t$ rad. Use a step $\Delta\theta = \frac{\pi}{180}$ rad. For variations of 20° of angle θ, list the parameters: φ_2, φ_3, φ_4, φ_5, ω_2, ω_3, ω_4, ω_5, ε_2, ε_3, ε_4, ε_5. Draw the variation of the angular velocity $\omega_5 = \omega_5(\theta)$.

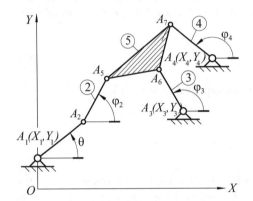

Figure 2.40 Mechanism with 6R triad.

Solution: The program incorporating the *Triad_6R* procedure is shown in the companion website of this book (the name of the program is *Mec_R6R*). To call the *Triad_6R* procedure it is necessary to determine the approximate position of the triad at time $t = 0$. For that reason, we use AutoCAD and, applying the method of the pattern (described in the Section 2.9.2), we position the triangle $A_5A_5A_7$ (constructed with the AutoCAD command **Pline**), using the commands **Move** and **Rotate**, on the circular arcs with their centres at points A_2 (0.3, 0), A_3 (0.9, 0.2), and A_4 (1.6, 0.4) and having radii of 0.5, 0.6, and 0.84, respectively.

The geometrical construction obtained after drawing the dimensions of the angles φ_2, φ_3, and φ_4 is given in Figure 2.41. In this way, the values φ_2^0, φ_3^0, φ_4^0 required for the initiation of the iterative process are obtained.

In a *For* loop, giving degree-by-degree increments to angle θ:

- Determine the coordinates of point A_2 using

$$X_2 = X_{A_1} + l_1 \cos\theta, \quad Y_2 = Y_{A_1} + l_1 \sin\theta \tag{2.211}$$

- Call the *Triad_6R* procedure, knowing the positions of points A_2, A_3, and A_4; the procedure returns the values of the angles φ_2, φ_3, and φ_4, with precision $\nu = 10^{-6}$.
- Determine the value of angle φ_5 using (2.206).
- The vectors previously formed (*Phi2Vect*, *Phi3Vect*, *Phi4Vect*, *Phi5Vect*) with the values of the angles φ_2, φ_3, φ_4, and φ_5, respectively are necessary.

The results obtained are shown in Table 2.3.

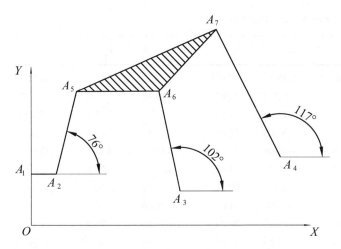

Figure 2.41 Geometrical construction of the 6R triad using AutoCAD.

Table 2.3 The values of the angles φ_2, φ_3, φ_4, and φ_5.

θ	φ_2	φ_3	φ_4	φ_5	θ	φ_2	φ_3	φ_4	φ_5
0	75.769	102.22	117.04	0.198	180	63.054	122.81	130.90	−4.756
20	47,033	82.760	106.49	8.979	200	70.090	128.44	134.91	−5.613
40	36.775	79.791	105.31	10.923	220	77.557	133.11	138.26	−6.183
60	34.711	82.380	106.33	9.216	240	85.235	136.70	140.86	−6.538
80	36.517	87.618	108.73	6.226	260	92.727	139.06	142.56	−6.731
100	40.323	94.356	112.33	3.095	280	99.382	139.90	143.17	−6.792
120	45.143	101.76	116.75	0.350	300	104.08	138.60	142.23	−6.696
140	50.581	109.23	121.57	−1.854	320	104.82	133.89	138.83	−6.266
160	56.548	116.34	126.38	−3.532	340	97.781	123.19	131.17	−4.820

The vectors previously formed are necessary to determine, by finite differences, the angular velocities ω_2, ω_3, and ω_4, and the angular accelerations ε_2, ε_3, and ε_4, using the relations established in the Section 2.3.4.

For this reason, a new *For* loop is used, in which the indexed components of the vectors are called, and, using (2.39) and (2.40), the values of the angular velocities ω_2, ω_3, ω_4, and ω_5, and the angular accelerations ε_2, ε_3, ε_4, and ε_5 are determined. The values obtained may be written in a text file, for later use. Based on such a file, we obtained the values in Table 2.4 and Figure 2.42.

If the linear velocities and accelerations of points A_5, A_6, and A_7 are also required, then in the calculation program (2.208) and (2.210) can be added in a second *For* loop.

2.9.5 The Assisted Graphical Method

Because there is no direct command in AutoCAD to position a geometrical figure on three arcs of a circle, we wrote an AutoLisp function to do this so. Using the approach of Section 2.9.2, the algorithm for the positioning of the triangle $A_5A_6A_7$ in Figure 2.39 is:

1) Choose a point A_5 on the circle (C_2).
2) At the intersection between the circle centred at point A_5 with radius l_5, and the circle centred at point A_3 with radius l_3 is obtained point A_6.

Table 2.4 The values of the angular velocities and accelerations.

θ	ω_2	ω_3	ω_4	ω_5	ε_2	ε_3	ε_4	ε_5
20	−18.942	−9.442	−4.077	5.737	1320.13	1068.53	544.35	−534.35
40	−5.391	0.402	0.149	−0.279	443.55	292.34	108.16	−202.56
60	0.092	4.023	1.666	−2.542	220.88	150.91	78.54	−73.23
80	2.933	6.069	2.981	−3.185	114.50	87.52	70.33	−7.45
100	4.370	7.159	4.040	−3.008	57.15	39.49	49.25	22.99
120	5.131	7.521	4.673	−2.509	34.45	4.04	23.33	31.67
140	5.677	7.366	4.874	−1.958	29.91	−20.27	0.56	30.57
160	6.207	6.860	4.724	−1.460	31.01	−36.66	−16.81	26.25
180	6.752	6.118	4.317	−1.045	30.84	−47.63	−29.05	21.29
200	7.255	5.218	3.734	−0.715	25.78	−55.07	−37.28	16.57
220	7.610	4.203	3.029	−0.463	13.62	−61.22	−43.33	12.47
240	7.675	3.069	2.218	−0.275	−8.00	−69.38	−49.95	9.26
260	7.250	1.739	1.258	−0.132	−43.69	−84.85	−61.35	7.41
280	6.010	0.012	0.009	−0.001	−103.85	−117.03	−84.63	8.25
300	3.352	−2.547	−1.842	0.197	−211.62	−184.39	−133.31	16.34
320	−1.927	−6.834	−4.931	0.701	−415.44	−322.82	−231.27	48.22
340	−12.263	−14.547	−10.307	2.359	−813.67	−584.60	−390.85	167.87
360	−31.122	−26.140	−16.581	8.170	−1089.94	−406.43	−30.64	498.52

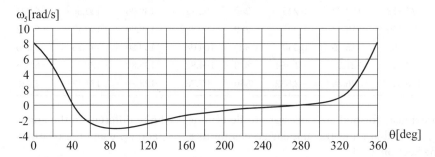

Figure 2.42 The variation of the angular velocity $\omega_5 = \omega_5\,(\theta)$.

3) At the intersection between the circle centred at point A_5 with radius l_7, and the circle centred at point A_6 with radius l_6 is obtained point A_7.
4) Determine the distance A_4A_7 and compare it to the value l_4.
5) Repeat steps 1–4 until an acceptable deviation for the difference $|A_4A_7 - l_4|$ is obtained.

Denoting by φ_2 the positional angle of element 2 relative to the axis OX (Figure 2.40), the position of point $A_5\left(X_{A_5}, Y_{A_5}\right)$ in the reference system OXY will be given by:

$$X_{A_5} = X_2 + l_2 \cos \varphi_2, \ \ Y_{A_5} = Y_2 + l_2 \sin \varphi_2. \tag{2.212}$$

In the last algorithm, the choice of point A_5 on the circle $\left(C_2\right)$ is similar to the choice of angle φ_2; the exact position of the triangle at which its apices are on the three circles is determined by varying angle φ_2 until the difference between the distance A_4A_7 and l_4 is, in absolute value, less than a required threshold value.

The algorithm for the variation of angle φ_2 is as follows:

1) Determine the distance $d_1 = A_4A_7$ for angle φ_2.
2) Determine the distance $d_2 = A_4A_7$ for angle $\varphi_1 + \Delta\varphi$.
3) Calculate the difference $diff = |l_4 - d_1|$, which, if it is not less than an imposed error ε, then vary angle φ_2 using:

$$\varphi_2 = \varphi_2 + \Delta\varphi\frac{l_4 - d_1}{d_2 - d_1}. \tag{2.213}$$

4) Repeat the previous steps, diminishing the step $\Delta\varphi$ (for instance $\Delta\varphi = \frac{\Delta\varphi}{10}$), until

$$|l_4 - d_1| < \varepsilon. \tag{2.214}$$

Later, we will present the AutoLisp functions that solve this problem. A first function is called *Calculations* and successively determines, as function of angle φ_2:

- the coordinates of point A_5 (X_{A_5}, Y_{A_5}), using (2.212)
- points P_1 and P_2, situated at the intersection between the circle of centre A_5 with radius L_2, and the circle of centre A_3 with radius l_5, using function *Int_2C*
- the coordinates of point A_5, by choosing either P_1 or P_2, depending on the position of the mechanism
- points P_1, P_2 at the intersection between the circle centred at point A_5 with radius l_7, and the circle centred at point A_6 with radius l_6, using function *Int_2C*
- the coordinates of point A_7, by choosing either P_1 or P_2, depending on the position of the mechanism
- the distance A_4A_7, using $d = A_4A_7 = \sqrt{(X_4 - X_{A_7})^2 + (Y_4 - Y_{A_7})^2}$, or using the AutoLisp function *Distance*.

The function is given below.

```
(Defun Calculations ()
(Setq xA5 (+ xA2 (* 12 (Cos phi))) yA5 (+ yA2 (* 12 (Sin phi))) al xA3
  bl yA3 rl 13 a2 xA5 b2 yA5 r2 15)
(Int_2C)
(Setq xA6 x2 yA6 y2)
(Setq al xA6 bl yA6 rl 16 a2 xA5 b2 yA5 r2 17)
(Int_2C)
(Setq xA7 x2 yA7 y2 d (Distance (List xA4 yA4) (List xA7 yA7)))
)
```

The calling of the two AutoLisp functions is performed inside the main body of the function called *Triad*, the content of which is presented below.

```
(Defun C:Triad ()
(Setq xA2 15.0 yA2 30.0 12 50.0 xA3 90.0 yA3 20.0 13 60.0 xA4 150.0 yA4
40.0 14 84.0 15 50.0 16 50.0 17 92.0 err 0.00001)
(Setq phi (/ (* 90.0 Pi) 180) step (/ (* 0.01 Pi) 180))
(Setq PC1 (List xA2 yA2) PC2 (List xA3 yA3) PC3 (List xA4 yA4))
(Command "Erase" "All" "" "Osnap" "Off" "Color" "1")
(Command "Circle" PC1 12 "Circle" PC2 13 "Circle" PC3 14)
(Calculations)
(Setq A5 (List xA5 yA5) A6 (List xA6 yA6) A7 (List xA7 yA7))
(Command "Pline" A5 "W" "0.5" "0.5" A6 A7 "Close")
```

```
(Setq d1 d phi (+ phi step))
(Calculations)
(Setq d2 d)
(Setq phi (+ phi (/ (* step (- 14 d1)) (- d2 d1))))
(Calculation)
(Setq d1 d diff (Abs(- 14 d1)))
(If (>= diff err)
(PROGN ;Then instructions
(Setq pas (/ step 10) phi (+ phi step))
(Calculations)
(Setq d2 d phi (+ phi (/ (* step (- 14 d1)) (- d2 d1))))
(Calculations)
(Setq d1 d diff (Abs(- 14 d1)))
) ; from PROGN
) ; end If
(Setq A5 (List xA5 yA5) A6 (List xA6 yA6) A7 (List xA7 yA7))
(Command "Color" "7" "Pline" A5 "W" "0.5" "0.5" A6 A7 "Close")
)
```

After the assignment of values (the function *Setq*) to the distances and the coordinates of the fixed points, the values 90° and 0.01° are given to the angles φ_2 and $\Delta\varphi$, respectively.

After erasing the existing constructions and the inhibition of the *Osnap* modes, three circles are constructed having their centres at points A_2, A_3, and A_4, and with radii l_2, l_3, and l_4, respectively. Function *Calculations*, is called and this determines the coordinates of points A_5, A_6, A_7, Using the AutoCAD command **Pline**, the triangle $A_5A_6A_7$ is constructed (approximate position), drawn in red. The algorithm continues, varying the angle φ_2 in a repetitive cycle, defined by the instruction *If*, until the deviation from the distance A_4A_7 is less than 0.00001. The function ends after the construction (in black) of the triangle $A_5A_6A_7$.

Using values of $X_2 = 15$, $Y_2 = 30$, $l_2 = 50$, $X_3 = 90$, $Y_3 = 20$, $l_3 = 60$, $X_4 = 150$, $Y_4 = 40$, $l_4 = 84$, $l_5 = 50$, $l_6 = 50$, and $l_7 = 92$, the function gives the construction in Figure 2.43.

The algorithm began from the approximate value of $\varphi_1 = 90°$ and reached the exact value $\varphi_1 = 75.77°$. In Figure 2.43 the approximate solution is drawn with a dotted line, while the exact

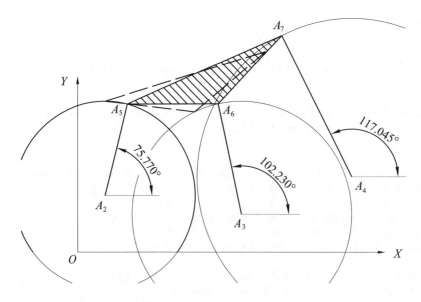

Figure 2.43 The four-bar mechanism.

solution is drawn with a solid line. The angles of position of elements 1, 2, and 3 were dimensioned using the AutoCAD command **Dimangular**. The values obtained are identical to those obtained with the analytical method outlined in the previous section.

The AutoLisp function presented here may be modified to be used exclusively from the keyboard. In the presented form, the user intervenes to choose one solution from points A_2 or A_3 (one of the two points of intersection of the two circles). The function may also be set so that the values of the coordinates of points O_1, O_2, O_3, the lengths O_1A_1, O_2A_2, O_3A_3, or the dimensions of the triangle $A_1A_2A_3$ can be introduced using keyboard or mouse.

2.10 Kinematic Analysis of Some Planar Mechanisms

2.10.1 Kinematic Analysis of the Four-Bar Mechanism

Perform a kinematic analysis for the four-bar mechanism in Figure 2.44. The following are known:

- the dimensions of the mechanism: $OA = 0.1$ m, $AB = 0.25$ m, $BC = 0.25$ m, $BM = 0.25$ m, $\alpha = 180°$
- the positions of the kinematic pairs linked to the base: $O(0,0)$, $C(0.20,0)$
- the constant angular velocity of element 1: $\omega_1 = 10$ rad/s.

The kinematic analysis is performed using a constant angular step of $10°$.

The kinematic analysis is to be performed using the analytical and the assisted graphical methods. For the analytical method, the method described in Section 2.4.2 can be used, the relations determined being a function of the positions of the kinematic pairs A and C. To avoid repetition, we will use the analytical method of the projections described in the Section 2.3.2.

The analytical method of projections
We choose the loop $OABC$. The vector equation $\mathbf{OA} + \mathbf{AB} + \mathbf{BC} = \mathbf{OC}$, projected onto the system of axes OXY, leads to the system of non-linear equations

$$\begin{cases} OA \cos \varphi_1 + AB \cos \varphi_2 + BC \cos \varphi_3 = X_C - X_O, \\ OA \sin \varphi_1 + AB \sin \varphi_2 + BC \sin \varphi_3 = Y_C - Y_O. \end{cases} \qquad (2.215)$$

The angles are measured in a trigonometric sense (relative to the axis OX) starting from the end of the element. In this way, although working in scalar mode, in the case of the velocities and accelerations, the obtained parameters are vectors.

For the determination of angle φ_2 angle φ_3 is eliminated by passing the expressions that contain angle φ_3 to the right-hand term and by squaring the obtained expressions

$$\begin{aligned} \left(AB \cos \varphi_2 + A_1\right)^2 &= \left(-BC \cos \varphi_3\right)^2, \\ \left(AB \sin \varphi_2 + A_2\right)^2 &= \left(-BC \sin \varphi_3\right)^2, \end{aligned} \qquad (2.216)$$

Figure 2.44 The position of the triad.

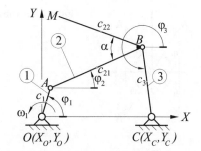

where A_1 and A_2 remain for the expressions

$$A_1 = OA \cos \varphi_1 - X_C + X_O, \ A_2 = OA \sin \varphi_1 - Y_C + Y_O. \tag{2.217}$$

Squaring and summing the expressions (2.216) gives:

$$A \cos \varphi_2 + B \sin \varphi_2 + C = 0, \tag{2.218}$$

where

$$A = 2A_1 AB, \ B = 2A_2 AB, \ C = AB^2 - BC^2 + A_1^2 + A_2^2. \tag{2.219}$$

The expression for angle φ_2 is

$$\varphi_2 = 2 \arctan \left(\frac{-B \pm \sqrt{A^2 + B^2 - C^2}}{C - A} \right). \tag{2.220}$$

For the determination of angle φ_3 angle φ_2 is eliminated by passing the expressions that contain angle φ_2 to the right-hand term and squaring:

$$\begin{aligned} \left(BC \cos \varphi_3 + A_1 \right)^2 &= \left(-AB \cos \varphi_2 \right)^2, \\ \left(BC \sin \varphi_3 + A_2 \right)^2 &= \left(-AB \sin \varphi_2 \right)^2. \end{aligned} \tag{2.221}$$

By summing, we obtain an expression in the form

$$A' \cos \varphi_3 + B' \sin \varphi_3 + C' = 0, \tag{2.222}$$

where

$$A' = 2A_1 BC, \ B' = 2A_2 BC, \ C' = BC^2 - AB^2 + A_1^2 + A_2^2. \tag{2.223}$$

The expression for angle φ_3 is

$$\varphi_3 = 2 \arctan \left(\frac{-B' \pm \sqrt{A'^2 + B'^2 - C'^2}}{C' - A'} \right). \tag{2.224}$$

For the correct use of the arctan function we make the following remarks regarding how to obtain the correct value of angle φ (this angle is measured in a trigonometric sense starting from the axis OX). Depending on the quadrant in which the segment ON is situated, we will need to correct the returned value using the function arctan; hence (see Figure 2.45):

- in quadrant I: $a > 0, b > 0$, and $\varphi = \arctan \left(\dfrac{b}{a} \right)$
- in quadrants II and III: $a < 0, \varphi = \pi + \arctan \left(\dfrac{b}{a} \right)$
- in quadrant IV: $a > 0, b < 0, \varphi = 2\pi + \arctan \left(\dfrac{b}{a} \right)$.

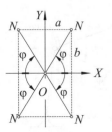

Figure 2.45 The arctan function.

With the angles φ_2 and φ_3 known, the positions of the kinematic pairs can be determined:

$$\begin{cases} X_A = X_O + OA \cos \varphi_1, \\ Y_A = Y_O + OA \sin \varphi_1, \end{cases} \begin{cases} X_B = X_A + AB \cos \varphi_2, \\ Y_B = Y_A + AB \sin \varphi_2, \end{cases}$$

$$\begin{cases} X_M = X_B + BM \cos \left(\varphi_2 + \pi - \alpha \right), \\ Y_M = Y_B + BM \sin \left(\varphi_2 + \pi - \alpha \right), \end{cases} \qquad (2.225)$$

as can the positions of the centres of mass of the elements:

$$\begin{cases} X_{C_1} = \dfrac{X_O + X_A}{2} \\ Y_{C_1} = \dfrac{Y_O + Y_A}{2} \end{cases} \begin{cases} X_{C_{21}} = \dfrac{X_A + X_B}{2}, \\ Y_{C_{21}} = \dfrac{Y_A + Y_B}{2}, \end{cases}$$

$$\begin{cases} X_{C_{22}} = \dfrac{X_B + X_M}{2}, \\ Y_{C_{22}} = \dfrac{Y_B + Y_M}{2}, \end{cases} \begin{cases} X_{C_3} = \dfrac{X_B + X_C}{2}, \\ Y_{C_3} = \dfrac{Y_B + Y_C}{2}. \end{cases} \qquad (2.226)$$

Considering that the elements are of the homogeneous bar type (the centres of mass are situated in the middle of the element), element 2 consists of the elements AB and AM.

For the analysis of the velocities, the functions of position (2.215) with respect to time are derived, giving:

$$\begin{cases} -OA\omega_1 \sin \varphi_1 - AB\omega_2 \sin \varphi_2 - BC\omega_3 \sin \varphi_3 = 0, \\ OA\omega_1 \cos \varphi_1 + AB\omega_2 \cos \varphi_2 + BC\omega_3 \cos \varphi_3 = 0, \end{cases} \qquad (2.227)$$

with the solution

$$\omega_2 = \frac{OA \sin \left(\varphi_1 - \varphi_3 \right)}{AB \sin \left(\varphi_3 - \varphi_2 \right)} \omega_1, \; \omega_3 = \frac{OA \sin \left(\varphi_2 - \varphi_1 \right)}{BC \sin \left(\varphi_3 - \varphi_2 \right)} \omega_1. \qquad (2.228)$$

Differentiating (2.225) and (2.226) with respect to time gives the expressions for the velocities of the kinematic pairs and of the centres of mass of the elements:

$$\begin{cases} v_{Ax} = \dot{X}_A = -OA\omega_1 \sin \varphi_1, \\ v_{Ay} = \dot{Y}_A = OA\omega_1 \cos \varphi_1, \end{cases} \begin{cases} v_{Bx} = \dot{X}_B = v_{Ax} - AB\omega_2 \sin \varphi_2, \\ v_{By} = \dot{Y}_B = v_{Ay} + AB\omega_2 \cos \varphi_2, \end{cases}$$

$$\begin{cases} v_{Mx} = \dot{X}_M = v_{Bx} - BM\omega_2 \sin \left(\varphi_2 + \pi - \alpha \right), \\ v_{My} = \dot{Y}_M = v_{By} + BM\omega_2 \cos \left(\varphi_2 + \pi - \alpha \right), \end{cases} \qquad (2.229)$$

$$\begin{cases} v_{C_1x} = \dot{X}_{C_1} = \dfrac{v_{Ax}}{2}, \\ v_{C_1y} = \dot{Y}_{C_1} = \dfrac{v_{Ay}}{2}, \end{cases} \begin{cases} v_{C_{22}x} = \dot{X}_{C_{22}} = \dfrac{v_{Bx} + v_{Mx}}{2}, \\ v_{C_{22}y} = \dot{Y}_{C_{22}} = \dfrac{v_{By} + v_{My}}{2}, \end{cases} \begin{cases} v_{C_3x} = \dot{X}_{C_3} = \dfrac{v_{Bx}}{2}, \\ v_{C_3y} = \dot{Y}_{C_3} = \dfrac{v_{By}}{2}. \end{cases} \qquad (2.230)$$

To determine the accelerations, the functions of position (2.215) are differentiated once more with respect to time, giving:

$$\begin{cases} -AB\varepsilon_2 \sin \varphi_2 - BC\varepsilon_3 \sin \varphi_3 = \overline{A}, \\ -AB\varepsilon_2 \cos \varphi_2 - BC\varepsilon_3 \cos \varphi_3 = \overline{B}, \end{cases} \qquad (2.231)$$

where

$$\begin{aligned} \overline{A} &= AB\omega_2^2 \cos \varphi_2 + BC\omega_3^2 \cos \varphi_3 + OA\omega_1^2 \cos \varphi_1, \\ \overline{B} &= AB\omega_2^2 \sin \varphi_2 + BC\omega_3^2 \sin \varphi_3 + OA\omega_1^2 \sin \varphi_1, \end{aligned} \qquad (2.232)$$

with the solution

$$\varepsilon_2 = \frac{\overline{A}\cos\varphi_3 + \overline{B}\sin\varphi_3}{AB\sin\left(\varphi_3 - \varphi_2\right)}, \ \varepsilon_3 = \frac{\overline{A}\sin\varphi_2 + \overline{B}\cos\varphi_2}{BC\sin\left(\varphi_3 - \varphi_2\right)}. \tag{2.233}$$

Differentiating (2.229) and (2.230) with respect to time gives the expressions for the accelerations

$$\begin{cases} a_{Ax} = \ddot{X}_A = -OA\omega_1^2\cos\varphi_1, \\ a_{Ay} = \ddot{Y}_A = -OA\omega_1^2\sin\varphi_1, \end{cases}$$

$$\begin{cases} a_{Bx} = \ddot{X}_B = a_{Ax} - AB\varepsilon_2\sin\varphi_2 - AB\omega_2^2\cos\varphi_2, \\ a_{By} = \ddot{Y}_B = a_{Ay} + AB\varepsilon_2\cos\varphi_2 - AB\omega_2^2\sin\varphi_2, \end{cases} \tag{2.234}$$

$$\begin{cases} a_{Mx} = \ddot{X}_M = a_{Bx} - BM\varepsilon_2\sin\left(\varphi_2 + \pi - \alpha\right) - BM\omega_2^2\cos\left(\varphi_2 + \pi - \alpha\right), \\ a_{My} = \ddot{Y}_M = a_{By} + BM\varepsilon_2\cos\left(\varphi_2 + \pi - \alpha\right) - BM\omega_2^2\sin\left(\varphi_2 + \pi - \alpha\right), \end{cases}$$

$$\begin{cases} a_{C_1x} = \ddot{X}_{C_1} = \dfrac{a_{Ax}}{2}, \\ a_{C_1y} = \ddot{Y}_{C_1} = \dfrac{a_{Ay}}{2}, \end{cases} \begin{cases} a_{C_{22}x} = \ddot{X}_{C_{22}} = \dfrac{a_{Bx} + a_{Mx}}{2}, \\ a_{C_{22}y} = \ddot{Y}_{C_{22}} = \dfrac{a_{By} + a_{My}}{2}, \end{cases} \begin{cases} a_{C_3x} = \ddot{X}_{C_3} = \dfrac{a_{Bx}}{2}, \\ a_{C_3y} = \ddot{Y}_{C_3} = \dfrac{a_{By}}{2}. \end{cases} \tag{2.235}$$

To determine numerical results the Turbo Pascal calculation program presented in the companion website of this book can be used. The values obtained using this program will be written to three files, which may be be imported into a text editor and transformed into tables, as in the companion website of this book.

In a repetitive *For* loop, values are given to angle φ_1, in the interval $[0, 2\pi]$, with an angular step of $\frac{\pi}{18}$ rad. Inside this loop this gives values for the expressions given in (2.219)–(2.235).

The program uses a procedure for the determination of the function $2\arctan(\cdot)$, the name of the procedure being *Inv_2Tan*. The procedure has as input data the parameters given by (2.219), and the sign in front of the radical, while the output data is the angle given by (2.220). The procedure is called twice, for the determination of the angles φ_2, and φ_3, respectively. For angle φ_2 the convenient solution is that with the + sign in front of the radical in the expression (2.220), while for angle φ_3, the convenient solution is the one having the − sign in front of the radical in (2.224). For the choice of the ± sign, the reference position in Figure 2.24 is taken into account. After the exit from the repetitive loop, the data files are closed.

The tables in the companion website of this book were created using the following steps:

1) In a new document insert the files *Pos.txt*, *Vel.txt*.
2) For each inserted file, after the selection of the all values, perform a text-to-table conversion, using ",", as the separator.

The input data of the program can be modified for any type of four-bar mechanism. Depending on the initial position of the mechanism, it is only necessary to intervene in the program to choose the sign ± when the procedure *Inv_2Tan* is called.

The assisted graphical method

Because the kinematic analysis of the mechanism in Figure 2.24 is performed in 36 positions, AutoLisp is used to automate the function, which is based on the *RRR* function, used in Section 2.4.4. The new function is called *R_3R* and its content is shown in the companion website of this book.

As for the previous method – the analytical method of projections – the values obtained after the kinematic analysis are written in three files: *Posit.txt*, *Veloc.txt*, and *Accel.txt*. Their main use is in verifying the correct functioning of the function, as well for comparing the values obtained to those obtained using the assisted analytical method.

The AutoLisp function starts by creating the three files and by writing the heads of tables in them. It also used a fourth script file *Curve.scr*, which contains the values necessary to represent the trajectory described by point M.

After the declaration of the values assigned to:

- the elements
- the kinematic pairs linked to the base
- the angular velocity ω_1 of the driving element,

the value of angle φ_1 is varied in the interval $[0, 360]$, with an angular step of $10°$ in a repetitive loop. Similar to the previous AutoLisp programs (Section 2.6.4), the angles are converted to radians. To obtain all the positions of the mechanism in a single picture, the constructions are created in different layers for each angle, the name of the layer matching the value of the angle. To obtain the ordered layers, we count them starting with 1000. Thus the first layer, which contains the position of the mechanism, and the polygons of velocities and accelerations for $\varphi_1 = 0°$, will be named '1000', while the last layer, named '1350', will contain the same constructions, but for angle $\varphi_1 = 350°$.

After the constructions have all been created in the layers, the layers will be closed in order to avoid the visual superposition of the all geometrical constructions. The layers open when the dimensions are added or listed. No matter the status of a layer (open/closed, frozen/unfrozen, visible/invisible and so on) all the entities will be saved when the file with the current drawing is saved.

Knowing the position of point A, the two solutions for point B can be obtained by intersecting, using the AutoLisp function *Int_2C*:

- the circle centred at point A (X_A, Y_A) and of radius AB
- the circle of centre at point C (X_C, Y_C) and of radius BC.

The first solution is retained and angles φ_2 and φ_3 are determined using the function *Angle*.

point M is obtained using the function *Polar*, starting from point B with an angle $\varphi_2 + \pi - \alpha$, and the distance BM. The coordinates of point M (X_M, Y_M) are written in the script file *Curve.scr*, which is then run using the AutoCAD command **Script**, in the layer 0. We get the construction in Figure 2.46, superimposed over the position of the mechanism.

Later, the function draws the loop of the mechanism, using a line with non-zero thickness, and then the axes of coordinates are drawn too.

For the notation of the kinematic pairs of the mechanism we wrote an AutoLisp function called *Writing*, the content of which is as follows.

Figure 2.46 The position of the four-bar mechanism at angle $\varphi_1 = 70°$.

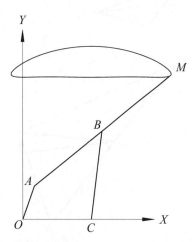

```
(Defun Writing (head tail nr_head nr_tail sign_head sign_tail)
(Command "Text" (Polar head (Angle tail head) 1) 1 "0"
   (Strcat (Chr nr_head) (Chr sign_cap)))
(Command "Text" (Polar tail (Angle head tail) 1) 1 "0"
   (Strcat (Chr nr_tail) (Chr sign_tail)))
)
```

The function has as parameters the coordinates of the points (*head* and *tail*) at which the text is to be placed. The next two parameters are for the ASCII codes of the text letters (integer numbers), while the last two parameters are for the second symbol of the text (when this is necessary). If no second character is wanted, then the number 32, the ASCII code for the *Space* key, is used. We used two characters for the annotations to allow us to add prime characters in the polygon of accelerations (for example, a', b', or m'). For instance, for the notation of the kinematic pairs of the element BC, the function is called with the coordinates of point B, the coordinates of point C, the number 66 (the ASCII code for the B letter), the number 67 (the ASCII code for the C letter) and the number 32 twice, for two *Space* characters. The characters are written at a certain distance from the element, in the direction of its extension, at a distance l.

After writing the characters A, M, C, and B, the representation in Figure 2.46 is obtained (for angle $\varphi_1 = 70°$); it can be seen after the selecting layer 1070 and setting it to 'visible'. To get the curve described by point M the script file mentioned above is run in layer 0.

By calling the function *Velocities*, presented in the Section 2.4.4, and completing with the functions *Arrows* and *Writing*, the polygon of velocities in Figure 2.47 is obtained (for angle $\varphi_1 = 70°$). We used, for the velocity of point A, $v_A = \omega_1 OA$ and angle $\alpha_v = \varphi_1 + \frac{\pi}{2}$, while for the velocity of point C, $v_C = 0$ and angle $\beta_v = 0$.

Using the functions already presented, we added the instructions necessary for the construction of point M (see the companion website of the book).

The accelerations are obtained in a similar way. The Autolisp function *Accelerations*, used in the Section 2.4.4 for the kinematic analysis of the RRR dyad, completes with the function *Writing* and the instruction necessary to obtain the acceleration of point M is added. In the companion website of the book, the modifications are annotated with comments. As input data we used: for the acceleration of point A, $a_A = \omega_1^2 OA$ and angle $\alpha_a = \varphi_1 + \pi$, while for the acceleration of point C, $a_C = 0$ and angle $\beta_a = 0$. For angle $\varphi_1 = 70°$ the polygon of accelerations in Figure 2.48 is obtained.

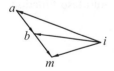

Figure 2.47 The polygon of velocities at angle $\varphi_1 = 70°$.

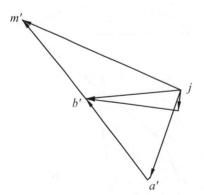

Figure 2.48 The polygon of accelerations at angle $\varphi_1 = 70°$.

The constructions may be dimensions to obtain values visually. These values are already written in the text files discussed above and presented in the companion website of this book. As expected, there is no difference between the results for the two methods.

When AutoCAD runs, an animation of the mechanism will be shown in the window declared for the position (the bottom left corner $-0.6, -0.2$, the up right corner $1, 1$). Different layers can be visualized to obtain the positions and the polygons of velocities and accelerations for the 36 angles. The animation of the polygon of velocities can then be viewed in a window of visualization of the velocities, and the animation of the polygon of accelerations in a window of visualization of the accelerations, but also reveals many details about changes in the magnitudes, senses, and velocities caused by increases and decreases in the vector parameters.

The animation is not only spectacular, but it also gives many details about the magnitudes, senses, and velocities of increasing and decreasing of the vector parameters. And because that the letter is 'linked' to the end of the vector, and moves with it, the didactic goal of the function is attained, too. As the dimensions of the mechanism are changed, a visual appreciation of the effect of the modifications can be gained.

The kinematic analysis can also be performed with an angular step of $1°$; it is only necessary to modify the function that assigns the values to angle φ_1. With AutoCAD it is an easy task to obtain precise geometric constructions in 360 layers.

2.10.2 Kinematic Analysis of the Crank-shaft Mechanism

Perform the kinematic analysis for the crank-shaft mechanism in Figure 2.49. The following are known:

- the dimensions of the mechanism (in metres): $OA = 0.0365, AB = 0.128, e = 0.0015$
- the position of the kinematic pair linked to the base $O(0, 0)$
- the constant angular velocity of element 1, $\omega_1 = 300$ rad/s.

The names of the elements are:

1) crank
2) shaft
3) piston.

The kinematic analysis is to be performed with an angular step of $10°$.

We will perform the kinematic analysis in a similar way to that used for the four-bar mechanism; that is, using the analytical method and the assisted graphical method. For the analytical method, the method described in Section 2.5.2 is used (with the length of the element $BC = 0$). Later on, we will use the analytical method of projections, which was described in the Section 2.3.2.

Figure 2.49 The crank-shaft mechanism.

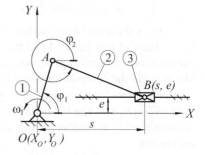

The analytical method of projections

We choose the loop OAB and the vector equation $\mathbf{OA} + \mathbf{AB} = \mathbf{OB}$. Projecting this equation onto the axes of the system OXY gives the system of non-linear equations of position

$$\begin{cases} OA\cos\varphi_1 + AB\cos\varphi_2 = s - X_O, \\ OA\sin\varphi_1 + AB\sin\varphi_2 = e - Y_O. \end{cases} \qquad (2.236)$$

From the second expression in (2.236):

$$\varphi_2 = \arcsin\left(\frac{e - Y_O - OA\sin\varphi_1}{AB}\right), \qquad (2.237)$$

which, substituted into the first expression in (2.236) gives the displacement of piston 3,

$$s = OA\cos\varphi_1 + AB\cos\varphi_2 + X_O. \qquad (2.238)$$

The coordinates of points A and B are obtained from:

$$\begin{cases} X_A = X_O + OA\cos\varphi_1, \\ Y_A = Y_O + OA\sin\varphi_1, \end{cases} \quad \begin{cases} X_B = s, \\ Y_B = e. \end{cases} \qquad (2.239)$$

The linear and angular velocities are obtained by deriving the system of the equations of position with respect to time,

$$\begin{cases} -OA\omega_1\sin\varphi_1 - AB\omega_2\sin\varphi_2 = \dot{s}, \\ OA\omega_1\cos\varphi_1 + AB\omega_2\cos\varphi_2 = 0, \end{cases} \qquad (2.240)$$

from which:

$$\omega_2 = \frac{-OA\omega_1\cos\varphi_1}{AB\cos\varphi_2},$$
$$\dot{s} = v_B = -OA\omega_1\sin\varphi_1 - AB\omega_2\sin\varphi_2. \qquad (2.241)$$

The derivation of (2.239) with respect to time gives the components of the velocities

$$\begin{cases} v_{Ax} = \dot{X}_A = -OA\omega_1\sin\varphi_1, \\ v_{Ay} = \dot{Y}_A = OA\omega_1\cos\varphi_1, \end{cases} \quad \begin{cases} v_{Bx} = \dot{X}_B = \dot{s}, \\ v_{By} = \dot{Y}_B = 0. \end{cases} \qquad (2.242)$$

Differentiating once again the equations of position with respect to time, we obtain

$$\begin{cases} -OA\omega_1^2\cos\varphi_1 - AB\omega_2^2\cos\varphi_2 - AB\varepsilon_2\sin\varphi_2 = \ddot{s}, \\ -OA\omega_1^2\sin\varphi_1 - AB\omega_2^2\sin\varphi_2 - AB\varepsilon_2\cos\varphi_2 0, \end{cases} \qquad (2.243)$$

from which

$$\varepsilon_2 = \frac{OA\omega_1^2\sin\varphi_1 + AB\omega_2^2\sin\varphi_2}{AB\cos\varphi_2},$$
$$\ddot{s} = a_B = -OA\omega_1^2\cos\varphi_1 - AB\omega_2^2\cos\varphi_2 - AB\varepsilon_2\sin\varphi_2. \qquad (2.244)$$

The components of the accelerations of the points are obtained from (2.242) by differentiation with respect to time

$$\begin{cases} a_{Ax} = \ddot{X}_A = -OA\omega_1^2\cos\varphi_1, \\ a_{Ay} = \ddot{Y}_A = -OA\omega_1^2\sin\varphi_1, \end{cases} \quad \begin{cases} a_{Bx} = \ddot{X}_B = \ddot{s}, \\ a_{By} = \ddot{Y}_B = 0. \end{cases} \qquad (2.245)$$

Based on these relations, we have written the calculation program presented in the companion website of this book. As in the case of the previous mechanism, the values obtained by running the calculation program are written to three files, which are transformed into tables like those in the companion website of this book.

In a repetitive *For* loop, with an angular step of $\frac{\pi}{18}$ rad, values are obtained for the expressions in (2.237)–(2.245). The function arctan is used to determine the value of angle φ_2. Taking into account the initial position of the mechanism, only the solutions situated in quadrants I or IV are chosen.

The assisted graphical method

For the kinematic analysis of the mechanism in Figure 2.49, the AutoLisp function *R_RRT* is used. This is shown in the companion website to this book. The function has no local variables. After being called in AutoCAD, it will

- determine the positions and the create the animation of the mechanism
- determine the velocities and accelerations
- construct the polygons of velocities and accelerations
- write the values to three text files.

The dimensions of the mechanism are declared at the beginning of the function. These can be modified to obtain the diagrams for other crank-shaft mechanisms.

The angular step is 10° (in a repetitive *While* loop) and each position of the mechanism is recorded in a dedicated layer, as in the previous example (Section 2.10.1; the assisted graphical method for the four-bar mechanism).

The kinematic pair at point *B* is obtained as the intersection between:

- the circle centred at point *A* with radius *AB*
- a straight line parallel to the axis *OX*, situated at the distance *e* from it.

The AutoLisp function *Int_CL*, previously defined (Section 2.5.1), is used. point P_1 is chosen according to the initial position of the mechanism.

The relations (2.237) and (2.238) given for the analytical method can also be used, but in this case when the position of the mechanism changes angle φ_2 has to be corrected, as in Figure 2.45.

Care must be taken, because the cosine of angle φ_2 was obtained as $\cos\varphi_2 = \sqrt{1 - \sin^2\varphi_2}$, so a solution is lost and the procedure for Figure 2.45 can no longer be used (the solutions in quadrants II and III were lost, the radical being always positive). The determination of the position of the kinematic pair at point *B*, using the function *Int_CL* has the advantage that, no matter which position the *R_RRT* mechanism is in, it is not necessary to manually intervene in the function. The only input required is to choose one of the two points of intersection between a circle and a straight line. It does not matter if the piston moves along the axis *OY*, along the negative portion of the axis *OX*, or along an inclined straight line situated, for instance, in the third quadrant.

After drawing the axes of coordinates and plotting the points with the function *Writing*, (Section 2.10.1), the representation in Figure 2.50 is obtained (for the position given by angle $\varphi_1 = 50°$).

To derive the velocities, the following vector relation is used:

$$\underline{\mathbf{v}_B} = \underline{\mathbf{v}_A} + \underline{\mathbf{v}_{BA}}$$
$$\parallel OX \quad \perp OA \quad \perp AB \tag{2.246}$$

where $v_A = \omega_1 OA$.

To obtain the polygon of velocities, the pole of velocities *i* is used. The velocity $\mathbf{v}_A$ (the vector **ia**) is obtained Using the *Polar* function. 'Departing' from point *i* at an angle $\varphi_1 + \frac{\pi}{2}$ and the

Figure 2.50 The position of the crank-shaft mechanism for angle $\varphi_1 = 50°$.

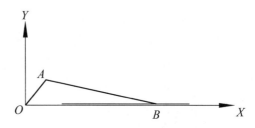

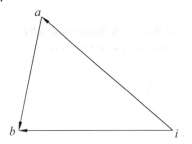

Figure 2.51 The polygon of velocities corresponding to angle $\varphi_1 = 50°$.

length $\omega_1 OA$, leads to a point a. From point a, along a perpendicular to AB (the angle is $\varphi_2 + \frac{\pi}{2}$), of length 1, leads to a point denoted by r.

In a similar way, constructing a parallel to AB passing through point i (using *Polar* from point i, with an angle of 0° and length 1) gives point s.

point b in the plane of velocities is obtained at the intersection of the straight lines defined by points a, r and i, s. The velocity v_B is the distance (*Distance*) between i and b, while ω_2 is the ratio between distances ab and AB.

The polygon of velocities is drawn using the command **Line**, and is completed with arrows using the function *Arrows*, and with letters using the function *Writing*. The construction in layer 1050 (corresponding to angle $\varphi_1 = 50°$) is as shown in Figure 2.51.

To obtain the accelerations the following vector relation is used:

$$\begin{array}{ccccc} \mathbf{a}_B & = & \mathbf{a}_A & + & \mathbf{a}_{BA}^v & + & \mathbf{a}_{BA}^\tau \, . \\ \| \, OX & & A \to O & & B \to A & & \perp AB \\ & & OA\omega_1^2 & & AB\omega_2^2 & & \end{array}$$

(2.247)

Starting from the pole of accelerations j, in *Polar* way, with angle $\varphi_1 + \pi$, and the length $\omega_1^2 OA$, gives point a' in the plane of accelerations (Figure 2.52). The point m' is obtained departing in *Polar* way from point a', with angle $\varphi_2 + \pi$, and the length $\omega_2^2 AB$.

point r' (situated on the perpendicular to AB) is obtained departing in *Polar* way from point m' with angle $\varphi_2 + \frac{\pi}{2}$ and an arbitrary length of 10, while point s' (situated on the parallel to OX) is obtained departing *Polar* from point j with an angle of 0° and length 10. point b' in the plane of accelerations is at the intersection of the straight lines defined by points a', r' and f, s', respectively. The acceleration a_B is the distance (*Distance*) between points j and b', while the angular acceleration ε_2 is the ratio between the distance $m'b'$ and AB.

The polygon of accelerations is drawn using the command **Line** through points j, a', b', m', j, and is completed with arrows (function *Arrows*) and letters (function *Writing*). If layer 1050, corresponding to angle $\varphi_1 = 50°$, is made visible, the representation in Figure 2.52 is obtained.

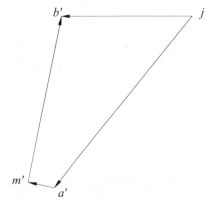

Figure 2.52 The polygon of accelerations corresponding to angle $\varphi_1 = 50°$.

The values for the positions (φ_2 and s), velocities (ω_2 and $\mathbf{v}_B$), and accelerations (ε_2 and $\mathbf{a}_B$) are written, depending on the position of the driving element (φ_1), in the files *Positions.txt*, *Velocities.txt*, and *Accelerations.txt*. The corresponding values are found in table form in the companion website of the book.

Similar to the case of the four-bar mechanism, there are no differences between the values obtained by the analytical and the assisted graphical methods.

2.10.3 Kinematic Analysis of the Crank and Slotted Lever Mechanism

Perform the kinematic analysis for the crank and slotted lever mechanism in Figure 2.53. The following are known:

- the dimensions of the mechanism: $OA = 0.030$ m, $BC = 0.100$ m
- the positions of the kinematic pairs linked to the base: $O(0, 0)$, $B(0.050, 0)$
- the constant angular velocity of the element 1: $\omega_1 = 10$ rad/s.

The kinematic analysis is performed using a constant angular step of $10°$. The analytical and the assisted graphical methods will be used, as was done for the four-bar and crank-shaft mechanisms.

The analytical method of projections
We choose loop OAB and the vector equation $\mathbf{OA} + \mathbf{AB} = \mathbf{OB}$ which, projected onto the axes of the system OXY, leads to the system of non-linear equations of position

$$\begin{cases} OA \cos \varphi_1 + s \cos \varphi_3 = X_B - X_O, \\ OA \sin \varphi_1 + s \sin \varphi_3 = Y_B - Y_O, \end{cases} \tag{2.248}$$

with the solution

$$\varphi_3 = \arctan \left(\frac{Y_B - Y_O - OA \sin \varphi_1}{X_B - X_O - OA \cos \varphi_1} \right), \ s = \frac{Y_B - Y_O - OA \sin \varphi_1}{\sin \varphi_3}. \tag{2.249}$$

The coordinates of points A and C are

$$\begin{cases} X_A = X_O + OA \cos \varphi_1, \\ Y_A = Y_O + OA \sin \varphi_1, \end{cases} \begin{cases} X_C = X_B + BC \cos \left(\varphi_3 - \pi \right), \\ Y_C = Y_B + BC \sin \left(\varphi_3 - \pi \right). \end{cases} \tag{2.250}$$

The differentiation of the equations of position (2.248) with respect to time leads to the system

$$\begin{cases} \dot{s} \cos \varphi_3 + s\omega_3 \sin \varphi_3 = OA\omega_1 \sin \varphi_1, \\ \dot{s} \sin \varphi_3 + s\omega_3 \cos \varphi_3 = -OA\omega_1 \cos \varphi_1. \end{cases} \tag{2.251}$$

with the solution

$$\dot{s} = OA\omega_1 \sin \left(\varphi_1 - \varphi_3 \right), \ \omega_3 = -\frac{OA \cos \left(\varphi_1 - \varphi_3 \right)}{s} \omega_1. \tag{2.252}$$

Figure 2.53 The crank and slotted lever mechanism.

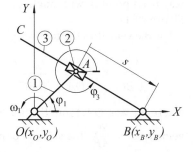

The velocities of points A and C are given by:

$$\begin{cases} v_{Ax} = \dot{X}_A = -OA\omega_1 \sin \varphi_1, \\ v_{Ay} = \dot{Y}_A = OA\omega_1 \cos \varphi_1, \end{cases} \quad \begin{cases} v_{Cx} = \dot{X}_C = -BC\omega_3 \sin(\varphi_3 - \pi), \\ v_{Cy} = \dot{Y}_C = BC\omega_3 \cos(\varphi_3 - \pi). \end{cases} \tag{2.253}$$

The accelerations are obtained by differentiating (2.251) with respect to time:

$$\begin{cases} \ddot{s} \cos \varphi_3 - s\varepsilon_3 \sin \varphi_3 = A_1, \\ \ddot{s} \sin \varphi_3 + s\varepsilon_3 \cos \varphi_3 = B_1, \end{cases} \tag{2.254}$$

where A_1 and B_1 are given by the expressions

$$\begin{cases} A_1 = 2\dot{s}\omega_3 \sin \varphi_3 + s\omega_3^2 \cos \varphi_3 + OA\omega_1^2 \cos \varphi_1, \\ B_1 = -2\dot{s}\omega_3 \cos \varphi_3 + s\omega_3^2 \sin \varphi_3 + OA\omega_1^2 \sin \varphi_1. \end{cases} \tag{2.255}$$

The solution of the system (2.254) is

$$\ddot{s} = A_1 \cos \varphi_3 + B_1 \sin \varphi_3, \quad \varepsilon_3 = \frac{-A_1 \sin \varphi_3 + B_1 \cos \varphi_3}{s}. \tag{2.256}$$

The accelerations of points A and C read

$$\begin{cases} a_{Ax} = \ddot{X}_A = -OA\omega_1^2 \cos \varphi_1, \\ a_{Ay} = \ddot{Y}_A = -OA\omega_1^2 \sin \varphi_1, \end{cases}$$

$$\begin{cases} a_{Cx} = \ddot{X}_C = -BC\omega_3^2 \cos(\varphi_3 - \pi) - BC\varepsilon_3 \sin(\varphi_3 - \pi), \\ a_{Cx} = \ddot{X}_C = -BC\omega_3^2 \cos(\varphi_3 - \pi) + BC\varepsilon_3 \cos(\varphi_3 - \pi). \end{cases} \tag{2.257}$$

The program used to obtain the numerical values is written in Pascal and is shown in the companion website of the book. It is written in a way similar to the programs discussed in previous sections, but using different equations, namely (2.249)–(2.257). The three files contain: the positions (φ_3, s), the velocities (ω_3, $\dot{s}$, and v_C), and the accelerations (ε_3, $\ddot{s}$, and a_C), as functions of the position of the driving element (angle φ_1). The corresponding tables can be found in the companion website of the book.

The assisted graphical method

The AutoLisp function for the kinematic analysis of the crank and slotted lever mechanism in Figure 2.53 is called *R_RTR* and it is shown in the companion website of this book. Proceeding as in Sections 2.10.1 and 2.10.2, we will obtain the positions and the polygons of velocities and accelerations in 36 distinct layers. The corresponding data are given in the companion website of the book.

The positions are easily obtained since, for a given angle φ_1 the position of point A is known. Using the function *Distance*, the distance between points A and B (Figure 2.53) is determined, and using the function *Angle*, angle φ_3 is determined. point C is obtained departing in *Polar* way from point B, with angle $\varphi_3 - \pi$ and length BC. The loop of the mechanism is shown in Figure 2.54a.

For the velocities the following vector relation is used:

$$\underset{\perp BC}{\underline{\mathbf{v}_{A_3}}} = \underset{\perp OA}{\underline{\mathbf{v}_{A_2}}} + \underset{\parallel BC}{\underline{\mathbf{v}_{A_3 A_2}}} \tag{2.258}$$

where $v_{A_2} = \omega_1 OA$. From the polygon of velocities (Figure 2.54b), $v_{A_3 A_2} = a_2 a_3$ and the angular velocity $\omega_3 = \dfrac{ia_3}{AB}$. The velocity of point C is determined using the theorem of similarity.

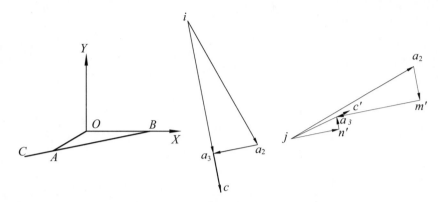

Figure 2.54 The position, polygon of velocities, and polygon of accelerations corresponding to $\varphi_1 = 210°$.

To determine the accelerations, the following vector relation is used:

$$\mathbf{a}_{A_3} = \underset{A \to O}{\mathbf{a}_{A_2}} + \underset{\perp BC}{\mathbf{a}^C_{A_3A_2}} + \underset{\parallel BC}{\mathbf{a}^r_{A_3A_2}},$$

$$= \underset{A \to B}{\mathbf{a}^\nu_{A_3}} + \underset{\perp AB}{\mathbf{a}^\tau_{A_3}}$$

(2.259)

giving $a^r_{A_3A_2} = m'a_3$ and $\varepsilon_3 = \dfrac{n'a_3}{AB}$ (Figure 2.54c). To determine the sense of the angular velocity ω_3, the sign of the expression $\cos(\varphi_1 - \varphi_3)$ must be examined.

3

Kinetostatics of Planar Mechanisms

Kinetostatics deals with the determination of the forces that act upon the elements of mechanisms. The kinetostatic analysis is preceded by the kinematic analysis; in determining the forces, the velocities and accelerations are considered as known. At the end of the analysis the reactions in the kinematic pairs of the mechanism are determined.

3.1   General Aspects: Forces in Mechanisms

The forces that act upon a mechanism can be divided into two categories

- given (known) forces
- unknown forces.

The latter are usually the reactions in the kinematic pairs and the forces (torques) of equilibration. The forces (torques) that act in the mechanisms are:

Driving forces (torques):   These forces (torques) act upon the driving element and maintain the motion in the mechanism. They are produced by driving (motor) sources such as thermal engines, and electrical, hydraulic, and electro-magnetic motors.

The weights of the elements: These are considered in calculations for mechanisms with small velocities. For mechanisms with high velocities, the weights become negligible with respect to inertial forces.

The technological forces (torques):   These are forces (torques) that act upon the driven elements and they produce negative work. In most cases the main destination of the mechanism is to create these forces (cutting forces in machine-tools, the useful forces at the brakes, weight forces in cranes, and so on).

The forces of friction: These are forces (torques) that appear in the kinematic pairs and which, in most cases, are coulombian (dry) friction forces. There are also cases in which the forces of friction are viscous ones.

The reactions: These are forces and torques that appear in the kinematic pairs, and are determined by kinetostatic calculation.

The forces of inertia: For the calculation of the forces in mechanisms, the d'Alembert principle, is used. From this comes the necessity to determine the inertial forces and torques. These are calculated from the results of the kinematic analysis.

Classical and Modern Approaches in the Theory of Mechanisms, First Edition.
Nicolae Pandrea, Dinel Popa and Nicolae-Doru Stănescu.
© 2017 John Wiley & Sons Ltd. Published 2017 by John Wiley & Sons Ltd.
Companion Website: www.wiley.com/go/pandmech17

The forces (torques) of equilibration: In the kinetostatic calculation the velocities and accelerations are assumed known from the kinematic analysis. The kinematic analysis is performed starting from the driving element, the velocity of which is considered to be a given (known) and constant. In these conditions, the force (torque) of equilibration represents the force (torque) that must act upon the driving element such that its velocity has the chosen value.

3.2 Forces of Inertia

3.2.1 The Torsor of the Inertial Forces

It is known that the torsor of inertia is calculated at the centre of rotation O for rigid solids in rotational motion, and at the centre of mass C for rigid solids in plane-parallel (translational) motion. The components of the torsor of inertia are

$$\boldsymbol{F} = -m\mathbf{a}_C, \ \boldsymbol{M}_O = -\dot{\mathbf{K}}_O \tag{3.1}$$

for a rigid body in rotational motion, and

$$\boldsymbol{F} = -m\mathbf{a}_C, \ \boldsymbol{M}_C = -\dot{\mathbf{K}}_C \tag{3.2}$$

for a rigid body in plane-parallel (translational) motion, where $\boldsymbol{F}$ is the inertial force and $\boldsymbol{M}$ is the resultant moment of the inertial forces.

Since, in general, the elements of planar mechanisms are bars or shells, the moments of the inertial forces reduce to the formulae

$$\boldsymbol{M}_O = -J_O \varepsilon, \ \boldsymbol{M}_C = -J_C \varepsilon. \tag{3.3}$$

Figure 3.1a,b shows the components of the torsor of inertia for rotational motion (Figure 3.1a), and for the plane-parallel motion (Figure 3.1b).

3.2.2 Concentration of Masses

This concept represents a particular case of the general concept of approximation by meshing; that is, a study of motion using the method of concentrated parameters. Essentially, in the present case, an element representing a continuous medium has to be replaced with concentrated masses situated at fixed distances from one another.

To ensure the conservation of the components of the torsor of inertia, it is necessary to conserve the mass, the position of the centre of mass, and the moment of inertia. For a bar (Figure 3.2) this gives:

$$\sum m_i = m, \ \sum m_i x_i = 0, \ \sum m_i x_i^2 = J_C. \tag{3.4}$$

In (3.4) the unknowns are the parameters m_i, x_i; some of the masses and their positions can be defined in advance, while only three can be determined from (3.4). Consider three concentrated

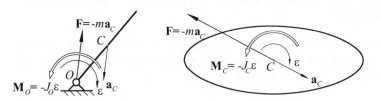

Figure 3.1 The components of the torsor of inertia.

Figure 3.2 Concentration of masses.

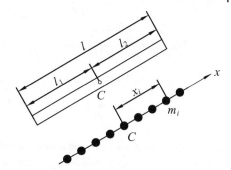

masses m_1, m_2, and m_3, such that m_2 is situated at the centre of mass, and $x_1 = l_1$, $x_2 = 0$, and $x_3 = -l_2$. The unknowns are the masses m_1, m_2, and m_3. From (3.4) we get

$$m_1 = \frac{J_c}{l_1 l}, \quad m_3 = \frac{J_c}{l_2 l}, \quad m_2 = m - \frac{J_c}{l_1 l_2}. \tag{3.5}$$

The problem can be solved if and only if

$$m l_1 l_2 > J_C. \tag{3.6}$$

3.3 Equilibration of the Rotors

3.3.1 Conditions of Equilibration

It is known from mechanics that, for a rigid solid in rotational motion, the conditions of dynamic equilibration are:

1) The rotational axis is a principal axis of inertia.
2) The centre of mass is situated on the rotational axis.

Using the notation in Figure 3.3, the equilibration conditions read

$$J_{xz} = J_{yz} = 0, \ \xi = 0 \tag{3.7}$$

A rigid solid in rotational motion can be said to be static equilibrated if its centre of mass is situated on the rotational axis; that is,

$$\xi = 0. \tag{3.8}$$

3.3.2 The Theorem of Equilibration

The theorem of equilibration states that a non-equilibrated rigid solid in rotational motion can always be equilibrated using two masses situated in two arbitrary planes perpendicular to the rotational axis. To demonstrate the theorem, consider the rigid solid in Figure 3.3, of mass m, which fulfills the conditions $J_{xz} \neq 0$, $J_{yz} \neq 0$, $\xi \neq 0$. We add the masses m_1 and m_2 in two planes

Figure 3.3 A rigid solid in rotational motion.

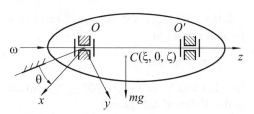

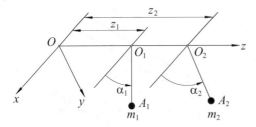

Figure 3.4 Equilibration masses and their positions.

perpendicular to the axis Oz and situated at the distances z_1 and z_2 relative to the Oxy plane. If we denote by x_1, y_1, z_1 and x_2, y_2, z_2 the coordinates of points A_1 and A_2 at which the masses m_1 and m_2, respectively, are situated, then the equilibration conditions are:

$$m_1 x_1 + m_2 x_2 + m\xi = 0, \tag{3.9}$$

$$m_1 y_1 + m_2 y_2 = 0, \tag{3.10}$$

$$m_1 x_1 z_1 + m_2 x_2 z_2 + J_{xz} = 0, \tag{3.11}$$

$$m_1 y_1 z_1 + m_2 y_2 z_2 + J_{yz} = 0. \tag{3.12}$$

These conditions form a system of four equations with eight unknowns ($x_1, y_1, z_1, x_2, y_2, z_2, m_1, m_2$). Choosing the equilibration planes (that is, the distances z_1 and z_2), from the equations (3.9), (3.11) and (3.10), (3.12), respectively, we deduce that

$$m_1 x_1 = \frac{m\xi z_2 - J_{xz}}{z_1 - z_2}, \quad m_2 x_2 = \frac{-m\xi z_1 + J_{xz}}{z_1 - z_2}, \tag{3.13}$$

$$m_1 y_1 = \frac{-J_{yz}}{z_1 - z_2}, \quad m_2 y_2 = \frac{J_{yz}}{z_1 - z_2}. \tag{3.14}$$

Let us denote by O_1 and O_2 the intersection points between:

- the perpendiculars to the Oz axis that pass through points A_1 and A_2, respectively
- the Oz axis (Figure 3.4)

Then let α_1 and α_2 be the angles between the straight lines $O_1 A_1$ and $O_2 A_2$, respectively, and the axis Ox (that is, the angles between the straight lines $O_1 A_1$ and $O_2 A_2$, and plane formed by the rotational axis and the centre of mass). If the planes are selected, then the angles α_1 and α_2 are given by:

$$\tan \alpha_1 = \frac{y_1}{x_1} = \frac{J_{yz}}{J_{xz} - m\xi z_2}, \quad \tan \alpha_2 = \frac{y_2}{x_2} = \frac{J_{yz}}{J_{xz} - m\xi z_1}. \tag{3.15}$$

Then, if $r_1 = O_1 A_1$ and $r_2 = O_2 A_2$, from the first parts of (3.13) and (3.14), and from the second parts of (3.13) and (3.14), respectively, we get the equilibration masses

$$m_1 = \frac{\sqrt{J_{yz}^2 + (J_{xz} - m\xi z_2)^2}}{r_1 |z_1 - z_2|}, \quad m_2 = \frac{\sqrt{J_{yz}^2 + (J_{xz} - m\xi z_1)^2}}{r_2 |z_1 - z_2|}. \tag{3.16}$$

If the rigid solid is static equilibrated ($\xi = 0$), then from (3.13)–(3.15) the angles α_1 and α_2 fulfill the relation

$$\alpha_2 = \alpha_1 + \pi. \tag{3.17}$$

In other words, points A_1 and A_2 are situated in a plane that contains the rotational axis and and in different half-planes.

This means:

- Dynamic equilibration may be brought about with two positive masses (the masses are added) or two negative masses (the masses are eliminated) situated in two arbitrary planes, but perpendicular to the rotational axis.
- The angles that define the positions of the equilibration masses relative to the rotational axis are given by (3.15).
- If the distances to the rotational axis are chosen, then the equilibration masses are given by (3.16).
- From the point of view of the equilibration, any rigid solid may be replaced by two concentrated masses, as given by (3.16), situated in two distinct planes perpendicular to the rotational axis, and in the directions given by (3.15).

3.3.3 Machines for Dynamic Equilibration

The dynamic equilibration of rotors (such as rotors for electrical motors, turbines, or bent axles) is vital nowadays, because working angular velocities are so high. For short cylinders (homogeneous discs), equilibration is performed in a single plane (the plane of the disc) and, in conclusion, dynamic equilibration reduces to static equilibration.

Modern machines for equilibration are equipped with transducers, amplifiers, systems for analog conversion, EDP systems, and systems to display the values of the equilibration masses and their positions. In Figure 3.5 we present the schema of a system of equilibration, which can be adapted to fulfill the previous requirements (Section 3.3.2).

The system consists of an oscillating frame OF, linked at point O to a fixed support frame FF. The OF frame is linked at point A to the fixed frame with an elastic rod ER, on which strain gauges SG are assembled. On the oscillating frame OF is a direct-current electric motor M, which is charged by a variator of angular velocity VAV. From the motor M the motion is transmitted using belt pulley S_1 and belt B to belt pulley S_2; this second belt pulley turns an axle, which is coupled, using a friction clutch and an axial bearing FC, to another axle, which is coupled, using fixed clutches C_1 and C_2, to rotor R, which has to be equilibrated. Planes P_I and P_{II} are chosen as the equilibriation planes, plane P_{II} passing through the kinematic pair O.

The signal from the strain gauge SG is processed by the tensometer T and is registered on paper by the signal register SR. The whole system may be modernized by processing the signal from the tensometer T, converting it to digital form and processing it in a computer, so that the maximum amplitudes are recording and subsequent calculations can be performed.

Motor M is switched on and the angular velocity is gradually increased using the variator VAV until the critical angular velocity is exceeded. At this point the motor is unplugged and the amplitude B_1 of the vibrations of the resonance is recorded. The motion of the system is stopped

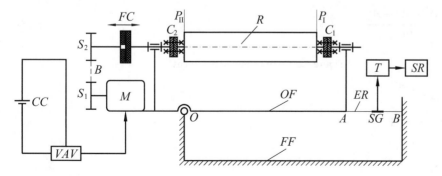

Figure 3.5 The schema of an equilibration machine.

and a mass m is added in plane P_I, at distance r from the centre of rotation. The direction is recorded and the operation is repeated, recording a new amplitude B_2. Again, the motion of the system is stopped. Then mass m is moved in the opposite direction, and the process is repeated, giving a third amplitude B_3.

The following calculations start from the hypothesis that the non-equilibrated masses m_I and m_{II} are situated in planes P_I and P_{II}, respectively. These masses produce inertial forces, which are decomposed into horizontal and vertical components. The horizontal components do not produce vibrations because their loadings are taken by the radial bearing situated at point O. The vertical component in plane P_{II} does not produce vibrations either, because its direction passes through point O. Hence, the component that produces the vibrations is the vertical one in plane P_I, the expression for which is $F_I \cos \omega t$. Here, F_I is the force of inertia,

$$F_I = m_I r_I \omega^2. \tag{3.18}$$

where ω is the angular velocity (at resonance, when the signal is the strongest possible), and r_I is the distance from mass m_I to the rotational axis.

The amplitude of the motion A_I is proportional to the magnitude of the force:

$$A_I = \lambda m_I r_I \omega^2. \tag{3.19}$$

Mass m produces an inertial force $F = mr\omega^2$ which, at resonance, produces an amplitude of $A = \lambda mr\omega^2$. By the decomposition of forces $\mathbf{F}_1$ and $\mathbf{F}$, a new force of magnitude F_2 is obtained, and this produces vibrations with amplitude A_2

$$A_2 = \lambda F_2. \tag{3.20}$$

Moving mass m in the opposite direction and compounding the inertial force $-\mathbf{F}$ produced by this mass with the force $\mathbf{F}_I$, a new force is obtained, the magnitude of which is F_3; this force produces vibrations of amplitude A_3:

$$A_3 = \lambda F_3. \tag{3.21}$$

The forces and the amplitudes (the differences of phase being $-\frac{\pi}{2}$) are shown in Figures 3.6 and 3.7, where α is the position angle of mass m_I relative to mass m (of known position). From the theorem of the median:

$$A = \sqrt{\frac{A_2^2 + A_3^2 - 2A_I^2}{2}}, \tag{3.22}$$

while angle α is obtained from:

$$\cos \alpha = \frac{A^2 + A_I^2 - A_3^2}{2AA_I}. \tag{3.23}$$

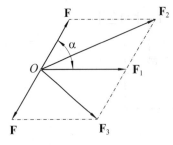

Figure 3.6 Composition of the forces.

Figure 3.7 Composition of the amplitudes of the vibrations.

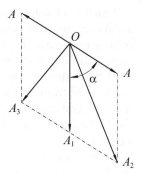

Since the amplitudes A_i are proportional to the registered amplitude B_i, one has

$$B = \sqrt{\frac{B_2^2 + B_3^2 - 2B_I^2}{2}}, \quad \cos \alpha = \frac{B^2 + B_I^2 - B_3^2}{2BB_I}. \tag{3.24}$$

Then, knowing angle α, we also know the position of mass m_I that produces the disequilibrium. In the direction of mass m_i, but in the opposite sense, we have to add the equilibration mass m_e at distance r_e, so that

$$m_e r_e = m_I r_I. \tag{3.25}$$

Taking into account

$$A = \lambda m r \omega^2, \quad A_I = \lambda m_I r_I \omega^2, \tag{3.26}$$

the equilibration mass is given by $m_e = \dfrac{mr_I A_I}{Ar_e}$; that is, by:

$$m_e = \frac{mr_I B_I}{Br_e}. \tag{3.27}$$

For equilibration in plane P_{II}, the axle is disassembled and re-assembled such that plane P_I passes through point O.

Figure 3.8 shows the schema for another machine for equilibration.

The part to be equilibrated, labelled 13, is assembled on the axle 12 using flanges F_I and F_{II}. From the direct-current electrical motor, driven by the variator 18, the motion is transmitted to belt pulley 2 and, using belt 4, which is rigidly linked to tube axle 5, to belt pulley 4. The motor is assembled on the oscillating frame 16, which is linked to the base with spring 17.

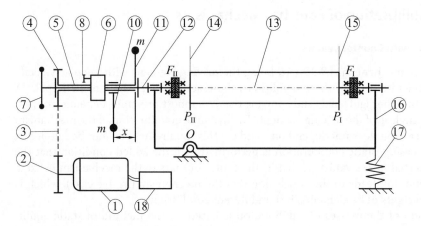

Figure 3.8 Mechanical machine for equilibration.

With the aid of differential 6 and by the action of screw 7, arms 10 and 11 – at the end of which are the masses m situated in the same plane – are constrained to have angular displacements relative to the axle that has to be equilibrated; this axle has an angular velocity ω, equal to the angular velocity of the arms.

By the action of screw 8, distance x between arms 10 and 11 may be varied. Motion is transmitted from differential 6 to axle 13, using axle 12. Planes P_I and P_{II} are the planes for the equilibration, plane P_{II} passing through the oscillating centre O. To increase the sensitivity, the system works at resonance; the angular velocity at resonance is obtained using the variator 18.

The disequilibrium masses m_I and m_{II}, situated in planes P_I and P_{II}, at distances r_I and r_{II} from the rotational axis, produce forces $F_I = m_I r_I \omega^2$ and $F_{II} = m_{II} r_{II} \omega^2$, respectively. The latter force does not produce vibrations because it is situated in plane P_{II}, which contains the axes of the radial bearing at point O. To reduce the level of vibration, arms 10 and 11 are rotated using screw 7, and distance x is varied using screw 8, until the frame no longer vibrates. This means that arms 10 and 11 arrive in the same plane as mass m_I and produce a torque of moment equal to the moment of the inertial force F_I. The moments are calculated with respect to point O_2 at which axle 12 intersects plane P_{II}. Denoting by d the distance between planes P_I and P_{II}, and by r the distance from the masses m situated at the end of the arms to axle 12 gives:

$$mr\omega^2 x = m_I r_I \omega^2 d. \tag{3.28}$$

For equilibration in plane P_I, in the direction and sense of arm 11, at a chosen distance r_e, we have to apply the equilibration mass m_e that verifies:

$$m_e r_e = m_I r_I. \tag{3.29}$$

From this and (3.28):

$$m_e = \frac{mrx}{r_e d}. \tag{3.30}$$

Since the parameters m, r, r_e, and d are constant, the rule for the measurement of distance x can be directly calibrated in grams, and in this way equilibration becomes a very simple process. By inverting the ends of flanges F_I and F_{II}, equilibration can easily be readily brought about in plane P_{II}.

3.4 Static Equilibration of Four-bar Mechanisms

3.4.1 Equilibration with Counterweights

Consider a four-bar mechanism $ABCD$ (AB being the driving element with angular velocity ω and angular acceleration ε). In these conditions, the reaction R_D at the kinematic pair D, calculated by isolating the rigid solids and applying the d'Alembert principle, depends on ε, ω^2, and the rotational angle φ of the driving element. The dynamic equilibration of the mechanism requires that the reaction does not depend on ε and ω^2; this requirement cannot be fulfilled in general. to In this case, a static equilibration is performed with the added condition that the resultant of the inertial forces vanishes; that is, the centre of mass of the mechanism is fixed. Obviously, this condition leads to the conclusion that the mechanism (in this case), which is acted only by the weights of its elements, is in indifferent equilibrium.

The determination of the masses of equilibration is based on the method of static equilibration of the elements, using two concentrated masses. As an example, consider the four-bar

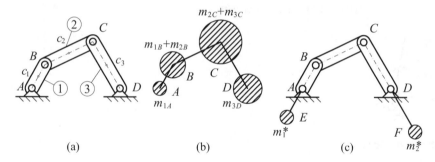

Figure 3.9 Static equilibration of a four-bar mechanism: (a) four-bar mechanism; (b) equivalent mechanism with concentrated masses; (c) equilibrated four-bar mechanism.

mechanism in Figure 3.9a. We distribute the masses at the ends of the elements using the following relations:

$$m_{1A} = m_1 \frac{BC_1}{l_1}, \; m_{1B} = m_1 \frac{AC_1}{l_1}, \; m_{2B} = m_2 \frac{C_2C}{l_2},$$

$$m_{2C} = m_2 \frac{AC_2}{l_2}, \; m_{3C} \frac{C_3D}{l_3}, \; m_{3D} = m_3 \frac{C_3C}{l_3}, \tag{3.31}$$

where m_1, m_2, and m_3 are the masses of the elements, l_1, l_2, and l_3 are their lengths, and C_1, C_2, and C_3 are the centres of weight.

This gives the equivalent mechanism with the masses concentrated at points A, B, C, and D (Figure 3.9b). For equilibration, we add the concentrated masses m_1^* and m_2^* at points E and F, respectively (Figure 3.9b), such that they equilibrate the masses at points B and C:

$$m_1^* = \frac{(m_{1B} + m_{2B}) l_1}{AE}, \; m_2^* = \frac{(m_{2C} + m_{3C}) l_3}{DF}. \tag{3.32}$$

In this way, the mechanism reduces to a system of concentrated masses at points A and D, the centre of mass of the mechanism being situated on the straight line AD.

For the crank-shaft mechanism in Figure 3.10a, distributing the masses of elements 1 and 2 at points A, B and B, C, respectively, we obtain the situation shown in Figure 3.10b. For equilibration, we put mass m_2^* at point D, so that the centre of mass of element 2 is at point B:

$$m_2^* = \frac{(m_{2C} + m_3) l_3}{DB}. \tag{3.33}$$

The mechanism now reduces to the concentrated mass at point B

$$m_B = m_{1B} + m_{2B} + m_2^* + m_{2C} + m_3$$

and to the concentrated mass m_{1A} at point A. For equilibration, we add mass m_1^* at point E, so that point A becomes the centre of mass of the mechanism.

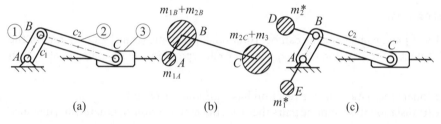

Figure 3.10 Static equilibration of a crank-shaft mechanism:(a) crank-shaft mechanism; (b) equivalent mechanism with concentrated masses; (c) equilibrated crank-shaft mechanism.

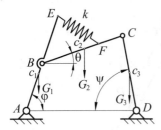

Figure 3.11 Equilibration with springs.

3.4.2 Equilibration with Springs

As we have seen, static equilibration with counterweights increases the weight of the mechanism; in some cases, this might be unacceptable. In these situations – mainly in the equilibration of weight forces – springs can be added, the characteristics of which (stiffness k and the length l_0 in non-deformed status) and the points at which they are fixed having to be determined. As an example, consider the mechanism $ABCD$ in Figure 3.11, situated in a vertical plane. We wish to determine points E and F ($AE = \lambda_1$, $BF = \lambda_2$), the stiffness k, and the length l_0 of the spring, such that the mechanism is equilibrated. The condition for the realization of an indifferent equilibrium states that the potential energy is constant; this requirement may be written as

$$V = \frac{k}{2}\left(EF - l_0\right)^2 + \sum_{i=1}^{3} G_i y_{C_i} = \text{const.} \tag{3.34}$$

If we denote the distances AE and BF with λ_1 and λ_2, respectively, then (3.34) becomes

$$\begin{aligned} V = \frac{k}{2}\Bigg\{ &\sqrt{\left[(\lambda_1 - l_1)\cos\varphi - \lambda_2\cos\theta\right]^2 + \left[(\lambda_1 - l_1)\sin\varphi - \lambda_2\sin\theta\right]^2} - l_0 \Bigg\}^2 \\ &+ G_1\frac{l_1}{2}\sin\varphi + G_2\left(l_1\sin\varphi + \frac{l_2}{2}\sin\theta\right) + G_3 l_3\sin\psi = C, \end{aligned} \tag{3.35}$$

where θ and ψ are parameters that depend (in a positional analysis) on angle φ. In general notation, (3.35) reads

$$V\left(\varphi, \lambda_1, \lambda_2, k, l_0\right) = C. \tag{3.36}$$

Because, in general, (3.36) does not hold true for any value of angle φ, we will choose five positions for which we can write

$$V\left(\varphi_i, \lambda_1, \lambda_2, k, l_0\right) = C, \ i = 1, 2, 3, 4, 5. \tag{3.37}$$

From these last equations, parameters λ_1, λ_2, k, l_0, and C can be determined.

3.5 Reactions in Frictionless Kinematic Pairs

3.5.1 General Aspects

Kinematic pairs of planar mechanisms can be: rotational, prismatic (translational), and of the fourth class (contact). The reactions that appear in each type of kinematic pair are shown in Figure 3.12:

- in each rotational kinematic pair, the reaction has components H and V
- in prismatic (translational) kinematic pairs there is a normal reaction N situated at a distance x (that is, a force and a moment)
- in contact kinematic pairs the reaction is a normal one and passes through the contact point.

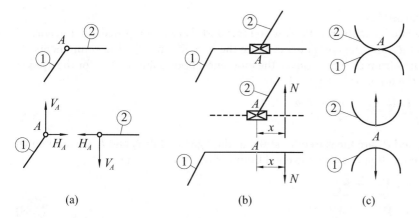

Figure 3.12 Reactions in frictionless kinematic pairs: (a) rotational kinematic pair; (b) prismatic (translational) kinematic pair; (c) fourth class kinematic pair.

The representation of the reactions on the two rigid bodies is created using the principle of reciprocal actions. For a dyad we have two elements and three class-5 kinematic pairs. Since for each element there are three equations of dynamic equilibrium, while for each kinematic pair there are two unknown parameters that define the reactions, the dyads are determined from the point of view of the forces. In more general terms, it can be shown that all the kinematic groups are determined from the point of view of the forces.

The unknowns that appear at the driving element are:

- the two parameters in the kinematic pair at the base
- the moment (force) of equilibrium that has to be applied to the driving element in order to produce the velocities and accelerations considered in the kinematic calculation,

3.5.2 Determination of the Reactions for the RRR Dyad

Formulation of the problem

Let us assume that known (given and inertial) forces $\mathbf{F}_2$ and $\mathbf{F}_3$, and torques of known (given and inertial) moments $\mathbf{M}_2$ and $\mathbf{M}_3$ act on dyad ABC, as shown in Figure 3.13a. We also assume that a kinematic analysis has been performed. In these conditions we want to determine the reactions in the kinematic pairs A, B, and C. We consider the components F_{2x}, F_{2y}, F_{3x} and F_{3y} as known, and want to determine components H_A, V_A, H_B, V_B, H_C and V_C of the reactions in the kinematic pairs A, B, and C.

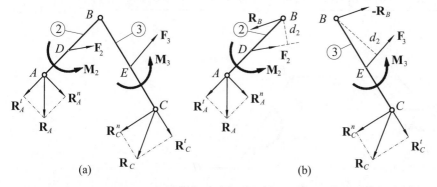

Figure 3.13 Kinetostatic analysis of the RRR dyad using the grapho-analytical method: (a) isolation of the RRR dyad; (b) isolation of the elements of the RRR dyad.

The grapho-analytical method

The elements are isolated, and the reactions at points A and C are decomposed into the normal components R_A^n and R_C^n and the components along the bars R_A^t and R_C^t. The components R_A^n and R_C^n are determined from the equations of the moments about point B. Thus for the dyad in Figure 3.13a, using the notation in Figure 3.13b:

$$R_A^n = -\frac{M_2 + F_2 d_2}{l_2}, \ R_C^n = \frac{M_3 + F_3 d_3}{l_3},$$

(3.38)

where distances d_2 and d_3 are measured (at scale) in the figure, while l_2 and l_3 are the known lengths of the elements. Then, writing the equilibrium of the forces for the dyad

$$
\begin{aligned}
&\underline{\underline{\mathbf{R}_A^n}} + \underline{\underline{\mathbf{F}_2}} + \underline{\underline{\mathbf{F}_3}} + \underline{\underline{\mathbf{R}_C^n}}\\
&\qquad + \underline{\mathbf{R}_C^t} + \underline{\mathbf{R}_A^t} = \mathbf{0}\\
&\qquad\ \ \| BC \quad\ \| AB
\end{aligned}
$$

(3.39)

and representing it at the scale chosen, we obtain the components $\mathbf{R}_A^t$ and $\mathbf{R}_C^t$. From the equilibrium of the forces that act upon element 2

$$\underline{\underline{\mathbf{R}_A^t}} + \underline{\underline{\mathbf{R}_A^n}} + \underline{\underline{\mathbf{F}_2}} + \underline{\underline{\mathbf{R}_B}} = \mathbf{0},$$

(3.40)

the reaction $\mathbf{R}_B$ can be determined.

The analytical method

Consider the isolation of the elements in Figure 3.14. From the equations of moments about points A and C for the two elements, the following system of equations is obtained:

$$
\begin{aligned}
&M_2 + F_{2y}\left(X_D - X_A\right) - F_{2x}\left(Y_D - Y_A\right) + V_B\left(X_B - X_A\right) - H_B\left(Y_B - Y_A\right) = 0,\\
&M_3 + F_{3y}\left(X_E - X_C\right) - F_{3x}\left(Y_E - Y_C\right) - V_B\left(X_B - X_C\right) + H_B\left(Y_B - Y_C\right) = 0.
\end{aligned}
$$

(3.41)

Then, if we denote

$$
\begin{aligned}
A_1 &= -M_2 - F_{2y}\left(X_D - X_A\right) + F_{2x}\left(Y_D - Y_A\right),\\
B_1 &= -M_3 - F_{3y}\left(X_E - X_C\right) + F_{3x}\left(Y_E - Y_C\right),\\
C_1 &= \left(X_B - X_A\right)\left(Y_B - Y_C\right) - \left(X_B - X_C\right)\left(Y_B - Y_A\right),
\end{aligned}
$$

(3.42)

we get

$$H_B = \frac{A_1\left(X_B - X_C\right) + B_1\left(X_B - X_A\right)}{C_1}, \ V_B = \frac{A_1\left(Y_B - Y_C\right) + B_1\left(Y_B - Y_A\right)}{C_1}.$$

(3.43)

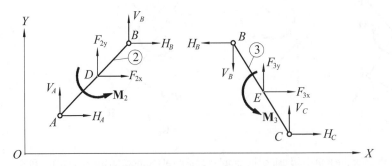

Figure 3.14 Kinetostatic analysis of the RRR dyad using the analytical method.

From the equations of projections for each element

$$H_A + F_{2x} + H_B = 0, \ V_A + F_{2y} + V_B = 0,$$
$$H_C + F_{3x} - H_B = 0, \ V_C + F_{3y} - V_B = 0,$$

(3.44)

we get the components H_A, V_A, H_C, and V_C, and the problem is solved.

Numerical example

Problem: Determine the reactions in the kinematic pairs for the dyad ABC shown in Figure 3.15, for which a kinematic analysis was performed in Sections 2.4.1 (grapho-analytical method), 2.4.2 (analytical method), 2.4.3 (assisted analytical method), and 2.4.4 (assisted graphical method). The dyad is acted on only by the inertial forces and moments, while the elements are homogeneous straight bars with masses $m_2 = 1$ kg and $m_3 = 1.25$ kg. For simplicity the weights are neglected.

Solutions:

The grapho-analytical method: The dyad is represented at a scale $k_l = 0.02 \left[\dfrac{m}{mm}\right]$ in Figure 3.15a, while the accelerations are represented at a scale $k_a = 4.5 \left[\dfrac{m/s^2}{mm}\right]$ in Figure 3.15b. The centres of weight D and E are situated at the middle of the bars and correspond, in the plane of accelerations (according to the theorem of similarity), to the accelerations $\mathbf{jd'}$ and $\mathbf{je'}$. The values of these accelerations, measured on the figure, are $\left[jd'\right] = 16$ mm and $\left[je'\right] = 26$ mm. The real values for the accelerations of points D and E are $a_D = 72$ m/s^2 and $a_E = 117$ m/s^2, while the magnitudes of the inertial forces are $F_2 = 72$ N and $F_3 = 146.25$ N (see Figure 3.15a).

The angular accelerations deduced from the kinematic analysis are $\varepsilon_2 = 450$ rad/s^2 and $\varepsilon_3 = 40.29$ rad/s^2, while the inertial moments $J_i = \dfrac{m_i l_i^2}{12}$, are $J_2 = 0.021$ kgm^2 and $J_3 = 0.047$ kgm^2; the moments of inertia are then $M_2 = 9$ Nm and $M_3 = 1.89$ Nm. Measuring on the figure, we obtain $\left[d_2\right] = 7$ mm and $\left[d_3\right] = 10$ mm, hence $d_2 = 0.14$ m and $d_3 = 0.2$ m.

From the equations of moments about point B for elements 2 and 3, we get

$$R_A^n = \frac{F_2 d_2 - M_2}{l_2} = 2.16 \text{ N}, \ R_C^n = \frac{F_3 d_3 - M_3}{l_3} = 40.83 \text{ N}.$$

Choosing the scale of the forces $k_F = 3 \left[\dfrac{N}{mm}\right]$ and representing the vector equation

$$\underline{\underline{\mathbf{R}_A^n}} + \underline{\underline{\mathbf{F}_2}} + \underline{\underline{\mathbf{F}_3}} + \underline{\underline{\mathbf{R}_C^n}}$$
$$+ \underline{\underline{\mathbf{R}_C^t}} + \underline{\underline{\mathbf{R}_A^t}} = \mathbf{0}.$$
$$\quad \| BC \quad \| AB$$

we obtained the polygon of forces, with the pole at point f (Figure 3.15c). We deduce that $R_A^t = 120$ N and $R_C^t = 159$ N. Considering the pole at point f_2 and using the vector relation (3.40) with the representation in Figure 3.15c, we obtain the value $R_B = 72$ N for the reaction in the kinematic pair at point B.

Finally, the results are

$$R_A = \sqrt{2.16^2 + 120^2} = 120.02 \text{ N}, \ R_C = \sqrt{40.83^2 + 159^2} = 164.16 \text{ N}.$$

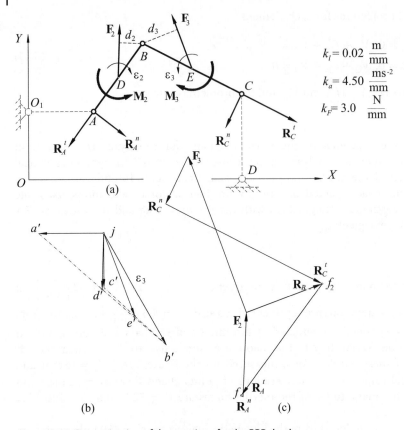

Figure 3.15 Determination of the reactions for the RRR dyad.

The analytical method First, we have to determine the coordinates of points D and E

$$X_D = \frac{1}{2}\left(X_A + X_B\right) = \frac{1}{2}\left(0.4 + 0.7\right) = 0.55\,\text{m},$$

$$Y_D = \frac{1}{2}\left(Y_A + Y_B\right) = \frac{1}{2}\left(0.4 + 0.8\right) = 0.6\,\text{m},$$

$$X_E = \frac{1}{2}\left(X_C + X_B\right) = \frac{1}{2}\left(1.3 + 0.7\right) = 1\,\text{m},$$

$$Y_E = \frac{1}{2}\left(Y_C + Y_B\right) = \frac{1}{2}\left(0.5 + 0.8\right) = 0.65\,\text{m}.$$

The components of the accelerations of points D and E are

$$a_{Dx} = a_{Ax} - \varepsilon_2\left(Y_D - Y_A\right) - \omega_2^2\left(X_D - X_A\right)$$
$$= -90 - 456.17 \cdot 0.2 - 5.46^2 \cdot 0.15 = -3.24\,\text{m/s}^2,$$

$$a_{Dy} = a_{Ay} + \varepsilon_2\left(X_D - X_A\right) - \omega_2^2\left(Y_D - Y_A\right)$$
$$= 0 - 457.17 \cdot 0.15 - 5.46^2 \cdot 0.2 = -74.54\,\text{m/s}^2,$$

$$a_{Ex} = a_{Cx} - \varepsilon_3\left(Y_E - Y_C\right) - \omega_3^2\left(X_E - X_C\right)$$
$$= 0 - 47.28 \cdot 0.15 + 12.75^2 \cdot 0.3 = 41.68\,\text{m/s}^2,$$

$$a_{Ey} = a_{Cy} + \varepsilon_3\left(X_E - X_C\right) - \omega_3^2\left(Y_E - Y_C\right)$$
$$= -72 - 47.28 \cdot 0.3 - 12.75^2 \cdot 0.15 = -110.56\,\text{m/s}^2.$$

Then the forces and moments of inertia are:

$$F_{Dx} = -m_2 a_{Dx} = -3.24\text{N}, \quad F_{Dy} = -m_2 a_{Dy} = 74.54 \text{ N},$$

$$F_{Ex} = -m_3 a_{Ex} = -52.1 \text{ N}, \quad F_{Ey} = -m_3 a_{Ey} = 138.2 \text{ N},$$

$$M_2 = -J_2 \varepsilon_2 = 0.021 \cdot 456.17 = 9.6 \text{ Nm},$$

$$M_3 = -J_3 \varepsilon_3 = -0.017 \cdot 47.28 = -2.22 \text{ Nm},$$

From (3.42) we deduce that $A_1 = -20$, $B_1 = 35.82$, and $C_1 = 0.33$, while from (3.43) we get $H_B = 69.13$ N, $V_B = 25.20$ N, and $R_B = 73.58$ N. From (3.44):

$$H_A = -F_{Dx} - H_B = -3.24 - 69.13 = -72.37 \text{ N},$$

$$V_A = -F_{Dy} - V_B = -74.54 - 25.2 = -99.7 \text{ N},$$

$$H_C = -F_{Ex} + H_B = 52.1 + 69.13 = 121.23 \text{ N},$$

$$V_C = -F_{Ey} + V_B = -138.2 + 25.2 = -113 \text{ N},$$

from which

$$R_A = \sqrt{H_A^2 + V_A^2} = 123.19 \text{ N}, \quad R_C = \sqrt{H_C^2 + V_C^2} = 165.72 \text{ N}.$$

Comparing these results with those obtained using the grapho-analytical method, it can be seen that the errors are less than 5%.

The assisted analytical method   Based on the relations established in Section 2.5.2, either a dedicated procedure or *Dyad_RRR* (see Section 2.4.3) can be used, with kinematic and kinetostatic relations that permit determination of the reactions in the kinematic pairs of the RRR dyad. We prefer the latter approach due to its simplicity; masses m_2 and m_3 and the moments of inertia J_2 and J_3, corresponding to elements 2, and 3, respectively are added to the input data of the procedure *Dyad_RRR*. The reactions $H_{Ar}, V_{Ar}, V_{Ar}, R_{Ar}, H_{Br}, V_{Br}, R_{Br}, H_{Cr}, V_{Cr}, R_{Cr}$ are added to the output data and as local variables it is also necessary to introduce $X_D, Y_D, X_E, Y_E, a_{Dx}, a_{Dy},$ $a_{Ex}, a_{Ey}, F_{Dx}, F_{Dy}, F_{Ex}, F_{Ey}, J_2, J_3, M_2, M_3, A_1, B_1$. The modified procedure can be found in the companion website of this book.

The procedure is called with the known data

Dyad_RRR $(0.4, 0.4, 1.3, 0.5, 0.5, 0.67, 0, -6, 6, 0, -90, 0, 0, -72, 1, 1.25,$

$J2, J3, Phi2, Phi3, XB, YB, Omega2, Omega3, vBx, vBy, vB, Epsilon2, Epsilon3,$

$aBx, aBy, aB, HAr, VAr, RAr, HBr, VBr, RBr, HCr, VCr, RCr)$

and in addition to to the results in Section 2.4.3, the following values are now obtained:

$$H_A = -72.500004, \quad V_A = -99.750351, \quad R_A = 123.314164,$$

$$H_B = 69.129797, \quad V_B = 25.223822, \quad R_B = 73.587839,$$

$$H_C = 121.167037, \quad V_C = -112.934339, \quad R_C = 165.637000.$$

The assisted graphical method   In Section 2.4.4, the kinematic analysis of the RRR dyad was performed using AutoCAD, giving the graphical construction in Figures 2.9–2.11. Now we will present the AutoLisp function that determines the reactions in the kinematic pairs and constructs the polygon of forces. The function is called *Forces_RRR* and is based on the kinematic calculation encapsulated in the function *Dyad_RRR.lsp*. For this reason it is called before the function RRR, and after the function *Accelerations* (*Forces_RRR*). The content of the function *Forces_RRR* is as follows:

```
(Defun Forces_RRR ()
(Command "Zoom" "30,80" "600,320")
(Command "-Layer" "N" "Forces" "C" "6" "Forces" "S" "Forces" "" ""
   "Color" "6")
(Setq m2 1.0 m3 1.25 fsmall(List 300 100) J2(/ (* m2 lAB lAB) 12) J3(/
(* m3 lBC lBC) 12))
(Setq xaprime(princ(car aprime)) yaprime(princ(car(cdr aprime)))
   xbprime(princ(car bprime))
ybprime(princ(car(cdr bprime))) xcprime(princ(car cprime)) ycprime
   (princ(car(cdr cprime)))
dprime(List (/ (+ xaprime xbprime) 2) (/ (+ yaprime ybprime) 2))
   eprime(List (/ (+ xcprime xbprime) 2) (/ (+ ycprime ybprime) 2))
   accD(Distance jsmall dprime)
accE(Distance jsmall eprime))
(Setq F2(* m2 accD) F3(* m3 accE) M2(* J2 eps2) M3(* J3 eps3))
(Setq angF2(- (Angle jsmall dprime) Pi) angF3(- (Angle jsmall eprime)
Pi) angRAn(+ phi2 (* Pi 1.5)) angRAt(+ phi2 Pi) angRCn(- phi3 (* Pi
0.5)) angRCt phi3 alpha(- phi2 angF2) beta(- phi3 angF3 Pi))
(Setq d2(* (/ lAB 2) (Sin alpha)) d3(* (/ lBC 2) (Sin beta)) RAn(/ (+ (*
F2 d2) M2) lAB -1) RCn(/ (- (* F3 d3) M3) lBC))
(Setq fg(Polar fsmall angRAn RAn) gh(Polar fg angF2 F2) hk(Polar gh
   angF3 F3)
kl(Polar hk angRCn RCn) lm(Polar kl phi3 10) fn(Polar fsmall phi2 10)
int(Inters kl lm fsmall fn Nil) RAt(Distance fsmall int) RCt(Distance
kl int) RA(Sqrt(+ (* RAn RAn) (* RAt RAt))) RC(Sqrt(+ (* RCn RCn) (* RCt
RCt))) RB(Distance int gh))
(Command "Line" fsmall fg gh hk kl int "C")
(Setq l(/ RA 40) g(* l 0.353))
(Arrows fg fsmall) (WritingF fsmall fg 32 82 102 65 32 110)
(Arrows gh fg) (WritingF fg gh 32 32 32 70 32 50)
(Arrows hk gh) (WritingF gh hk 32 32 32 70 32 51)
(Arrows kl hk) (WritingF hk kl 32 82 32 67 32 110)
(Arrows int kl) (WritingF kl int 32 82 32 67 32 116)
(Arrows fsmall int)(WritingF int fsmall 32 82 32 65 32 116)
(Command "PLine" gh "W" "0.4" "" int "")
(Arrows int gh) (WritingF gh int 32 32 32 82 32 66)
)
```

The function constructs the polygon of forces in Figure 3.16 in a new window of visualization and a dedicated layer called *Forces*, the colour of the layer being violet (6). Values from the previous applications (Section 3.5.2, the grapho-analytical method) are assigned to the masses and the torsor of the inertial forces that act at the centres of weight of elements 2 and 3 is determined. To do this, the centres of segments $a'b'$ and $b'c'$ in the plane of accelerations, denoted by d' and e', respectively, are determined. The distances jd' and je' are the accelerations of the centres of weight of the homogeneous bars 2 and 3 (a_D and a_E, respectively). To use the vector relation (3.39) the following are determined:

- the angles made by the inertial forces ($\mathbf{F}_2$, $\mathbf{F}_3$) and the normal reactions ($\mathbf{R}_A^n$, $\mathbf{R}_C^n$) with the axis OX
- the distances d_2 and d_3 in Figure 3.15
- the values of the normal components $\mathbf{R}_A^n$, $\mathbf{R}_C^n$ using (3.38).

The points of the polygon of forces in Figure 3.16 are determined using function *Polar*, and the intersection point between the parallel to the straight line *AB*, and the parallel to the straight

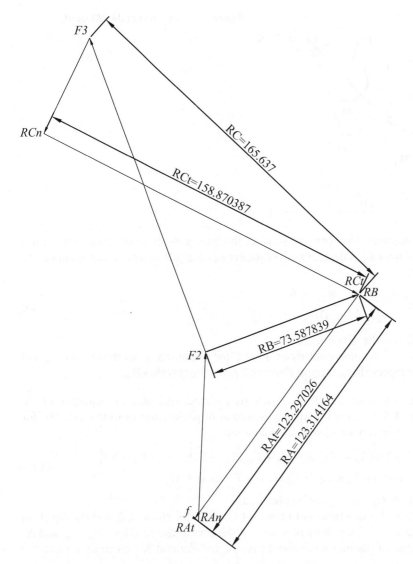

Figure 3.16 The forces for the RRR dyad obtained using the assisted graphical method.

line *BC* is found using the function *Inters*. This gives the values of the tangential components $(\mathbf{R}_A^t, \mathbf{R}_C^t)$ of the reactions in the kinematic pairs *A* and *C*. The reaction at point *B* is determined using (3.34).

Using the command **Pline**, the loop of the polygon of forces passing through the previously drawn points is drawn and the polygon is completed with arrows and notations. After the dimension, the representation in Figure 3.16 is obtained.

3.5.3 Determination of the Reactions for the RRT Dyad

Formulation of the problem

Assume that forces $\mathbf{F}_2$ and $\mathbf{F}_3$ and torques of moments $\mathbf{M}_2$ and $\mathbf{M}_3$ act upon dyad *ABC* in Figure 3.17. We want to determine the reactions in the kinematic pairs *A*, *B* and *C*. A kinematic analysis has already been performed.

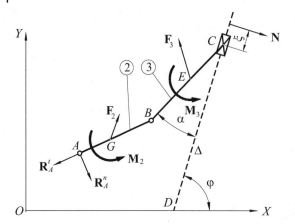

Figure 3.17 The forces at the RRT dyad.

The grapho-analytical method Element 2 is isolated, the equations of moments about point B are set out, and component R_A^n is determined. From the equilibrium of forces that act upon the dyad

$$\underline{\underline{\mathbf{R}_A^n}} + \underline{\underline{\mathbf{F}_2}} + \underline{\underline{\mathbf{F}_3}} + \underset{\perp \Delta}{\underline{\mathbf{N}}} + \underset{\parallel AB}{\underline{\mathbf{R}_A^t}} = \mathbf{0},$$

(3.45)

the reactions $\mathbf{N}$ and $\mathbf{R}_A^t$ can be determined.

Writing the equation of the moments about point B for element 3, gives the distance ξ, and the equilibrium of the forces that act upon element 2 gives the reaction $\mathbf{R}_B$.

The analytical method The elements (Figure 3.18) are isolated, and with the equation of the moments about point A for element 2 and the equation of projections onto the axis Dx' for element 3, the following system of equations is obtained:

$$-H_B\left(Y_B - Y_A\right) + V_B\left(X_B - X_A\right) = -M_2 + F_{2x}\left(Y_D - Y_A\right) - F_{2y}\left(X_D - X_A\right),$$
$$H_B \cos\varphi + V_B \sin\varphi = F_{3x}\cos\varphi + F_{3y}\sin\varphi + M_3$$

(3.46)

from which the reactions H_B and V_B can be obtained.

Writing the equations of projections onto the axes OX, OY for element 2, and the equation of projection onto the axis Dy' for element 3, we can determine the reactions H_A, V_A, and N, while from the equation of the moments about point C for element 3, we can determine the distance ξ.

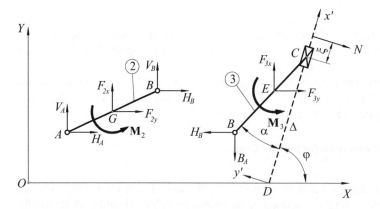

Figure 3.18 The forces at the RRT dyad (analytical method).

Numerical example

For the dyad ABC in Figure 3.17, the kinematic analysis of which was performed in the Sections 2.5.1–2.5.4, determine the reactions in the kinematic pairs A, B, and C, assuming that the dyad is acted on only by inertial forces and moments. The elements are homogeneous straight bars with masses $m_2 = 1$ kg and $m_3 = 1.25$ kg. For simplicity, the weight of the bars can be neglected.

Solution:

The assisted graphical method As noted in the previous examples (Section 3.5.2), the grapho-analytical method follows the same steps as the assisted graphical method. The difference is only in the scale representation of the linear or vector parameters. Here we will only present the assisted graphical method based on AutoLisp functions; the other constructions result from it. We will also present the parameters calculated, to verify the function and show that it is correct. The parameters used in the function will be drawn and annotated.

Figure 3.19 isolates the elements of the RRT dyad in Section 2.5.1. This uses the representation in Figure 2.14b and the polygon of accelerations in Figure 2.14d. To determine the torsor of the forces of inertia, we will have to determine the accelerations of the centres of weight of elements 2 and 3; the elements being homogeneous, their centres of weight are situated at the centre of the bars.

The acceleration of point G is determined using the theorem of similarity, point g' in the plane of accelerations being situated at the middle of the segment jb'_2. Therefore, $a_G = \dfrac{a_{B_2}}{2} = 101.96928$ m/s². The acceleration of point E, a point in element 3, is determined from the motion of the points in plane 3 relative to the points in plane 4 (the plane containing the slider)

$$\mathbf{a}_{E_3} = \mathbf{a}_{E_4} + \mathbf{a}^C_{E_3 E_4} + \mathbf{a}_{E_3 E_4}. \tag{3.47}$$

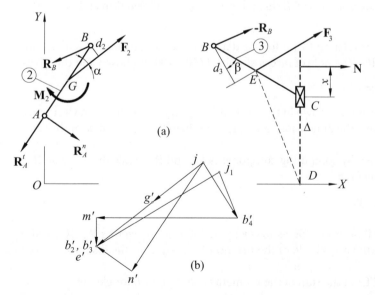

(a)

(b)

Figure 3.19 The isolation of the elements of the dyad RRT and the polygon of accelerations (the graphical method): (a) isolation of the RRR dyad; (b) polygon of accelerations.

The relative motion between the planes is the same, no matter the points of the plane: $\mathbf{a}_{E_3 E_4}^C =$ $\mathbf{a}_{B_3 B_4}^C$, $\mathbf{a}_{E_3 E_4} = \mathbf{a}_{B_3 B_4}$; the acceleration of point E in plane 4 is determined using the *Rivals'* relation:

$$\mathbf{a}_E = \mathbf{a}_D + \mathbf{a}_{ED}^v + \mathbf{a}_{ED}^\tau, \tag{3.48}$$

where $\mathbf{a}_D = \mathbf{0}$ and $\mathbf{a}_{ED}^\tau = \mathbf{0}$. This gives the vector relation for the determination of the acceleration of point E

$$\mathbf{a}_{E_3} = \mathbf{a}_{ED}^v + \mathbf{a}_{B_3 B_4}^C + \mathbf{a}_{B_3 B_4}. \tag{3.49}$$

Since in the polygon of acceleration we have already represented the accelerations $\mathbf{a}_{B_3 B_4}^C$ ($\mathbf{b}'_4 \mathbf{m}'$) and $\mathbf{a}_{B_3 B_4}$ ($\mathbf{m}' \mathbf{b}'_2$), we will mark a new pole j_1, situated at a distance $\omega_4^2 ED$ from point b'_4, in the direction of the straight line ED, as in the polygon of the bottom part of Figure 3.19. This gives $a_E = j_1 b'_2 = 217.2841$ m/s^2.

The magnitudes of the inertial forces can now be determined:

$$F_2 = m_2 a_G = 101.969 \text{ N}, \ F_3 = m_3 a_{E_3} = 217.284 \text{ N},$$

as can the moments of the forces of inertia:

$$M_2 = J_2 \varepsilon_2 = 2.70863 \text{ Nm}, \ M_3 = J_3 \varepsilon_4 = 0.$$

For element 2, the equation of the moments about point B gives:

$$R_A^n = \frac{M_2 - F_2 d_2}{AB}, \tag{3.50}$$

distance d_2 being the distance between point B and the support of the force F_2.

To determine distance d_2 the following AutoLisp procedure is followed:

1) Determine angle α as the difference between the angle made by the straight line AB and the horizontal direction, and the angle made by the force $\mathbf{F}_2$ and the horizontal direction (that is, the angle between the vector $\mathbf{b}'_4 \mathbf{j}$ in the plane of accelerations and the horizontal direction); this gives $\alpha = 0.324419$ rad $= 18.5878°$.
2) Determine the distance d_2 from $d_2 = GB \sin \alpha$; this gives $d_2 = 0.0796895$ m and hence $R_A^n = -10.8345$ N.

Now we will use the vector relation (3.45) for the determination of the normal reaction and the tangential component $\mathbf{R}_A^t$. The pole of forces at point f (400, 100) is chosen and this gives:

$$N = 169.488 \text{ N}, \ R_A^t = 245.919 \text{ N}.$$

The vector polygon of forces is as shown in Figure 3.20. The polygon is completed with arrows and notations. We have drawn the dimensions and noted on them the significance of the values obtained for the reactions.

The reaction $\mathbf{R}_A$ is obtained by computing components $\mathbf{R}_A^n$ and $\mathbf{R}_A^t$, while the reaction $\mathbf{R}_B$ is obtained using the vector relation

$$\underline{\underline{\mathbf{R}_A^t}} + \underline{\underline{\mathbf{R}_A^n}} + \underline{\underline{\mathbf{F}_2}} + \mathbf{R}_B = \mathbf{0}, \tag{3.51}$$

which comes from the equilibrium of forces for element 2. Since the vectors $\mathbf{R}_A^t$, $\mathbf{R}_A^n$, and $\mathbf{F}_2$ are already in the polygon, the magnitude of the reaction $\mathbf{R}_B$ is equal to the distance between $\mathbf{R}_A^t$ and $\mathbf{F}_2$.

Distance x is determined from the sum of the moments about point B for element 3

$$x = y_B - y_C - \frac{F_3 d_3}{N}. \tag{3.52}$$

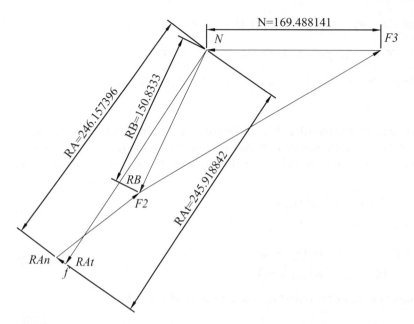

Figure 3.20 The polygon of forces for the RRT dyad obtained using the assisted graphical method.

We have denoted by d_3 the length of the straight line passing through point B and perpendicular to the support of the force $\mathbf{F}_3$. Proceeding as in the case of the distance d_2, we first determine angle $\beta = 0.523858$ rad $= 30.0148°$, and then the distance d_3, using $d_3 = BE \cos \beta = 0.259769$ m. This gives the value $x = -0.11628$ m.

The values were obtained using the AutoLisp function *Forces_RRT*, which is presented below. The function is called (*Forces_RRT*) before the last line of the function *Dyad_RRT.lsp*, which was presented in Section 2.5.4. It is positioned before the *RRT* function.

```
(Defun Forces_RRT ()
(Command "Zoom" "-100,80" "800,500")
(Command "-Layer" "N" "Forces" "C" "6" "Forces" "S" "Forces" "" "")
(Setq m2 1.0 m3 1.25 fsmall(List 400 100) J2(/ (* m2 AB AB) 12) J3(/
   (* m3 BC BC) 12) GCapital(List (/ (+ xa xb) 2) (/ (+ ya yb) 2))
   ECapital(List (/ (+ xb xc) 2) (/ (+ yb yc) 2)) DCapital (List xD yD))
(Setq jsmall1(Polar bprime4 (Angle DCapital ECapital) (* om4 om4
   (Distance ECapital DCapital))) accE(Distance jsmall1 bprime23) accG(/
   (Distance jsmall bprime23) 2))
(Setq F2(* m2 accG) F3(* m3 accE) M2(* J2 eps2) M3 0)
(Setq angF2(Angle bprime23 jsmall) angF3(Angle bprime23 jsmall1)
   angRAn(+ phiAB (* Pi 1.5)) angRAt(+ phiAB Pi) alpha(- phiAB angF2)
   beta(- (Angle (List xB yB) (List xC yC)) angF3 (* Pi 1.5)))
(Setq d2(* (/ AB 2) (Sin alpha)) d3(* (/ BC 2) (Cos beta)) RAn(/ (- M2
   (* F2 d2)) AB))
(Setq pRAn(Polar fsmall angRAn RAn) pF2(Polar pRAn angF2 F2) pF3(Polar
pF2 angF3 F3) perD(Polar pF3 (+ phiDelta (/ Pi 2)) 10) parAB(Polar
fsmall phiAB 10) int(Inters pF3 perD fsmall parAB Nil) RAt(Distance
fsmall int) N(Distance pF3 int) RB(Distance int pF2) RA(Distance int
pRAn) x(- (- yB yC) (/ (* F3 d3) N)))
(Command "Line" fsmall pRAn pF2 pF3 int "C")
(Setq l(/ RAt 30) g(* l 0.353))
```

```
(Arrows pRAn fsmall) (WritingF fsmall pRAn 32 82 102 65 32 110)
(Arrows pF2 pRAn)
(Arrows pF3 pF2) (WritingF pF2 pF3 32 32 70 70 50 51)
(Arrows int pF3)
(Arrows fsmall int) (WritingF int fsmall 32 82 78 65 32 116)
)
```

The assisted analytical method For the RRT dyad in Figure 2.14 (Section 2.5.1) the isolation in Figure 3.21 is obtained. For the determination of the torsor of the inertial forces we proceed as in the case of the graphical method and determine the accelerations of points G and E using the *Rivals'* relations

$$\begin{cases} a_{Gx} = a_{Ax} - \varepsilon_2 \left(y_G - y_A \right) - \omega_2^2 \left(x_G - x_A \right), \\ a_{Gy} = a_{Ay} + \varepsilon_2 \left(x_G - x_A \right) - \omega_2^2 \left(y_G - y_A \right), \end{cases} \tag{3.53}$$

$$\begin{cases} a_{Ex} = a_{Bx} - \varepsilon_4 \left(y_E - y_B \right) - \omega_4^2 \left(x_E - x_B \right), \\ a_{Ey} = a_{By} + \varepsilon_4 \left(x_E - x_B \right) - \omega_4^2 \left(y_E - y_B \right). \end{cases} \tag{3.54}$$

The sum of the projections of forces onto the axis Δ gives for element 2

$$V_B = F_{3y}, \tag{3.55}$$

while from the sum of moments about point A, for the same element, we deduce

$$H_B = \frac{V_B \left(x_B - x_A \right) + M_2 - F_{2x} \left(y_G - y_A \right) + F_{2y} \left(x_G - x_A \right)}{y_B - y_A}. \tag{3.56}$$

The equilibrium of the forces onto the axes OX and OY, for element 2, gives

$$H_A = H_B - F_{2x}, \tag{3.57}$$

$$V_A = -V_B - F_{2y}, \tag{3.58}$$

and from the sum of forces onto the axis OX, for element 3, we get

$$N = H_B - F_{3x}. \tag{3.59}$$

The distance x is obtained from the sum of the moments about point C for element 3:

$$x = \frac{M_3 + V_B \left(x_C - x_B \right) + H_B \left(y_B - y_C \right) - F_{3x} \left(y_E - y_C \right) - F_{3y} \left(x_C - x_E \right)}{N}. \tag{3.60}$$

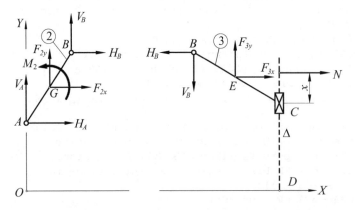

Figure 3.21 The isolation of the elements of the RRT dyad (the analytical method).

Based on these relations we have writtern the procedure given in the companion website of this book and which is called from the *Dyad_RRT* procedure. Calling the procedure with the known parameters

$$\text{Forces\_RRT}\left(Pi/2, 2 * Pi/3, 0.5, 0.6, 0.8, 0, 0, 0.4, Lambda, 0, 0, aBx, aBy,\right.$$
$$\left.Omega2, -10, Epsilon2, 0, HaA, VaA, RaA, HaB, VaB, RaB, Na, xa\right)$$

the following values are obtained

$$H_A = -146.8745, \quad V_A = -197.5382, \quad R_A = 2461.1574$$
$$H_B = 65.7640, \quad V_B = 135.7416, \quad R_B = 150.8333$$
$$N = -169.4881, \quad x = -0.1163$$

As expected, there are no differences between the two methods. This again highlights the simplicity and similarity to well-known vector methods of the assisted graphical method.

3.5.4 Determination of the Reactions for the RTR Dyad

Formulation of the problem
Determine the reactions in the kinematic pairs of the RTR dyad in Figure 3.22, knowing that forces $\mathbf{F}_2$ and $\mathbf{F}_3$ and the torques of moments $\mathbf{M}_2$ and $\mathbf{M}_3$ act upon the elements. Assume that the kinematic analysis has been performed.

The grapho-analytical method   At point A, the orthogonal axes Ax' and Ay' are constructed so that the axis Ax' is parallel to the axis Δ of the slider B. The elements are isolated (Figure 3.22). The equation of projections onto the direction of the axis Δ for the two components gives the reactions $\mathbf{R}_A^t$ and $\mathbf{R}_C^t$. The equation of the moments about point A for the entire dyad gives the normal component $\mathbf{R}_C^n$, while the equation of projections onto the direction of the axis Ay' for element 3 gives the reaction $\mathbf{N}$.

The distance ξ is obtained from the equation of the moments about point B for element 3, and the reaction $\mathbf{R}_A^n$ is determined from the equation of projections onto the axis Ay' for element 2.

The analytical method   In this method uses the isolation of the elements in Figure 3.23. The equations of the moments about point A for element 2, and about point C for element 3 are written, giving a system of two linear equations:

$$N\left[(X_B - X_A)\cos\varphi + (Y_B - Y_A\sin\varphi)\right] + N\xi = -M_2 + F_{2x}\left(Y_D - Y_A\right) - F_{2y}\left(X_D - X_A\right),$$
$$-N\left[(X_B - X_A)\cos\varphi + (Y_B - Y_A)\sin\varphi\right] - N\xi = -M_3 + F_{3x}\left(Y_E - Y_C\right) - F_{3y}\left(X_E - X_C\right),$$

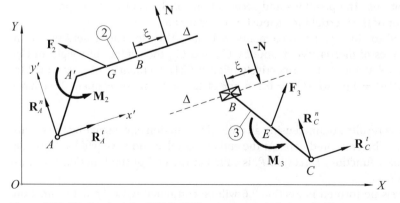

Figure 3.22 The forces at the RRT dyad (the grapho-analytical method).

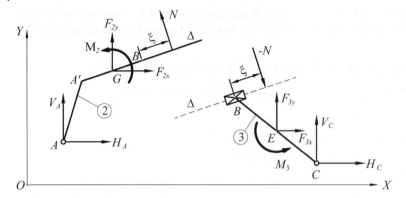

Figure 3.23 The forces at the RTR dyad (the analytical method).

from which:

$$N = \frac{-M_2 - M_3 + F_{2x}\left(Y_D - Y_A\right) + F_{3x}\left(Y_E - Y_C\right) - F_{2y}\left(X_D - X_A\right) - F_{2y}\left(X_E - X_C\right)}{\left(X_C - X_A\right)\cos\varphi + \left(Y_C - Y_A\right)\sin\varphi},$$

$$\xi = \frac{-M_2 + F_{2x}\left(Y_D - Y_A\right) - F_{2y}\left(X_D - X_A\right) - N\left[\left(X_D - X_A\right)\cos\varphi + \left(Y_B - Y_A\right)\sin\varphi\right]}{N}.$$

The reactions at points A and C result from the equations of equilibrium for each element

$$H_A = N\sin\varphi - F_2, \quad V_A = -N\cos\varphi - F_{2y},$$
$$H_C = -N\sin\varphi - F_{3x}, \quad V_C = N\cos\varphi - F_{3y}. \tag{3.61}$$

Numerical example

Problem: For the mechanism with the RTR dyad in Figure 2.21, and considering the numerical and kinematic data in Section 2.6.1, determine the reactions in the kinematic pairs. Assume that the elements are homogeneous, having masses $m_{21} = 0.2$ kg (the mass of the bend AA'), $m_{22} = 1$ kg (the mass of the slider $A'A^*$), and $m_3 = 1$ kg (the mass of element BC). The dyad is assumed to be acted on only by the forces and moments of inertia.

Solution:

The assisted graphical method: The kinematic analyses in Sections 2.6.1 (the analytical method) and 2.6.4 (the assisted graphical method) give the positions of points A and B, and the angular velocities and accelerations. The positions and accelerations of the centres of weight for the calculation of the torsor of the inertial forces must be determined.

Element 2 is assumed to consist of two elements, AA' and $A'A^*$, for easier determination of the torsor of the forces of inertia. We denote by C_{21} and C_{22} the centres of weight of the homogeneous elements AA' and $A'A^*$, respectively (Figure 3.24). Element $A'A^*$ is assumed to have length equal to $A'A^* = 1$ m, so $AC_{21} = 0.1$ m and $A'C_{22} = 0.5$ m; for element 3 we have $CC_3 = \dfrac{0.4}{\sqrt{3}}$ m.

Now, we will show the results obtained using an AutoLisp function encapsulating the calculation algorithm that will be presented here. At the end of the subsection, we will list the code of the function. This new function, *Forces_RTR*, is called at the end of the function that was presented in Section 2.6.4.

The kinematic analysis performed in Section 2.6 will be continued with the determination of the centres of weight. For element 2, which has plane-parallel motion, we apply the *Rivals'*

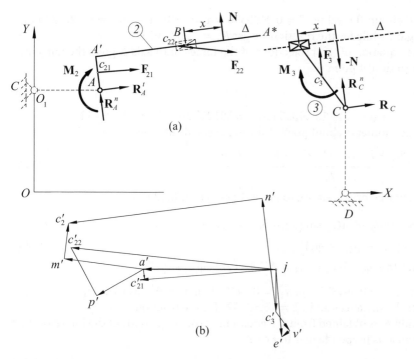

Figure 3.24 The isolation of the elements of the RTR dyad and the polygon of accelerations (the graphical method): (a) isolation of the RTR dyad; (b) polygon of accelerations.

relation and obtain

$$\mathbf{a}_{C_{21}} = \mathbf{a}_A + \mathbf{a}^{\nu}_{C_{21}A} + \mathbf{a}^{\tau}_{C_{21}A}, \ \mathbf{a}_{C_{22}} = \mathbf{a}_A + \mathbf{a}^{\nu}_{C_{22}A} + \mathbf{a}^{\tau}_{C_{22}A}. \tag{3.62}$$

The bottom part of the Figure 3.24 shows the completion of the polygon of accelerations in Figure 2.21 with the polygons $ja't'c'_{21}$ for the determination of the acceleration $\mathbf{a}_{C_{21}}$ ($\mathbf{jc'}_{21}$) and $ja'p'c'_{22}$ for the determination of the acceleration $\mathbf{a}_{C_{22}}$ ($\mathbf{jc'}_{22}$). Similarly, the absolute acceleration of point C_3 can be determined from

$$\mathbf{a}_{C_3} = \mathbf{a}_E = \mathbf{a}_C + \mathbf{a}^{\nu}_{EC} + \mathbf{a}^{\tau}_{EC}. \tag{3.63}$$

The polygon $jc'_3v'e'$ gives $\mathbf{je'}$; that is, the acceleration $\mathbf{a}_{C_3}$. This gives the values $a_{C_{21}} = 164.175$ m/s², $a_{C_{22}} = 230.455$ m/s², and $a_{C_3} = 79.7741$ m/s², from which the torsor of the inertial forces can be derived:

$$F_{21} = m_{21}a_{C_{21}}, \ F_{22} = m_{22}a_{C_{22}}, \ F_3 = m_3 a_{C_3},$$
$$M_2 = (J_{21} + J_{22})\, \varepsilon_2, \ M_3 = J_3 \varepsilon_2, \tag{3.64}$$

where $J_{21} = \dfrac{m_{21}AA'^2}{12}, J_{22} = \dfrac{m_{22}A'A^{*2}}{12}$, and $J_3 = \dfrac{m_3 CB^2}{12}$.

After the isolation of the elements of the dyad (Figure 3.24), the equation of the projections of the forces that act upon element 2 onto axis Δ gives the tangential component of the reaction at point A:

$$R^t_A = -F_{21} \cos \delta - F_{22} \cos \chi, \tag{3.65}$$

where the angles δ and χ are $\delta = \varphi_2 - \varphi_{F_{21}}$ and $\chi = 2\pi - \varphi_{F_{22}} + \varphi_2$, respectively.

We denote by $\varphi_{F_{21}}$ and $\varphi_{F_{22}}$ the angles with the horizontal axis made by the inertial forces $\mathbf{F}_{21}$ and $\mathbf{F}_{22}$, respectively. From now on, we will use the notation φ_F for the angle made by force

F with axis OX. We obtain the values $\delta = 0.0495396$ rad $= 2.83841°$ and $\chi = 0.227529$ rad $= 13.0365°$ for the angles, while for the reaction we get $R_A^t = -262.165$ N.

For element 3, the equation of the projections of the forces onto axis Δ gives the tangential component of the reaction at point C

$$R_C^t = -F_3 \cos \nu, \tag{3.66}$$

where $\nu = \varphi_{F_3} - \varphi_2$. This gives $\nu = 1.534355$ rad $= 87.9121°$ and $R_C^t = -2.90645$ N.

The equation of the moments about point A for the entire dyad gives

$$R_C^n = \frac{M_{21} + M_{22} + F_{21}d_{21} + F_{22}d_{22} - F_3 d_3 - R_C^t d_4}{d_5}, \tag{3.67}$$

where the distances of the forces relative to point A are determined from:

$$\begin{aligned}
&d_{21} = \frac{AA'}{2} \cos \delta, \quad d_{22} = AC_{22} \sin \left(2\pi - \varphi_{F_{22}} - \varphi_{AC_{22}}\right), \\
&d_3 = AC_3 \sin \left(\varphi_{F_3} - \varphi_{AC_3} + \varphi_2\right), \\
&d_4 = AC \sin \left(2\pi - \varphi_{AC} + \varphi_2\right), \quad d_5 = AC \cos \left(2\pi - \varphi_{AC} + \varphi_2\right).
\end{aligned} \tag{3.68}$$

This gives the values $d_{21} = 0.0998773$ m, $d_{22} = 0.247534$ m, $d_3 = 0.665113$ m, $d_4 = 0.2$ m, and $d_5 = 0.781025$ m for the distances and $R_C^n = 16.8873$ N for the reaction.

The normal reaction N is obtained from the equation of the projections of the forces that act upon element 3 on an axis perpendicular to axis Δ:

$$N = R_C^n + F_3 \sin \nu, \tag{3.69}$$

giving the value $N = 96.6085$ N.

For element 3, the equation of the moments about point B gives distance x:

$$x = \frac{F_3 d_6 + R_C^n d_7 + R_C^t d_8 - M_3}{N}. \tag{3.70}$$

The distances to point B are as follows:

$$d_6 = \frac{BC}{2} \sin \left(\varphi_{CB} - \varphi_{F_3}\right), \quad d_7 = BC \cos \beta, \quad d_8 = BC \sin \beta. \tag{3.71}$$

This gives the values $d_6 = 0.12268$ m, $d_7 = 0.23094$ m, $d_8 = 0.4$ m, and $x = 0.119979$ m.

The normal component of the reaction R_A is obtained from the equation of the projections of the forces that act upon element 3 onto an axis that is perpendicular to axis Δ:

$$R_A^n = F_{22} \sin \chi - F_{21} \sin \delta - N, \tag{3.72}$$

and its value is $R_A^n = -75.8912$ N.

The normal and tangential components give the values

$$R_A = 272.0417 \text{ N}, \quad R_C = 17.1356 \text{ N}.$$

The content of the AutoLisp function is given below.

```
(Defun Forces_RTR ()
(Setq m21 0.2 m22 1.0 m3 1.0 lAAsec 1.0)
(Setq xAstar(+ xAsec (* lAAsec (Cos phi))) yAstar(+ yAsec (* lAAsec
(Sin phi))) pctAstar(List xAstar yAstar) xC22(/ (+ xAsec xAstar) 2)
yC22(/ (+ yAsec yAstar) 2) pctC22(List xC22 yC22) xC3(/ (+ xB xC) 2)
yC3(/ (+ yB yC) 2) pctC3(List xC3 yC3) pctA(List xA yA) pctB(List xB
yB) pctC(List xC yC) pctAsec(List xAsec yAsec) tprime(Polar aprime
(Angle pctAsec pctA) (* om2 om2 (/ lAAp 2))) c21prime(Polar tprime (+
```

```
(Angle tprime aprime) (/ Pi 2)) (* eps2 (/ lAAp 2))) pprime(Polar aprime
(Angle pctC22 pctA) (* om2 om2 (Distance pctC22 pctA))) c22prime(Polar
pprime (+ (Angle pprime aprime) (/ Pi 2)) (* eps2 (Distance pctC22
pctA))) vprime(Polar cprime3 (Angle pctC3 pctC) (* om2 om2 (/ lBC 2)))
eprime(Polar vprime (+ (Angle vprime cprime3) (/ Pi 2)) (* eps2 (/
lBC 2))) aC21(Distance c21prime jsmall) aC22(Distance c22prime jsmall)
aC3(Distance eprime jsmall))
(Command "Line" aprime tprime c21prime jsmall "")
(Arrows tprime aprime) (Arrows c21prime jsmall)
(Command "Line" aprime pprime c22prime jsmall "")
(Arrows pprime aprime) (Arrows c22prime jsmall)
(Command "Line" cprime3 vprime eprime jsmall "")
(Arrows vprime cprime3) (Arrows eprime jsmall)
(Arrows eprime vprime)
(Setq F21(* m21 ac21) F22(* m22 aC22) F3(* m3 aC3) J21(/ (* m21 lAAp
lAAp) 12) J22(/ (* m22 lAAsec lAAsec) 12) J2(+ J21 J22) J3(/ (* m3 lBC
lBC) 12) M2(* j2 eps2) M3(* J3 eps2))
(Setq angF21(Angle c21prime jsmall) angF22(Angle c22prime jsmall) angF3
(Angle eprime jsmall))
(Setq alphaF(- phi angF21) betaF(+ phi (- (* 2 Pi) angF22))
  gammaF(- angF3 phi))
(Setq RAt(+ (* F21 (Cos alphaF) -1) (* F22 (Cos betaF) -1)) RCt(* f3
(Cos gammaF) -1))
(Setq d21(* (/ lAAp 2.0) (Cos alphaF)) d22(* (Distance pctA pctC22) (Sin
(+ (Angle pctA pctC22) (* 2 Pi) (* -1 angF22)))) d3(* (Distance pctA
pctC3) (Sin (- angF3 (Angle pctA pctC3)))) d4(* (Distance pctA pctC)
(Sin (+ (* 2 Pi) (* -1 (Angle pctA pctC)) phi))) d5(* (Distance pctA
pctC) (Cos (+ (* 2 Pi) (* -1 (Angle pctA pctC)) phi))) RCn(/ (+ M2 M3
(* F21 d21) (* F22 d22) (* -1 F3 d3) (* -1 RCt d4)) d5) N(+ RCn (* F3
(Sin gammaF))) RC(Sqrt(+ (* RCt Rct) (* RCn RCn))) d6(* lBC 0.5 (Sin
(- (Angle pctC pctB) angF3))) d7(* lBC (Cos betaRad)) d8(* lBC (Sin
betaRad)) x(/ (+ (* F3 d6) (* RCn d7) (* RCt d8) (* M3 -1) ) N)
RAn(- (* F22 (Sin betaF)) (* F21 (Sin alphaF)) N) RA(Sqrt(+ (* RAt RAt)
(* RAn RAn))))
)
```

Comparing to the other two AutoLisp function used for the determination of the reactions, this function has more instructions. These are easily written because they are called more often: for the determination of the angles of the forces relative to the horizontal direction (*Angles*), or for the determination of the distances from a point to a force (*Distance*). From this point of view, in the absence of the polygons of forces, things are simpler, the analysis consisting of solving some simple problems of planar geometry.

The assisted analytical method   As in the case of the graphical method, for the solution of the problem, we will firstly determine the coordinates of point A'

$$x_{A'} = x_A + AA' \cos\left(\frac{\pi}{2} + \varphi_2\right), \ y_{A'} = y_A + AA' \sin\left(\frac{\pi}{2} + \varphi_2\right), \tag{3.73}$$

and the coordinates of the centres of weight

$$x_{C_{21}} = \frac{x_A + x_{A'}}{2}, \ y_{C_{21}} = \frac{y_A + y_{A'}}{2},$$

$$x_{C_{22}} = x_{A'} + \frac{AA'}{2} \cos\varphi_2, \ y_{C_{22}} = y_{A'} + \frac{AA'}{2} \sin\varphi_2, \tag{3.74}$$

$$x_{C_3} = x_C + \frac{BC}{2} \cos\left(\pi - \beta + \varphi_2\right), \ y_{C_3} = y_C + \frac{BC}{2} \sin\left(\pi - \beta + \varphi_2\right).$$

The determination of the accelerations of the centres of weight is made using:

$$\begin{cases} a_{C_{21}x} = a_{Ax} - \varepsilon_2 \left(y_{C_{21}} - y_A\right) - \omega_2^2 \left(x_{C_{21}} - x_A\right), \\ a_{C_{21}y} = a_{Ay} + \varepsilon_2 \left(x_{C_{21}} - x_A\right) - \omega_2^2 \left(y_{C_{21}} - y_A\right), \end{cases}$$
$$\begin{cases} a_{C_{22}x} = a_{Ax} - \varepsilon_2 \left(y_{C_{22}} - y_A\right) - \omega_2^2 \left(x_{C_{22}} - x_A\right), \\ a_{C_{22}y} = a_{Ay} + \varepsilon_2 \left(x_{C_{22}} - x_A\right) - \omega_2^2 \left(y_{C_{22}} - y_A\right), \end{cases} \qquad (3.75)$$
$$\begin{cases} a_{C_3x} = a_{Cx} - \varepsilon_2 \left(y_{C_3} - y_C\right) - \omega_2^2 \left(x_{C_3} - x_C\right), \\ a_{C_3y} = a_{Cy} + \varepsilon_2 \left(x_{C_3} - x_C\right) - \omega_2^2 \left(y_{C_3} - y_C\right). \end{cases}$$

The forces and the moment of inertia can now be determined from:

$$\begin{aligned} & F_{21x} = -m_{21}a_{C_{21}x}, \ F_{21y} = -m_{21}a_{C_{21}y}, \\ & F_{22x} = -m_{22}a_{C_{22}x}, \ F_{22y} = -m_{22}a_{C_{22}y}, \\ & F_{3x} = -m_3 a_{C_3x}, \ F_{3y} = -m_3 a_{C_3y}, \\ & M_2 = - \left(J_{21} + J_{22}\right) \varepsilon_2, \ M_3 = -J_3 \varepsilon_2. \end{aligned} \qquad (3.76)$$

The elements are isolated, as in Figure 3.25. For element 2, from the equations of the moments about point A, and for element 3, the equations of the moments about point C give a system of two equations with two unknowns, N and x. The solutions of the system are

$$N = \frac{A_1 + B_1}{BC \cos \beta + \lambda}, \ x = \frac{N \cdot BC \cos \beta - B_1}{N}, \qquad (3.77)$$

where

$$\begin{aligned} A_1 & = -M_2 + F_{22x} \left(y_{C_{22}} - y_A\right) - F_{22y} \left(x_{C_{22}} - x_A\right), \\ & + F_{21x} \left(y_{C_{21}} - y_A\right) + F_{21y} \left(x_A - x_{C_{21}}\right), \\ B_1 & = -M_3 + F_{3x} \left(y_{C_3} - y_C\right) + F_{3y} \left(x_C - x_{C_3}\right). \end{aligned} \qquad (3.78)$$

The reactions at point A result from the equations of equilibrium for element 2

$$H_A = N \sin \varphi_2 - F_{21x} - F_{22x}, \ V_A = -N \cos \varphi_2 - F_{21y} - F_{22y}, \qquad (3.79)$$

while the reactions at point C is obtained from the equations of equilibrium for element 3

$$H_C = -N \sin \varphi_2 - F_{3x}, \ V_C = N \cos \varphi_2 - F_{3y}. \qquad (3.80)$$

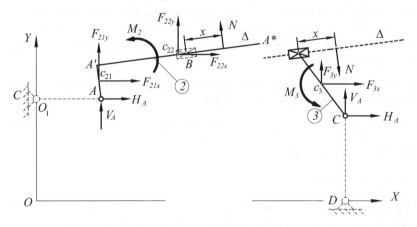

Figure 3.25 The isolation of the elements of the RTR dyad (the assisted analytical method).

For the numerical results we have written a procedure in Turbo Pascal. The procedure is called *Forces_RTR* and the code can be found in the companion website of this book. The procedure works inside the procedure *Dyad_RTR* (see the Section 2.6.2), and it is called before the last line of the program.

The procedure uses (3.73)–(3.80) and it is called with the data

Forces_RTR $(0.4, 0.6, 0.2, 0.8/\text{sqrt}(3), 1.2, 0.5, Pi/3, Phi2, -160, 0, 0, -50,$

$Lambda, Omega2, Epsilon2, Nr, xr, HAr, VAr, HCr, VCr)$

The calculation program gives the values:

$N = 96.608459, \ x = 0.119979,$

$H_A = -250.922984, \ V_A = -105.092133, \ R_A = 272.041725,$

$H_C = -5.011017, \ V_C = 16.386507, \ R_C = 17.135575.$

3.5.5 Determination of the Reactions for the TRT Dyad

Formulation of the problem
Determine the reactions in the kinematic pairs of the TRT dyad in Figure 3.26, knowing that the elements are acted up by forces $\mathbf{F}_2$ and $\mathbf{F}_3$ and torques of moments $\mathbf{M}_2$ and $\mathbf{M}_3$. Assume that the kinematic analysis has been performed.

The grapho-analytical method   The vector equation of equilibrium for the dyad

$$\underset{=}{\mathbf{F}_2} + \underset{=}{\mathbf{F}_3} + \underset{\perp \Delta_4}{\mathbf{N}_3} + \underset{\perp \Delta_1}{\mathbf{N}_2} = \mathbf{0}, \tag{3.81}$$

gives the normal reactions $\mathbf{N}_2$ and $\mathbf{N}_3$, while the equilibrium equation of element 2 gives the reaction at point B :

$$\mathbf{R}_B = -\mathbf{F}_2 - \mathbf{N}_2. \tag{3.82}$$

Now, from the equations of the moments about points A and C for elements 2, and 3, respectively, the distances ξ_2 and ξ_3 can be found.

The analytical method   Projecting the equation (3.81) onto the axes gives the equations

$$N_2 \sin \varphi_1 + N_3 \sin \varphi_4 = F_{2x} + F_{3x},$$
$$N_2 \cos \varphi_1 + N_3 \cos \varphi_4 = -F_{2y} - F_{3y}, \tag{3.83}$$

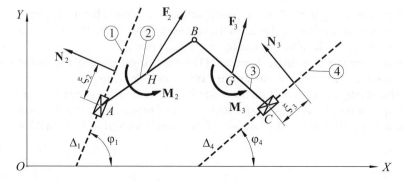

Figure 3.26 The forces in the TRT dyad.

which gives the reactions

$$N_2 = \frac{\left(F_{2x} + F_{3x}\right)\cos\varphi_4 + \left(F_{2y} + F_{3y}\right)\sin\varphi_4}{\sin\left(\varphi_1 - \varphi_4\right)}, \tag{3.84}$$

$$N_3 = \frac{-\left(F_{2x} + F_{3x}\right)\cos\varphi_1 - \left(F_{2y} + F_{3y}\right)\sin\varphi_1}{\sin\left(\varphi_1 - \varphi_4\right)}. \tag{3.85}$$

If we denote by H_B and V_B the reactions in the rotational kinematic pair at point B, (3.82) gives

$$H_B = N_2 \sin\varphi_1 - F_{2x}, \quad V_B = -N_2 \cos\varphi_1 - F_{2y}. \tag{3.86}$$

Finally, the equations of the moments about points A and C, for elements 2, and 3, respectively, give the distances ξ_2 and ξ_3:

$$\begin{aligned}
\xi_2 &= \frac{1}{N_2}\left[F_{2x}\left(Y_H - Y_A\right) - F_{2y}\left(X_H - X_A\right)\right.\\
&\quad \left. + H_B\left(Y_B - Y_A\right) - V_B\left(X_B - X_A\right) - M_2\right],\\
\xi_3 &= \frac{1}{N_3}\left[F_{3x}\left(Y_G - Y_C\right) - F_{3y}\left(X_G - X_C\right)\right.\\
&\quad \left. - H_B\left(Y_B - Y_C\right) + V_B\left(X_B - X_C\right) - M_3\right].
\end{aligned} \tag{3.87}$$

Numerical example

Problem: Knowing the dimensions and the kinematic analysis for the mechanism with the TRT dyad in Section 2.7, determine the reactions in the kinematic pairs assuming that the dyad is acted upon only by the forces and moments of inertia. The elements of the dyad are homogeneous, with masses $m_2 = m_3 = 1$ kg.

Solution:

The assisted graphical method: The assisted graphical method presented in Section 2.7.4 is completed with the determination of the coordinates and accelerations of the centres of weight. Since element 2 is in translational motion, the acceleration $\mathbf{a}_H = \mathbf{a}_{C_2}$. Element 3 has plane-parallel motion, so

$$\mathbf{a}_{C_3} = \mathbf{a}_{G_3} = \mathbf{a}_{G_4} + \mathbf{a}_{G_3 G_4}^C + \mathbf{a}_{G_3 G_4}. \tag{3.88}$$

Consider the relative motion between point G in plane 3, and point G in plane 4. Since the relative motions between the planes are equal, no matter the point, and point G in plane 4 has the acceleration $\mathbf{a}_G = \mathbf{a}_E + \mathbf{a}_{GE}^\nu + \mathbf{a}_{GE}^\tau$, with $\mathbf{a}_E = \mathbf{a}_{GE}^\tau = 0$ and $a_{GE}^\nu = \omega_4^2 GE$:

$$\mathbf{a}_{C_3} = \mathbf{a}_{GE}^\nu + \mathbf{a}_{B_3 B_4}^C + \mathbf{a}_{B_3 B_4}. \tag{3.89}$$

The polygon of accelerations in Figure 2.32 is completed with a new pole j_1, from which the vector polygon given by (3.89) can be drawn. This gives the representation on the left-hand side of Figure 3.27 and the values $a_{C_2} = 2351.49$ m/s^2 and $a_{C_3} = 2372.50$ m/s^2.

In the figure, we have also shown the isolated elements of the dyad. These elements are acted on only by the forces of inertia; the inertial moments are null for both elements.

The values were obtained using an AutoLisp function added to the kinematic calculation performed with the AutoLisp function *Dyad_TRT.lsp* presented in the Section 2.7.4. The function is called *Forces_TRT* and it is called before the last line of the function given in Section 2.7.4 (*Forces_TRT*).

```
(Defun Forces_TRT ()
(Setq m2 1.0 m3 1.0 fsmall(List 7000 8000))
```

```
(Setq xC2 (/ (+ xA xB) 2) yc2 (/ (+ yA yB) 2) xC3 (/ (+ xC xB) 2) yC3 (/
(+ yC yB) 2) pctC3 (List xC3 yC3) pctE (list xE yE) pctC (List xC yc)
pctA (list xA yA))
(Setq jsmall1 (Polar bprime (Angle pctE pctC3) (* om4 om4 (Distance pctE
pctC3))) acc3 (Distance jsmall1 bprime23) acc2 (Distance jsmall bprime23))
(Command "Line" bprime jsmall1 bprime23 "")
(Setq F2 (* m2 acc2) F3 (* m3 acc3))
(Setq pF2 (Polar fsmall (Angle bprime23 jsmall) F2) pF3 (Polar pF2 (Angle
bprime23 jsmall1) F3) pmf (Polar pF3 (Angle pctC BCapital) 100) prf (Polar
fsmall 0 100) int (Inters fsmall prf pF3 pmf Nil) N2 (Distance fsmall int)
N3 (Distance pF3 int))
(Command "Zoom" "-1000,3550" "12000,8000")
(Command "-Layer" "N" "Forcea" "S" "Forces" "" "")
(Command "Color" "6")
(Command "Line" fsmall pF2 pF3 int "C")
(Setq l (/ F2 20) g (* l 0.353))
(Arrows pF2 fsmall) (Arrows pF3 pF2)
(Arrows fsmall int) (Arrows int pf3)
(Writing fsmall pF2 102 70 32 50) ; f - F2
(Writing pF3 int 70 78 51 51) ; F3 - N3
(Writing int fsmall 32 78 32 50) ; - N2
(Setq prB (Polar int (Angle bprime23 jsmall) F2) RB (Distance int pF2))
(Command "Line" int fsmall pF2 "C")
(Arrows int pF2)
(Writing pF2 int 32 82 32 66) ; - RB
(Setq d2 (* lBC (Sin (- (Angle int pF2) (Angle BCapital pctC)))) d3 (* 0.5
lBC (Sin (- (Angle BCapital pctC) (Angle pF2 pF3)))) x3 (/ (- (* RB d2)
(* F3 d3)) N3) d1 (* lAB (Sin (- (Angle BCapital pctA) (Angle pF2 int))))
x2 (/ (- (* RB d1) (* F2 xC2)) N2))
(Command "Color" "7")
)
```

After the declaration of the values assigned to the masses of the elements and to the coordinates of pole f of the forces, f (7000, 8000), the coordinates of the centres of elements 2 and 3 (the centres of weight) are determined. The vector polygon given by (3.89) is found for the

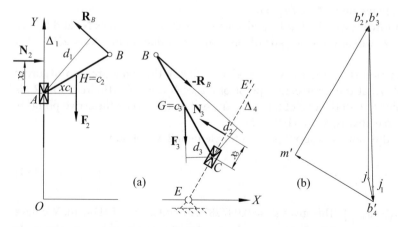

Figure 3.27 The isolation of the elements of the TRT dyad and the polygon of accelerations (the assisted graphical method): (a) isolation of the TRT dyad; (b) polygon of accelerations.

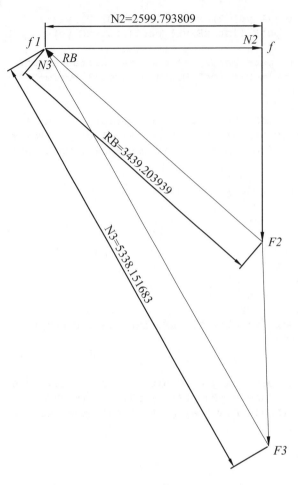

Figure 3.28 The polygon of forces for the TRT dyad.

determination of the acceleration $\mathbf{a}_{C_3}$. This is completed with arrows, as in Figure 3.27. Now, the inertial forces are found using:

$$F_2 = -m_2 a_{C_2}, \ F_3 = -m_3 a_{C_3}. \tag{3.90}$$

The moments of inertia being null, $M_2 = M_3 = 0$.

From the polygon of forces given by (3.81), and starting from pole f, forces $\mathbf{N}_2$ and $\mathbf{N}_3$ are determined. In Figure 3.28 we show the polygon of forces and the values found: $N_2 = 2599.79$ N and $N_3 = 5338.15$ N.

To determine the reaction $\mathbf{R}_B$ using the vector relation (3.82), the same polygon of forces is used with a new pole f_1, at the intersection point of forces $\mathbf{N}_2$ and $\mathbf{N}_3$. The reaction $\mathbf{R}_B$ is obtained by uniting point f_1 with the end of force $\mathbf{F}_2$. The distance between these two points is the magnitude of the force; that is, $R_B = 3439.20$ N.

From the equation of the moments about point A for element 2, we obtain

$$x_2 = \frac{R_B d_1 - F_2 x_{C_2}}{N_2}, \tag{3.91}$$

where $d_1 = AB \sin\left(\varphi_{\mathbf{BA}} - \varphi_{\mathbf{R}_{B_2}}\right)$. This gives $d_1 = 0.0872872$ m and $x_2 = 0.404145$ m. We have maintained the notational conventions for the angles used in Section 3.5.4 ($\varphi_{\mathbf{BA}}$ is the angle between the vector $\mathbf{BA}$ and the horizontal, while $\varphi_{\mathbf{R}_{B_2}}$ is the angle between the horizontal and force $\mathbf{R}_B$ situated on element 2). Similarly, for element 3, from the sum of the moments about point C, we obtain

$$x_3 = \frac{R_B d_2 - F_3 d_3}{N_3}, \tag{3.92}$$

where $d_2 = BC \sin\left(\varphi_{R_{B_3}} - \varphi_{BC}\right)$ and $d_3 = \frac{BC}{2} \sin\left(\varphi_{BC} - \varphi_{F_3}\right)$.

This gives the values $d_2 = 0.226779$ m, $d_3 = 0.164371$ m, and $x_3 = 0.073053$ m.

The AutoLisp function described above gives the polygons of accelerations and forces, as well as the numerical values presented above. As in the previous cases, the vector polygon is completed with arrows and notations. The polygon of forces is constructed in a dedicated layer, *Forces*, and using colour 6 (violet).

The drawing of the polygons is optional, the numerical values being sufficient for the kinematic and kinetostatic analyses. But from a didactical point of view, as well for the confirmation of the correctness of the algorithm, the graphical representation is very useful.

The assisted analytical method   To calculate the forces of inertia and to compare them to the values calculated in Section 2.7.2, we will also have determine:

- the coordinates of points A and C

$$\begin{aligned} x_A &= 0, \; y_A = \lambda_1, \\ x_C &= x_D + \lambda_4 \cos \varphi_4, \; y_C = \lambda_4 \sin \varphi_4 \end{aligned} \tag{3.93}$$

- the coordinates of the centres of weight

$$\begin{aligned} x_{C_2} &= \frac{x_A + x_B}{2}, \; y_{C_2} = \frac{y_A + y_B}{2}, \\ x_{C_3} &= \frac{x_C + x_B}{2}, \; y_{C_3} = \frac{y_C + y_B}{2} \end{aligned} \tag{3.94}$$

- the accelerations of the centres of weight

$$\begin{aligned} a_{C_2 x} &= a_{Bx}, \; a_{C_2 y} = a_{By}, \\ a_{C_3 x} &= a_{Bx} - \omega_4^2 \left(x_{C_3} - x_B\right), \; a_{C_3 y} = a_{By} - \omega_4^2 \left(y_{C_3} - y_B\right) \end{aligned} \tag{3.95}$$

- the forces of inertia

$$\begin{aligned} F_{2x} &= -m_2 a_{C_2 x}, \; F_{2y} = -m_2 a_{C_2 y}, \\ F_{3x} &= -m_3 a_{C_3 x}, \; F_{3y} = -m_3 a_{C_3 y} \end{aligned} \tag{3.96}$$

- the moments of inertia, which are null, $M_2 = M_3 = 0$.

Figure 3.29 shows the isolated elements of the TRT dyad. Taking into account the data in the Section 2.7.2, (3.83) becomes:

$$\begin{aligned} N_3 &= \frac{-\left(F_{2y} + F_{3y}\right)}{\sin \varphi_4}, \\ N_2 &= F_{2x} + F_{3x} - N_3 \cos \varphi_4. \end{aligned} \tag{3.97}$$

The components H_B and V_B of the reaction $\mathbf{R}_B$ are

$$H_B = N_2 - F_{2x}, \; V_B = -F_{2y}. \tag{3.98}$$

Using the notation in Figure 3.29, (3.87) becomes

$$\begin{aligned} x_2 &= \frac{F_{2x}\left(y_{C_2} - y_A\right) - F_{2y}\left(x_{C_2} - x_A\right) - H_B\left(y_B - y_A\right) - V_B\left(x_B - x_A\right)}{N_2}, \\ x_3 &= \frac{F_{3x}\left(y_{C_3} - y_C\right) + F_{3y}\left(x_C - x_{C_3}\right) - H_B\left(y_B - y_C\right) - V_B\left(x_C - x_B\right)}{N_3}. \end{aligned} \tag{3.99}$$

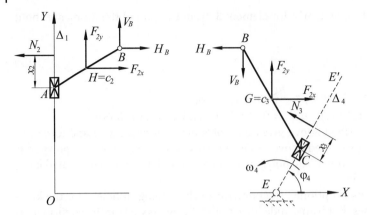

Figure 3.29 The isolation of the elements of the TRT dyad (the assisted analytical method).

To obtain numerical results we incorporated these equations in a procedure written in Turbo Pascal, called *Forces_TRT*. This procedure is called at the end of the procedure *Dyad_TRT*, which was presented in the Section 2.7.3. The code for the procedure can be found in the companion website of this book.

Calling the procedure with the known data:

$$\text{Forces\_TRT} \left(\lambda_1, \lambda_4, 0.6, 0, \frac{\pi}{3}, x_B, y_B, 20, 0, a_{Bx}, a_{By}, N_3, N_2, H_B, V_B, x_2, x_3 \right)$$

gives the following results:

$$N_2 = -2599.7938, \ N_3 = 5338.1517,$$
$$H_B = -2599.7938, \ V_B = 2251.4875, \ R_B = 3439.2039,$$
$$x_2 = 0.404145, \ x_3 = 0.073053.$$

3.5.6 Determination of the Reactions for the RTT Dyad

Formulation of the problem
We want to determine the reactions in the kinematic pairs of the RTT dyad in Figure 3.30, knowing that the elements of the dyad are acted on only by forces $\mathbf{F}_2$ and $\mathbf{F}_3$ and moments $\mathbf{M}_2$ and $\mathbf{M}_3$. We assume that the kinematic analysis has already been performed.

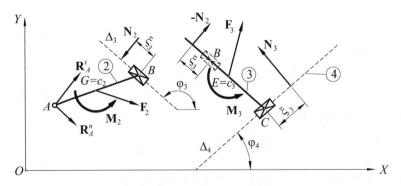

Figure 3.30 The forces in the RTT dyad.

The grapho-analytical method Isolating elements 2 and 3 (Figure 3.30), we can write the vector equation of equilibrium for element 3

$$\underset{\perp\ \Delta_4}{\mathbf{N}_3} + \underset{\perp\ \Delta_3}{\mathbf{N}_2} + \underset{=}{\mathbf{F}_3} = \mathbf{0}. \tag{3.100}$$

Representing this equation in the polygon of forces gives the reactions $\mathbf{N}_2$ and $\mathbf{N}_3$. The reaction at point A decomposes into component $\mathbf{R}_A^t$, parallel to Δ_3, and component $\mathbf{R}_A^n$, normal to Δ_3. The reactions $\mathbf{R}_A^t$ and $\mathbf{R}_A^n$ are obtained from the equation of equilibrium of element 2:

$$\underset{\parallel\ \Delta_3}{\mathbf{R}_A^t} + \underset{\perp\ \Delta_3}{\mathbf{R}_A^n} \underset{=}{\mathbf{F}_2} + \underset{=}{\mathbf{N}_2} = \mathbf{0}. \tag{3.101}$$

The equations of the moments about point B for element 2, and about point C for element 3, gives the distances ξ_2 and ξ_3.

The analytical method Using the notation in Figure 3.30 and projecting the vector relation (3.100) onto the axes Ox and Oy gives the scalar equations

$$N_2 \cos\left(\varphi_3 - \frac{\pi}{2}\right) + N_3 \cos\left(\varphi_4 + \frac{\pi}{2}\right) = -F_{3x},$$
$$N_2 \sin\left(\varphi_3 - \frac{\pi}{2}\right) + N_3 \sin\left(\varphi_4 + \frac{\pi}{2}\right) = -F_{3y}, \tag{3.102}$$

from which:

$$N_2 = \frac{F_{3x}\cos\varphi_4 + F_{3y}\sin\varphi_4}{\sin\left(\varphi_4 - \varphi_3\right)}, \quad N_3 = \frac{F_{3x}\cos\varphi_3 + F_{3y}\sin\varphi_3}{\sin\left(\varphi_4 - \varphi_3\right)}. \tag{3.103}$$

Denoting by H_A and V_A the projections of the reaction at point A onto the axes Ox and Oy, respectively, from the vector equation of equilibrium

$$\mathbf{R}_A + \mathbf{F}_2 + \mathbf{N}_2 = \mathbf{0}, \tag{3.104}$$

gives

$$H_A = -F_{2x} + N_2\sin\varphi_3, \quad V_A = -F_{2y} - N_2\cos\varphi_3. \tag{3.105}$$

Writing the equations of the moments about point B for element 2, and about point C for element 3, we get

$$\xi_2 = \frac{1}{N_2}\left[F_{2x}\left(Y_D - Y_A\right) - F_{2y}\left(X_D - X_A\right)\right.$$
$$\left. + H_A\left(Y_A - Y_B\right) - V_A\left(X_A - X_B\right) - M_2\right],$$
$$\xi_3 = \frac{1}{N_3}\left[N_2\left(\lambda_3 + \xi_2\right) + F_{3x}\left(Y_E - Y_C\right) - F_{3y}\left(X_E - X_C\right) - M_3\right], \tag{3.106}$$

where λ_3 is the distance BC, which was calculated in the kinematic analysis.

Numerical example

Problem: For the mechanism with the RTT dyad in Section 2.8, and knowing the dimensions and the kinematic analysis, determine the reactions in the kinematic pairs, assuming that the dyad is acted upon only by the inertial forces and moments. The elements of the dyad are homogeneous, having masses $m_2 = 0.5$ kg and $m_3 = 1.5$ kg, while the element is of length $CC' = 1.4$ m.

Solution:

The assisted graphical method: To the relations obtained in the assisted graphical method presented in Section 2.8.4 one has to add the relations for the determination of the coordinates and accelerations of the centres of mass. Figure 3.31 shows the isolation of the elements of the RTT dyad shown in Figure 2.34, the numerical data being those set out in Section 2.8.1. The polygon of accelerations is completed with the accelerations of the centres of weight. For element 2 we have

$$\mathbf{a}_{C_2} = \mathbf{a}_A + \mathbf{a}^v_{C_2A} + \mathbf{a}^\tau_{C_2A}, \tag{3.107}$$

with $\mathbf{a}_A = \mathbf{0}$ and $\mathbf{a}^v_{C_2A} = \mathbf{0}$. This gives the magnitude of the acceleration $a_{C_2} = \omega_4^2 C_2 A$, its sense being from C_2 to A. The acceleration $\mathbf{a}_{C_3}$ is the absolute acceleration of point E in plane 3. Between the accelerations of points E_3 and E_2 there is a relation

$$\mathbf{a}_{C_3} = \mathbf{a}_{E_3} = \mathbf{a}_{E_2} + \mathbf{a}^C_{E_3E_2} + \mathbf{a}_{E_3E_2}. \tag{3.108}$$

The relative motions between the planes are equal, no matter which point is considered; point E in plane 2 has the acceleration $\mathbf{a}_E = \mathbf{a}_A + \mathbf{a}^v_{EA} + \mathbf{a}^\tau_{EA}$, with $\mathbf{a}_A = \mathbf{a}^\tau_{EA} = \mathbf{0}$ and $a^v_{EA} = \omega_4^2 EA$. This gives

$$\mathbf{a}_{C_3} = \mathbf{a}^v_{EA} + \mathbf{a}^C_{A_3A_2} + \mathbf{a}_{A_3A_2}. \tag{3.109}$$

The polygon of accelerations is completed with a new pole j_1. The magnitude of the acceleration $\mathbf{a}_{C_3}$ is equal to the distance $j_1 a_{C_3}$, as shown in Figure 3.31.

The AutoLisp function for the determination of the reactions in the kinematic pairs is called *Forces_RTT* and is called before the last instruction of the AutoLisp function *Dyad_RTT.lsp* (see Section 2.8.4). It gives the values $a_{C_2} = 84.8528$ m/s^2 and $a_{C_3} = 703.8386$ m/s^2.

The kinetostatic analysis gives the magnitudes of the inertial forces:

$$F_2 = m_2 a_{C_2}, \quad F_3 = m_3 a_{C_3}. \tag{3.110}$$

The inertial moments being null, $M_2 = M_3 = 0$. The vector relation (3.100), and the pole of the forces having the coordinates $f(800, 800)$, gives the values $N_2 = 593.513$ N, and $N_3 = 1218.94$ N.

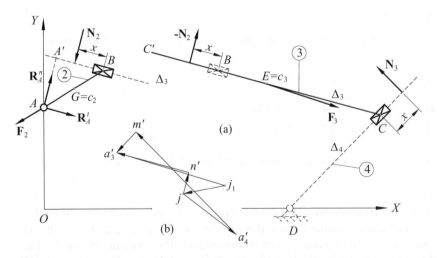

Figure 3.31 The isolation of the elements of the RTT dyad and the polygon of accelerations (the assisted graphical method): (a) isolation of the RTT dyad; (b) polygon of accelerations.

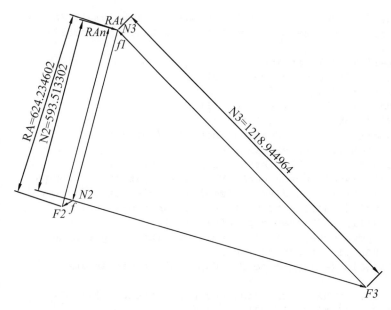

Figure 3.32 The polygon of forces for the RTT dyad.

Because we want a graphical representation of the polygon of forces, it was constructed in a new layer called *Forces*, the colour being violet. Figure 3.32 shows this polygon of forces completed using (3.101), which gives $R_A^n = 623.513$ N and $R_A^t = 30.0$ N.

To use the existing constructions, the new pole f_1 is situated at the intersection of forces $\mathbf{N}_2$ and $\mathbf{N}_3$. After the composition of forces $\mathbf{R}_A^n$ and $\mathbf{R}_A^t$ we get $R_A = 624.235$ N.

From the equation of the moments about point B for element 2 we obtain the expression for distance x_2

$$x_2 = \frac{R_A^n d_1 - R_A^t d_2}{N_2}, \tag{3.111}$$

where $d_1 = AB \sin\left(\varphi_{R_A^n} - \varphi_{AB}\right)$ and $d_2 = AB \sin \alpha$. This gives the values $d_1 = 0.3$ m, $d_2 = 0.3$ m, and $x_3 = 0.3$ m.

For element 3, from the equations of the moments about point C, we obtain the expression for distance x_3

$$x_3 = \frac{N_2\left(BC + x_2\right) - F_3 d_3}{N_3}, \tag{3.112}$$

where $d_3 = EC \sin\left(\varphi_{F_3} - \varphi_{BC}\right)$. This gives $d_3 = 0.0105816$ m and $x_3 = 0.6501$ m.

In the AutoLisp function, described above, the polygons of accelerations and forces are completed with arrows, notations, and dimensions as in Figure 3.32. The content of the AutoLisp function is:

```
(Defun Forces_RTT ()
(Setq m2 0.5 m3 1.5 CCp 1.4 fsmall(List 800 800) xc2(/ (+ xA xB) 2)
yC2(/ (+ yA yB) 2) pctA(List xA yA) pctB(List xB yB) pctC2(List xC2 yC2)
pctCp(Polar CCapital (+ Pi phiG (* -1 betaG)) CCp) xCp(princ(car pctCp))
yCp(princ(car(cdr pctCp))) xC(princ(car CCapital)) yC(princ(car(cdr
CCapital))) xc3(/ (+ xCp xC) 2) yC3(/ (+ yCp yC) 2) pctC3(List xC3 yC3))
```

```
(Setq acc2(* om4 om4 (Distance pctA pctC2)) jsmall1(Polar jsmall (Angle
pctA pctC3) (* om4 om4 (Distance pctC3 pctA))) acc3(Distance jsmall1
aprime3))
(Command "Line" jsmall jsmall1 aprime3 "")
(Arrows jsmall jsmall1) (Arrows aprime3 jsmall1)
(Setq F2(* m2 acC2) F3(* m3 acC3))
(Setq pF3(Polar fsmall (Angle aprime3 jsmall1) F3))
(Setq int1(Polar pF3 (+ phiG (/ Pi 2)) 100))
(Setq int2(Polar fsmall (- (Angle CCapital pctC3) (/ Pi 2)) 100))
(Setq fsmall1(Inters fsmall int2 pF3 int1 Nil))
(Setq N3(Distance pF3 fsmall1) N2(Distance fsmall1 fsmall))
(Setq pF2(Polar fsmall (Angle pctC2 pctA) F2) int1(Polar pF2 (- (Angle
CCapital pctC3) (/ Pi 2)) 10) int2(Polar fsmall1 (Angle CCapital pctC3)
10) int(Inters pF2 int1 fsmall1 int2 Nil) RAn(Distance pF2 int)
RAt(Distance int fsmall1) RA(Distance pf2 fsmall1))
(Setq d1(* (Distance pctA pctB) (Sin (- (Angle pF2 int) (Angle pctA
pctB)))) x2(/ (- (* RAn d1) (* RAt d2)) N2))
(Setq d3(* (/ CCp 2) (Sin (- (Angle pctC3 CCapital) (Angle aprime3
jsmall1)))))
(Setq x3(/ (+ (* F3 d3) (* N2 (+ x2 (Distance pctB CCapital)))) N3))
(Command "Zoom" "-450,250" "2800,1700")
(Command "-Layer" "N" "Forces" "S" "Forces" "" "")
(Command "Color" "6")
(Command "Line" fsmall pF3 fsmall1 "C")
(Setq l(/ F2 2) g(* l 0.353))
(Arrows pF3 fsmall) (Arrows fsmall1 pF3)
(Arrows fsmall fsmall1)
(Command "Line" fsmall pF2 int fsmall1 "")
(Arrows pF2 fsmall) (Arrows int pF2)
(Arrows fsmall1 int)
(WritingF fsmall pF3 32 32 32 70 102 51) ;f - F3
(WritingF fsmall1 int 32 82 78 65 51 110) ;N3 - RAn
(WritingF pF2 int 32 82 70 65 50 116) ;F2 - RAt
(Command "Color" "7")
)
```

The assisted analytical method The kinematic analysis in Section 2.8.2 is completed with the relations for the determination of the coordinates of points C, B, and C'

$$x_C = x_D + \lambda_4 \cos \varphi_4, \ y_C = y_D + \lambda_4 \sin \varphi_4,$$
$$x_B = x_C + \lambda_3 \cos (\pi - \beta + \varphi_4), \ y_B = y_C + \lambda_3 \sin (\pi - \beta + \varphi_4), \tag{3.113}$$
$$x_{C'} = x_C + CC' \cos (\pi - \beta + \varphi_4), \ y_{C'} = y_C + CC' \sin (\pi - \beta + \varphi_4),$$

and the coordinates of the centres of weight

$$x_{C_2} = \frac{x_A + x_B}{2}, y_{C_2} = \frac{y_A + y_B}{2}, \ x_{C_3} = \frac{x_C + x_{C'}}{2}, \ y_{C_3} = \frac{y_C + y_{C'}}{2}. \tag{3.114}$$

Taking into account (3.107) and (3.109) and the comments made for the grapho-analytical method, gives the accelerations of the centres of weight:

$$a_{C_2 x} = -\omega_4^2 (x_{C_2} - x_A), \ a_{C_2 y} = -\omega_4^2 (y_{C_2} - y_A),$$
$$a_{C_3 x} = -\omega_4^2 (x_{C_3} - x_A) - 2\omega_4 v_{A_1 A_2} \cos \left(\frac{\pi}{2} - \beta + \varphi_4 \right) + a_{A_1 A_2} \cos (\pi - \beta + \varphi_4), \tag{3.115}$$
$$a_{C_3 y} = -\omega_4^2 (y_{C_3} - y_A) - 2\omega_4 v_{A_1 A_2} \sin \left(\frac{\pi}{2} - \beta + \varphi_4 \right) + a_{A_1 A_2} \sin (\pi - \beta + \varphi_4).$$

The torsor of the inertial forces is

$$F_{2x} = -m_2 a_{C_2 x}, \; F_{2y} = -m_2 a_{C_2 y}, \; M_2 = 0,$$
$$F_{3x} = -m_3 a_{C_3 x}, \; F_{3y} = -m_3 a_{C_3 y}, \; M_3 = 0. \tag{3.116}$$

Equations 3.103, 3.105, and 3.106 give the components of the reactions, and the distances ξ_2 and ξ_3 of the reactions $\mathbf{N}_2$ and $\mathbf{N}_3$, respectively.

Based on the previous kinematic and kinetostatic relations, we wrote a Turbo Pascal procedure called *Forces_RTT*, which is described in the companion website of the book. It is called just before the last instruction of the procedure *Dyad_RTT*, which was presented in Section 2.8.3.

Running the program with the data as follows:

Forces_RTT $\left(x_A, y_A, x_D, y_D, \lambda_3, \lambda_4, \varphi_4, \beta, v_{A_3 A_2}, a_{Ax}, a_{Ay}, a_{A_3 A_2}, \omega_4, \varepsilon_4, \right.$
$\left. N_{f2}, N_{f3}, H_{fA}, V_{fA}, x_{f2}, x_{f3} \right)$

gives the values

$$N_2 = -593.5133, \; N_3 = 1218.9450,$$
$$H_A = -190.3549, \; V_A = -594.5030, \; R_A = 624.2346,$$
$$x_2 = 0.300, \; x_3 = -0.6501.$$

3.5.7 Determination of the Reactions at the Driving Element

Formulation of the problem
Consider a driving element with rotational motion (Figure 3.33a). Let us denote by $\mathbf{F}_1$ and $\mathbf{M}_1$ the force and the moment, respectively, that act upon the element. Let $\mathbf{M}_e$ be the equilibration moment (the moment that must act upon the driving element to give the status of velocities and accelerations considered in the kinematic calculation).

We want to know the reactions at point O and the equilibration moment. If the driving element is a translational one (Figure 3.33b), we want to know the reaction $\mathbf{N}$, the distance ξ, and the equilibration force $\mathbf{F}_e$ (the reaction $\mathbf{R}_A$ is known).

The grapho-analytical method The reaction $\mathbf{R}_{O_1}$ is determined from the vector equation

$$\mathbf{R}_{O_1} + \mathbf{F}_1 + \mathbf{R}_A = \mathbf{0}, \tag{3.117}$$

while the equilibration moment is determined from the equation of the moments about point O_1. For translational motion, the equilibrium equation

$$\underset{}{\underline{\underline{\mathbf{F}_A}}} + \underset{}{\underline{\underline{\mathbf{R}_A}}} + \underset{\perp \Delta_1}{\underline{\mathbf{N}}} + \underset{\| \Delta_1}{\underline{\mathbf{F}_e}} = \mathbf{0} \tag{3.118}$$

gives the forces $\mathbf{F}_e$ and $\mathbf{N}$, while the equation of the moments about point O_1 gives the distance ξ.

Figure 3.33 The forces at the driving element: (a) driving element in rotational motion; (b) driving element in translational motion.

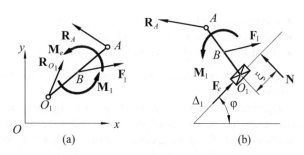

(a)   (b)

The analytical method Denoting by H_{O_1} and V_{O_1} the components of the reaction $\mathbf{R}_{O_1}$, from the equation (3.117) we deduce

$$H_{O_1} = -F_{1x} - H_A, \; V_{O_1} = -F_{1y} - V_A,$$ (3.119)

and from the equation of the moments about point O_1 we get

$$M_e = F_{1y}(X_B - X_A) - F_{1x}(Y_B - Y_A) + V_A(X_A - X_1) - H_A(Y_A - Y_1).$$ (3.120)

For translational motion, from (3.118) we get

$$F_e \cos\varphi - N\sin\varphi = -F_{1x} - H_A,$$
$$F_e \sin\varphi + N\cos\varphi = -F_{1y} - V_A,$$ (3.121)

which leads to

$$F_e = -(F_{1x} + H_A)\cos\varphi - (F_{1y} + V_A)\sin\varphi,$$
$$N = (F_{1x} + H_A)\sin\varphi - (F_{1y} + V_A)\cos\varphi.$$ (3.122)

The equations of the moments about point O_1 give the distance

$$\xi = \frac{1}{N}\left[-M_1 + F_{1x}(Y_B - Y_1) - F_{1y}(X_B - X_1)\right.$$
$$\left. + H_A(Y_A - Y_1) - V_A(X_A - X_1)\right].$$ (3.123)

3.5.8 Determination of the Equilibration Force (Moment) using the Virtual Velocity Principle

According to the virtual velocity principle, for a planar mechanical system in equilibrium (static or dynamic) the following relation holds:

$$\sum (\mathbf{F}\cdot\mathbf{v} + M\omega) = 0,$$ (3.124)

where $\mathbf{v}$ stands for the velocities of the points of application of forces $\mathbf{F}$ (given forces, forces of weight, inertial forces and forces of equilibration); ω stands for the angular velocities of the elements, and M stands for the moments (given, inertial, and of equilibration). The reactions do not appear in (3.124).

If the driving element has rotational motion, then from (3.124), if we denote by M_e the equilibration moment and by ω_1 the angular velocity of the driving element, we deduce

$$M_e = -\frac{1}{\omega_1}\sum(\mathbf{F}\cdot\mathbf{v} + M\omega).$$ (3.125)

Similarly, if the driving element is in translational motion, then we deduce:

$$F_e = -\frac{1}{v_1}\sum(\mathbf{F}\cdot\mathbf{v} + M\omega),$$ (3.126)

where v_1 is the velocity of the driving element.

As an example, consider the mechanism in Figure 3.34a, for which the following are known:

- the distance $OA = r$
- the masses m_2 and m_3 of the elements
- the stiffness k of the spring; for the position characterized by $\theta = 0$ the spring is at rest.

Determine the equilibration moment knowing that the driving element has a constant angular speed ω_1.

Figure 3.34b shows the forces of weight, the force in the spring $\mathbf{F}_{spring}$, and the inertial forces. There are no moments of inertia in this case because elements 2 and 3 have translational motion, while the driving element does not have an angular acceleration. With the notation in

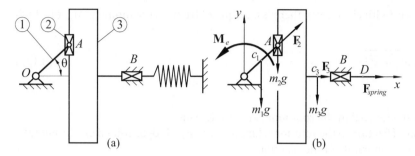

Figure 3.34 The calculation of the equilibration moment: (a) the mechanism; (b) the forces in mechanism.

Figure 3.34b and taking into account (3.125):

$$M_e = -\frac{1}{\omega_1}\left(-m_1 g v_{C_1 y} - m_2 g v_{Ay} - m_3 g v_{C_3 y}\right.$$
$$\left. + F_{2x} v_{Ax} + F_{2y} v_{Ay} + F_{3x} v_{C_3 x} + F_{spring} v_{Dx}\right).$$

The kinematic relations

$$y_{C_1} = \frac{r}{2}\sin\theta, \ v_{C_1 y} = \frac{r}{2}\omega_1\cos\theta,$$
$$x_A = r\cos\theta, \ v_{Ax} = -r\omega_1\sin\theta, \ y_A = r\sin\theta, \ v_{Ay} = r\omega_1\cos\theta,$$
$$v_{C_3 x} = v_{Ax} = v_{Dx} = -r\omega_1\sin\theta, \ v_{C_1 y} = 0,$$

and the dynamic relations

$$F_2 = m_1 r\omega_1^2, \ F_{2x} = m_2 r\omega_1^2\cos\theta, \ F_{2y} = m_2 r\omega_1^2\sin\theta,$$
$$F_{3x} = m_3 r\omega_1^2\cos\theta, \ F_{3y} = m_3 r\omega_1^2\sin\theta, \ F_{spring} = kr(1-\cos\theta),$$

give the moment of equilibration

$$M_e = \left[\frac{m_1 g}{2} + m_2 g + m_3 r\omega_1^2\cos\theta + kr(1-\cos\theta)\right] r\sin\theta.$$

If the driving element is acted upon by a moment that varies as function of θ after the last expression, then the driving element has uniform rotational motion with angular velocity ω_1.

3.6 Reactions in Kinematic Pairs with Friction

3.6.1 Friction Forces and Moments

Friction in a rotational kinematic pair
Consider two elements linked to one another by a rotational kinematic pair (Figure 3.35a), and let ω_{21} be the angular velocity of element 2 relative to element 1.

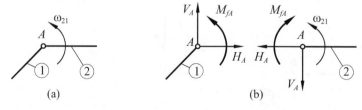

Figure 3.35 Friction in a rotational kinematic pair: (a) rotational kinematic pair; (b) isolation of the elements of the rotational kinematic pair.

In the kinematic pair (with clearance), there is a torque of friction, the moment M_{fA} of which is given by:

$$M_{fA} = \mu_0 r_0 \sqrt{H_A^2 + V_A^2}, \tag{3.127}$$

where

- r_0 is the radius of the journal in the rotational kinematic pair
- μ_0 is the coefficient of friction in the rotational kinematic pair; it is determined experimentally
- H_A and V_A are the components of the reaction.

The moment of friction that acts upon the element 1 has the sense of the angular velocity ω_{21}, while the friction moment that acts upon element 2 (according to the principle of action and reaction) has an opposite sense to the angular velocity ω_{21}.

Equation (3.127) shows that the moment of friction varies proportionally to the magnitude of the reaction in the kinematic pair; since this magnitude varies depending on the position, the friction moment depends on the position too.

For a rotational kinematic pair without clearance, it is assumed that the moment of friction is constant.

Friction in a prismatic (translational) kinematic pair

Consider two elements linked to one another by a prismatic (translational) kinematic pair (Figure 3.36b) and let $\mathbf{v}_{21}$ be the velocity of element 2 relative to element 1. In a prismatic (translational) kinematic pair there is a force of friction given by:

$$F_f = \mu N, \tag{3.128}$$

where μ is the coefficient of sliding friction, while N is the normal reaction. The friction force that acts upon element 1 has the sense of the relative velocity $\mathbf{v}_{21}$, while the friction force that acts upon element 2, according the principle of action and reaction, has opposite sense to the relative velocity $\mathbf{v}_{21}$.

For prismatic (translational) kinematic pairs, the lock phenomenon may appear. Consider element 2 of the kinematic pair (Figure 3.37) and let B be the point through which passes the resultant $\mathbf{F}$ of the forces (given forces, inertial forces, forces of weight and forces of constraint), unlike the prismatic kinematic pair at point A. Under the action of force $\mathbf{F}$, the contact between the two elements takes place at points A_1 and A_2, where forces N_1 and N_2 and the friction forces μN_1 and μN_2 appear (μ being the coefficient of friction).

If φ is the friction angle, and taking into account the sense of motion, then constructing the straight lines $A_3 A_1$ and $A_3 A_2$ that pass through points A_1 and A_2, respectively, and make angles of φ with the reactions $\mathbf{N}_1$ and $\mathbf{N}_2$, gives the *friction zone* described by the angle $A_1 A_3 A_2$.

The condition of motion states that the support of force $\mathbf{F}$ does not intersect the hatched zone contained between the straight lines $A_1 A_3$ and $A_3 A_2$. If one denotes by α the angle between force

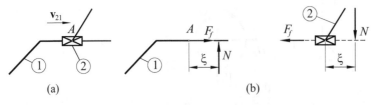

(a)   (b)

Figure 3.36 Friction in a prismatic (translational) kinematic pairs: (a) prismatic kinematic pair with friction; (b) isolation of the elements of the prismatic kinematic pair with friction.

Figure 3.37 The forces in a prismatic (translational) kinematic pair.

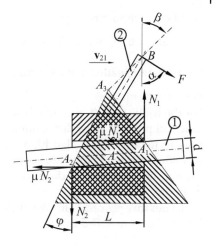

F and the normal to the axis of the kinematic pair, and by β the angle between the straight line BA_3 and the same normal, then the condition of motion reads

$$\pi - \beta > \alpha > \varphi. \tag{3.129}$$

If this condition does not hold true – that is, if $0 < \alpha < \varphi$ or $\pi > \alpha > \pi - \beta$ – then the lock phenomenon appears.

Friction in higher kinematic pairs

Consider the higher (contact) kinematic pair between elements 1 and 2. Let ω_{21} be the angular velocity of element 2 relative to element 1 (Figure 3.38a), and let $\mathbf{v}_{21}$ be the relative velocity at the contact point (the velocity of point A_2 relative to point A_1). At the contact point A, the torsor of the friction forces consists of the force of sliding friction $\mathbf{F}_f$ situated on the common tangent of the two bodies, and the rolling friction moment $\mathbf{M}_r$. These components of the torsor of the friction forces, according to the principle of action and reaction, lead to the representation in Figure 3.38b, where N stands for the normal reaction. In general, the relative motion of body 2 relative to body 1 is a rolling-with-sliding motion ($\omega_{21} \neq 0$, $v_{21} \neq 0$); in this case

$$F_f = \mu N, \quad M_r = sN, \tag{3.130}$$

where μ is the coefficient of sliding friction and s is the coefficient of rolling friction. If the relative motion is rolling without sliding:

$$\mathbf{v}_{21} = \mathbf{0}, F_f < \mu N, M_r = sN. \tag{3.131}$$

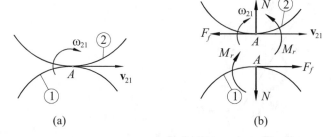

(a) (b)

Figure 3.38 Friction forces and moments in a higher (contact) kinematic pair: (a) higher kinematic pair with friction; (b) isolation of the elements of the higher kinematic pair with friction.

For sliding without rolling:

$$\omega_{21} = 0, \ F_f = \mu N, \ M_r < sN. \tag{3.132}$$

3.6.2 Determination of the Reactions with Friction

The working method

For the friction effect, the equilibrium equations are, in general, non-linear and thus, from the practical point of view, they cannot be solved by analytical methods. The numerical solution uses an iterative linear algorithm. The first step of the algorithm is the determination of the reactions, without taking into account the friction forces. These are used as the initial values for the non-linear system. In the next step the friction forces and moments and then the resulting reactions are determined.

The iterative process continues until the values of the reactions change between iterative steps by less than a specified value.

Numerical example

Problem: For the *ABC* dyad in Figure 3.39, determine the reactions in the kinematic pairs assuming that the dyad is acted on only by the inertial forces and moments. The elements are homogeneous straight bars with masses $m_2 = 1$ kg and $m_3 = 1.25$ kg; the coefficients of friction in the rotational kinematic pairs are $\mu_A = \mu_B = \mu_C = 0.4$ and the radii of the journals are $r_{0A} = r_{0B} = r_{0C} = 0.02$ m.

Solution: The kinematic analysis of this dyad was performed in Section 2.4.2, and the reactions (in the case without friction) were determined in Section 3.5.2. The results were:

$$X_A = 0.4 \,\text{m}, \ Y_A = 0.4 \,\text{m}, \ X_B = 0.7 \,\text{m}, \ Y_B = 0.8 \text{m}, \ X_C = 1.3 \,\text{m}, \ Y_C = 0.5 \,\text{m},$$

$$X_D = 0.55 \,\text{m}, \ Y_D = 0.6 \,\text{m}, \ X_E = 1 \,\text{m}, \ Y_E = 0.65 \,\text{m},$$

$$\omega_{10} = -15 \,\text{rad/s}, \ \omega_{20} = -5.46 \,\text{rad/s}, \ \omega_{30} = 12.75 \,\text{rad/s}, \ \omega_{40} = 12 \,\text{rad/s},$$

$$F_{Dx} = -3.04 \,\text{N}, \ F_{Dy} = 74.54 \,\text{N}, \ F_{Ex} = -52.1 \,\text{N}, \ F_{Ey} = 138.2 \,\text{N},$$

$$M_2 = 9.6 \,\text{Nm}, \ , M_3 = -2.22 \,\text{Nm}.$$

The relative angular velocities are

$$\omega_{21} = \omega_{20} - \omega_{10} = 9.54 \,\text{rad/s}, \ \omega_{32} = \omega_{30} - \omega_{20} = 18.21 \,\text{rad/s},$$

$$\omega_{43} = \omega_{40} - \omega_{30} = -0.75 \,\text{rad/s}.$$

These are drawn in the figure and are then used to draw the friction moments M_{fA}, M_{fB}, and M_{fC}. Using the results obtained in the case without friction ($R_A = 120$ N, $R_B = 75.43$ N, $R_C =$

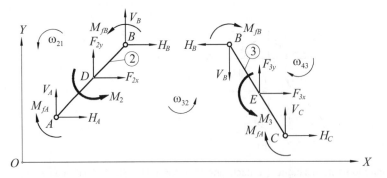

Figure 3.39 The determination of the reactions in the RRR dyad taking into account the friction forces.

168.26 N), relation (3.127) gives the following values for the friction moments

$$M_{fA} = 0.008R_A = 0.96 \text{ Nm}, \ M_{fB} = 0.008R_B = 0.60 \text{ Nm},$$
$$M_{fC} = 0.008R_C = 1.35 \text{ Nm}.$$

The influence of the friction torques can be found in the calculation of the parameters A_1 and B_1 in (3.42); here we have to replace M_2 and M_3 by $M_2^* = M_2 + M_{fB} - M_{fA}$ and $M_3^* = M_3 - M_{fB} - M_{fC}$, respectively. This means that the new values of the parameters A_1 and B_1 are

$$A_1 \leftarrow A_1 - M_{fB} + M_{fA} = -21.03, \ B_1 \leftarrow B_1 + M_{fB} + M_{fC} = 37.715.$$

Taking into account that $C_1 = 0.33$, from (3.43) we deduce

$$H_B = \frac{-0.6A_1 + 0.3B_1}{C_1}, \ V_B = \frac{0.3A_1 + 0.4B_1}{C_1}.$$

These relations ensure the calculation can be easily continued. In this case we have

$$H_B = 72.52 \text{ N}, \ V_B = 26.6 \text{ N}.$$

The reactions at points A and C are directly deduced from:

$$H_A = -H_B - F_{Dx}, \ V_A = -V_B - F_{Dy}, \ H_C = H_B - F_{Ex}, \ V_C = V_B - F_{Ey},$$

from which

$$H_A = 69.48 \text{ N}, \ V_A = 101.14 \text{ N}, \ H_C = 129.62 \text{ N}, \ V_C = -111.6 \text{ N}.$$

This gives the reactions

$$R_A = 122.71 \text{ N}, \ R_B = 77.24 \text{ N}, \ R_C = 171.04 \text{ N}.$$

The new values of the friction moments are

$$M_{fA} = 0.98 \text{ Nm}, \ M_{fB} = 0.62 \text{ Nm}, \ M_{fC} = 1.37 \text{ Nm}.$$

The actual values of the parameters A_1 and B_1 are $A_1 = -21.02$ and $B_1 = 37.855$; the new values of the components of the reaction at point B are then:

$$H_B = 72.63 \text{ N}, \ V_B = 26.77 \text{ N}.$$

These values are very close to those calculated in the previous iterations.

3.7 Kinetostatic Analysis of some Planar Mechanisms

3.7.1 Kinetostatic Analysis of Four-bar Mechanism

For the four-bar mechanism in Figure 3.40 the following are known: the dimensions of the elements of the mechanism, the position of the kinematic pairs linked to the base, the kinematic analysis performed in Section 2.10.1, the masses of the elements – $m_1 = 0.05$ kg, $m_{21} = 0.125$ kg, $m_{22} = 0.125$ kg, and $m_3 = 0.125$ kg – and the force that acts at point M

$$F = \begin{cases} 0 \text{ for } \varphi \in [0°, 80°), \\ 50 \text{ N for } \varphi \in [80°, 280°], \\ 0 \text{ for } \varphi \in [280°, 360°). \end{cases}$$

We want to determine the reactions in the kinematic pairs and the equilibration moment using the virtual velocities principle.

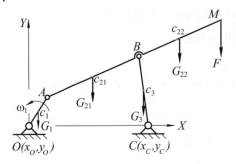

Figure 3.40 Four-bar mechanism.

Solution: To obtain numerical results we will use the assisted analytical method. To avoid duplication, the assisted graphical method will be not presented, since this would be almost identical to the procedure in Section 3.5.2; the AutoLisp function *Forces_RRR* presented there could be easily modified to call function *Mechanism_R-3R.lsp*, which was described in Section 2.10.1.

Since the analytical method involves determining the positions, velocities, and accelerations of the centres of weight, the magnitudes of the components of the inertial forces torsors can be determined using:

$$F_{i,1x} = -m_1 a_{C_1 x}, \ F_{i,1y} = -m_1 a_{C_1 y}, \ M_{i,1} = 0,$$

$$F_{i,21x} = -m_{21} a_{C_{21} x}, \ F_{i,21y} = -m_{21} a_{C_{21} y}, \ M_{i,21} = -J_{21} \varepsilon_2, \ \text{where } J_{21} = \frac{m_{21} AB^2}{12},$$

$$F_{i,22x} = -m_{22} a_{C_{22} x}, \ F_{i,22y} = -m_{22} a_{C_{22} y}, \ M_{i,22} = -J_{22} \varepsilon_2, \ \text{where } J_{22} = \frac{m_{22} MB^2}{12}, \quad (3.133)$$

$$F_{i,3x} = -m_3 a_{C_3 x}, \ F_{i,3y} = -m_3 a_{C_3 y}, \ M_{i,3} = -J_3 \varepsilon_3, \ \text{where } J_3 = \frac{m_3 BC^2}{3}.$$

The elements of the mechanism (Figure 3.41) are isolated; writing the equations $\sum M_A = 0$ and $\sum M_C = 0$ for elements 2 and 3, respectively, gives the system

$$\begin{cases} R_{Bx} (y_B - y_A) + R_{By} (x_B - x_A) = A, \\ R_{Bx} y_B - R_{By} (x_B - l_4) = B, \end{cases} \quad (3.134)$$

where

$$A = (G_{21} - F_{i,21y}) (x_{C_{21}} - x_A) + F_{i,21x} (y_{C_{21}} - y_A) - M_{i,21}$$
$$+ (G_{22} - F_{i,22y}) (x_{C_{22}} - x_A) + F_{i,22x} (y_{C_{22}} - y_A) - M_{i,22} + F (x_M - x_A), \quad (3.135)$$
$$B = G_3 (x_{C_3} - l_4) - M_{i,3}.$$

Solving the system, we obtain the solution

$$R_{Bx} = \frac{-A (x_B - x_C) - B (x_B - x_A)}{\Delta},$$

$$R_{By} = \frac{-B (y_B - y_A) - A (y_B - y_C)}{\Delta}, \quad (3.136)$$

where

$$\Delta = (y_B - y_A) (x_B - x_C) - (x_B - x_A) (y_B - y_C). \quad (3.137)$$

The magnitude of the reaction $\mathbf{R}_B$ is given by:

$$R_B = \sqrt{R_{Bx}^2 + R_{By}^2}. \quad (3.138)$$

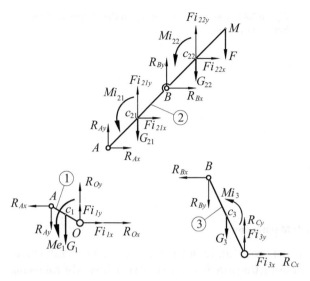

Figure 3.41 The isolation of the elements of the four-bar mechanism.

Equations $\sum F_x = 0$ and $\sum F_y = 0$ for element 2 give the components of the reaction at point A and its magnitude

$$R_{Ax} = -R_{Bx} - F_{i,21x} - F_{i,22x}, \quad R_{Ay} = G_{21} + G_{22} - R_{By} - F_{i,21y} - F_{i,22y},$$
$$R_A = \sqrt{R_{Ax}^2 + R_{Ay}^2}. \tag{3.139}$$

Equations $\sum F_x = 0$ and $\sum F_y = 0$ for element 3 give the components of the reaction at point C and its magnitude

$$R_{Cx} = R_{Bx} - F_{i,3x}, \quad R_{Cy} = G_3 + R_{By} - F_{i,3y}, \quad R_C = \sqrt{R_{CX}^2 + R_{Cy}^2}. \tag{3.140}$$

Equations $\sum F_x = 0$ and $\sum F_y = 0$ for element 1 give the component and the magnitude of the reaction at point O

$$R_{Ox} = R_{Ax} - F_{i,1x}, \quad R_{Oy} = G_1 + R_{Ay} - F_{i,1y}, \quad R_O = \sqrt{R_{Ox}^2 + R_{Oy}^2}, \tag{3.141}$$

while from equation $\sum M_O = 0$ we will deduce the expression of the equilibration moment at element 1

$$M_{e_1} = G_1 x_{C_1} - R_{Ax} y_A + R_{Ay} x_A. \tag{3.142}$$

To apply the principle of virtual velocities we will take into account the forces in Figure 3.42 that act on the four-bar mechanism. Applying (3.125), we obtain the expression for the equilibration moment at element 1

$$M_{e_1} = -\frac{1}{\omega_1} \left[-G_1 v_{C_1 y} + F_{i,21x} v_{C_{21x}} + \left(F_{i,21y} - G_{21} \right) v_{C_{21y}} + \left(M_{i,21} + M_{i,22} \right) \omega_2 \right.$$
$$\left. - F v_{My} + F_{i,22x} v_{C_{22x}} + \left(F_{i,22y} - G_{22} \right) v_{C,22y} - G_3 v_{C_3 y} + M_{i,3} \omega_3 \right]. \tag{3.143}$$

In the companion website of the book, we present the numerical results. These were obtained with the Turbo Pascal program that was described in Section 2.10.1, although we did not mention there that the program also included the code for calculating the reactions in the kinematic pairs. Similar to the case of of the kinematic analysis, the values obtained using (3.133)–(3.143) are saved a text file. These values can then be imported into a text editor and transformed into a table. The values obtained for M_{e_1} using the two methods are identical, thus validating the calculation.

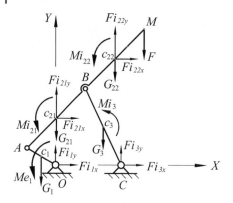

Figure 3.42 The forces that act upon the elements of the four-bar mechanism.

3.7.2 Kinetostatic Analysis of Crank-shaft Mechanism

For the crank-shaft mechanism in Figure 3.43 we want to determine the reaction in the kinematic pairs and the equilibration moment using the principle of virtual velocities. The following are known:

- the dimensions of the mechanism and the kinematic analysis performed in Section 2.10.2
- the masses of the elements: $m_1 = 0.250$ kg, $m_2 = 0.500$ kg, $m_3 = 0.300$ kg
- the positions of the centres of weight: $OC_1 = 0.020$ m, $AC_2 = 0.045$ m
- the moment of the inertia of coupler 2: $J_2 = 0.001$ kgm^2
- force F that acts upon piston 3, the sense of which is given in Figure 3.43

$$F = \begin{cases} 0 \text{ for } \varphi \in [0°, 180°) \\ k\,(s - s^*) \text{ for } [180°, 360°), \end{cases}$$

where $k = 100000$ Nm and $s^* = 0.0915$ m.

Solution: As in the previous application, the assisted analytical method is used. The calculation formulae presented in Section 2.10.2 is completed with the kinematic relations that permit the determination of the position, velocities, and accelerations of the centres of weight:

$$x_{C_1} = OC_1 \cos \varphi_1, \; y_{C_1} = OC_1 \sin \varphi_1$$
$$v_{C_1 x} = -OC_1 \omega_1 \sin \varphi_1, v_{C_1 y} = OC_1 \omega_1 \cos \varphi_1$$
$$a_{C_1 x} = -OC_1 \omega_1^2 \cos \varphi_1, \; a_{C_1 y} = -OC_1 \omega_1^2 \sin \varphi_1$$
$$x_{C_2} = x_A + AC_2 \cos \varphi_2, \; y_{C_2} = y_A + AC_2 \sin \varphi_2 \tag{3.144}$$
$$v_{C_2 x} = v_{Ax} - AC_2 \omega_2 \sin \varphi_2, \; v_{C_2 y} = v_{Ay} + AC_2 \omega_2 \cos \varphi_2$$
$$a_{C_2 x} = a_{Ax} - AC_2 \varepsilon_2 \sin \varphi_2 - AC_2 \omega_2^2 \cos \varphi_2$$
$$a_{C_2 y} = a_{Ay} + AC_2 \varepsilon_2 \cos \varphi_2 - AC_2 \omega_2^2 \sin \varphi_2$$

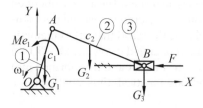

Figure 3.43 The crank-shaft mechanism.

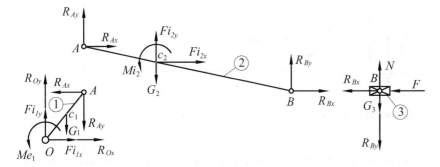

Figure 3.44 The isolation of the elements of the crank-shaft mechanism.

The torsor of the inertial forces reads:

$$F_{i,1x} = -m_1 a_{C_1,x}, \ F_{i,1y} = -m_1 a_{C_1 y}, \ M_{i,1} = 0$$
$$F_{i,2x} = -m_2 a_{C_2 x}, \ F_{i,2y} = -m_2 a_{C_2 y}, \ M_{i,2} = -J_2 \varepsilon_2 \tag{3.145}$$
$$F_{i,3x} = -m_3 a_{C_3 x}, \ F_{i,3y} = -m_3 a_{C_3 y}, \ M_{i,3} = 0$$

The elements of the mechanism are isolated, resulting the representation in Figure 3.44. Applying the d'Alembert principle, equations $\sum F_x = 0$ and $\sum M_A = 0$ for elements 2 and 3, respectively, give the components R_{Bx} and R_{By} of the reaction at point B

$$R_{Bx} = F_{i,3x} - F,$$
$$R_{By} = -\frac{R_{Bx}\left(y_A - y_B\right) + \left(F_{i,2y} - G_2\right)\left(x_{C_2} - x_A\right) + F_{i,2x}\left(y_A - y_{C_2}\right) - M_{i,2}}{x_B - x_A}, \tag{3.146}$$
$$R_B = \sqrt{R_{Bx}^2 + R_{By}^2}.$$

Similarly, equations $\sum F_x = 0$ and $\sum F_y = 0$ for element 2 give the components of the reaction at point A

$$R_{Ax} - R_{Bx} - F_{i,1x}, \ R_{Ay} = G_2 - R_{By} - F_{i,2y}, \ R_A = \sqrt{R_{Ax}^2 + R_{Ay}^2}. \tag{3.147}$$

The normal reaction N is obtained from $\sum F_y = 0$ for element 3

$$N = G_3 + R_{By}. \tag{3.148}$$

The reaction at the driving element is determined from the equations $\sum F_x = 0$ and $\sum F_y = 0$ for element 1

$$R_{Ox} = R_{Ax} - F_{i,1x}, \ R_{Oy} = G_1 + R_{Ay} - F_{i,1y}, \ R_O = \sqrt{R_{Ox}^2 + R_{Oy}^2}, \tag{3.149}$$

while $\sum M_O$ gives the expression for the equilibration moment at element 1

$$M_{e_1} = G_1 x_{C_1} - R_{Ax} y_A + R_{Ay} x_A. \tag{3.150}$$

Figure 3.45 shows the forces that act upon the elements of the crank-shaft mechanism. The equilibration moment, determined with (3.125), reads

$$M_{e_1} = -\frac{1}{\omega_1}\left[-G_1 v_{C_1} + F_{i,2x} v_{C_2 x} + \left(F_{i,2x} - G_2\right) v_{C_2 y} + M_{i_2} \omega_2 + \left(F_{i,3x} - F\right) v_B\right]. \tag{3.151}$$

In the companion website of this book we show the numerical results obtained using a Turbo Pascal program that is itself set out in the companion website of this book. We did not specify in Section 2.10.2 that the program contained both the code for the kinematic analysis and

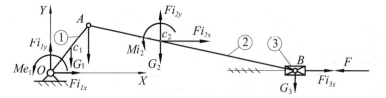

Figure 3.45 The forces that act upon the elements of the crank-shaft mechanism.

code encapsulating (3.144)–(3.151) for the kinetostatic analysis. This program outputs a text file called *React.txt*. The values obtained for M_{e_1} using the principle of virtual velocities are identical to those obtained via the isolation of the elements, thus confirming the accuracy of the methods and calculation.

3.7.3 Kinetostatic Analysis of Crank and Slotted Lever Mechanism

Determine the reactions in the kinematic pairs for the crank and slotted lever mechanism in Figure 3.46. The kinematic analysis is known: it resulted from the numerical application in Section 2.10.3. The following are also known:

- the masses of the elements: $m_1 = 0.150$ kg, $m_2 = 0.200$ kg, and $m_3 = 0.350$ kg
- the moment of inertia of slider 2: $J_2 = 10^{-5}$ kg m^2
- force F that acts at point C

$$F = \begin{cases} 100 \text{ N for } \varphi \in [0°, 180°,) \\ -100 \text{ N for } \varphi \in [180°, 360°,) \end{cases}$$

Verify the values obtained for the equilibration moment M_{e_1} using the principle of virtual velocities.

Solution: The kinematic analysis in Section 2.10.3 is continued with the determination of the positions, velocities, and accelerations of the centres of weight:

$$x_{C_1} = \frac{x_D + x_A}{2}, \ y_{C_1} = \frac{y_D + y_A}{2},$$

$$x_{C_2} = x_A, \ y_{C_2} = y_A,$$ (3.152)

$$x_{C_3} = \frac{x_B + x_C}{2}, \ y_{C_3} = \frac{y_B + y_C}{2},$$

$$v_{C_1 x} = \frac{v_{Ax}}{2}, \ v_{C_1 y} = \frac{v_{A2}}{2}, \ v_{C_2 x} = v_{A_x}, \ v_{C_2 y} = v_{A_y}, \ v_{C_3 x} = \frac{v_{Cx}}{2}, \ v_{C_3 y} = \frac{v_{Cy}}{2},$$ (3.153)

$$a_{C_1 x} = \frac{a_{Ax}}{2}, \ a_{C_1 y} = \frac{a_{A2}}{2}, \ a_{C_2 x} = a_{A_x}, \ a_{C_2 y} = a_{A_y}, \ a_{C_3 x} = \frac{a_{Cx}}{2}, \ a_{C_3 y} = \frac{a_{Cy}}{2}.$$ (3.154)

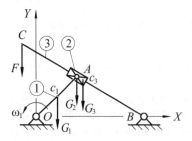

Figure 3.46 Crank and slotted lever mechanism.

The components of the torsor of inertia are:

- for element 1

$$F_{i,1x} = -m_1 a_{C_1x}, \ F_{i,1y} = -m_1 a_{C_1y}, \ M_{i,1} = 0; \tag{3.155}$$

- for element 2

$$F_{i,2x} = -m_2 a_{C_2x}, \ F_{i,2y} = -m_2 a_{C_2y}, \ M_{i,2} = -J_2 \varepsilon_2; \tag{3.156}$$

- for element 3

$$F_{i,3x} = -m_3 a_{C_3x}, \ F_{i,3y} = -m_3 a_{C_3y}, \ M_{i,3} = -J_3 \varepsilon_3. \tag{3.157}$$

Figure 3.47 shows the isolated elements of the mechanism. The equations $\sum M_B = 0$ and $\sum M_a = 0$ for elements 3 and 2, respectively, give

$$N = -\frac{G_3 \left(x_B - x_{C_3}\right) - M_{i,3} - M_{i,2} - F\left(x_B - x_C\right)}{s}, \ x = \frac{M_{i,2}}{N}. \tag{3.158}$$

The reaction R_B is determined from the equations $\sum F_x = 0$ and $\sum F_y = 0$ for element 3

$$R_{Bx} = -F_{i,3x} - N \cos\left(\varphi_3 - \frac{\pi}{2}\right), \ R_{By} = -F_{i,3y} + G_3 - N \sin\left(\varphi_3 - \frac{\pi}{2}\right) + F,$$
$$R_B = \sqrt{R_{Bx}^2 + R_{By}^2}. \tag{3.159}$$

The reaction R_A is determined from $\sum F_x = 0$ and $\sum F_y = 0$ for element 2

$$R_{Ax} = -F_{i,2x} - N \cos\left(\varphi_3 - \frac{3\pi}{2}\right), \ R_{Ay} = -F_{i,2y} + G_2 - N \sin\left(\varphi_3 - \frac{3\pi}{2}\right),$$
$$R_A = \sqrt{R_{Ax}^2 + R_{Ay}^2}. \tag{3.160}$$

For element 1, the reaction R_O is obtained from the equations $\sum F_x = 0$ and $\sum F_y = 0$

$$R_{Ox} = R_{Ax} - F_{i,1x}, \ R_{Oy} = G_1 + R_{Ay} - F_{i,1y}, \ R_O = \sqrt{R_{Ox}^2 + R_{Oy}^2}, \tag{3.161}$$

while the equilibration moment M_{e_1} is deduced from the equation $\sum M_O = 0$

$$M_{e_1} = G_1 x_{C_1} - R_{Ax} y_A + R_{Ay} x_A. \tag{3.162}$$

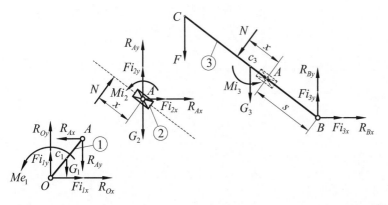

Figure 3.47 The isolation of the elements of the crank and slotted lever mechanism.

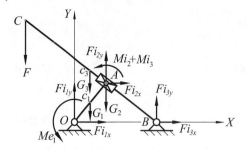

Figure 3.48 The forces that act upon the elements of the crank and slotted lever mechanism.

Using the principle of virtual velocities, Figure 3.48 gives the expression for the equilibration moment M_{e_1}

$$M_{e_1} = -\frac{1}{\omega_1} \left[-G_1 v_{C_1 y} + F_{i,2x} v_{Ax} + \left(F_{i,2y} - G_2 \right) v_{Ay} \right. \\ \left. -F v_{Cy} - G_3 v_{C_3 y} + \left(M_{i,2} + M_{i,3} \right) \omega_3 \right]$$

(3.163)

In the companion website of the book, we show the results obtained with a calculation program, the code for which is also found in the companion website. This calculation program was used for the kinematic analysis too and it was completed with (3.152)–(3.163). The values obtained for the equilibration moment M_{e_1} using the two methods are identical, so the calculation is correct.

4

Dynamics of Machines

In this chapter we study the motion of the mechanisms and machines under the action of the forces and moments that act upon them. For this purpose the dimensions, masses, moments of inertia, forces, and moments that act upon the elements are all assumed as known. We want to determine the motion of the mechanism or machine.

4.1 Dynamic Model: Reduction of Forces and Masses

4.1.1 Dynamic Model

In the dynamic study of a mechanism or machine a simple model is used. This represents an element in rotational or translational motion. The element's motion is identical to that of a chosen element of the mechanism; usually the driving or initial element. It is referred to as the element at which the reduction is made.

To determine the dynamic equivalence, it is assumed that the element has a variable moment of inertia J_{red} depending on the position, and it is acted on by a torque of variable moment M_{red}. Figure 4.1a shows a four-bar mechanism $OABC$ acted on by a system of forces and moments, while Figure 4.1b shows the dynamic model of this mechanism, where J_{red} is the reduced moment of inertia, and M_{red} is the reduced moment. The calculation model for the parameters J_{red} and M_{red} is described below.

If the element at which the reduction is made is in translational motion, then:

- instead of the reduced moment of inertia J_{red}, the reduced inertial mass m_{red} is calculated
- instead of the reduced moment M_{red}, the reduced force F_{red} is calculated.

4.1.2 Reduction of Forces

The reduction of the forces is based on the equivalence between the instantaneous power given by the forces that act upon the machine, and the instantaneous power given by the reduced moment M_{red} that acts upon the driving element. Denoting with $\boldsymbol{\omega}$ the angular velocity of the element at which the reduction is made, with $\mathbf{F}_j$ the force that acts at point A_j, with $\mathbf{v}_j$ the velocity of point A_j, and with $\mathbf{M}_k$ the moment of the torque that acts upon element k of angular velocity $\boldsymbol{\omega}_k$, gives:

$$\mathbf{M}_{red} \cdot \boldsymbol{\omega} = \sum_j \mathbf{F}_j \cdot \mathbf{v}_j + \sum_k \mathbf{M}_k \cdot \boldsymbol{\omega}_k. \tag{4.1}$$

For planar mechanisms we get

$$M_{red} = \frac{1}{\omega} \left(\sum_j F_j v_j \cos \alpha_j + \sum_k M_k \omega_k \right), \tag{4.2}$$

Classical and Modern Approaches in the Theory of Mechanisms, First Edition.
Nicolae Pandrea, Dinel Popa and Nicolae-Doru Stănescu.
© 2017 John Wiley & Sons Ltd. Published 2017 by John Wiley & Sons Ltd.
Companion Website: www.wiley.com/go/pandmech17

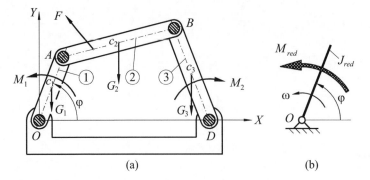

Figure 4.1 The four-bar mechanism and its equivalent dynamic model: (a) four-bar mechanism; (b) equivalent dynamic model.

where α_k is the angle between force $\mathbf{F}_j$ and the velocity $\mathbf{v}_j$. Projecting onto the axes of a fixed reference system Oxy gives:

$$M_{red} = \frac{1}{\omega} \left[\sum_j \left(F_{jx} v_{jx} + F_{jy} v_{jy} \right) + \sum_k M_k \omega_k \right]. \tag{4.3}$$

If the element at which the reduction is made is in translational motion with velocity v, then the reduced force is obtained in a similar way:

$$F_{red} = \frac{1}{v} \left(\sum_j \mathbf{F}_j \cdot \mathbf{v}_j + \sum_k \mathbf{M}_k \cdot \boldsymbol{\omega}_k \right). \tag{4.4}$$

For the mechanism in Figure 4.2, acted on by moments M_1 and M_3 and by the weights of the bars G_1, G_2, and G_3, this gives

$$M_{red} = \frac{1}{\omega} \left(M_1 \omega_1 + G_1 v_1 \cos \alpha_1 + G_2 v_2 \cos \alpha_2 + G_3 v_3 \cos \alpha_3 - M_3 \omega_3 \right), \tag{4.5}$$

where $\omega = \omega_1$.

Figure 4.2 Four-bar mechanism.

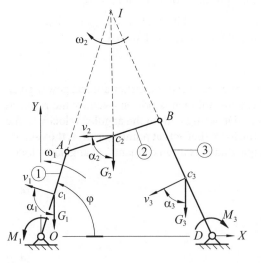

Figure 4.3 Crank-shaft mechanism.

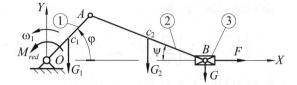

In the following numerical case:

$$G_1 = 0.14 \text{ kg}, G_2 = 0.16 \text{ kg}, G_3 = 0.20 \text{ kg}, M_1 = 0.2 \text{ Nm}, M_2 = 0.1 \text{ Nm},$$
$$OA = 0.14 \text{ m}, AB = 0.16 \text{ m}, BC = 0.2 \text{ m}, OC = 0.28 \text{ m}, \varphi = 72°, \omega = 100 \text{ rad/s}$$

and using the grapho-analytical method, the following values are obtained:

$$v_1 = 7 \text{ m/s}, v_A = 14 \text{ m/s}, \omega_2 = \frac{v_A}{IA} = 70 \text{ rad/s}, IC_2 = 0.17 \text{ m}, v_2 = 12 \text{ m/s},$$
$$IB = 0.18 \text{ m}, v_B = 12.6 \text{ m/s}, v_3 = 6.3 \text{ m/s}, \omega_3 = 63 \text{ rad/s},$$
$$\alpha_1 = 108°, \alpha_2 = 86°, \alpha_3 = 62°.$$

Substituting in (4.5), for the position given by $\varphi = 72°$, gives the reduced moment $M_{red} = 0.142$ Nm.

Another example is shown in Figure 4.3. This mechanism consists of two homogeneous straight bars OA and OB, having lengths r and l and weights G_1 and G_2, respectively; the mechanism is acted on by force F at point B. The reduced moment is expressed by:

$$M_{red} = \frac{1}{\omega} \left(-G_1 \dot{y}_1 - G_2 \dot{y}_2 + F \dot{x}_B \right); \tag{4.6}$$

and since

$$y_1 = y_2 = r \sin \varphi, \ \dot{\varphi} = \omega, \ x_B = r \cos \varphi + l \cos \psi, \ r \sin \varphi = l \sin \psi, \ \dot{\psi} = \frac{r\omega \cos \varphi}{\cos \psi},$$

this gives

$$M_{red} = - \left(G_1 + G_2 \right) r \cos \varphi - \frac{Fr \sin \left(\varphi + \psi \right)}{\cos \psi}. \tag{4.7}$$

Based on (4.7), the function $M_{red} \left(\varphi \right)$ can be assessed for different values of angle φ.

4.1.3 Reduction of Masses

The reduction of the masses is based on the equivalence between the kinetic energy of the real machine and the kinetic energy of the dynamic model. For the case of planar mechanisms, we used the following notation:

- ω, the angular velocity of the element in rotational motion at which the reduction is made;
- ω_i, the angular velocity of the elements in rotational motion;
- J_i, the moment of inertia of element i in rotational motion relative to the rotational centre;
- v_j, the velocity of the centre of mass of element j in translational motion;
- m_j, the mass of element j in translational motion;
- ω_k, the angular velocity of element k in plane-parallel motion;
- v_k, the velocity of the centre of mass of element k in plane-parallel notion;
- m_k, the mass of element k in plane-parallel motion;
- J_k, the moment of inertia of element k in plane-parallel motion relative to the centre of mass.

This gives the equality

$$\frac{1}{2}J_{red}\omega^2 = \frac{1}{2}\left[\sum_i J_i\omega_i^2 + \sum_j m_j v_j^2 + \sum_k \left(m_k v_k^2 + J_k\omega_k^2\right)\right], \tag{4.8}$$

which gives the reduced moment of inertia

$$J_{red} = \frac{1}{\omega^2}\left[\sum_i J_i\omega_i^2 + \sum_j m_j v_j^2 + \sum_k \left(m_k v_k^2 + J_k\omega_k^2\right)\right]. \tag{4.9}$$

If the element at which the reduction is made is in translational motion with velocity v, then the reduced mass is:

$$m_{red} = \frac{1}{v^2}\left[\sum_i J_i\omega_i^2 + \sum_j m_j v_j^2 + \sum_k \left(m_k v_k^2 + J_k\omega_k^2\right)\right]. \tag{4.10}$$

For the example in Figure 4.2, $J_{red} = \frac{1}{\omega^2}\left(J_1\omega^2 + J_2\omega_2^2 + m_2 v_2^2 + J_3\omega_3^2\right)$; since the bars are homogeneous, $J_1 = \frac{G_1 l_1^2}{3g} = 9.32 \cdot 10^{-5}$ kgm^2, $J_2 = \frac{G_2 l_2^2}{12g} = 3.48 \cdot 10^{-5}$ kgm^2, and $J_3 = \frac{G_3 l_3^2}{3g} = 27.18 \cdot 10^{-5}$ kgm^2. For angle $\varphi = 72°$:

$$J_{red} = J_1 + J_3\left(\frac{63}{100}\right)^2 + \frac{G_1}{g}\left(\frac{12}{100}\right)^2 + J_2\left(\frac{70}{100}\right)^2 = 4.53 \cdot 10^{-4}\ \text{kgm}^2.$$

For the example in Figure 4.3, performing the reduction at element OA gives the reduced moment of inertia

$$J_{red} = \frac{1}{\omega^2}\left(J_1\omega^2 + J_2\omega_2^2 + m_2 v_2^2 + m_3 v_B^2\right); \tag{4.11}$$

the calculation being performed using:

$$J_1 = \frac{G_1 r^2}{3g},\ J_2 = \frac{G_2 l^2}{12g},\ m_2 = \frac{G_2}{g},$$
$$v_2^2 = \dot{x}_2^2 + \dot{y}_2^2,\ v_B^2 = \dot{x}_B^2,$$
$$\sin\psi = \frac{r}{l}\sin\varphi,\ x_2 = r\cos\varphi + \frac{l}{2}\cos\psi,\ y_2 = r\sin\varphi,\ x_B = r\cos\varphi + l\cos\psi.$$

If $r = l$ and $G_1 = G_2 = G$, then $\psi = \varphi$, $\omega_2 = \omega$, $v_2^2 = \frac{r^2\omega^2}{4}\left(1 + 8\sin^2\varphi\right)$, $v_B^2 = 4r^2\omega^2\sin^2\varphi$, and the periodic function $J_{red}(\varphi)$, of period $T = \pi$, is obtained

$$J_{red} = \frac{2Gr^2}{3g} + \frac{2\left(G + 2G_3\right)r^2}{g}\sin^2\varphi, \tag{4.12}$$

This is shown in Figure 4.4.

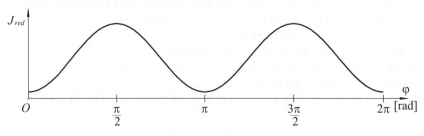

Figure 4.4 The representation of the reduced moment of inertia.

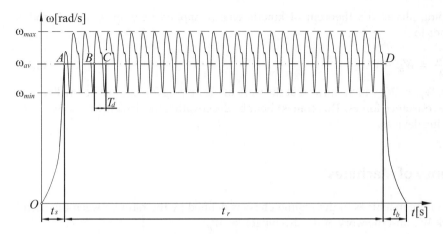

Figure 4.5 The phases of motion of the machine.

4.2 Phases of Motion of a Machine

During the operation of a machine (Figure 4.5) the following phases are distinguished:

- the starting phase, of duration t_S, in which the angular velocity ω increases from $\omega = 0$ to an average value ω_{av};
- the regime phase, of duration t_R, in which the angular velocity ω oscillates around an average value ω_{av};
- the stopping phase, of duration t_B, in which the angular velocity decreases from ω_{av} to $\omega = 0$.

In most cases $t_S \ll t_R$, $t_B \ll t_R$. For an energetic analysis of the phases of motion, the following are denoted:

- by T_0 and T, the initial and the final energy, respectively
- by W_m and W_r, the work of the driving and resistance forces, respectively.

Applying the theorem of kinetic energy on segment OA (Figure 4.5), gives $T - T_0 = W_m - W_r$; that is,

$$J_{red} \frac{\omega_{av}^2}{2} = W_m - W_r; \tag{4.13}$$

Since $J_{red} \frac{\omega_{av}^2}{2} > 0$, the work of the driving forces in the starting phase must be greater than the work of the resistance forces.

During the regime phase, the angular velocity ω has a periodic variation, of period T_d (the *dynamic cycle*), which represents the time necessary to return the dynamic parameters (forces, moments) to their initial values. In the general case, the dynamic cycle T_d and the kinematic cycle T_k are different, the latter representing the time necessary to return the kinematic parameters to their initial values. For instance, consider a thermal engine four-stroke engine, in which the kinematic cycle of the crank-shaft mechanism is $T_k = \frac{2\pi}{\omega}$ (ω being the angular velocity of the shaft), while the dynamic cycle is given by $T_d = \frac{4\pi}{\omega} = 2T_k$. Applying the theorem of kinetic energy in the regime phase for a dynamic cycle (between points B and C in Figure 4.5) and taking into account that $T_B = T_C$, gives $W_m = W_r$; this shows that the work of the driving forces is used to overcome the resistance in the system.

In the stopping phase, the theorem of kinetic energy applied between points D and E (Figure 4.5) leads to

$$-J_{red}\frac{\omega_{av}^2}{2} = W_m - W_r,$$ (4.14)

In other words, $W_m < W_r$, which shows that the work of the driving forces must be smaller than the work of the resistance forces. This comes about by deactivating the driving forces ($W_m = 0$) and using braking devices.

4.3 Efficiency of Machines

The efficiency of the machines in the regime phase is defined by the ratio between the useful resistance work W_{ru} and the work of the driving forces W_m:

$$\eta = \frac{W_{ru}}{W_m}$$ (4.15)

We take into account $W_r = W_{ru} + W_{rf}$, where W_r is the work done by all the resistance forces, while W_{rf} is the work of the friction forces. Since $W_m = W_r$ for a dynamic cycle, and denoting the coefficient of looses by ψ

$$\psi = \frac{W_{rf}}{W_m},$$ (4.16)

we get

$$\eta = 1 - \psi;$$ (4.17)

so

$$0 < \eta < 1.$$ (4.18)

For machines connected in series (Figure 4.6), we get

$$\eta = \frac{W_{ru_n}}{W_{m_1}} = \frac{W_{ru_1}}{W_{m_1}} \cdot \frac{W_{ru_2}}{W_{m_2}} \cdots\cdots \frac{W_{ru_n}}{W_{m_n}},$$

which gives $\eta = \eta_1 \cdot \eta_2 \cdots\cdots \eta_n$.

For machines connected in parallel (Figure 4.7), we define

- ξ the coefficient of repartition as $\xi_i = \dfrac{W_{m_i}}{W_m}$
- η_i, the efficiency as $\eta_i = \dfrac{W_{ru_i}}{W_{m_i}}$,

This gives the expressions $W_{m_i} = \xi_i W_m$, $W_{ru_i} = \xi_i W_{m_i}$, and $W_{ru_i} = \eta_i \xi_i W_m$, and the total efficiency reads

$$\eta = \frac{W_{ru}}{W_m} = \frac{\sum\limits_{i=1}^{n} W_{ru_i}}{W_m} = \sum\limits_{i=1}^{n} \xi_i \eta_i,$$

or

$$\eta = \sum\limits_{i=1}^{n} \eta_i \xi_i.$$ (4.19)

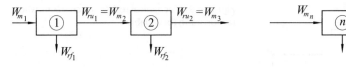

Figure 4.6 The connection of machines in series.

Figure 4.7 The connection of machines in parallel.

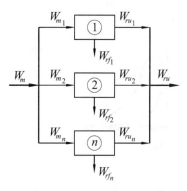

When $\xi_i = \dfrac{1}{n}$ this gives

$$\eta = \frac{1}{n} \sum_{i=1}^{n} \eta_i. \tag{4.20}$$

4.4 Mechanical Characteristics of Machines

The forces and torques that act upon mechanisms and machines are the driving forces (torques) and the useful resistant forces (torques). These forces and torques are defined by the mechanical characteristics that represent the variations of the (driving or resistant) forces as functions of the position and velocity of the element upon which they act.

Denoting by M_d and M_r the moments of the driving and resistant torques, respectively:

$$M_d = M_d\,(\varphi, \omega)\,,\ M_r = M_r\,(\varphi, \omega)\,. \tag{4.21}$$

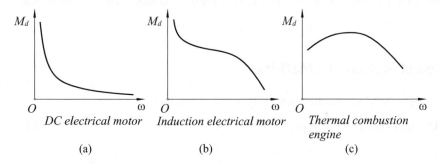

Figure 4.8 Mechanical driving characteristics: (a) for a DC electrical motor; (b) for an induction electrical motor; (c) for a thermal combustion engine.

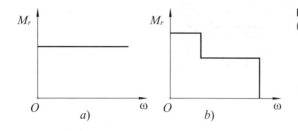

Figure 4.9 Mechanical resistance characteristics: (a) constant; (b) variable.

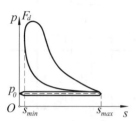

Figure 4.10 Pressure versus piston position in an internal combustion engine.

In more general terms, these moments depend on time too, so they may expressed as

$$M_d = M_d(\varphi, \omega, t), \quad M_r = M_r(\varphi, \omega, t). \tag{4.22}$$

The mechanical characteristics are obtained experimentally. In some machines, these characteristics depend on only the position, while in others they depend only on the velocity. Figure 4.8 shows the mechanical (driving) characteristics and Figure 4.9 the mechanical resisting characteristics for a few everyday machines.

Analytically, the mechanical characteristics may be approximated by second-degree polynomials:

$$M_d = a_d + b_d\omega + c_d\omega^2, \quad M_r = a_r + b_r\omega + c_r\omega^2, \tag{4.23}$$

where a_d, b_d, c_d, a_r, b_r, and c_r are constants.

In Figure 4.10 we show the diagram for an internal combustion engine; that is, the variation of the pressure of the gases inside the cylinder as function of the position of the piston (parameter s). Taking into account

$$F_d = (p - p_0)\frac{\pi D^2}{4}, \tag{4.24}$$

where F_d is the force of the gases inside the cylinder and D is the diameter of the piston, the diagram represents (at a different scale) the variation of the driving force as a function of position $F_d(s)$.

4.5 Equation of Motion of a Machine

Consider the general case of a dynamic model (Figure 4.11) in which

$$M_{red} = M_{red}(\varphi, \omega, t), \quad J_{red} = J_{red}(\varphi), \tag{4.25}$$

where

$$\omega = \frac{d\varphi}{dt}. \tag{4.26}$$

Figure 4.11 The dynamic model.

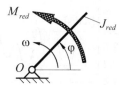

The theorem of kinetic energy is $dT = dW$, where T is the kinetic energy, $T = \frac{1}{2}J_{red}(\varphi)\omega^2$ and dW is the elementary work, $dW = M_{red}d\varphi$. This means that $\dfrac{dT}{dt} = \dfrac{dW}{dt}$, or

$$J_{red}(\varphi)\,\omega\frac{d\omega}{dt} + \frac{1}{2}\omega^2\frac{dJ_{red}}{dt} = M_{red}\frac{d\varphi}{dt};$$

Taking into account (4.26) and the expressions

$$\frac{d\omega}{dt} = \frac{d^2\varphi}{dt^2}, \quad \frac{dJ_{red}}{dt} = \omega\frac{dJ_{red}}{d\varphi},$$

gives the second-order non-linear equation

$$F_{red}(\varphi)\frac{d^2\varphi}{dt^2} + \frac{1}{2}\left(\frac{d\varphi}{dt}\right)^2\frac{dJ_{red}}{d\varphi} = M_{red}\left(t, \varphi, \frac{d\varphi}{dt}\right). \qquad (4.27)$$

If M_{red} does not depend on t, then M_{red} may be considered a function of the angle φ and angular velocity ω. Knowing that

$$\frac{d^2\varphi}{dt^2} = \frac{d\omega}{dt} = \frac{d\omega}{d\varphi}\cdot\frac{d\varphi}{dt} = \omega\frac{d\omega}{d\varphi},$$

the relation (4.27) gives the first-order differential equation

$$J_{red}(\varphi)\frac{d\omega}{d\varphi} + \frac{1}{2}\omega\frac{dJ_{red}}{d\varphi} = \frac{1}{\omega}M_{red}(\varphi, \omega). \qquad (4.28)$$

4.6 Integration of the Equation of Motion

4.6.1 General Case

Consider as known the functions $J_{red}(\varphi)$, $\dfrac{dJ_{red}}{d\varphi}$, $M_{red}\left(t, \varphi, \dfrac{d\varphi}{dt}\right)$, and the initial conditions

$$t = 0 \rightarrow \begin{cases} \varphi = \varphi_0, \\ \omega = \dfrac{d\varphi}{dt} = \omega_0. \end{cases} \qquad (4.29)$$

Using the notation

$$\varphi = Y_1, \quad \frac{d\varphi}{dt} = Y_2, \qquad (4.30)$$

equation (4.27) becomes

$$J_{red}(Y_1)\frac{dY_2}{dt} + \frac{1}{2}Y_2^2\frac{dJ_{red}}{dY_1} = M_{red}(t, Y_1, Y_2), \qquad (4.31)$$

which gives

$$\frac{dY_2}{dt} = \frac{1}{J_{red}(Y_1)}\left[M_{red}(t, Y_1, Y_2) - \frac{1}{2}Y_2^2\frac{dJ_{red}}{dY_1}\right]. \qquad (4.32)$$

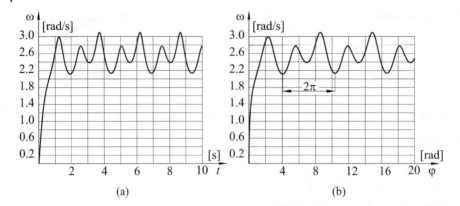

Figure 4.12 Variation of the angular velocity: (a) $\omega = \omega(t)$ and $\omega = \omega(t)$.

This then gives the system of equations

$$
\begin{cases}
\dfrac{dY_1}{dt} = Y_2, \\
\dfrac{dY_2}{dt} = \dfrac{1}{J_{red}(Y_1)} \left[M_{red}(t, Y_1, Y_2) - \dfrac{1}{2} Y_2^2 \dfrac{dJ_{red}}{dY_1} \right],
\end{cases}
\tag{4.33}
$$

with the initial conditions

$$
t = 0 \rightarrow \begin{cases} Y_1 = \varphi_0, \\ Y_2 = \omega_0. \end{cases}
\tag{4.34}
$$

The system of differential equations (4.33) can be solved using the fourth-order Runge–Kutta method. As an example, consider the case in which the reduced moment of inertia is $J_{red} = 4 + \sin(2\varphi)$ kgm^2 and the reduced moment is $M_{red} = 2\left(10e^{-\omega t} + \cos\varphi\right)$ Nm; in addition $\varphi = 0$ and $\omega = 0$ at $t = 0$. Choosing the integration step $\Delta t = 0.01$ s, we obtain a table of values from which the graphs in Figure 4.12 have been drawn.

4.6.2 The Regime Phase

In general, the regime phase is characterized by two aspects:

- the reduced moment depends on only angle φ, $M_{red} = M_{red}(\varphi)$,
- the work produced in a dynamic cycle φ_d is null; that is

$$
\int_{\varphi_0}^{\varphi_0 + \varphi_d} M_{red}(\varphi)\, d\varphi = 0.
\tag{4.35}
$$

In these conditions, the angular velocity $\omega(\varphi)$ has periodic variation of period φ_d, oscillating between the values ω_{min} and ω_{max}. To determine the function $\omega(\varphi)$, $J_{red}(\varphi)$, $M_{red}(\varphi)$, and φ_d are assumed as known, and the average angular velocity ω_{av} is defined by:

$$
\omega_{av} = \frac{1}{\varphi_d} \int_{\varphi_0}^{\varphi_0 + \varphi_d} \omega(\varphi)\, d\varphi.
\tag{4.36}
$$

The initial angular velocity ω_0 is not known. Applying the theorem of kinetic energy gives

$$
J_{red}(\varphi) \frac{\omega^2}{2} - J_{red}(\varphi_0) \frac{\omega_0^2}{2} = \int_{\varphi_0}^{\varphi_0 + \varphi_d} M_{red}(\varphi)\, d\varphi,
\tag{4.37}
$$

where angle φ_0 can be chosen as $\varphi_0 = 0$. Using the notation

$$L(\varphi) = \int_0^\varphi M_{red}(\varphi)\, d\varphi, \quad J_{red}(0) = J_0,$$

gives

$$\omega = \sqrt{\frac{2L(\varphi) + J_0\omega_0^2}{J_{red}(\varphi)}}. \tag{4.38}$$

The parameter ω_0 in expression (4.38), which is unknown, is determined from the equation

$$\omega_{av} = \frac{1}{\varphi_d} \int_0^{\varphi_d} \sqrt{\frac{2L(\varphi) + J_0\omega_0^2}{J_{red}(\varphi)}}\, d\varphi. \tag{4.39}$$

Solution using the grapho-analytical method

Because, in general, the solving of (4.39) is laborious, a numerical approach is used. In order to determine the function $\tilde{\omega}(\omega_0)$, a value is given to the parameter ω_0:

$$\tilde{\omega}(\omega_0) = \frac{1}{\varphi_d} \int_0^{\varphi_d} \sqrt{\frac{2L(\varphi) + J_0\omega_0^2}{J_{red}(\varphi)}}\, d\varphi. \tag{4.40}$$

For this purpose, for a chosen value of the parameter ω_0, degree-by-degree increments are made to the parameter φ, and the integral (4.40) is numerically calculated from:

$$\tilde{\omega}(\omega_0) = \frac{1}{\varphi_d} \sum_{i=1}^{n} \sqrt{\frac{2L(\varphi_i) + J_0\omega_0^2}{J_{red}(\varphi_i)}}, \tag{4.41}$$

where $\varphi_i = \dfrac{i\varphi_d}{n}$, n being the number of equal intervals into which the dynamic cycle is divided (usually, if $\varphi_d = 360°$, then $n = 360$). For further values of ω_0 around the value ω_{av}, a graphic similar to that in Figure 4.13 is obtained. The value of ω_0 is obtained by drawing in the figure a straight line passing through ω_{av} and parallel to the horizontal axis (Figure 4.13).

Solution using the iterative method

Values are given to ω_0 around the value ω_{av}, and the values $\tilde{\omega}$ and ω_{av} are compared. The following iterative formula is used:

$$\omega_0^{(i)} = \omega_0^{(i-1)} + \Delta\tilde{\omega}^{(i-1)}, \tag{4.42}$$

where

$$\Delta\tilde{\omega}^{(i)} = \omega_{av} - \tilde{\omega}^{(i)}. \tag{4.43}$$

We have used the following notation:

Figure 4.13 Graph for obtaining the value of ω_0.

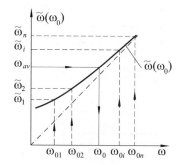

- $\omega_0^{(i)}$, the initial angular velocity after i iterations
- $\tilde{\omega}^{(i)}$, the angular velocity after i iterations
- $\Delta\tilde{\omega}^{(i)}$, the deviation of the calculated value for the average angular velocity from ω_{av}.

The iterative process ends when

$$\left|\Delta\tilde{\omega}^{(i)}\right| \le \rho, \tag{4.44}$$

where ρ stands for the imposed maximum error. This process is quickly convergent; after a maximum of 10 iterations the process ends, the precision being $\frac{1}{1000}$ of the average angular velocity.

Numerical example

Problem: Consider the case in which $\varphi_d = 2\pi$, $J_{red} = 1$ kgm², $\omega_{av} = 3.5$ rad/s, and

$$M_{red}(\varphi) = \begin{cases} M_0, & \text{for } 0 \le \varphi < \pi, \\ -M_0, & \text{for } \pi \le \varphi < 2\pi. \end{cases} \quad \text{where } M_0 = \frac{4}{\pi} \text{ Nm.}$$

The condition $\int_0^{2\pi} M_{red}(\varphi)\,d\varphi = 0$ being fulfilled, we get

$$L(\varphi) = \begin{cases} M_0\varphi & \text{for } 0 \le \varphi < \pi, \\ 2M_0\pi - M_0\varphi & \text{for } \pi \le \varphi < 2\pi, \end{cases}$$

$$\tilde{\omega} = \frac{1}{2\pi\sqrt{J}} \int_0^{2\pi} \sqrt{2L(\varphi) + J\omega_0^2}\,d\varphi;$$

after the calculation we get

$$\tilde{\omega} = \frac{1}{2\pi\sqrt{J}} \left[\int_0^\pi \sqrt{2M_0\varphi + J\omega_0^2}\,d\varphi + \int_\pi^{2\pi} \sqrt{4\pi M_0 - 2M_0\varphi + J\omega_0^2}\,d\varphi \right]$$

or

$$\tilde{\omega} = \frac{1}{3\pi M_0 \sqrt{J}} \left[\left(2M\pi + J\omega_0^2\right)^{\frac{3}{2}} - J^{\frac{3}{2}}\omega_0^3 \right].$$

The values for the graph of $\tilde{\omega}(\omega_0)$ shown in Figure 4.14 can be found in Table 4.1.

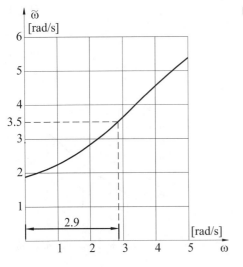

Figure 4.14 The function $\tilde{\omega} = \tilde{\omega}(\omega_0)$.

Table 4.1 Values for the curve $\tilde{\omega}(\omega_0)$.

ω_0	0	1	2	3	4	5
$\tilde{\omega}$	1.88	2.17	2.79	3.59	4.46	5.38

Figure 4.15 The function $\omega = \omega(\varphi)$.

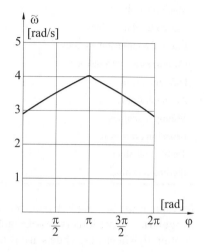

Figure 4.16 The function $\tilde{\omega} = \tilde{\omega}(\omega_0)$.

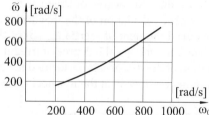

Drawing the straight line $\tilde{\omega} = \omega_{av}$ on the graph of the function $\tilde{\omega}(\omega_0)$ (Figure 4.14) gives the value $\omega_0 = 2.9$ rad/s, for which the expressions for the angular velocity are:

$$\omega = \begin{cases} \sqrt{8.41 + 2.55\varphi} \text{ for } 0 \leq \varphi < \pi, \\ \sqrt{24.41 - 2.55\varphi} \text{ for } \pi \leq \varphi < 2\pi. \end{cases}$$

Figure 4.15 shows the function $\omega(\varphi)$.

In another application in which $\varphi_d = 4\pi$, $J_{red}(\varphi) = 0.02(2 - \cos\varphi)$ kgm^2, $M_{red}(\varphi) = 10\sin\left(\frac{\varphi}{2}\right)$ Nm, and $L(\varphi) = 20\left[1 - \cos\left(\frac{\varphi}{2}\right)\right]$ J, and using a calculation program to determine the integral in (4.40), we obtained the graph for $\tilde{\omega}(\omega_0)$ shown in Figure 4.16.

4.7 Flywheels

4.7.1 Formulation of the Problem: Definitions

The technological and dynamical requirements of a machine (aggregate) demand that variations in the angular velocity during a dynamic cycle are bounded.

The *degree of irregularity* δ of the motion of a machine is defined as the ratio

$$\delta = \frac{\omega_{max} - \omega_{min}}{\omega_{av}}. \tag{4.45}$$

Table 4.2 The admissible degree of irregularity.

Type of machine	Degree of irregularity
Pumps	0.033–0.200
Machine tools	0.020–0.050
Agricultural machines	0.020–0.200
Internal combusion engines	0.006–0.012
Compressors with pistons	0.010–0.020
DC generators	0.005–0.010
AC generators	0.003–0.005
Milling machines	0.050–0.200
Broaching machines	0.160–0.180
Textile machines	0.100–0.350
Etching machines	0.300–0.400

Its value must be kept between certain limits. In Table 4.2 we set out the admissible degree of irregularity for various types of machine. To keep the machines within the limits, they usually contain flywheels. Flywheels are rotating bodies that have large moments of inertia J_f, so that they reduce the degree of irregularity.

For the calculation of the flywheel we assume as known: the reduced moment $M_{red}(\varphi)$, the reduced moment of inertia $J_{red}(\varphi)$ of the pieces of the aggregates (without the flywheel), the dynamic cycle φ_d, the degree of irregularity δ, and the average angular velocity ω_{av}. The calculation of the flywheel is performed for the regime phase and it reduces to the calculation of the moment of inertia of the flywheel J_f (reduced at the element at which the reduction is made).

4.7.2 Approximate Calculation

Adding the moment of inertia of the flywheel J_f at $J_{red}(\varphi)$, the angular velocity ω is expressed by the known relation in which $J_{red}(\varphi)$ is replaced with $J_{red}(\varphi) + J_f$

$$\omega = \sqrt{\frac{2L(\varphi) + (J_0 + J_f)\,\omega_0^2}{J_{red}(\varphi) + J_f}}. \tag{4.46}$$

In the approximate calculation, it is assumed that $J_f \gg J_{red}(\varphi)$. From (4.46) and neglecting $J_{red}(\varphi)$, we get

$$\omega^2 = \frac{2L(\varphi) + J_f\omega_0^2}{J_f}. \tag{4.47}$$

The approximation

$$\omega_{av} = \frac{\omega_{max} + \omega_{min}}{2}; \tag{4.48}$$

gives $\omega_{max} + \omega_{min} = 2\omega_{av}$ and $\omega_{max} - \omega_{min} = \delta\omega_{av}$, from which:

$$\omega_{max} = \omega_{av}\left(1 + \frac{\delta}{2}\right), \quad \omega_{min} = \omega_{av}\left(1 - \frac{\delta}{2}\right). \tag{4.49}$$

Squaring the expressions in (4.49) and neglecting the terms that contain δ^2 gives

$$\omega_{max}^2 = \omega_{av}^2(1 + \delta), \quad \omega_{min}^2 = \omega_{av}^2(1 - \delta). \tag{4.50}$$

If $L(\varphi)$ is replaced by L_{max} and L_{min} in (4.47), we get

$$\omega_{max}^2 = \frac{2L_{max} + J_f \omega_0^2}{J_f}, \quad \omega_{min}^2 = \frac{2L_{min} + J_f \omega_0^2}{J_f};$$

Then subtracting these two expressions from each other gives

$$\omega_{max}^2 - \omega_{min}^2 = \frac{2\left(L_{max} - L_{min}\right)}{J_f}; \tag{4.51}$$

Taking into account (4.50), we obtain

$$2\omega_{av}^2 = \frac{2\left(L_{max} - L_{min}\right)}{J_f}.$$

We thus obtain the moment of inertia of the flywheel

$$J_f = \frac{L_{max} - L_{min}}{\delta\omega_{av}^2}. \tag{4.52}$$

As a numerical example, we consider the first numerical example in Section 4.6.2, where $L_{max} = 4$ J, $L_{min} = 0$; if we choose $\delta = 0.05$, then $J_f = 6.53$ kgm^2.

4.7.3 Exact Calculation

Using the grapho-analytical method
For the exact calculation of the flywheel we start from (4.46). The steps are as follows:

1) Choose values for J_f: $J_{f_1}, J_{f_2}, J_{f_3}, ..., J_{f_i}, ...$
2) For a value J_{f_i} draw the curve $\tilde{\omega}(\omega_0)$ and determine the value ω_0, following the procedure described for the analysis of the regime phase (Figure 4.17).
3) From the calculation of the angular velocity with (4.46), in which J_f and ω_0 are replaced by J_{f_i} and ω_{0_i}, respectively, deduce the values ω_{max_i} and ω_{min_i} and then δ_i.
4) Repeat this procedure for different values of the parameter J_{f_i}, obtaining the numerical values for $\tilde{\delta}(J_f)$
5) On a graph of $\tilde{\delta}(J_f)$, draw the straight line $\tilde{\delta} = \delta$ (Figure 4.18) and determine the moment of inertia J_f of the flywheel.

We will outline a numerical example in Section 4.6.2. We obtain:

$$\tilde{\omega} = \frac{1}{3\pi M_0 \sqrt{J + J_f}} \left\{ \left[2M\pi + \left(J + J_f\right)\omega_0^2\right]^{\frac{3}{2}} - \left(J + J_f\right)^{\frac{1}{2}}\omega_0^{\frac{3}{2}} \right\};$$

Figure 4.17 The curves $J_f = $ const.

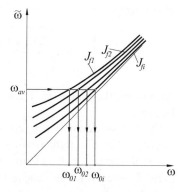

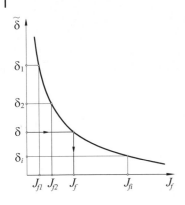

Figure 4.18 The function $\tilde{\delta} = \tilde{\delta}\left(J_f\right)$.

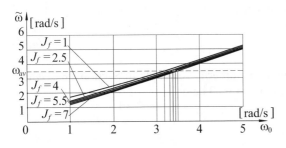

Figure 4.19 The graph of the function $\tilde{\omega} = \tilde{\omega}(\omega_0)$ for different values of J_f.

Table 4.3 The values of the angular velocity ω_0 as a function of J_f.

J_f	1	2.5	4	5.5	7
ω_0	3.194	3.323	3.374	3.401	3.418

Taking for J_f the values $J_f = 1$ kgm^2, $J_f = 2.5$ kgm^2, $J_f = 4$ kgm^2, $J_f = 5.5$ kgm^2, and $J_f = 7$ kgm^2, we obtain the curves in Figure 4.19.

If we find the intersections of the curves in Figure 4.19 and the straight line $\tilde{\omega} = \omega_{av} = 3.5$ s^{-1}, we obtain the values in Table 4.3. The curves can be drawn using AutoCAD for accurate determination of the intersection points. In the case considered, the values ω_{min} and ω_{max} are given by

$$\omega_{min} = \omega_0, \quad \omega_{max} = \sqrt{\frac{2M_0\pi + \left(J + J_f\right)\omega_0^2}{J + J_f}};$$

Therefore

$$\delta = \frac{1}{\omega_{av}}\left[\omega_0 + \sqrt{\frac{2M_0\pi + \left(J + J_f\right)\omega_0^2}{J + J_f}}\right].$$

We obtain the values in Table 4.4, and drawing these using AutoCAD gives the graph in Figure 4.20.

The intersection between the curve and the straight line $\delta = 0.05$ gives the exact value for the moment of inertia of the flywheel $J_f = 5.57$ kgm^2. The difference between this and the approximate value of $J_{f_{approx}} = 6.53$ kgm^2, determined in Section 4.7.2, is a result of the moment of inertia of the flywheel J_f being close to the value of the reduced moment of inertia J_{red}.

Table 4.4 The values of the degree of irregularity δ as a function of J_f.

J_f	1	2.5	4	5.5	7
δ	0.16455	0.09388	0.06569	0.05053	0.05105

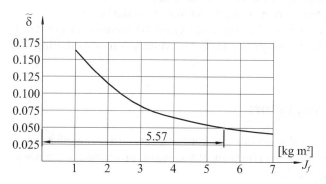

Figure 4.20 The graph of the function $\tilde{\delta} = \tilde{\delta}(J_f)$.

Using the iterative method

For the iterative method for the exact calculation of the moment of inertia of a flywheel, we need an approximate value for the first iteration. This is the value from the approximate calculation (4.52). The iterative formula used is:

$$J_f^{(i)} = J_f^{(i-1)} \frac{\delta_c^{(i-1)}}{\delta_{imposed}},$$

(4.53)

the deviation from the imposed value being

$$\left|\Delta\delta_i\right| = \left|\delta_{imposed} - \delta_c^{(i)}\right|.$$

(4.54)

The iterative procedure ends when

$$\left|\Delta\delta_i\right| \leq \rho,$$

(4.55)

where

- J_f is the value of the moment of inertia of the flywheel after i iterations
- $\delta_{imposed}$ is the maximum admissible degree of irregularity
- $\delta_c^{(i)}$ is the calculated value for the degree of irregularity after i iterations.

This procedure is quickly convergent; in a maximum of 10 iterations it ends at an error value equal to $\frac{1}{100}$ of the admissible degree of irregularity.

We now introduce the procedure *Eq_Motion*, which contains code for the the iterative solution of the equation of motion in the regime phase, and the exact calculation of the moment of inertia of the flywheel. It has as input data the arrays of 360 components M_{red} and J_{red}, and as output data the vector angular velocity ω and the exact value of the moment of inertia of the flywheel J_f.

The procedure first verifies if the system is in the regime phase by calculating the work W_{cr} produced by the reduced moment of inertia M_{red} during a dynamic cycle. This uses (4.35), the integral being calculated using the method of rectangles. Then two repetitive loops are used. The first determines the exact value of the moment of inertia of the flywheel for an imposed degree of irregularity δ. Inside the first loop, the second loop determines the values for the angular velocity ω_0, ω_{max}, ω_{min}, maximum work $W_{cr_{max}}$, and minimum work $W_{cr_{min}}$.

After the angular velocity ω_0 has been determined iteratively using (4.41)–(4.44), the error being equal to ρ, the variation of the angular velocity is determined using (4.46). The values of the angular velocity are written in the array *Omegastel*. The exact calculation of the moment of inertia of the flywheel uses (4.53) and (4.54), the approximate value for the first iteration being given by (4.52). The iterative procedure ends when (4.55) holds true.

The code for the procedure can be found in the companion website of the book.

Calling the procedure with the values from Section 4.6.2, gives the moment of inertia of the flywheel as $J_f = 5.56838$ kgm^2, which is very close to that previously determined using the graphical approach and AutoCAD.

4.8 Adjustment of Motion Regulators

Consider a thermal engine and an electrical generator that must work at constant angular velocity. Constant angular velocity is maintained with centrifugal regulators. From the point of view of static forces, there are two types of regulator:

- gravitational (for example, the Watt and Porter types), in which the static forces are provided only by the weights of the bodies
- with springs (for example, the Hartnell and Tolle types), in which the static forces are provided by both elastic and gravitational forces.

Figure 4.21 shows a Tolle-type regulator with a spring, which adjusts the angular velocity of an aggregate consisting of thermal engine 1 and electric generator 2, joined to one another by clutch EC. Axle 3 of the regulator, which rotates with angular velocity ω, is set in motion by generator 2. Together with axle 3, levers 4, 4', 5, and 5', the balls of weight G, and slider 6 of weight Q, all rotate with the same angular velocity.

The inertial (centrifugal) forces F_C are equilibrated by the weights G and Q, and the force in the spring BB'. At the displacement of slider 6, the displacement of lever 7 is produced with the aid of a fork; using lever 8, slider 6 sets clack 9 in motion, which, using pipe 10, can either close or open the fuel–air inlet of the engine. In this way, if the angular velocity increases over

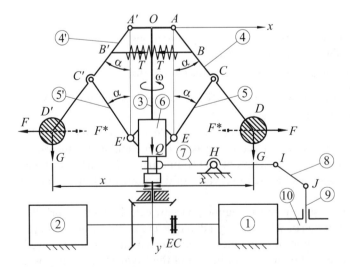

Figure 4.21 Centrifugal regulator with spring.

a pre-established value, then forces F_C increase, bar 4 rotates upwards, moving slider 6, lever 7, and crossbar 8, which produce a downward motion of clack 9, which closes the fuel-air mixture inlet, producing a reduction in the angular velocity of the engine.

To perform the kinetostatic calculation we use the following notation:

- $OA = OA' = a, AB = AB' = b, AC = A'C' = CE = C'E' = d, AD = A'D' = l$
- k, the stiffness of the spring
- l_0, the length of the spring at rest
- x, the distance between the axis of the regulator and weight G
- α, the angle between lever AC and the frame OA.

A mobile system of reference $Oxyz$ is chosen (Figure 4.21), and forces G, Q, and T (the force in the spring) are reduced to forces F^*, at points D and D'. Writing the virtual work, gives

$$F^*\delta x_{D'} - F^*\delta x_D = Q\delta y_E + G\delta y_{D'} + G\delta y_D + T\delta x_{B'} - T\delta x_B; \tag{4.56}$$

Since

$$x_{B'} = -a - b\sin\alpha, \ x_B = a + b\sin\alpha, \ y_E = 2d\cos\alpha,$$
$$y_{D'} = y_D = l\cos\alpha, \ x_{D'} = -a - l\sin\alpha, \ x_D = a + l\sin\alpha,$$

we get the expression

$$-2F^*l\cos\alpha\delta\alpha = (-2Qd\sin\alpha - 2Gl\sin\alpha - 2Tb\cos\alpha)\,\delta\alpha$$

or

$$F^* = \left(G + Q\frac{d}{l}\right)\tan\alpha + T\frac{b}{l}. \tag{4.57}$$

Force T in the spring is expressed by:

$$T = k\left(2a + 2b\sin\alpha - l_0\right); \tag{4.58}$$

so the reduced force F^* becomes

$$F^* = \left(G + Q\frac{d}{l}\right)\tan\alpha + \frac{kb}{l}\left(2a + 2b\sin\alpha - l_0\right). \tag{4.59}$$

Expressing the trigonometric functions $\sin\alpha$ and $\tan\alpha$ as functions of the parameter x:

$$\sin\alpha = \frac{x - a}{l}, \quad \tan\alpha = \frac{x - a}{\sqrt{l^2 - (x - a)^2}},$$

and the expression for force F^* becomes

$$F^* = \left(G + Q\frac{d}{l}\right)\frac{x - a}{\sqrt{l^2 - (x - a)^2}} + \frac{2kb^2}{l^2}(x - a) + \frac{kb}{l}\left(2a - l_0\right). \tag{4.60}$$

The centrifugal force F_C (Figure 4.21) is given by:

$$F_C = \frac{G}{g}\omega^2 x, \tag{4.61}$$

while the condition for dynamic equilibrium reads

$$F^* = F_C. \tag{4.62}$$

The *characteristic* of the regulator is defined by (4.60), which represents the variation of the reduced force F^* as a function of the parameter x. The function is drawn in Figure 4.22.

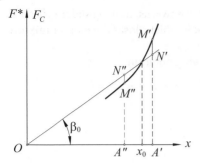

Figure 4.22 The characteristic of the regulator.

The variation of the centrifugal force as a function of the parameter x, for angular velocity in the regime phase $\omega = \omega_0$, is given by the straight line OM (Figure 4.22). The intersection between this straight line and the characteristic of the regulator defines the point of functioning M that corresponds to distance x_0. For the study of the stability of the regulator it is assumed that the regulator is not at the equilibrium position, but the angular velocity is unmodified; this new position corresponds to point M'. Since the variation of the centrifugal force is along the same straight line ($\tan \beta_0 = \dfrac{G}{g}\omega_0^2$), from the inequality $A'M' > A'N'$ it follows that $F'^* > F'_C$. This leads to the diminishing of the parameter x, which tends to x_0, the position of dynamic equilibrium.

If the position of the regulator corresponds to point M'', then from the inequality $BP'' > BP'$ it follows that $F''_C > F''^*$, so x increases to a value of x_0, which corresponds to the equilibrium position. Therefore a characteristic of the shape shown in Figure 4.22 ensures the stability of the regulator. This is referred to as a *static* characteristic. Characteristics that do not ensure the stability of the regulator referred to as *non-static*, while those that have the equality $F_C = F$ at any position are called *astatic* (the characteristic coincides with the straight line OM).

Taking into account (4.60) and (4.61), from (4.62) it follows that:

$$\omega^2 = \frac{g}{Gx}\left[\left(G + Q\frac{d}{l}\right)\frac{x - a}{\sqrt{l^2 - (x - a)^2}} + \frac{2kb^2}{l^2}(x - a) + \frac{kb}{l}(2a - l_0)\right]. \tag{4.63}$$

A graphical representation of this function (see Figure 4.23) gives the *characteristic of equilibrium $\omega(x)$* of the regulator.

Equation (4.63) for the equilibrium position x_0 gives the angular velocity in the regime phase ω_0. Taking into account the friction, then gives an interval $\left[\omega''_0, \omega'_0\right]$ of dynamic equilibrium for the regulator (Figure 4.23). Different values of the parameter x give the hatched equilibrium zone in Figure 4.23.

To highlight the insensitivity of the regulator, we define the coefficient of insensitivity ν using:

$$\nu = \frac{\omega' - \omega''}{\omega_0}. \tag{4.64}$$

The regulator does not react to variations of the angular velocity during a dynamic cycle if $\nu \geq \delta$, where δ is the degree of irregularity of the aggregate.

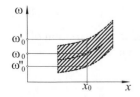

Figure 4.23 The characteristic of equilibrium of the regulator.

4.9 Dynamics of Multi-mobile Machines

Constantinescu's use of dynamic convertors as adaptive gearboxes in automobiles in the 1920s led to a need for the dynamic study of multi-mobile mechanisms. Consider a planar mechanism with two degrees of mobility, defined by the generalized coordinates q_1 and q_2. The kinetic energy is expressed by

$$T = \frac{1}{2} \sum_j \left(m_j v_j^2 + J_j \omega_j^2 \right),$$

$$(4.65)$$

where v_j, and ω_j represent the velocity of the centre of mass, and the angular velocity, respectively, for the jth element.

For holonomic and scleronomic constraints, and denoting by $\mathbf{r}_j$ and θ_j the vector of position of the centre of mass and the position angle, respectively, of the element, we get

$$\mathbf{v}_j = \frac{\partial \mathbf{r}_j}{\partial q_1} \dot{q}_1 + \frac{\partial \mathbf{r}_j}{\partial q_2} \dot{q}_2, \quad \omega_j = \frac{\partial \theta_j}{\partial q_1} \dot{q}_1 + \frac{\partial \theta_j}{\partial q_2} \dot{q}_2$$

$$(4.66)$$

and (4.65) can be written in the form

$$T = \frac{1}{2} \left(A_{11} \dot{q}_1 + A_{22} \dot{q}_2 \right) + A_{12} \dot{q}_1 \dot{q}_2,$$

$$(4.67)$$

where A_{ij} are functions that depend on q_1 and q_2.

The generalized forces Q_1 and Q_2 are obtained from the expression for the power

$$P = Q_1 \dot{q}_1 + Q_2 \dot{q}_2 = \sum_j \left(M_j \omega_j + \mathbf{F}_j \cdot \mathbf{v}_j \right),$$

$$(4.68)$$

where M_j and $\mathbf{F}_j$ are the forces and moments that act upon the system. Then, taking into account (4.67), gives:

$$Q_1 = \sum_j \left(M_j \frac{\partial \theta_j}{\partial q_1} + \mathbf{F} \frac{\partial \mathbf{r}_j}{\partial q_1} \right), \quad Q_2 = \sum_j \left(M_j \frac{\partial \theta_j}{\partial q_2} + \mathbf{F} \frac{\partial \mathbf{r}_j}{\partial q_2} \right).$$

$$(4.69)$$

Using (4.67) and (4.69) and the Lagrange equations

$$\frac{\mathrm{d}}{\mathrm{d}t} \left(\frac{\partial T}{\partial \dot{q}_k} \right) - \frac{\partial T}{\partial q_k} = Q_k, \; k = 1, \, 2,$$

$$(4.70)$$

gives the differential equations of the motion, which are solved numerically using the Runge–Kutta method.

As an example, consider the Constantinescu torque convertor, the constructive schema of which is shown in Figure 4.24. This convertor consists of principal axle 1, to which is jointed, at point A, mobile lever 2 which, through multiple joint B, acts upon the bars of connection 3 and $\tilde{3}$, which, through the revolute joints D and D', act upon the bars 4 and $\tilde{4}$, jointed to fixed point E. In this way, the rotational motion of principal axle 1 is transformed into the oscillatory motion of lever 2, and the latter is transformed into oscillatory motion in bars 4 and $\tilde{4}$. Now, the oscillatory motion of bars 4 and $\tilde{4}$ is transformed, using various coupling systems, into rotational motion of the same sense for the secondary axle 5.

In Figure 4.24 the coupling system of lever 4 consists of ratchet 8, rigidly linked to the secondary axle 5 using chock 9, latches 6, and springs 7. An identical coupling system is used in the back plane to couple lever $\tilde{4}$.

Constantinescu patented many technical inventions, some of them involving blocks that act by friction. In our example we assume that the masses of the bars are negligible, and the only non-zero mass is the concentrated mass m at the end F of lever 2 (Figure 4.24).

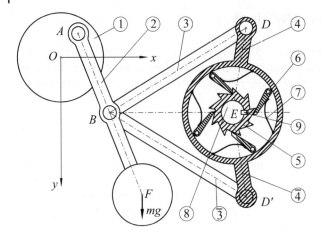

Figure 4.24 The constructive schema of the torque converter.

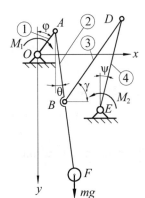

Figure 4.25 The kinematic schema of the torque converter.

For the dynamic study of the convertor, consider the model in Figure 4.25, which represents a mechanism with two degrees of mobility, the generalized coordinates being φ and θ. Choosing the system of coordinates in Figure 4.25, denoting the lengths of the segments OA, AB, AE, BD, and ED by a, b, l, c, and d, respectively, and denoting the coordinates of the fixed point E by x_E, y_E, then taking the vector relation

$$\mathbf{OA} + \mathbf{AB} + \mathbf{BD} + \mathbf{DE} = \mathbf{OE}, \tag{4.71}$$

and projecting it onto the axes and taking into account the notation in Figure 4.25, gives

$$
\begin{aligned}
a \sin \varphi + b \sin \theta + c \cos \gamma - d \sin \psi &= x_E, \\
-a \cos \varphi + b \cos \theta - c \sin \gamma + d \cos \psi &= y_E
\end{aligned}
\tag{4.72}
$$

or

$$
\begin{aligned}
a \sin \varphi + b \sin \theta - d \sin \psi - x_E &= -c \cos \gamma, \\
-a \cos \varphi + b \cos \theta + d \cos \psi - y_E &= c \sin \gamma.
\end{aligned}
\tag{4.73}
$$

Squaring the equations in (4.73), summing the results and using the notation:

$$
\begin{aligned}
A &= 2d \left(-a \cos \varphi + b \cos \theta - y_E \right), \quad B = 2d \left(a \sin \varphi + b \sin \theta - x_E \right), \\
C &= \left(a \sin \varphi + b \sin \theta - x_E \right)^2 + \left(-a \cos \varphi + b \cos \theta - y_E \right)^2 + d^2 - c^2,
\end{aligned}
\tag{4.74}
$$

gives

$$A \cos \psi - B \sin \psi + C = 0.$$

Taking into account the formulae

$$cos\psi = \frac{1 - \tan^2\left(\frac{\psi}{2}\right)}{1 + \tan^2\left(\frac{\psi}{2}\right)}, \quad \sin\psi = \frac{2\tan\left(\frac{\psi}{2}\right)}{1 + \tan^2\left(\frac{\psi}{2}\right)},$$

gives the second-order equation

$$(C - A)\tan^2\left(\frac{\psi}{2}\right) - @B\tan\left(\frac{\psi}{2}\right) + C + A = 0,$$

from which

$$\tan\left(\frac{\psi}{2}\right) = \frac{B + \sqrt{A^2 + B^2 - C^2}}{C - A}. \tag{4.75}$$

Differentiating (4.72) with respect to time gives

$$a\dot{\varphi}\cos\varphi + b\dot{\theta}\cos\theta - c\dot{\gamma}\sin\gamma - d\dot{\psi}\cos\psi = 0,$$
$$a\dot{\varphi}\sin\varphi - b\dot{\theta}\sin\theta - c\dot{\gamma}\cos\gamma - d\dot{\psi}\sin\psi = 0, \tag{4.76}$$

which gives

$$\dot{\psi} = A_4\dot{\varphi} + B_4\dot{\theta}, \tag{4.77}$$

where

$$A_4 = \frac{a\cos(\varphi + \gamma)}{d\cos(\psi + \gamma)}, \quad B_4 = \frac{b\cos(\theta - \gamma)}{d\cos(\psi + \gamma)}. \tag{4.78}$$

The kinetic energy reads

$$T = \frac{1}{2}mv_F^2 = \frac{1}{2}m\left(\dot{x}_F^2 + \dot{y}_F^2\right); \tag{4.79}$$

and since

$$x_F = a\sin\varphi + l\sin\theta, \quad y_F = -a\cos\varphi + l\cos\theta,$$
$$\dot{x}_F = a\dot{\varphi}\cos\varphi + l\dot{\theta}\cos\theta, \quad \dot{y}_F = a\dot{\varphi}\sin\varphi - l\dot{\theta}\sin\theta,$$

we get

$$T = \frac{1}{2}m\left[a^2\dot{\varphi}^2 + l^2\dot{\theta}^2 + 2al\dot{\varphi}\dot{\theta}\cos(\varphi + \theta)\right]. \tag{4.80}$$

The generalized forces are obtained from the expression for the power; that is, from the expression:

$$P = M_1\dot{\varphi} + mg\dot{y}_F - M_2\dot{\psi} = M_1\dot{\varphi} + mg\left(a\dot{\varphi}\sin\varphi - l\dot{\theta}\sin\theta\right) - M_2\left(A_4\dot{\varphi} + B_4\dot{\theta}\right)$$

or

$$P = \left(M_1 + mga\sin\varphi - M_2A_4\right)\dot{\varphi} - \left(mgl\sin\theta + M_2B_4\right)\dot{\theta}.$$

This gives

$$Q_\varphi = M_1 + mga\sin\varphi - M_2A_4, \quad Q_\theta = -mgl\sin\theta - M_2B_4. \tag{4.81}$$

By calculating the expressions

$$\frac{d}{dt}\left(\frac{\partial T}{\partial\dot{\varphi}}\right) = m\left[a^2\ddot{\varphi} + al\ddot{\theta}\cos(\varphi + \theta) - al\dot{\theta}\left(\dot{\varphi} + \dot{\theta}\right)\sin(\varphi + \theta)\right],$$

$$\frac{d}{dt}\left(\frac{\partial T}{\partial\dot{\theta}}\right) = m\left[l^2\ddot{\theta} + al\ddot{\varphi}\cos(\varphi + \theta) - al\dot{\varphi}\left(\dot{\varphi} + \dot{\theta}\right)\sin(\varphi + \theta)\right],$$

$$\frac{\partial T}{\partial\varphi} = -mal\dot{\varphi}\dot{\theta}\sin(\varphi + \theta), \quad \frac{\partial T}{\partial\theta} = -mal\dot{\varphi}\dot{\theta}\sin(\varphi + \theta),$$

the Lagrange equations

$$\frac{\mathrm{d}}{\mathrm{d}t}\left(\frac{\partial T}{\partial \dot{\varphi}}\right) - \frac{\partial T}{\partial \varphi} = Q_\varphi, \quad \frac{\mathrm{d}}{\mathrm{d}t}\left(\frac{\partial T}{\partial \dot{\theta}}\right) - \frac{\partial T}{\partial \theta} = Q_\theta \tag{4.82}$$

become

$$m\left[a^2\ddot{\varphi} + al\ddot{\theta}(\varphi+\theta) - al\dot{\theta}^2\sin(\varphi+\theta)\right] = Q_\varphi,$$
$$m\left[al\ddot{\varphi}\cos(\varphi+\theta) + l^2\ddot{\theta} - al\dot{\varphi}^2\sin(\varphi+\theta)\right] = Q_\theta. \tag{4.83}$$

To integrate the system using the Runge–Kutta method, $\ddot{\theta}$ is eliminated from (4.83), giving the expression for $\ddot{\varphi}$; similarly, eliminating $\ddot{\varphi}$ gives the expression for $\ddot{\theta}$.

Then, denoting

$$Y_1 = \varphi, \ Y_2 = \dot{\varphi}, \ Y_3 = \theta, \ Y_4 = \dot{\theta},$$

gives:

$$\frac{\mathrm{d}Y_1}{\mathrm{d}t} = Y_2, \quad \frac{\mathrm{d}Y_3}{\mathrm{d}t} = Y_4, \tag{4.84}$$

to which can be added the equations obtained from (4.83), which can be put in the form

$$\frac{\mathrm{d}Y_2}{\mathrm{d}t} = F_1\left(t, Y_1, Y_2, Y_3, Y_4\right), \quad \frac{\mathrm{d}Y_4}{\mathrm{d}t} = F_2\left(t, Y_1, Y_2, Y_3, Y_4\right). \tag{4.85}$$

We consider the case in which

$$\dot{\varphi} = \omega = \text{const}, \ \varphi = \omega t + \tfrac{\pi}{2}, \ m = 3\,\text{kg},$$
$$a = 0.02\,\text{m}, \ b = 0.1\,\text{m}, \ c = 0.2\,\text{m}, \ d = 0.3\,\text{m},$$
$$d = x_E = \sqrt{b^2 - a^2}, \ y_E = \sqrt{c^2 - d^2};$$

The initial conditions are

$$t = 0 \rightarrow \begin{cases} \theta = \arctan\left(\dfrac{a}{d}\right), \\ \dot{\theta} = 0, \end{cases}$$

the integration step being $\Delta t = 0.001$ s.

For

$$M_2 = \begin{cases} M_5 \text{ for } \dot{\psi} \geq 0, \\ -M_5 \text{ for } \dot{\psi} < 0 \end{cases}$$

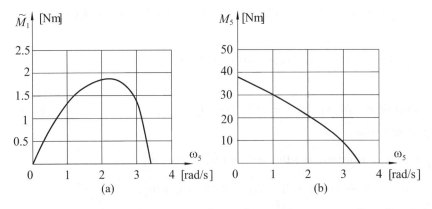

Figure 4.26 The variation of the moments $\tilde{M}_1$ and $\tilde{M}_5$ as functions of ω_5: (a) $\tilde{M}_1 = \tilde{M}_1(\omega_5)$; $M_5 = M_5(\omega)$.

and the values $\omega = 10$ rad/s, $\omega = 20$ rad/s, $\omega = 40$ rad/s, $\omega = 60$ rad/s, $M_5 = 5$ Nm, $M_5 = 10$ Nm, and $M_5 = 20$ Nm, the system (4.83) and (4.85), integrated using the Runge–Kutta method, gives the results in Figure 4.26.

The parameters $\tilde{M}_1$ and $\tilde{\omega}$ in Figure 4.26 represent average values over the period $T_\omega = \dfrac{2\pi}{\omega}$ calculated for $\omega = 20$ rad/s.

These diagrams are in full agreement with the characteristic obtained by Constantinescu: $M_5^2 = C_1^2 \omega^2 \left(\omega^2 - C_2 \tilde{\omega}_5^2 \right)$, where C_1 and C_2 are constants that depend on the inertial characteristics of the convertor. The diagrams express the *adaptive property* of the convertor; that is, how the output angular velocity $\tilde{\omega}_5$ varies as a function of the resistant moment M_5.

5

Synthesis of Planar Mechanisms with Bars

The term 'synthesis' covers the establishment of design criteria, and the design of mechanisms that satisfy them. In the most cases, the requirements imposed by the theme (geometric, kinematic, dynamic, or any combination thereof) can be fulfilled by four-bar mechanisms.

We will now discuss the synthesis of four-bar mechanisms that satisfy pure geometric requirements.

The majority of these mechanisms are designed to solve problems in machines in which a point in the plane of the coupler must approximate a certain curve. This *coupler curve* may be a straight line or a circle. The first of these applies in, for example, boring or digging machines and devices for guiding movie cameras; the second is used in stop mechanisms. These kinds of problem characterize the synthesis of path-generating four-bar mechanisms.

Another kind of mechanism is for situations in which a certain segment in the plane of the coupler must occupy a defined series of successive positions. An example is found in the feed mechanisms of the transporters. This kind of problem is called a *multi-positional* synthesis.

Other mechanisms are for situations where there must be a certain function between the input and output parameters. These are called *function-generating* mechanisms.

5.1 Synthesis of Path-generating Four-bar Mechanism

5.1.1 Conditions for Existence of the Crank

Let us consider the four-bar mechanism $OABC$ (Figure 5.1), with dimensions $OA = a$, $AB = b$, and $CD = d$, and let us determine the conditions in which the driving element OA (a crank) can perform a complete revolution without locking the mechanism.

If we denote the diagonal AC by f, then from the triangle ABC we have the double inequality

$$|b - c| \leq f \leq |b + c|. \tag{5.1}$$

Edge f can be expressed using the cosine theorem, then from triangle OAC comes

$$f^2 = a^2 + d^2 - 2ad \cos \varphi, \tag{5.2}$$

from which

$$(a - d)^2 = f_{min}^2, \quad (a + d)^2 = f_{max}^2. \tag{5.3}$$

Writing (5.1) in the form

$$(b - c)^2 \leq f_{min}^2 \leq f_{max}^2 \leq (b + c)^2, \tag{5.4}$$

Classical and Modern Approaches in the Theory of Mechanisms, First Edition.
Nicolae Pandrea, Dinel Popa and Nicolae-Doru Stănescu.
© 2017 John Wiley & Sons Ltd. Published 2017 by John Wiley & Sons Ltd.
Companion Website: www.wiley.com/go/pandmech17

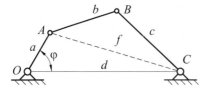

Figure 5.1 Four-bar mechanism.

gives the sequence of inequalities

$$(b-c)^2 \le (a-d)^2 \le (a+d)^2 \le (b+c)^2 \,, \tag{5.5}$$

which separates into the inequalities

$$(b-c)^2 \le (a-d)^2 \,, \quad (a+d)^2 \le (b+c)^2 \,. \tag{5.6}$$

This gives the system of inequalities

$$\begin{cases} (b-c-a+d)(b-c+a-d) \le 0, \\ \qquad\qquad a+d \le b+c, \end{cases} \tag{5.7}$$

which is equivalent to the systems

$$\begin{cases} a+d \le b+c, \\ b+d-c-a \le 0, \\ b+a-c-d \ge 0, \end{cases} \tag{5.8}$$

$$\begin{cases} a+d \le b+c, \\ b+d-c-a \ge 0, \\ b+a-c-d \le 0, \end{cases} \tag{5.9}$$

that is, the systems

$$\begin{cases} a+d \le b+c, \\ b+d \le c+a, \\ c+d \le a+b, \end{cases} \tag{5.10}$$

$$\begin{cases} a+d \le b+c, \\ a+c \le b+d, \\ a+b \le c+d. \end{cases} \tag{5.11}$$

The expressions in (5.10) are symmetrical in a and c, and so they represent the requirement that the mechanism be a double-crank one. For this case, the following also apply: $d \le a, d \le b, d \le c$; that is, the base is the shortest element.

The relations (5.11) are not symmetrical in a and c, and represent the requirement that the mechanism be a crank-rocker one. For this case we have $a \le d, a \le b, a \le c$; that is, the crank is the shortest element.

When neither (5.10) nor (5.11) are fulfilled, the mechanism either has no crank, or it is a double-rocker one, or the mechanism can not be constructed.

5.1.2 Equation of the Coupler Curve

The coupler curve is the geometric locus of a point in the plane of the coupler (Figure 5.2). If $M(x, y)$ is such a point and the dimensions of the edges are $O_1A = a$, $AM = e$, $MB = f$ and

Figure 5.2 The coupler curve.

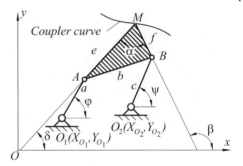

$O_2B = c$, then, using the notations in Figure 5.2, one may write:

$$\begin{cases} x = x_{O_1} + a \cos\varphi + e \cos\gamma, \\ y = y_{O_1} + a \sin\varphi + e \sin\gamma, \end{cases} \tag{5.12}$$

$$\begin{cases} x = x_{O_2} + c \cos\psi + f \cos\beta, \\ y = y_{O_2} + c \sin\psi + f \sin\beta. \end{cases} \tag{5.13}$$

Eliminating the angles φ and ψ, then (5.12) and (5.13) give the equalities

$$\left(x - x_{O_1} - e \cos\gamma\right)^2 - \left(y - y_{O_1} - e \sin\gamma\right)^2 = a^2, \tag{5.14}$$

$$\left(x - x_{O_2} - f \cos\beta\right)^2 + \left(y - y_{O_2} - f \sin\beta\right)^2 = c^2. \tag{5.15}$$

Now, taking into account $\beta = \alpha + \gamma$ and using the notation

$$\begin{aligned} & A_1 = 2e\left(x - x_{O_1}\right), B_1 = 2e\left(y - y_{O_1}\right), \ C_1 = \left(x - x_{O_1}\right)^2 + \left(y - y_{O_1}\right)^2 + e^2 - a^2, \\ & A_2 = 2f\left[\left(x - x_{O_2}\right)\cos\alpha + \left(y - y_{O_2}\right)\sin\alpha\right], \\ & B_2 = 2f\left[-\left(x - x_{O_2}\right)\sin\alpha + \left(y - y_{O_2}\right)\cos\alpha\right], \\ & C_2 = \left(x - x_{O_2}\right)^2 + \left(y - y_{O_2}\right)^2 + f^2 + c^2, \end{aligned} \tag{5.16}$$

gives the expressions

$$\begin{aligned} & A_1 \cos\gamma + B_1 \sin\gamma = C_1, \\ & A_2 \cos\gamma + B_2 \sin\gamma = C_2, \end{aligned} \tag{5.17}$$

from which

$$\cos\gamma = \frac{C_1 B_2 - C_2 B_1}{A_1 B_2 - A_2 B_1}, \quad \sin\gamma = \frac{-C_1 A_2 + C_2 A_1}{A_1 B_2 - A_2 B_1}. \tag{5.18}$$

Based on the equality $\cos^2\gamma + \sin^2\gamma = 1$, the equation of the coupler curve is obtained:

$$\left(C_1 B_2 - C_2 B_1\right)^2 + \left(-C_1 A_2 + C_2 A_1\right)^2 - \left(A_1 B_2 - A_2 B_1\right)^2 = 0. \tag{5.19}$$

The equation of the coupler curve is a sixth-degree polynomial, the sixth-degree terms in x and y forming the expression $\lambda\left(x^2 + y^2\right)^3$, where λ is a constant. This function, symbolically written in the form

$$F\left(x, y, x_{O_1}, y_{O_1}, x_{O_2}, y_{O_2}, a, c, e, f, \alpha\right) = 0, \tag{5.20}$$

shows that it depends on nine parameters; therefore a four-bar mechanism can be constructed with a coupler curve that passes through nine given points.

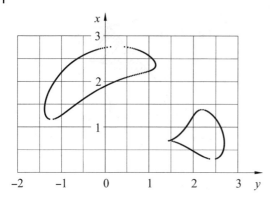

Figure 5.3 The coupler curves obtained as solutions.

If the nine points are defined by their coordinates (x_i, y_i), $i = 1, 2, \ldots, 9$, then the nine parameters $x_{O_1}, y_{O_1}, x_{O_2}, y_{O_2}, a, c, e, f, \alpha$ are obtained from the solution of the non-linear system

$$F\left(x_i, y_i, x_{O_1}, y_{O_1}, x_{O_2}, y_{O_2}, a, c, e, f, \alpha\right) = 0, \ i = 1, 2, \ldots, 9. \tag{5.21}$$

We wish to draw the graph of function (5.20) in AutoCAD. For that purpose we assume as known the dimensions of the mechanism and the positions of the kinematic pairs at the base. We use the following algorithm:

1) Give values to x, in the interval $[x_{min}, x_{max}]$, using a constant step Δx.
2) For each value of x, solve (5.20).
3) Writes the value of the solution in a script file.
4) Draw in AutoCAD the pairs of points recorded in the script file.

A classical method – namely the bisection method – is preferred for solving (5.20). To verify the solution, the Turbo Pascal calculation program also draws the coupler curves (as a result of the positional analysis of the four-bar mechanism R–$3R$). The solutions are marked as circles. The code of the program is shown in the companion website of this book.

Consider the data $x_{O_1} = 0$, $y_{O_1} = 0$, $x_{O_2} = 1.4$, $y_{O_2} = 0$, $a = 0.6$, $e = 2.2$, $c = 1.5$, $f = 2.2$, and $\alpha = 60°$. Running the script file *FCapital.scr*, gives the construction in Figure 5.3.

5.1.3 Triple Generation of the Coupler Curve

Consider the four-bar mechanism $OABC$ in Figure 5.4. Let us choose a point M in the plane of the coupler, and let $OA = a$, $AB = b$, $OC = d$, $MA = e$, $MB = f$, $\beta = \widehat{MAC}$, $\gamma = \widehat{ABM}$. We construct the parallelograms $OAMD$ and $CBME$ on edges MD and ME, respectively, and we construct the triangles MDF and MEG; these are similar to the triangle ABM. Finally, we construct the parallelogram $FMGH$.

We have thus constructed a desmodromic mechanism which, as we will prove, has the property that point H is fixed. In the reference system in Figure 5.4, the coordinates of point H read

$$\begin{aligned} x_H &= OD\cos(\beta + \theta) + DF\cos(\beta + \varphi) + FH\cos(\psi - \beta), \\ y_H &= OD\sin(\beta + \theta) + DF\sin(\beta + \varphi) - FH\sin(\psi - \beta). \end{aligned} \tag{5.22}$$

From the similarity of the triangles DFM and ABM, and MEG and ABM, respectively, we have:

$$DF = \frac{ea}{b}, \ FH = \frac{ec}{b}, \tag{5.23}$$

while from the equations of projections for the loop $OABC$ we have

$$\begin{aligned} a\cos\varphi + b\cos\theta + c\cos\psi &= d, \\ a\sin\varphi + b\sin\theta - c\sin\psi &= 0. \end{aligned} \tag{5.24}$$

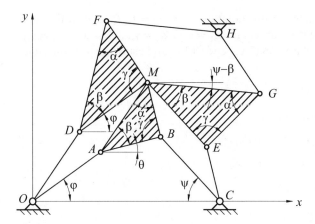

Figure 5.4 Triple generation of the coupler curve.

Taking into account (5.23), the coordinates of point H read

$$x_H = \frac{e}{b} \left[\cos \beta \left(b \cos \theta + a \cos \varphi + c \cos \psi \right) - \sin \beta \left(b \sin \theta + a \sin \varphi + c \sin \psi \right) \right],$$

$$y_H = \frac{e}{b} \left[\sin \beta \left(b \cos \theta + a \cos \varphi + c \cos \psi \right) - \cos \beta \left(b \sin \theta + a \sin \varphi + c \sin \psi \right) \right]; \tag{5.25}$$

Then, using (5.24) we obtain

$$x_H = \frac{ed}{b} \cos \beta, \quad y_H = \frac{ed}{b} \sin \beta. \tag{5.26}$$

The coordinates of point H being constant, this point is fixed and it can be considered a revolute joint. We thus obtain another two mechanisms $ODFH$ and $CEGH$, for which point M describes the same coupler curve. In the particular case in which point M is situated on the coupler AB ($\beta = 0$ or $\beta = \pi$), from (5.26) point H is situated on the base OC at a distance $\frac{ed}{b}$ from point O.

In the companion website of the book, we present the code for the AutoLisp function for construction of the three four-bar mechanisms that generate the same coupler curve. The coordinates of the points used in the function are $O\,(0,0)$, $A\,(42, 28)$, $C\,(115, 0)$, and $M\,(71, 70)$.

5.1.4 Analytic Synthesis

As we have already shown in Section 5.1.2, the problem admits an approximate solution (Figure 5.5); that is, we can determine a four-bar mechanism so that a point M of the coupler passes through nine chosen points $M_1, M_2, \ldots, M_9$ situated on the given path.

Figure 5.5 The coupler curve that approximates a given path.

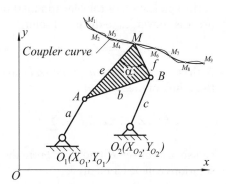

To establish the calculation algorithm we use the notation $(x_i, y_i), i = 1, 2, \ldots, 9$ for the coordinates of the points chosen on the given curve, and $Z_1 = x_{O_1}, Z_2 = y_{O_1}, Z_3 = x_{O_2}, Z_4 = y_{O_2}, Z_5 = a, Z_6 = c, Z_7 = e, Z_8 = f, Z_9 = \alpha$. Based on (5.16) and (5.19), and using the previous notation, we have

$$A_{1i} = 2Z_7 \left(x_i - Z_1\right), \quad B_{1i} = 2Z_7 \left(y_i - Z_2\right),$$

$$C_{1i} = \left(x_i - Z_1\right)^2 + \left(y_i - Z_2\right)^2 + Z_7^2 - Z_5^2,$$

$$A_{2i} = 2Z_8 \left[\left(x_i - Z_3\right) \cos Z_9 + \left(y_i - Z_4\right) \sin Z_9\right], \tag{5.27}$$

$$B_{2i} = 2Z_8 \left[-\left(x_i - Z_3\right) \sin Z_9 + \left(y_i - Z_4\right) \cos Z_9\right],$$

$$C_{2i} = \left(x_i - Z_3\right)^2 + \left(y_i - Z_4\right)^2 + Z_8^2 - Z_6^2,$$

$$F_i\left(Z_1, Z_2, \ldots, Z_9\right) = \left(C_{1i}B_{2i} - C_{2i}B_{1i}\right)^2$$

$$+ \left(-C_{1i}A_{2i} + C_{2i}A_{1i}\right)^2 - \left(A_{1i}B_{2i} - A_{2i}B_{1i}\right)^2. \tag{5.28}$$

Now let us denote

$$\{\mathbf{Z}\} = \begin{bmatrix} Z_1 & Z_2 & \cdots & Z_9 \end{bmatrix}^{\mathrm{T}}, \quad \{\mathbf{\Delta Z}\} = \begin{bmatrix} \Delta Z_1 & \Delta Z_2 & \cdots & \Delta Z_9 \end{bmatrix}^{\mathrm{T}},$$

$$\{\mathbf{F}\} = \begin{bmatrix} F_1 & F_2 & \cdots & F_9 \end{bmatrix}^{\mathrm{T}}, \tag{5.29}$$

and let

$$\{\mathbf{Z_0}\} = \begin{bmatrix} Z_1^0 & Z_2^0 & \cdots & Z_9^0 \end{bmatrix}^{\mathrm{T}} \tag{5.30}$$

be an approximate value for the solution of the problem. According to the Newton–Raphson method, if we denote by $[\mathbf{J}]$ the Jacobian matrix having as entries the partial derivatives $F_{ij} = \dfrac{\partial F_i}{\partial Z_j}$, then the deviation $\{\mathbf{\Delta Z}\}$ is given by:

$$\{\mathbf{\Delta Z}\} = - [\mathbf{J}]^{-1} \{\mathbf{F}\}, \tag{5.31}$$

where the Jacobian matrix $[\mathbf{J}]$ and the vector matrix $\{\mathbf{F}\}$ are calculated for the values $Z_1^0, Z_2^0, \ldots, Z_9^0$. Then, we make the actualization

$$\{\mathbf{Z}\} = \{\mathbf{Z_0}\} + \{\mathbf{\Delta Z}\} \tag{5.32}$$

and continue the iterative process until reaching the required level of precision

$$\left|\Delta Z_i\right| < \varepsilon. \tag{5.33}$$

With the solution obtained and using the Roberts theorem, we deduce the other two mechanisms that generate the same coupler curve.

The application of the Newton method must be made with care because, in most cases, the process converges to the banal solution

$$Z_1 = Z_3, \quad Z_2 = Z_4, \quad Z_5 = Z_6, \quad Z_7 = Z_8, \quad Z_9 = 0. \tag{5.34}$$

In these cases, the *gradient method* can be used. This involves determining the points at which the values of the function

$$F\left(Z_1, Z_2, \ldots, Z_9\right) = \frac{1}{9} \sum_{i=1}^{9} F_i^2 \tag{5.35}$$

vanish in a convenient domain. From the solutions obtained, the points that represent the banal solutions can be eliminated.

5.1.5 Mechanisms for which Coupler Curves Approximate Circular Arcs and Segments of Straight Lines

Figure 5.6a shows a four-bar mechanism, the coupler curve of which, in some portions ($M_{12}M_2$, M_3M_4), may be approximated by circular arcs. This property can be used, by attaching the dyad MPO, to construct a stop mechanism. In this case, the rocker PO is at rest when point M describes the arcs M_1M_2 and M_3M_4. Figure 5.6b shows a mechanism for which the coupler curve approximates a circle. The mechanisms for which the coupler curves approximate segments of straight lines are represented in Figure 5.6c–e.

This property may be used in designing certain cranes (horizontal displacement, Figure 5.6d), digging machines (vertical displacement, Figure 5.6e), and machinery for automatic transport of work-pieces and so on.

5.1.6 Method of Reduced Positions

Graphical methods have been abandoned over the last 25 years because the need to read distances with a ruler makes them very imprecise. That has changed with the advent of AutoCAD, which permits the determination of the distances with any required level of precision. We will now describe Hain's graphical method for designing a mechanism so that a point of the coupler approximates a given curve.

Approximation by three points

Consider an arc of curve C_1C_3. We want to design a mechanism A_0ABB_0 so that a point C of the coupler passes through point C_1, C_2, and C_3 (where C_2 is chosen to be on the arc C_1C_3).

For that purpose (Figure 5.7) we can choose the fixed revolute joints A_0, B_0, the radius r_0 of the circle on which point A moves, and the length l of the segment CA. Then the circular arcs

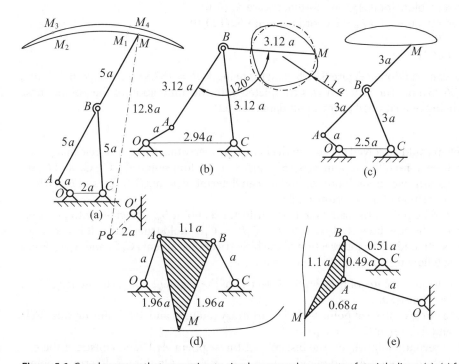

Figure 5.6 Coupler curves that approximate circular arcs and segments of straight lines: (a)–(c) four-bar mechanisms for which the coupler curves approximate circular arcs; (d) , (e) four-bar mechanisms for which the coupler curves approximate segments of straight lines.

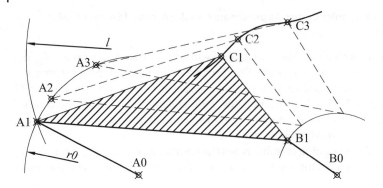

Figure 5.7 Synthesis of a four-bar mechanism so that a point of the coupler passes through three given points.

with their centres at points C_1, C_2, and C_3 and with radii equal to l can be constructed; these arcs intersect the circle centred at point A_0 and with radius r_0, at points A_1, A_2, and A_3, respectively.

For the determination of point B_1, note that if element $C_1A_1B_1$ is fixed, then point B_0 describes a circle centred at point B_1. The second and the third positions of point B_0 are denoted by B'_{02} and B'_{03}. Observing the equalities $\Delta C_3A_2B_0 = \Delta C_1A_1B'_{02}$ and $\Delta C_3A_3B_0 = \Delta C_1A_1B'_{03}$, points B'_{02} and B'_{03} can be constructed using the pattern method, while point B_1 is situated at the intersection of the bisectrices of the segments $B_0B'_{02}$ and $B_0B'_{03}$.

In the companion website of the book, we set out the code for the AutoLisp function that determines the dimensions of the mechanism that approximates the path by three points, following the previous methodm. This function has the following input data (in mm):

- for the coordinates: C_1 (50, 70), C_2 (60, 80), C_3 (90, 90)
- for the fixed revolute joint A_0 the coordinates are A_0 (0, 0)
- for the fixed revolute joint B_0 the coordinates are B_0 (120, 0)
- radius $r_0 = 70$
- length $l = 118.81$.

This gives the construction in Figure 5.7, the lengths being $A_1B_1 = 152.8$ mm, $B_1C_1 = 63.65$ mm, and $B_1B_0 = 36.35$ mm. The function was completed with the annotation of the points, these annotations being repositioned as the input data changes.

5.1.6.1 Approximation by four points

The last problem could be solved because of the known property that there is a circle that passes through any three non-collinear points (B_0, B'_{02}, and B'_{03}). For four points C_1, C_2, C_3, and C_4 we proceed analogously; we choose point B_0 and we can determine points B'_{02}, B'_{03}, and B'_{04}, which in general are not situated on the same circle.

The points can be put on the same circle if, for instance, point B'_{04} coincides with point B_0. For that purpose the equalities $A_4B_0 = A_1B_0$ and $C_4B_0 = C_1B_0$ hold true; hence, it is necessary that point B_0 is situated at the intersection of the bisectrices of segments C_1C_4 and A_1A_4. From this results the following series of constructions:

1) Choose the fixed revolute joint B_0 on the bisectrix of the segment C_1C_4; choose length l of segment CA (Figure 5.8).
2) Construct a circle centred at point B_0 and of arbitrary radius y, and determine on this circle points A_1 and A_4 so that $C_1A_1 = C_4A_4 = l$.
3) Choose the revolute joint A_0 on the bisectrix of the segment A_1A_4 and construct the circular arc centred at point A_0 and of radius $r_0 = A_0A_1 = A_0A_4$. On this arc determine points A_2 and A_3 by intersecting the circle centred at point A_0 and of radius A_0A_1 with the circles of radii equal to l with their centres at points C_2, and C_3, respectively.

Figure 5.8 Synthesis of a four-bar mechanism so that a point of the coupler passes through four points.

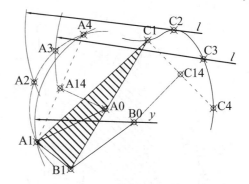

point B in position 1 of the mechanism (B_1) is obtained using the *method of associated positions* (also known as the the method of rigidity) in which triangle $C_2A_2B_2$ is equal to the triangle $C_1A_1B'_{02}$, while triangle $C_3A_3B_3$ is equal to triangle $C_1A_1B'_{03}$.

Points B'_{02} and B'_{03} are obtained from the equalities of the previous triangles as follows:

1) Overlap edge C_2A_2 of triangle $C_2A_2B_2$ over edge C_1A_1, giving point B'_{02}.
2) Overlap edge C_3A_3 of triangle $C_3A_3B_3$ over edge C_1A_1, giving point B'_{03}.

Points B_0, B'_{02}, and B'_{03} are situated on the circle centred at point B_1 (this circle circumscribes the triangle $B_0B'_{02}B'_{03}$). The mechanism we have searched is $A_0A_1B_1B_0$.

In the companion website of the book, we present the code for the AutoLisp function that creates the approximation of a curve by four points. For efficiency we used points defined inside the function. These points have the coordinates (in mm): C_1 $(75, 60)$, C_2 $(83, 63)$, C_3 $(92, 54.5)$, and C_4 $(95, 40)$. For the determination of points A_0 and B_0 we assume as known: $C_{14}B_0 = 20$ mm, and $A_{14}A_0 = 15$ mm, C_{14}, and A_{14} being the middles of segments C_1C_4, and A_1A_4, respectively (Figure 5.8). We also know $l = 45$ mm and $y = 30$ mm. This data, when processed through the AutoLisp function, gave the construction in Figure 5.8. The dimensions were:

$$A_0A_1 = 22.91 \text{ mm}, \quad A_1B_1 = 12.93 \text{ mm}, \quad B_1B_0 = 23.73 \text{ mm}, \quad B_1C_1 = 44.61 \text{ mm}$$

and the coordinates of the fixed joints (in mm) were:

$$A_0 \, (62.1886, 39.7305) \,, \quad B_0 \, (70.8279, 35.8579) \,.$$

As in the previous case, we annotated the points, repositioning the annotations as the input data changed.

Approximation by five points
This problem may be reduced to the four-point one if $B'_{01} = B'_{04}$ and $B'_{02} = B'_{03}$ (see Figure 5.9). We construct the bisectrices of segments C_1C_5 and C_2C_3; these bisectrices intersect at the fixed joint B_0. We then choose the straight line D that passes through point B_0, and construct the straight lines D_1 and D_2, which make angles of $\beta_{15} = \frac{1}{2}\widehat{C_1B_0C_5}$ and $\beta_{23} = \frac{1}{2}\widehat{C_2B_0C_3}$ with axis D, respectively. Then we choose length $l = CA$ and construct points A_1 and A_2, situated on the straight lines D_1 and D_2, respectively, so that $C_1A_1 = C_2A_2$. The bisectrix of segment A_1A_2 intersects straight line D at point A_0, which is a fixed joint too.

We construct the circular arc with its centre at point A_0 that passes through points A_1 and A_2 and then determine on this arc points A_3, A_4, and A_5, so that $C_3A_3 = C_4A_4 = C_5A_5 = l$. With the aid of these constructions one ensures the symmetry of points A_1, A_5 and A_2, A_3, respectively, relative to axis D.

We construct triangles $C_1A_1B'_{02}$ and $C_1A_1B'_{04}$ so that $C_1B'_{02} = C_2B_0$, $A_1B'_{02} = A_2B_0$, $C_1B'_{04} = C_4B_0$, and $A_1B'_{04} = A_4B_0$, and determine points B'_{02} and B'_{04}. The symmetries

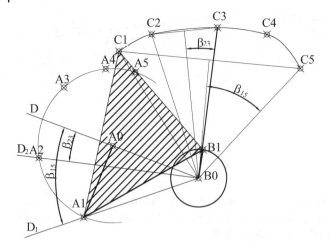

Figure 5.9 Synthesis of a four-bar mechanism so that a point on the coupler passes through five given points.

ensure the coincidence of points B'_{02}, B'_{03} and B'_{01}, B'_{04}, respectively. point B_1 is the centre of the circle circumscribing the triangle $B_0 B'_{02} B'_{04}$. The mechanism $A_0 A_1 B_1 B_0$ is the solution that we searched for, point C in the plane of the coupler successively passing through points C_1, C_2, C_3, C_4, and C_5.

In the companion website of this book we set out the code for the AutoLisp function that encapsulates this algorithm. Using this function we obtained the representation in Figure 5.9. The function had as input data the coordinates (in mm) of the points: C_1 (45, 125), C_1 (55, 135), C_3 (75, 132), C_4 (90, 130), C_5 (100, 120), the distance $l = A_1 C_1 = 50$ mm, and the angle of the straight line D equal to $160°$.

These values can be modified in order to obtain other mechanisms. The input data is introduced directly in the program because it is simpler to write the code and, later, to use the function. The coordinates of the points could also be entered using the keyboard with minimal modifications to the code.

Using the data above, we obtained

- coordinates (in mm) of the joints linked to the base: A_0 (43.1993, 97.1787), B_0 (69.3333, 87.6667)
- dimensions of the four-bar mechanism: $A_1 A_0 = 22.80$ mm, $A_1 B_1 = 41.36$ mm, $B_1 B_0 = 8.50$ mm, and $C_1 B_1 = 38.68$ mm.

5.2 Positional Synthesis

5.2.1 Formulation of the Problem

Technical requirements sometimes demand that the plane of the coupler occupy particulary positions. For instance, in the case of a bus, the doors, which are rigidly linked to the couplers of the driving mechanisms (Figure 5.10) must occupy two defined positions: closed and open. The plane of the wheel of a car must be guided by the suspension mechanism (Figure 5.11) so that it remains, as far as possible, parallel to its initial position.

The general formulation which may be discerned from these examples is that a four-bar mechanism can be constructed when n positions of the plane of the coupler are known or, in other words, when n positions of a segment CD situated in the plane of the coupler are known.

Figure 5.10 The doors of a bus in the closed/open positions.

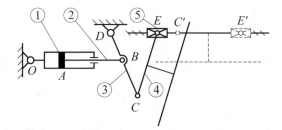

Figure 5.11 Independent suspension mechanism for a motor car.

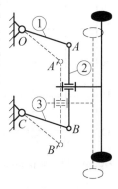

Figure 5.12 The pole of finite rotation.

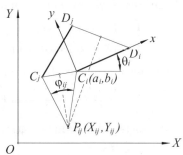

5.2.2 Poles of Finite Rotation

The position of index i of a segment C_iD_i in the plane of the coupler may be given relative to a reference system OXY by the coordinates (a_i, b_i) of point C_i and angle θ_i made by segment C_iD_i with axis OX. Segment CD may arrive from the position C_iD_i at position C_jD_j (Figure 5.12) by a rotation of angle θ_{ij} around a point P_{ij}, referred to as the *pole of finite rotation*. This point is situated at the intersection of the bisectrices of segments C_iC_j and D_iD_j.

We will consider a local reference system with Cxy, C as the origin, while Cx is the axis that overlaps segment CD. If a point M situated in the mobile plane has local coordinates x and y, then, in the position of index i, it has the coordinates X_i and Y_i relative to the reference system OXY:

$$X_i = a_i + x \cos\theta_i - y \sin\theta_i,$$
$$Y_i = b_i + x \sin\theta_i - y \cos\theta_i. \tag{5.36}$$

As functions of X_i, Y_i, the coordinates x and y are given by:

$$x = (X_i - a_i) \cos\theta_i + (Y_i - b_i) \sin\theta_i,$$
$$y = -(X_i - a_i) \sin\theta_i + (Y_i - b_i) \cos\theta_i. \tag{5.37}$$

The pole of finite rotation P_{ij} having the same coordinates x and y relative to the local reference systems $C_ix_iy_i$ and $C_jx_jy_j$, and the same coordinates $X_i = X_j = X_{ij}$ and $Y_i = Y_j = Y_{ij}$ relative to

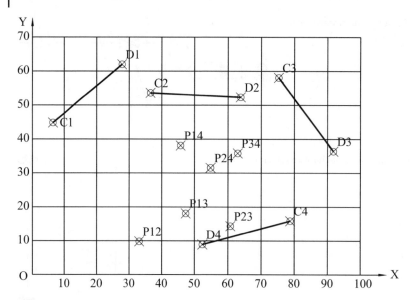

Figure 5.13 The poles of finite rotation for four positions of the plane of the coupler.

the reference system OXY, from (5.37) we obtain the system

$$\begin{cases} (X_{ij} - a_i)\cos\theta_i + (Y_{ij} - b_i)\sin\theta_i = (X_{ij} - a_j)\cos\theta_j + (Y_{ij} - b_j)\sin\theta_j, \\ -(X_{ij} - a_i)\sin\theta_i + (Y_{ij} - b_i)\cos\theta_i = -(X_{ij} - a_j)\sin\theta_j + (Y_{ij} - b_j)\cos\theta_j, \end{cases}$$ (5.38)

from which we deduce the coordinates

$$X_{ij} = \frac{a_i + a_j}{2} - \frac{b_i - b_j}{2}\cot\left(\frac{\theta_i - \theta_j}{2}\right), \quad Y_{ij} = \frac{b_i + b_j}{2} - \frac{a_i - a_j}{2}\cot\left(\frac{\theta_i - \theta_j}{2}\right).$$ (5.39)

The angle of finite rotation φ_{ij} is given by the equality

$$\varphi_{ij} = \theta_i - \theta_j.$$ (5.40)

In the companion website of the book, we set out the code for an AutoLisp function *Poles.lsp*, which determines, for four positions of the plane of the coupler, the poles of finite rotation P_{12}, P_{13}, P_{14}, P_{23}, P_{24}, and P_{34}. First of all, using AutoCAD, we draw the four positions, as in Figure 5.13; the function returns the six poles, drawing them in the figure, and positioning the corresponding annotations near each point.

5.2.3 Bipositional Synthesis

For bipositional synthesis, we will mainly use graphical methods, which, through the use of AutoCAD, can give numerical results of the same precision as analytical methods. For the bipositional synthesis it is necessary to use the *theorem of isovisibility* for the elements jointed to the base; the theorem states that elements jointed to the base are seen from the pole of finite rotation P_{12} at the same angle $\frac{\varphi_{12}}{2}$ (φ_{12} being the angle of finite rotation).

The proof is a simple one; taking into account that pole P_{12} is situated at the intersection of the bisectrices of the segments A_1A_2 and B_1B_2 (Figure 5.14a), which also pass through the fixed joints A_0, and B_0, respectively.

For the crank-shaft mechanism, the crank A_0A_1 and the semi-displacement B_0B_{12} are seen under the same angle $\frac{\varphi_{12}}{2}$.

Based on these properties, one may construct either a four-bar mechanism (Figure 5.15a), or a crank-shaft mechanism so that a segment from the plane of the coupler occupy the positions C_1D_1 and C_2D_2.

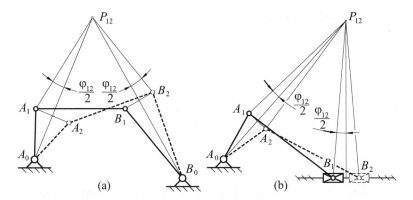

Figure 5.14 The equalities of the angles at which the elements jointed to the base and the linear semi-displacement are seen: (a) four-bar mechanism; (b) crank-shaft mechanism.

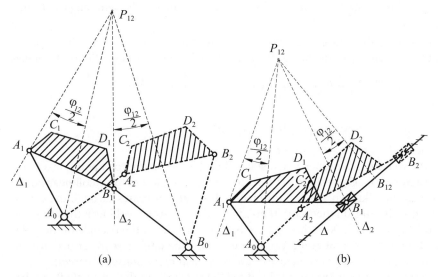

Figure 5.15 The syntheses of a four-bar mechanism and a crank-shaft mechanism for two positions of the plane of the coupler: (a) four-bar mechanism; (b) crank-shaft mechanism.

For the four-bar mechanism $A_0A_1B_1B_0$ (Figure 5.15a) one chooses the fixed joints A_0 and B_0, and points A_1, and B_1 situated on the straight lines Δ_1, and Δ_2, which make angle $\frac{\varphi_{12}}{2}$ with the directions $P_{12}A_0$, and $P_{12}B_0$, respectively.

For the crank-shaft mechanism $A_0A_1B_1$ in Figure 5.15b) we construct points A_0 and A_1 as in the previous situation, choose axis Δ, and construct the perpendicular P_1B_{12} to it. Then we construct the straight line Δ_2 that makes angle $\frac{\varphi_{12}}{2}$ with the straight line $P_{12}B_{12}$; the straight line Δ_2 intersects the straight line Δ at point B_1. The mechanism we are seeking is $A_0A_1B_1$.

5.2.4 Three-positional Synthesis

Triangle of poles: fundamental point, properties

When there are given three positions C_1D_1, C_2D_2, and C_3D_3 of a segment CD in the plane of the coupler, there are three poles of finite rotation, P_{12}, P_{23}, and P_{13}, that define the triangle of poles (Figure 5.16).

If A is a point in the plane of the coupler, and A_1 and A_2 are its first two positions, then angle $\widehat{A_1P_{12}A_2}$ is exactly the angle of finite rotation φ_2, and point A_{123} (the *fundamental point*) is both

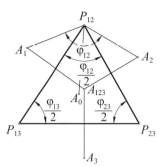

Figure 5.16 The triangle of the poles.

the symmetric point of point A_1 with respect to $P_{12}P_{13}$, and the symmetric point of point A_2 with respect to $P_{12}P_{23}$. In addition, angle $\widehat{P_{13}P_{12}P_{23}}$ is equal to the semi-angle of finite rotation $\frac{\varphi_{12}}{2}$. We also deduce that the fundamental point A_{123} is also the symmetric of point A_3 with respect to $P_{13}P_{23}$, and the angles $\widehat{P_{12}P_{23}P_{13}}$ and $\widehat{P_{12}P_{13}P_{23}}$ are equal to the semi-angles of finite rotation $\frac{\varphi_{23}}{2}$ and $\frac{\varphi_{13}}{2}$.

Now, if we denote by A_0 the centre of the circle circumscribing triangle $A_1A_2A_3$, then $P_{12}A_0$ is the bisectrix of the segment A_1A_2 and passes through point P_{12} too. The equalities of the angles

$$\widehat{A_1P_{12}P_{13}} = \widehat{P_{13}P_{12}A_{123}}, \quad \widehat{A_1P_{12}A_0} = \widehat{A_0P_{12}A_2} = \frac{\varphi_{12}}{2}$$

give

$$\widehat{P_{13}P_{12}A_0} = \widehat{A_{123}P_{12}P_{23}};$$

so the straight lines A_0P_{12} and $A_{123}P_{12}$ are *isogonal*.

Starting from the well-known property that states that the symmetries of a point on a circle circumscribed around a triangle, relative to the edges of the triangle, are three collinear points situated on a straight line (known as the Simson straight line) that passes through the ortho-centre of the triangle (the Simson theorem). If points A_1, A_2, and A_3 are collinear, then they are situated on a straight line that passes through the orthocentre (H) of the triangle of the poles (Figure 5.17), and the fundamental point A_{123} is situated on the circle circumscribed around the triangle of the poles (this is a consequence of *the reciprocal of the Simson theorem*).

This is the *theorem of collinearity* and it is used for the synthesis of the crank-shaft mecha-nism. It also means that points A_1, A_2, and A_3 are situated on *Carnot circles* $P_{12}P_{13}H$, $P_{12}P_{23}H$ and $P_{13}P_{23}H$ (that is, the symmetries of the circle circumscribed around the triangle of poles $P_{12}P_{13}P_{23}$ with respect with the edges of the triangle), and the straight line that passes through

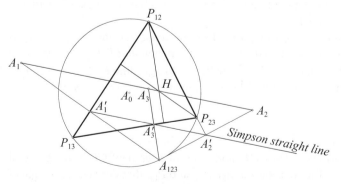

Figure 5.17 The collinearity of points A_1, A_2, and A_3 when the fundamental point A_{123} is situated on the circle circumscribed around the triangle of the poles.

points A_1, A_2, and A_3 is the geometric locus of the symmetric of the fundamental point A_{123} with respect to the Simson straight line $A'_1A'_3A'_2$ (Figure 5.17).

Synthesis of the four-bar mechanism

From the previous discussion the synthesis of the four-bar mechanism reduces to the following steps (Figure 5.18):

- Construct the triangle of the poles;
- Choose the fixed joints A_0 and B_0;
- Construct points A_{123} and B_{123} using the theorem of the isogonality

$$\widehat{A_0P_{12}P_{23}} = \widehat{A_{123}P_{12}P_{13}} \text{ and } \widehat{A_0P_{13}P_{12}} = \widehat{A_{123}P_{13}P_{23}},$$

$$\widehat{B_0P_{12}P_{23}} = \widehat{B_{123}P_{12}P_{13}} \text{ and } \widehat{B_0P_{13}P_{12}} = \widehat{B_{123}P_{13}P_{23}};$$

- one determines the symmetries A_1, and B_1 of points A_{123}, and B_{123}, respectively, with respect to the straight line $P_{12}P_{13}$, obtaining the quadrilateral $A_0A_1B_1B_0$.

Synthesis of the crank-shaft mechanism

W choose the fixed joint A_0, and, proceeding as for the synthesis of the four-bar mechanism, determine point A_1. Then we construct:

- the orthocentre H of the triangle of the poles $P_{12}P_{13}P_{23}$
- the Carnot circle that passes through points P_{13}, H, and P_{12}
- a horizontal straight line that passes through point H and intersects the Carnot circle at point B_1^*.

The solution are seeking is the crank-shaft mechanism for which slider B_1^* moves on the straight line B_1^*H.

For the three-positional synthesis we wrote an AutoLisp function called *Synthesis_3P.lsp*, the code for which is shown in the companion website of the book. It performs the synthesis previously presented and represented in Figure 5.18.

After calling the function, the coordinates of the three poles of finite rotation and the coordinates of points A_0 and B_0 must be specified. These are saved. Then the triangle of the poles,

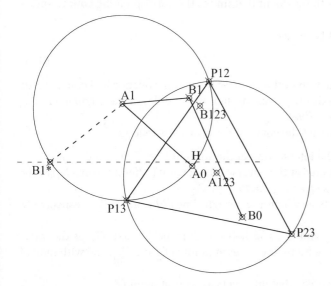

Figure 5.18 Three-positional synthesis of the four-bar and crank-shaft mechanisms.

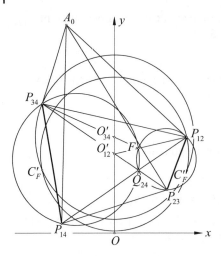

Figure 5.19 The isovisibility of segments $P_{12}P_{23}$ and $P_{34}P_{14}$.

and the positions and the notations of the points are constructed. Next, using the theorem of isogonality, points A_{123} and B_{123} are determined and, using the algorithm outlined above, the symmetries A_1 and B_1 of points A_{123} and B_{123}, respectively, with respect to the straight line $P_{12}P_{13}$ are constructed.

In Figure 5.18, the following data were used: P_{12} (40, 60), P_{12} (15, 25), P_{23} (65, 15), A_0 (35, 35), and B_0 (50, 20). This gave the coordinates of the points as follows: A_1 (13.84, 53.43), B_1 (34.10, 55.15) for the four-bar mechanism, and B_1^* (−8.18, 36.25) for the crank-shaft mechanism. The lengths of the elements are: $A_0A_1 = 28.06$ mm, $A_1B_1 = 20.33$ mm, $B_0B_1 = 38.58$ mm, and $A_1B_1^* = 27.93$ mm.

5.2.5 Four-positional Synthesis

For four-positional synthesis the fixed joint A_0 is situated on a curve called the *curve of the centres*. The *theorem of the isovisibility* states that from the centre A_0 of the circle on which are situated points $A_1, A_2, A_3,$ and A_4, the segment that unites a pair of poles (for instance, $P_{12}P_{13}$) is seen as being at the same angle as the segment that unites the corresponding counter-poles ($P_{34}P_{14}$) (see Figure 5.19).

From the theorem of the isovisibility we get:

$$\widehat{P_{14}A_0P_{34}} = \widehat{P_{12}A_0P_{23}} = \varphi; \tag{5.41}$$

from which we can establish the equation of the curve of the centres relative to a conveniently chosen reference system, in which axis Oy passes through the middles of the segments $P_{14}P_{23}$ and $P_{34}P_{12}$ (the *Newton-Gauss straight line*).

Using AutoCAD, it is easy to construct the curve of the centres as follows:

1) Construct point Q_{24} situated at the intersection of the straight lines $P_{14}P_{12}$ and $P_{23}P_{34}$.
2) Construct the *Miquel point F* situated at the obverse intersection (different from Q_{24}) of the circles circumscribed on triangles $Q_{24}P_{14}P_{34}$ and $Q_{24}P_{23}P_{12}$.
3) Denote by O'_{12} and O'_{34} the points at which the straight lines FP_{12} and FP_{34}, respectively, intersect axis Oy.
4) Construct the principal points C'_F and C''_F, symmetric with respect to axis Oy, as the intersection points between the circles with their centres at points O'_{12} and O'_{34}, and with radii of $O'_{12}P_{12}$ and $O'_{34}P_{34}$, respectively.
5) Construct an arbitrary straight line (Δ) that intersects axis Oy at point O'.

6) Construct the circle (C) centred at point O' that passes through points C'_F, and C''_F.
7) The points M and P where the straight line (Δ) and the circle (C) intersect belong to the curve of the centres.
8) By constructing different positions of the straight line (Δ), a sufficient number of points is obtained to draw the curve of the centres.

In general, when the principal points exist and are distinct, the curve of the centres has two branches. If the principal points coincide, then the curve of the centres becomes a focal one, with two foci.

In the case when the circles with the centres at points O'_{12} and O'_{34}, and of radii equal to $O'_{12}P_{12}$ and $O'_{34}P_{34}$, respectively, do not intersect, then the following method can be used:

1) Denotes by S and T the points at which the circle centred at point O'_{12} intersects axis OY.
2) Construct a circle that passes through points S and T, and which intersects the circle centred at point O'_{34} at points S' and T'.
3) Denotes by D the intersection point of the straight line $S'T'$ and axis Oy.
4) Construct the circle (C_0) centred at point D, of radius the length DD' of the tangent at point D to the circle centred at point O'_{12}.
5) At point F construct an arbitrary straight line (Δ) which intersects axis OY at point O'.
6) Construct the circle (C) centred at point O' of radius the length $O'O''$ of the tangent from point O' to the circle (C_0).
7) Points M and P where the circle (C) intersects straight line (Δ) belong to the curve of centres.

In both cases, the curve of the centre passes through the six poles: through point F, through points C'_F and C''_F (if they exist), and through points Q_{24}, and Q_{13} of intersection between the straight lines $P_{14}P_{12}$ and $P_{34}P_{23}$, and the straight lines $P_{14}P_{34}$ and $P_{12}P_{23}$, respectively.

The great number of graphical constructions demands the use of an AutoLisp function to obtain the curve of the centres. This function is called *Curve* and the code is set out in the companion website of the book. The function generates two script files *Curve1.scr* and *Curve2.scr*, from which the two branches of the curve of the centres can be drawn in AutoCAD. The input data are the four poles of finite rotation P_{12}, P_{23}, P_{34}, and P_{12}. The number of points used to obtain the curve of the centres (*nrsteps*) is also specified. The values of the ends of the interval (y_{min}, y_{max}) for the curve will be obtained. The function will construct two different curves of the centres: a curve with two branches, and a curve with one branch, depending on the intersection points of the circles with the centres at points O'_{12} and O'_{34}.

Using the symbolization:

- $\overline{AB} \cap \overline{CD} \Rightarrow M$, point M is obtained at the intersection of the straight lines AB and CD
- $C(M, N, P)$, the circle that passes through points M, N, and P
- $C(O)_{R=a}$, the circle centred at point O with radius a

and taking into account the previous discussion, the construction of the curve of the centre uses the following algorithm:

1) $\overline{P_{14}P_{12}} \cap \overline{P_{23}P_{34}} \Rightarrow Q_{24}$.
2) $C\left(Q_{24}, P_{14}, P_{34}\right) \cap C\left(Q_{24}, P_{23}, P_{12}\right) \Rightarrow P_1, P_2$.
3) if $P_1 = Q_{24}$, then $F = P_2$, else $F = P_1$.
4) $\overline{FP_{12}} \cap \overline{Oy} \Rightarrow O'_{12}$.
5) $\overline{FP_{34}} \cap \overline{Oy} \Rightarrow O'_{34}$.
6) $C\left(O'_{21}\right)_{R=O'_{12}P_{12}} \cap C\left(O'_{34}\right)_{R=O'_{34}P_{34}} \Rightarrow C'_F, C''_F$.

If points C'_F and C''_F do exist (that is, there is a curve with two branches), then:

Start a repetitive loop of *nrsteps* steps in which:

7') $y_{O'} \in [y_{min}, y_{max}]$ or $y_{O'} = y_{min} + \dfrac{y_{max} - y_{min}}{nrsteps} step$, where $step \in [0, nrsteps]$.

8') $C(O')_{R=C'F} \cap \overline{FO'} \Rightarrow M, N.$

9') Write in two script files the values found for M and N.

End repetitive loop

If points C'_F and C''_F do not exist (that is, there is a curve with one branch), then

7) If $O'P_{12} > O'_{34}P_{34}$, then $C(O'_{12}) \to C(O'_{12})$ and $C(O'_{34}) \to C(O'_{34})$, else $C(O'_{12}) \to C(O'_{34})$ and $C(O'_{34}) \to C(O'_{12})$.

8) $C(O'_{12}) \cap \overline{Oy} \Rightarrow S, T$ or, directly, $y_S = y_{O'_{12}} + O'_{12}P_{12}, y_T = y_{O'_{12}} - O'_{12}P_{12}.$

9) $C(O^*)_{R=y_{O'_{12}}\sqrt{(5)}} \cap C(O'_{34})_{R=O'_{34}P_{34}} \Rightarrow T', S'$, where $x_{O^*} = x_{O'_{12}}, y_{O^*} = 2O'_{12}P_{12}.$

10) $\overline{T'S'} \cap \overline{Oy} \Rightarrow D.$

11) $tangent = \sqrt{\left(y_D - y_{O'_{12}}\right)^2 + \left(O'_{12}P_{12}\right)^2}.$

Start a repetitive loop with *nrsteps* steps in which

12') $y_{O'} \in [y_{min}, y_{max}]$ or $y_{O'} = y_{min} + \dfrac{y_{max} - y_{min}}{nrsteps} step$, where $step \in [0, nrsteps]$.

13') $C(D)_{R=tangent} \cap \overline{F'O'} \Rightarrow P, R.$

14') Write in two script files the values found for P and R.

End repetitive loop

The function *Int_2C* is the function which specifies the construction in the two cases, by assigning the value 2 or 1 to the variable *branches*. The construction produced by the algorithm comes either from steps 7'–9' for *branches* = 2, or from steps 7–14' for *branches* = 1; the constructions from steps 1–6 are common.

When values of P_{12} (−65, 42) and P_{23} (−11, 108) are selected for poles P_{12} and P_{23}, and values of P_{34} (70, 49) and P_{14} (11, 163) for the counter-poles P_{34} and P_{14}, respectively, the curve of the centres being constructed through 2000 points in the interval $y \in [-100, 150]$, the curve of the centres with two branches shown in Figure 5.20a is obtained. When values of P_{12} (−71, 37) and P_{23} (−16, 113) are selected for poles P_{12} and P_{23}, and values of P_{34} (72, 44) and P_{14} (15, 171) for the counter-poles P_{34} and P_{14}, respectively, the curve of the centres being constructed through 2000 points in the interval $y \in [-250, 0]$, the curve of the centres with one branch shown in Figure 5.20b is obtained.

Based on the discussion above, we can now pass to the synthesis of the four-bar mechanism $A_0A_1B_1B_0$. We construct the curve of the centres using the procedure above, choose point A_0 on the curve, construct point A_{123} (applying the theorem of isogonality in triangle $P_{12}P_{13}P_{23}$) and then point A_1 (the symmetric of point A_{123} with respect to the straight line $P_{12}P_{13}$). In a similar way, we then choose point B_0 and construct point B_1.

For the crank-shaft mechanism $A_0A_1B_1$ we first construct point A_1 using the procedure above. point B_1 is situated at the intersection of the Carnot circles constructed on the segments $P_{12}P_{13}$

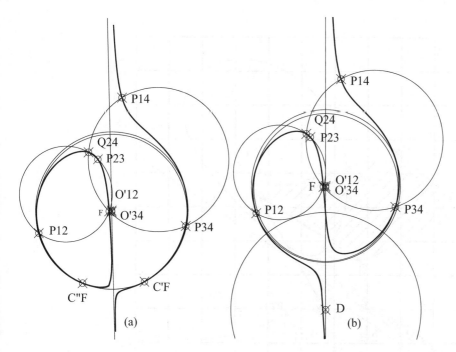

Figure 5.20 Curves of the centres: (a) curve with two branches; (b) curve with one branch.

and $P_{13}P_{14}$ for triangles $P_{12}P_{13}P_{23}$ and $P_{12}P_{14}P_{24}$, respectively; the support of slider B_1 is the straight line that unites point B_1 and the orthocentre of triangle $P_{12}P_{13}P_{23}$.

Let us consider the practical case in Figure 5.21, in which four positions of the plane of coupler $(C_1D_1, C_2D_2, C_3D_3, C_4D_4)$ are given. Let us construct the four-bar mechanism $A_0A_1B_1B_0$. We use the following steps:

1) Determine the position of poles P_{12}, P_{13}, P_{23}, P_{14}, P_{24}, and P_{34}.
2) Determine the cubic curve C_{1234}.
3) Choose points A_0 and B_0, and, by isogonality in the triangle $P_{12}P_{13}P_{23}$, determine points A_{123} and B_{123}.
4) Construct the symmetries of points A_{123} and B_{123} with respect to the straight line $P_{12}P_{13}$, determining points A_1 and B_1.

The quadrilateral $A_0A_1B_1B_0$ is the solution of the problem.

The construction in Figure 5.21 was created using an AutoLisp function called *Synthesis_4p*, the code for which is set out in the companion website of the book. The function has as input data the four positions of the plane of the coupler. For this example we used the construction in Figure 5.13.

The function calls several functions that have been described in Section 5.2.4. Function *Poles* determines the poles P_{12}, P_{13}, P_{14}, P_{23}, P_{24}, and P_{34}. Function *Curve* constructs the curve of the centres. The instructions necessary to construct the curve of the centres are written to two script files called *Curve1.scr* and *Curve2.scr*. Then the constructions created are drawn and labelled with the points determined. To obtain the curve of the centres, one the two script files are called in AutoCAD using the command **Script**. This gives the curve in Figure 5.21.

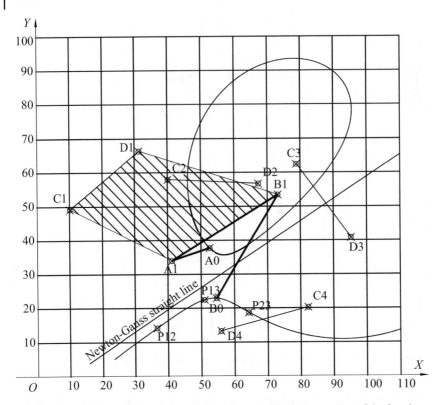

Figure 5.21 The positions of the plane of the coupler and the construction of the four-bar mechanism.

The four-bar mechanism is obtained using the AutoLisp function called *Synthesis_3p.lsp*, the code for which is shown in the companion website of this book. After the function has been called it is necessary to specify in the figure, the coordinates of the poles P_{12}, P_{13}, P_{23}, and the coordinates of points A_0 and B_0, which are on the curve of the centres, as in Figure 5.12. This gives the four-bar mechanism $A_0A_1B_1B_0$. Then the loop $A_1C_1D_1B_1$ can be drawn.

If the solution is not a convenient one, then different points A_0 and B_0 on the curve of the centres can be chosen until an optimal variant is obtained. Then, we obtain, by drawing the dimensions, the information about the coordinates of the points and the lengths of the elements.

5.2.6 Five-positional Synthesis

For a five-positional synthesis, it is necessary to construct the curves of the centres C_{1234} and C_{1245} which intersect at seven points at finite distance; three of these points are the poles of rotation P_{12}, P_{14}, and P_{24}. The other four points are the *Burmester points* and they can be real (four or two points) or imaginary. When at least two points are real, these points are chosen as fixed joints and the synthesis of the four-bar mechanism can be solved (Figure 5.22). The construction in Figure 5.22 was created using functions that were described in Sections 5.2.4 and 5.2.5. First the function *Synthesis_4p* is called to construct the curve C_{1234} in a dedicated layer. The curve C_{1245} is constructed in a similar way. A_0 and B_0 denote the common points of the two curves. Finally, after calling function *Synthesis_3p* and selecting poles P_{12}, P_{13}, P_{23} and

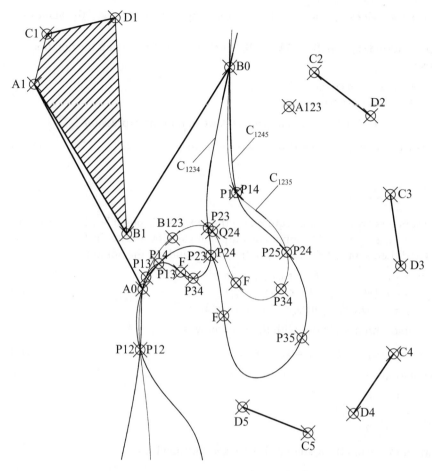

Figure 5.22 The Burmester points.

points A_0, B_0, the construction in Figure 5.22 will be obtained. The curve C_{1235} is also drawn. Note that three poles are common to each pair of curves, while points A_0 and B_0 are common to all the curves.

5.3 Function-generating Mechanisms

Some machines require the determination of a four-bar mechanism (Figure 5.23) so that there is a given correlation between the input angle φ and the output angle ψ; when the

Figure 5.23 Function-generating four-bar mechanism.

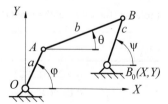

input angle φ takes the values φ_i, $i = 1$, 2, 3 the output angle ψ should takes the values ψ_i, $i = 1$, 2, 3.

Using the loop equation $\mathbf{OB}_0 + \mathbf{B}_0\mathbf{B} = \mathbf{OA} + \mathbf{AB}$, which, taking into account the notations in Figure 5.23, reads

$$b\cos\theta = x + c\cos\psi - a\cos\varphi,$$
$$b\sin\theta = y + c\sin\psi - a\sin\varphi. \tag{5.42}$$

We may take $a = 1$; from (5.42), by eliminating angle θ, we obtain the expression

$$x\cos\varphi + y\sin\varphi + c\cos(\psi - \varphi) + d = xc\cos\psi + yc\sin\psi, \tag{5.43}$$

where

$$d = \frac{b^2 - x^2 - y^2 - c^2 - 1}{2}. \tag{5.44}$$

In (5.43), four parameters (x, y, c, d) are required, so the synthesis is possible with four values φ_i and ψ_i, $i = 1$, 2, 3, 4, assigned to the input and output angles. If we write the four equations for $i = 1$, 2, 3, 4 and then subtract the first from the others, and use the notations

$$A_i = \cos\varphi_{i+1} - \cos\varphi_1, \; B_i = \sin\varphi_{i+1} - \sin\varphi_1,$$
$$C_i = \cos\left(\psi_{i+1} - \varphi_{i+1}\right) - \cos\left(\psi_i - \varphi_i\right), \tag{5.45}$$
$$D_i = \cos\psi_{i+1} - \cos\psi_i, \; E_i = \sin\psi_{i+1} - \sin\psi_i, \; i = 1, \; 2, \; 3,$$

we obtain the system of three equations with three unknowns

$$A_i x + B_i y + C_i c = D_i xc + E_i yc, \; i = 1, \; 2, \; 3. \tag{5.46}$$

Then, using the notation

$$|A_i \; B_i \; C_i| = \begin{vmatrix} A_1 & B_1 & C_1 \\ A_2 & B_2 & C_2 \\ A_3 & B_3 & C_3 \end{vmatrix} \tag{5.47}$$

and expressing x, y and C from the system (5.46), we get the equalities

$$\frac{1}{C}|A_i \; B_i \; C_i| = |D_i \; B_i \; C_i| + \frac{y}{x}|E_i \; B_i \; C_i|$$
$$= \frac{y}{x}|A_i \; D_i \; C_i| + |A_i \; E_i \; C_i|. \tag{5.48}$$

From the system (5.48), we obtain a second-order equation in $\frac{y}{x}$, so the problem has two solutions.

In a numerical application, with the input data shown in Table 5.1, we obtain the values

$$x = 2.3333333, \; y = -0.0000000, \; b = 3.6666667, \; c = 2.5000000.$$

The values in Table 5.1 were obtained by running a program of kinematic analysis for a four-bar mechanism with dimensions (Figure 5.23) as follows:

$$OA = 0.06, \; AB = 0.22, \; BC = 0.15 \; x_O = y_O = y_C = 0, \; x_C = 0.14.$$

Table 5.1 The values of the input data.

Nr. crt.	φ	ψ
1	0.06283185307	0.58131434789
2	0.12566370614	0.55239498924
3	0.18849555922	0.53491345981
4	0.25132741229	0.52754665107

Taking into account the convention $a = 1$, the mechanism which would be obtained would have the dimensions:

$$a = \frac{OA}{0.06} = 1, \; b = \frac{AB}{0.06} = 3.666, \; c = \frac{BC}{0.06} = 2.50, \; x = x_C = 2.333, \; y = y_C = 0.$$

The mechanism with these data gives the same values as those presented in Table 5.1.

6

Cam Mechanisms

Cams are simple mechanisms, their goal being the realization of a certain law of motion by the driven element (the follower). In most cases, such mechanisms have only two elements: the cam and the follower.

6.1 Generalities. Classification

In cam mechanisms, the motion is transmitted (Figure 6.1) by direct contact (higher kinematic pair) from the cam, which is the driving element, to the follower, which is the driven element. The cam is the element that imposes the particular law of motion on the follower. In some cases, when it is necessary to reduce the effect of friction, the motion is transmitted from the cam to the follower through a roller (Figure 6.2). Cams can act directly (Figures 6.1, 6.2) or they can act via a four-bar mechanism (Figure 6.3). There are also cam mechanisms that are fixed (Figure 6.4).

The use of cam mechanisms is justified by their many advantages: simplicity, small size, and the possibility of generating any form of motion in the driven element, providing the profile of the cam is suitable. Their disadvantages are the wear produced by the high pressures of contact in the higher pair, and the complicated technology required for their manufacture.

The classification of the cam mechanisms can be made according to various criteria.

1) The motion of the elements:
   - planar (Figures 6.1–6.4)
   - spatial (Figure 6.5)
2) The location of the cam's profile:
   - exterior (Figures 6.1 and 6.2)
   - interior (Figures 6.3 and 6.4)
3) The cam's motion:
   - fixed (Figure 6.4)
   - rotating (Figure 6.1, 6.2, 6.5, and 6.6)
   - translational (Figure 6.7)
   - cam mechanisms at which the cam has plane-parallel motion (Figure 6.8)
4) The shape of the follower:
   - with plunger tip (Figure 6.9)
   - with flat tappet (Figure 6.10)
   - with curvilinear follower (Figure 6.1)
   - with roller tappet (Figures 6.2, 6.7, and 6.8)

Classical and Modern Approaches in the Theory of Mechanisms, First Edition.
Nicolae Pandrea, Dinel Popa and Nicolae-Doru Stănescu.
© 2017 John Wiley & Sons Ltd. Published 2017 by John Wiley & Sons Ltd.
Companion Website: www.wiley.com/go/pandmech17

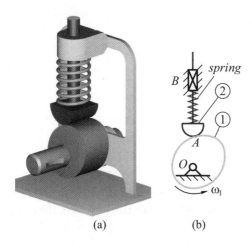

Figure 6.1 Cam mechanism with rotational cam: (a) constructive schema; (b) kinematic schema.

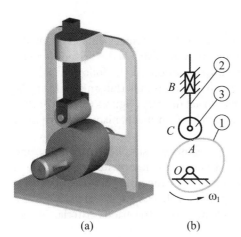

Figure 6.2 Cam mechanism with roller tappet: (a) constructive schema; (b) kinematic schema.

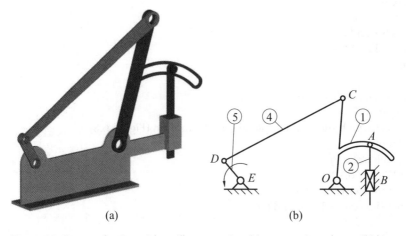

Figure 6.3 Cam mechanism with oscillatory motion: (a) constructive schema; (b) kinematic schema.

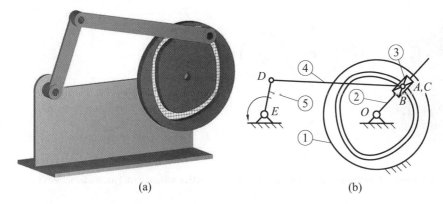

Figure 6.4 Cam mechanism with fixed cam: (a) constructive schema; (b) kinematic schema.

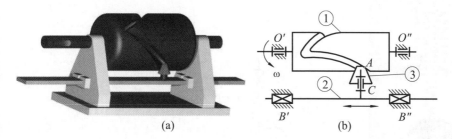

Figure 6.5 Cam mechanism with spatial motion: (a) constructive schema; (b) kinematic schema.

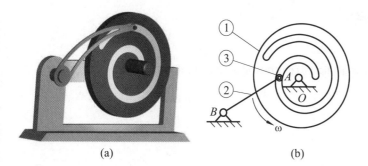

Figure 6.6 Cam mechanism with cam with interior profile: (a) constructive schema; (b) kinematic schema.

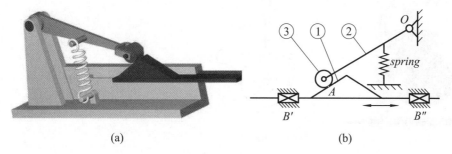

Figure 6.7 Cam mechanism with cam in translational motion: (a) constructive schema; (b) kinematic schema.

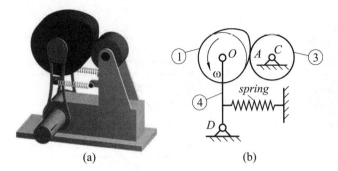

(a)   (b)

Figure 6.8 Cam mechanism with cam in rotational motion: (a) constructive schema; (b) kinematic schema.

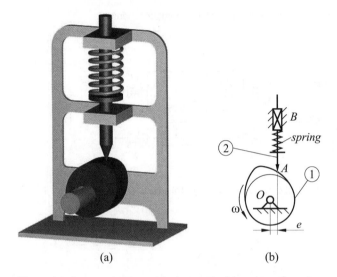

(a)   (b)

Figure 6.9 Cam mechanism with plunger tip: (a) constructive schema; (b) kinematic schema.

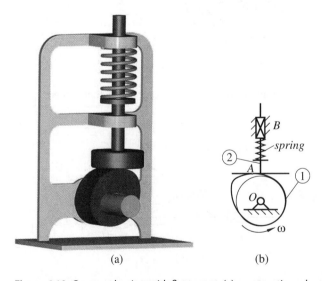

(a)   (b)

Figure 6.10 Cam mechanism with flat tappet: (a) constructive schema; (b) kinematic schema.

Figure 6.11 Cam mechanism with multiple cam.

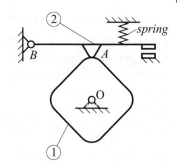

Figure 6.12 Cam mechanism with double contact.

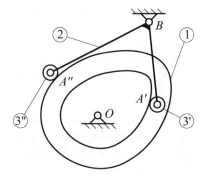

5) The motion of the follower:
- with fixed follower (Figure 6.8)
- follower in translational motion (Figures 6.1, 6.2, 6.3, 6.5, 6.9, and 6.10)
- follower in rotational motion (Figures 6.6, 6.7, and 6.8)

6) The number of courses of the follower int one revolution of the cam
- with simple cam (Figures 6.1, 6.2, 6.5, 6.7, 6.9, and 6.10)
- with multiple cams (Figure 6.11)

7) Position of the follower with respect to the cam:
- with axial follower (Figures 6.1, and 6.2)
- with non-axial follower (Figure 6.9)

8) after the way in which cam-follower contact is maintained:
- with spring (Figures 6.1, 6.7, 6.8, 6.9, and 6.10)
- with interior profile (Figures 6.3, 6.4, and 6.6)
- with double exterior contact (Figure 6.12)

In the kinematic schemata considered in earlier sections, we used the following notation:

1) Cam
2) Follower
3) Roller
4,5) Elements of the four-bar mechanism.

6.2 Analysis of Displacement of Follower

6.2.1 Formulation of the Problem

Consider as known: the profile of the cam, the law of motion of the cam, and the profile of the follower. We want to determine the law of motion of the follower.

The profile of the cam can be given by:

- the equation, in polar coordinates, $\rho = \rho(\psi)$
- the parametric equations $x = x(\lambda)$, $y = y(\lambda)$
- points (a table with values ψ_i, ρ_i)
- a drawing.

Analytical methods are used to determine the law of motion of the follower in the first three cases, with graphical methods only used for the last.

6.2.2 The Analytical Method

In this section we will first discuss some particular problems, before considering the general case.

Cam mechanism with axial translational plunger tip and rotational cam

The profile of the cam in Figure 6.13 is given in polar coordinates:

$$\rho = \rho(\psi), \tag{6.1}$$

where $\rho = OA$ and the reference axis for angle ψ is Ox. If the rotation of the cam starts from the position at which point A_0 coincides with point A^*, where $OA^* = \rho_{min} = r_0$, then, with the notation in Figure 6.13, the rotational angle of the cam φ is $\varphi = \psi$, and the displacement of the plunger tip reads

$$s = \rho(\varphi) - r_0. \tag{6.2}$$

Numerical example If, for distances expressed in mm:

$$\rho = \begin{cases} 10, & \text{for } \psi \in \left[0, \frac{\pi}{2}\right], \\ 10 + 8\cos^2 \psi, & \text{for } \psi \in \left(\frac{\pi}{2}, \frac{3\pi}{2}\right), \\ 10, & \text{for } \psi \in \left[\frac{3\pi}{2}, 2\pi\right], \end{cases} \tag{6.3}$$

we get $r_0 = \rho_{min} = 10$, $\varphi = \psi$ and

$$s = \begin{cases} 0, & \text{for } \psi \in \left[0, \frac{\pi}{2}\right], \\ 8\cos^2 \psi, & \text{for } \psi \in \left(\frac{\pi}{2}, \frac{3\pi}{2}\right), \\ 0, & \text{for } \psi \in \left[\frac{3\pi}{2}, 2\pi\right]. \end{cases}$$

In this case, the scale representation of the cam's profile and the law of motion are as shown in Figure 6.14.

Figure 6.13 Cam mechanism with axial plunger tip and rotational cam.

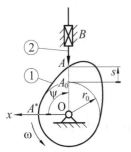

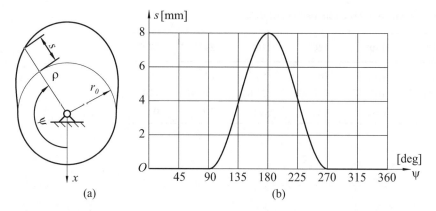

Figure 6.14 The determination of the law of motion of the plunger tip: (a) profile of the cam; (b) law of motion of the plunger tip.

Figure 6.15 Cam mechanism with non-axial plunger tip.

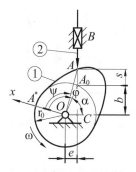

If the cam's profile is given by the parametric equations $x = x(\lambda)$, $y = y(\lambda)$, then φ and s are obtained from:

$$\tan \varphi = \frac{y}{x}, \ \rho = \sqrt{x^2 + y^2}, \ s = \rho - r_0. \tag{6.4}$$

Cam mechanism with non-axial translational plunger tip and rotational cam

The profile of the cam in Figure 6.15 is given by the equation in polar coordinates $\rho = \rho(\psi)$. We denote by e the eccentricity of the plunger tip.

If $OA^* = \rho_{min} = r_0$, then $\widehat{A^*OA_0} = \varphi$ is the rotational angle of the cam and $s = A_0A$ is the displacement of the plunger tip. Denoting $OA = \rho$, $\widehat{A^*OA} = \psi$, $\beta = \varphi - \psi$, and $\alpha = \widehat{A_0OC}$, gives

$$\alpha = \arccos\left(\frac{e}{r_0}\right), \ \alpha + \beta = \arccos\left(\frac{e}{\rho}\right), \ \varphi = \psi + \beta, \ s = AC - A_0C,$$

from which

$$\varphi = \psi + \arccos\left(\frac{e}{\rho}\right) - \arccos\left(\frac{e}{r_0}\right), \tag{6.5}$$

$$s = \sqrt{\rho^2 - e^2} - \sqrt{r_0^2 - e^2}. \tag{6.6}$$

To determine the law of motion of the plunger tip, we identify r_0, give values to angle ψ, for each value of which we calculate ρ, φ, and s with (6.5) and (6.6), obtaining thus the law of motion $s = s(\varphi)$.

Table 6.1 The numerical values for the cam defined by (6.3).

ψ	90°	120°	150°	180°	210°	240°	270°
ρ	10	12	16	18	16	12	10
φ	90°	126.87°	164.85°	197.40°	224.85°	246.87°	270°
s	0	2.39	6.83	8.97	6.83	2.39	0

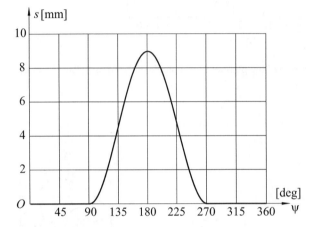

Figure 6.16 The law of motion of the plunger tip.

Numerical example Consider the cam defined by (6.3), with eccentricity $e = 6\,mm$ and $r_0 = 10\,mm$. The values in Table 6.1 are obtained, from which Figure 6.16 can be drawn.

If the profile of the cam is given by the parametric equations $x = x(\lambda)$, $y = y(\lambda)$, then the parameters ψ and ρ are obtained from (6.4), while the parameters φ and s are deduced from (6.5) and (6.6).

Cam mechanism with flat tappet and rotational cam

The profile of the cam in Figure 6.17 is given by the equation, in polar coordinates, $\rho = \rho(\psi)$. We denote $r_0 = \rho_{min}$ and $\rho = OA$, and construct the perpendicular OT to the base of the flat tappet.

Figure 6.17 Cam mechanism with flat tappet.

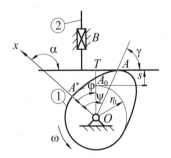

Figure 6.18 The angle between the polar radius and the tangent.

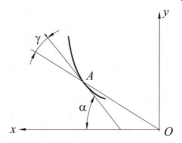

We know that the tangent to an arbitrary curve forms an angle γ with the polar radius (Figure 6.18) so that $\tan \gamma = \dfrac{\rho}{\dfrac{d\rho}{d\psi}}$; from here:

$$\gamma = \begin{cases} \arctan\left(\dfrac{\rho}{\dfrac{d\rho}{d\psi}}\right) & \text{for } \dfrac{d\rho}{d\psi} > 0, \\[2ex] \dfrac{\pi}{2} \text{ for } \dfrac{d\rho}{d\psi} = 0, \\[2ex] \pi + \arctan\left(\dfrac{\rho}{\dfrac{d\rho}{d\psi}}\right) & \text{for } \dfrac{d\rho}{d\psi} < 0. \end{cases} \tag{6.7}$$

Knowing angle γ, we can calculate the rotation angle φ and the displacement $s = OT - r_0$ using:

$$\varphi = \psi + \gamma - \frac{\pi}{2}, \tag{6.8}$$

$$s = \rho \sin \gamma - r_0. \tag{6.9}$$

Numerical example For the cam defined by the expressions

$$\rho = \begin{cases} 10 \text{ for } \psi \in \left[0, \dfrac{\pi}{2}\right], \\[1.5ex] 10 + 4\cos^2 \psi \text{ for } \psi \in \left(\dfrac{\pi}{2}, \dfrac{3\pi}{2}\right), \\[1.5ex] 10 \text{ for } \psi \in \left[\dfrac{3\pi}{2}, 2\pi\right], \end{cases}$$

applying (6.7)–(6.9) gives the values in Table 6.2, and we obtain the representation of the function $s = s(\varphi)$ shown in Figure 6.19.

Table 6.2 Numerical values for the representation of the function $s = s(\varphi)$.

ψ	90°	120°	150°	180°	210°	240°	270°
γ	90°	72.52°	75.08°	90°	104.92°	107.48°	90°
φ	90°	102.52°	135.08°	180°	224.92°	257.48°	270°
s	0	0.49	2.56	4	2.56	0.49	0

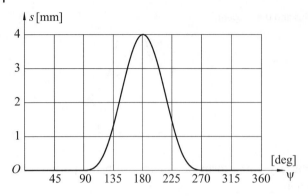

Figure 6.19 The law of motion of the flat tappet.

Figure 6.20 Cam mechanism with circular cam.

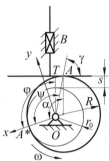

If the profile of the cam is given by the parametric coordinates $x = x(\lambda)$ and $y = y(\lambda)$, then denoting by x' and y' the derivatives with respect to λ, we successively obtain:

$$\tan \alpha = \frac{y'}{x'}, \quad \varphi = \alpha - \frac{\pi}{2}, \quad \tan \psi = \frac{y}{x}, \tag{6.10}$$

from which we can deduce the law of motion $s = s(\varphi)$.

As an example, consider the circular cam in Figure 6.20, with equations

$$x = r_0 - R + R \cos \lambda, \quad y = R \sin \lambda;$$

This gives:

$$\tan \alpha = \tan\left(\frac{\pi}{2} + \lambda\right), \quad \alpha = \frac{\pi}{2} + \lambda, \quad \varphi = \lambda,$$

$$\cos \psi = \frac{x}{\rho} = \frac{r_0 - R + R \cos \lambda}{\rho}, \quad \sin \psi = \frac{y}{\rho} = \frac{R \sin \lambda}{\rho},$$

$$\sin(\gamma) = \cos(\varphi - \psi) = \cos \varphi \cos \psi + \sin \varphi \sin \psi = \frac{R - (R - r_0) \cos \lambda}{\rho},$$

$$s = \rho \sin \gamma - r_0 = (R - r_0)(1 - \cos \varphi).$$

Remark: From the technical point of view, to construct a cam mechanism with a flat tappet and continuous motion it is necessary for the cam's profile to be represented as a curve without inflexion points; that is, the tangent to the cam does not intersect it at any points other than the tangency point. Mathematically, this condition is set out in the condition which states that the expression $\rho \rho'' - 2\rho'^2 - \rho^2$ has a constant sign for all values of θ.

If we consider the cam of equation

$$\rho = \begin{cases} r_0, & \text{if } \psi \in \left[0, \frac{\pi}{2}\right], \\ r_0 + b \cos^2 \psi, & \text{if } \psi \in \left(\frac{\pi}{2}, \frac{3\pi}{2}\right), \\ r_0, & \text{if } \psi \in \left[\frac{3\pi}{2}, 2\pi\right], \end{cases} \tag{6.11}$$

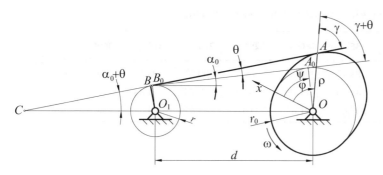

Figure 6.21 Cam mechanism with oscillatory follower.

then, from the condition of a constant sign for the expression $\rho\rho'' - 2\rho'^2 - \rho^2$, which is negative for $\psi = \frac{\pi}{2}$, the result is that $b < \frac{r_0}{2}$.

Cam mechanism with oscillatory follower and rotational cam

Consider the mechanism with oscillatory follower in Figure 6.21, O_1BA ($\widehat{O_1BA} = 90°$), and denote by φ and θ the rotation angles of the cam and follower, respectively. In the non-displaced position the edge A_0B_0 of the follower forms an angle α_0 with the straight line O_1O; the angle is defined by:

$$\alpha_0 = \arcsin\left(\frac{r_0 - r}{d}\right), \tag{6.12}$$

where $d = O_1O$, $r_0 = \rho_{min}$, and $r = O_1B$. In an arbitrary position of the follower, the edge AB makes an angle γ with the polar radius OA, so that $\tan\gamma = \dfrac{\rho}{\dfrac{d\rho}{d\psi}}$ or

$$\gamma = \begin{cases} \arctan\left(\dfrac{\rho}{\dfrac{d\rho}{d\psi}}\right) & \text{if } \dfrac{d\rho}{d\psi} > 0, \\[4mm] \dfrac{\pi}{2} \text{ if } \dfrac{d\rho}{d\psi} = 0, \\[4mm] \pi + \arctan\left(\dfrac{\rho}{\dfrac{d\rho}{d\psi}}\right) & \text{if } \dfrac{d\rho}{d\psi} < 0. \end{cases} \tag{6.13}$$

From the theorem of sine in the triangle COA, $\dfrac{CO}{\sin\gamma} = \dfrac{OA}{\sin C}$, we obtain $\sin(\theta + \alpha_0) = \dfrac{\rho\sin\gamma - r}{d}$, from which

$$\theta = \arcsin\left(\frac{\rho\sin\gamma - r}{d}\right) - \alpha_0. \tag{6.14}$$

Taking into account that $\widehat{A_0OA} = \frac{\pi}{2} - \theta - \gamma$, we obtain $\varphi = \psi - \left(\frac{\pi}{2} - \theta - \gamma\right)$; that is,

$$\varphi = \psi + \gamma + \theta - \frac{\pi}{2}; \tag{6.15}$$

We thus deduce the law of motion $\theta = \theta(\varphi)$.

Table 6.3 Numerical values for the representation of the function $\theta = \theta\,(\varphi)$.

ψ	0°	45°	90°	135°	180°	225°	270°	315°	360°
γ	90°	81.9°	80.0°	83.7°	90°	96.3°	100.0°	98.1°	90°
θ	0°	1.38°	5.25°	9.75°	11.79°	9.75°	5.25°	1.38°	0°
φ	0°	38.3°	85.2°	138.4°	191.8°	241.1°	285.3°	324.5°	360°

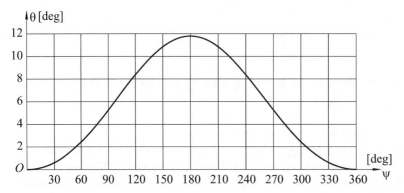

Figure 6.22 The law of motion of the follower.

Numerical example: If the profile of the cam is given by the equation $\rho = 7 + 3\sin^2\left(\frac{\psi}{2}\right)$, and $d = 16$, $r = 2$ (in mm), then applying (6.12) and (6.15) gives $\alpha_0 = 18.21°$ and the values in Table 6.3, from the law $\theta = \theta\,(\varphi)$, shown in Figure 6.22, is obtained.

Remark: In this case too, the profile of the cam must represent a curve that fulfills the condition that the expression $\rho\rho'' - 2\rho'^2 - \rho^2$ has a constant sign (in this case, negative).

The general case

In the most general case of planar motion of the cam and follower, their profiles (C_1) and (C_2) (se e Figure 6.24) may be defined with respect to the local reference systems $O_1x_1y_1$ and $O_2x_2y_2$ (to which they are rigidly linked) by the parametric equations

$$(C_1): \begin{cases} x_1 = x_1\,(\lambda_1), \\ y_1 = y_1\,(\lambda_1), \end{cases} \quad (C_2): \begin{cases} x_2 = x_2\,(\lambda_2), \\ y_2 = y_2\,(\lambda_2), \end{cases} \tag{6.16}$$

The positions of the reference systems $O_1x_1y_1$, and $O_2x_2y_2$ of the cam and follower, respectively, are defined with respect to the general reference system $OXYZ$ by the coordinates X_1, Y_1 and X_2, Y_2, and by the rotation angles φ_1 and φ_2, respectively.

The common tangent to the curves (C_1) and (C_2), at the contact point A, is inclined with respect to the axes OX, O_1x_1, and O_2x_2 with the angles α, α_1, and α_2, respectively, which verifies (see Figure 6.23):

$$\alpha = \varphi_1 + \alpha_1 = \varphi_2 + \alpha_2$$

or

$$\varphi_1 - \varphi_2 = \alpha_2 - \alpha_1; \tag{6.17}$$

The latter relation is the tangency condition.

Figure 6.23 The general case of the cam-follower contact.

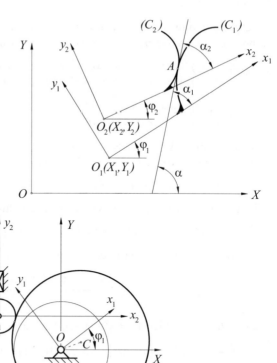

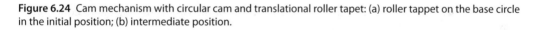

Figure 6.24 Cam mechanism with circular cam and translational roller tapet: (a) roller tappet on the base circle in the initial position; (b) intermediate position.

Now, taking into account $\tan \alpha_1 = \frac{y'_1}{x'_1}$ and $\tan \alpha_2 = \frac{y'_2}{x'_2}$ (the derivatives being calculated with respect to the parameters λ_1 and λ_2), the tangency condition reads

$$\tan \left(\varphi_1 - \varphi_2\right) = \frac{x'_1 y'_2 - x'_2 y'_1}{x'_1 x'_2 + y'_1 y'_2}. \tag{6.18}$$

By expressing the relations between the coordinates of point A relative to the reference systems OXY, $O_1 x_1 y_1$, and $O_2 x_2 y_2$, the geometric condition of contact between the two curves (C_1) and (C_2), at the contact point A, may be transcribed as

$$X_1 + x_1 \cos \varphi_1 - y_1 \sin \varphi_1 = X_2 + x_2 \cos \varphi_2 - y_2 \sin \varphi_2, \tag{6.19}$$

$$Y_1 + x_1 \sin \varphi_1 + y_1 \cos \varphi_1 = Y_2 + x_2 \sin \varphi_2 + y_2 \cos \varphi_2. \tag{6.20}$$

The laws of motion of the follower can be determined from (6.18)–(6.20). For instance, for the cam mechanism with roller tappet and rotational cam, the following conditions are necessary: $X_1 = X_{10}$, $Y_1 = Y_{10}$, $X_2 = X_{20}$ (X_{10}, Y_{10}, X_{20} being constant values), $\varphi_2 = 0$. For a given value of the parameter φ_1, from the system (6.18)–(6.20) we obtain the values λ_1 and Y_2, which give the law of motion of the roller tappet $s\left(\varphi_1\right) = Y_2 - Y_{2min}$.

As a first example, consider the cam mechanism with circular cam and roller tappet, shown in Figure 6.24a in the initial position, and in Figure 6.24b in an arbitrary position. Curve (C_1) is the circle of radius R with its centre at point C, so that $CO = R - r_0$ (Figure 6.24a). Curve (C_2) is the circle centred at point O_2 and of radius r. This gives the parametric equations

$$(C_1): \begin{cases} x_1 = (R - r_0) \cos\alpha - R\cos(\lambda_1 + \alpha), \\ y_1 = -(R - r_0)\sin\alpha + R\cos(\lambda_1 + \alpha), \end{cases}$$

$$(C_2): \begin{cases} x_2 = r\cos(\lambda_2 - \alpha), \\ y_2 = r\sin(\lambda_2 - \alpha) \end{cases}$$

and the distances $X_{10} = Y_{10} = 0$ and $X_{20} = -(r_0 + r)\cos\alpha$.

From (6.18), in which $\varphi_2 = 0$, we deduce that:

$$\tan\varphi_1 = \tan(\lambda_1 + \lambda_2) \text{ or } \varphi_1 = \lambda_1 + \lambda_2. \tag{6.21}$$

Then, in (6.19) we replace λ_2 with $\varphi_1 - \lambda_1$ obtaining

$$(R + r_0)\cos(\varphi_1 - \lambda_1 - \alpha) = (R - r_0)\cos(\varphi_1 - \alpha) + (r + r_0)\cos\alpha;$$

From which we can derive the parameter λ_1

$$\lambda_1 = \varphi_1 - \alpha + \arcsin\left(\frac{(R - r_0)\cos(\varphi_1 - \alpha) + (r_0 + r)\cos\alpha}{R + r}\right). \tag{6.22}$$

From (6.20), in which $Y_2 = s + (r + r_0)\sin\alpha$:

$$s = (r - r_0)\sin(\varphi_1 - \alpha) - (R + r)\sin(\varphi_1 - \alpha - \lambda_1) - (r + r_0)\sin\alpha. \tag{6.23}$$

For the determination of the law of motion the following steps must be followed:

1) Give values to angle φ_1.
2) Determine the value of λ_1 using (6.22).
3) Determine the value of λ_2 using (6.21).
4) Determine the value of s using (6.23).

A second example is given in Figure 6.25. The profile of the cam is the astroid of equations

$$x_1 = 2d\sin^3\lambda_1, \ y_1 = 2d\cos^3\lambda_1,$$

while the follower is a segment of the straight line of equations

$$x_2 = 0, \ y_2 = \lambda_2.$$

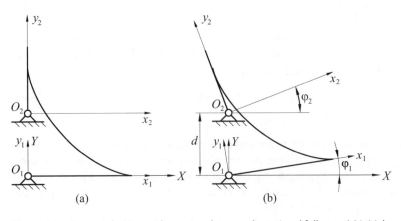

(a) (b)

Figure 6.25 Cam mechanism with rotational cam and rotational follower: (a) initial position; (b) intermediate position.

Equation (6.18) becomes $\tan(\varphi_1 - \varphi_2) = -\tan\lambda_1$ or $\varphi_1 - \varphi_2 = -\lambda_1$. Then, since $X_1 = Y_1 = 0$, $X_2 = 0$, and $Y_2 = d$, from (6.19) and (6.20) we get

$$2d\sin^3\lambda_1\cos\varphi_1 - 2d\cos^3\lambda_1\sin\varphi_1 = -\lambda_2\sin\varphi_2,$$
$$2d\sin^3\lambda_1\sin\varphi_1 + 2d\cos^3\lambda_1\cos\varphi_1 = d + \lambda_2\cos\varphi_2.$$

Eliminating the parameter λ_2 gives

$$2d\sin^3\lambda_1\cos(\varphi_1 - \varphi_2) - 2d\cos^3\lambda_1\sin(\varphi_1 - \varphi_2) = d\sin\varphi_2$$

or

$$2d\sin^3\lambda_1\cos\lambda_1 + 2d\cos^3\lambda_1\sin\lambda_1 = d\sin\varphi_2;$$

and from the last expression we successively obtain $\sin(2\lambda_1) = \sin\varphi_2$, $\varphi_2 = 2\lambda_1$, $\varphi_2 = 2\varphi_1$. This gives the law of motion $\varphi_2 = 2\varphi_1$.

6.2.3 The Graphical Method

Consider the scale drawing of the cam and the position and type of the follower as known. Determining the law of motion involves studying the relative motion of the follower with respect to the cam. The following graphical constructions can be created using AutoCAD.

Cam mechanism with axial plunger tip and rotational cam
To graphically determine the law of motion involves the following process:

1) Divide the arc of radius $r_0 = \rho_{min}$ (the circle of base) into n equal parts, obtaining points A_0, A_1, A_2, A_3,... (see Figure 6.26, in which $n = 12$).
2) Construct the radii OA_1, OA_2, OA_3,... that intersect the profile of the cam at points B_1, B_2, B_3,...
3) Consider the system of axes $A_0\psi s$; on axis $A_0\psi$ divide the angle of $360°$ into n equal parts, obtaining points A_0, A_1, A_2,... (see Figure 6.27, in which $n = 12$).
4) On the parallels to axis $A_0 s$ that pass through points A_1, A_2, A_3,..., consider the segments $s_1 = A_1B_1$, $s_2 = A_2B_2$,..., obtaining points B_1, B_2, B_3,..., as shown in Figure 6.27.
5) Connecting points B_0, B_1, B_2,... gives the graph of the function $s = s(\psi)$.

To draw Figures 6.26 and 6.27 using AutoCAD, the steps are as follows:

1) Create the exterior profile of the cam and draw the base circle

        Circle;0,0,15;

2) Construct a straight-line segment

        Line;0,0;@30,0;;

starting from point O, the length being greater than the distance OB_0.

Figure 6.26 The relative positions of the follower on the cam.

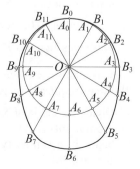

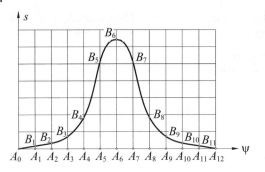

Figure 6.27 Graphical determination of the law of motion.

3) In polar mode, duplicate the segment from step 2 twelve times over, giving a total of 12 segments. The centre of this duplication is point O

   –Array;L;;P;0,0;12;360;Y;

4) Cut the segment that overlap the cam, the last one being the cutting edge

   Trim;\;\\\;

   This gives Figure 6.26.

5) Cuts the segments $A_1O, A_2O,..., A_{11}O$ using the circle as the cutting edge

   Trim;\;\\\;

   In this way only the segments $A_1B_1,..., A_{11}B_{11}$ will remain.

6) Declares as block the segment A_1B_1 using point A_1 as the reference point

   –Block;s_030;Int;\

   Proceed in a similar way for the other 11 segments.

7) Choose a system of axes on which $s = s(\psi)$ will be drawn. Divide axis $O\psi$

   Line;50,0;@100,0;;

   into 12 equal parts

   Divide;\12;PDmode;35;PDsize;35

   Alternatively, an increment can be used.

8) Insert block A_1B_1 at point A_1, which is on axis $O\psi$, by rotating this block through 30° and then multiplying it by a scaling factor (usually, this factor is the same for both axes)

   –Insert;s_030;Node;\2;2;30;

   Then insert blocks $A_2B_2,..., A_{11}B_{11}$ by rotating them through 60°,..., 330° and then multiplying by the same scale factor.

9) Draw a poly-line starting from point O and passing through the upper extremities of the inserted blocks

   Pline;50,0;End;\\\;

   This gives Figure 6.27. The curve can be created using interpolation with spline functions.

Cam mechanism with non-axial translational plunger tip and rotational cam
The steps required are as follows:

1) Divide the circle of base $(r_0 = \rho_{min})$ into n equal parts using points $A_0, A_1, A_2,...$
2) From points $A_1, A_2,...$, construct the tangents to the circle centred at point O of radius e (e being the eccentricity); these tangents intersect the profile of the cam at points $B_1, B_2,...$ (see Figure 6.28).

Figure 6.28 The relative positions of the plunger tip with respect to the cam.

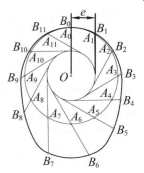

Figure 6.29 The graphical determination of the law of motion.

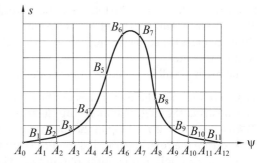

3) On axis $A_0 \psi$, divide the angle of 360° into n equal parts using points $A_0, A_1, A_2,...$
4) Construct the displacements $s_1 = A_1 B_1,..., s_{12} = A_{12} B_{12}$, obtaining points $B_1,..., B_{12}$.
5) Unite points $B_0, B_1, B_2,...$, obtaining the law of motion $s = s(\psi)$.

Figures 6.28 and 6.29 are drawn, using AutoCAD, as in the previous situation, through the following steps:

1) Create the exterior profile of the cam, and draw the base circle and the circle of radius e (e being the eccentricity of the plunger tip).
2) Construct a straight line segment (**Line**) from point O with length equal to the radius of the circle of the base.
3) Using the command **Array**, duplicate the segment 11 times, giving 12 segments in all, the centre of the multiplication being point O.
4) Construct lines (**Line**) starting from the points $A_1, A_2,..., A_{11}$ previously obtained, and tangential to the circle of radius e.
5) Extends (**Extend**) the segments obtained until they meet the cam.
6) Cuts (**Trim**) the segments obtained using the circle as the cutting edge.
7) Erases (**Erase**) the lines that do not intersect, leaving the cam, the circle of the base, the circle of radius e, and the segments $A_1 B_1,..., A_{11} B_{11}$, as in Figure 6.28.
8) Construct a circle centred at point A_1 and of radius $A_1 B_1$, and a straight line of length equal to the radius of the constructed circle, with one end at point A_1 and the other vertically above it. Declare as block (**Block**) the obtained segment using point A_1 as the reference point. Proceed in a similar way for the other 11 segments.
9) Choose a system of axes on which the graph $s = s(\psi)$ will be drawn. Divide axis $O\psi$ into 12 equal parts (**Divide**). Alternatively, use an increment.
10) Insert the block $A_1 B_1$ (**Insert**) at point A_1, which is on axis $O\psi$, without any rotation (the segment which was saved using the command **Block** is a vertical one), but possibly multiplied by a scaling factor (the same for the both axes). Then insert blocks $A_2 B_2,..., A_{11} B_{11}$ multiplied by the same scaling factor.

11) Draw a poly-line (**Pline**) starting from point O and passing through the upper ends (**End**) of the inserted blocks. The poly-line obtained, represented by points, my be interpolated using spline functions (command **Pedit**), giving Figure 6.29.

6.2.4 Analysis of Displacement of Follower using AutoLisp

In earlier sections, we presented the AutoLisp functions that we used for the analysis and synthesis of planar mechanisms with bars. These functions can also be used for the determination of the displacements of followers.

Because most cam mechanisms are spatial ones, we now present a working algorithm and an AutoLisp function for the determination of the displacements of followers, no matter what the shape of the cam or the follower. The AutoLisp functions for working with solids are used. This is easy in AutoCAD and AutoLisp, and is just as precise as for planar systems. The algorithm involves the stepwise movement of the solids that make up the cam and the follower until contact is made.

The method will be called *by approach* and the steps are as follows:

1) Construct the solid that makes up the cam and assign a name to it (*Cam*).
2) Construct the solid that makes up the follower and and assign a name to it (*Follower*).
3) In a repetitive *while* loop in the interval $[0, 2\pi]$:
   - create the rotation/translation of the cam with a constant angular/linear step
   - using a Boolean operation, obtain the solid of intersection between the cam and the follower (the AutoCAD command **Intersect**); if there is no intersection, then use a second repetitive *while* loop in which:
     – the follower is moved with a constant step (*step*) that is at least 100 times smaller than the previous one
     – the solid of intersection is obtained
     – if the solid of intersection does not exist, the follower is moved another step; if the solid of intersection does exist, then the repetitive loop is exited
   - write the value obtained for the displacement in a file
   - pass to a new position.

The time necessary to obtain the displacement of the follower depends on the value given to the parameter *step*, which, for correct results, must have a value less than 0.0001 mm (the profile of the cams for the motor car is given with a precision of a tenth of a micron).

As an example, we consider the cam mechanism with rotational circular cam and translational flat tappet in Figure 6.20. We used this cam mechanism because it has a simple analytical solution and we can compare the results obtained by the two methods (analytical and graphical). The numerical data are: the centre of rotation of the cam is at $O\,(0, 0, 0)$, the radius of the circle of base $r_0 = 25$ mm, and $R = 30$ mm. The solid that represents the cam is a cylinder centred at the point $C\,(0, -5, -10)$, of radius $R = 30$ mm and thickness $g = 20$ mm. The solid which represents the follower is also a cylinder, but in a plane perpendicular to the current plane. It is centred at the point $A\,(0, 25, 0)$, and has radius $R_T = 30$ mm and height $g_T = 30$ mm.

The function used (*Kine_cam.lsp*) is given in the companion website of the book. It also incorporates the code for the determination of the displacement of the follower; the underlying relation was determined in Section 6.2.2.

Table 6.4 shows the results obtained by both methods. There are no differences.

In the *Kine_cam* function, the shapes of the follower or cam are not important. The cam and the follower may be directly constructed in AutoCAD; in this case, the instructions used for their constructions will vanish from the AutoLisp function. The instructions which lead to the constructions described in function *Kine_cam* have the comments: 'construction of the cam'

Table 6.4 The numerical values obtained for the displacement of the follower $s = s(\psi)$.

ψ	s_{ACAD}	$s_{analytic}$	ψ	s_{ACAD}	$s_{analytic}$	ψ	s_{ACAD}	$s_{analytic}$	ψ	s_{ACAD}	$s_{analytic}$
0	0.076	0.076	90	5	5	180	10	10	280	4.1318	4.1318
10	0.3015	0.3015	100	5.8682	5.8682	200	9.924	9.924	300	3.2899	3.2899
20	0.6699	0.6699	110	6.7101	6.7101	210	9.6985	9.6985	310	2.5	2.5
30	1.1698	1.1698	120	7.5	7.5	220	9.3301	9.3301	320	1.7861	1.7861
40	1.7861	1.7861	130	8.2139	8.2139	230	8.8302	8.8302	330	1.1698	1.1698
50	2.5	2.5	140	8.8302	8.8302	240	8.2139	8.2139	340	0.6699	0.6699
60	3.2899	3.2899	150	9.3301	9.3301	250	7.5	7.5	350	0.3015	0.3015
70	4.1318	4.1318	160	9.6985	9.6985	260	6.7101	6.7101	360	0.076	0.076
80	0.076	0.076	170	9.924	9.924	270	5.8682	5.8682			

and 'construction of the follower'. After the cam's construction in AutoCAD, the entity obtained is named using the AutoLisp function *Setq* (*Setq Cam(EntLast)*) and similarly, for the follower, *Setq Follower(EntLast)*. To run an AutoLisp function in AutoCAD, we have to put the name of the function in parentheses in the command line. The function can be easily modified when the follower is in rotational motion, or if another law of motion is chosen for the cam.

6.3   Analysis of Velocities and Accelerations

6.3.1   Analytical Method

Velocity and acceleration of the follower
Knowing the function $s = s(\varphi)$, differentiation with respect to time gives the velocity and acceleration of the follower

$$v = \frac{ds}{d\varphi}\frac{d\varphi}{dt}, \quad a = \frac{d^2s}{d\varphi^2}\left(\frac{d\varphi}{dt}\right)^2 + \frac{ds}{d\varphi}\frac{d^2\varphi}{dt^2};$$

Since

$$\frac{d\varphi}{dt} = \omega, \quad \frac{d^2\varphi}{dt^2} = \varepsilon,$$

this gives

$$v = \omega\frac{ds}{dt}, \quad a = \omega^2\frac{d^2s}{d\varphi^2} + \varepsilon\frac{ds}{d\varphi}. \tag{6.24}$$

From (6.24), in order to obtain the velocity and acceleration it is first necessary to determine the derivatives $\frac{ds}{d\varphi}$ (the *reduced velocity*) and $\frac{d^2s}{d\varphi^2}$ (the *reduced acceleration*). If the function $s = s(\varphi)$ is given analytically, then the two derivatives will be calculated analytically too. If the function $s = s(\varphi)$ is given in a table, such as Table 6.5, then it is first necessary to calculate the derivatives $\frac{ds}{d\psi}, \frac{d^2s}{d\psi^2}, \frac{d\varphi}{d\psi}, \frac{d^2\varphi}{d\psi^2}$ at point $\psi = \psi_0$ while the derivatives $\frac{ds}{d\varphi}, \frac{d^2s}{d\varphi^2}$ are calculated using:

$$\frac{ds}{d\varphi} = \frac{\dfrac{ds}{d\psi}}{\dfrac{d\varphi}{d\psi}}, \quad \frac{d^2s}{d\varphi^2} = \frac{\dfrac{d^2s}{d\psi^2}\dfrac{d\varphi}{d\psi} - \dfrac{ds}{d\psi}\dfrac{d^2\varphi}{d\psi^2}}{\left(\dfrac{d\varphi}{d\psi}\right)^3}. \tag{6.25}$$

Table 6.5 The form of the function $s = s(\varphi)$.

ψ	0	...	$(i-1)\Delta\psi$	$i\Delta\psi$	$(i+1)\Delta\psi$	...	$n\Delta\psi$
φ	φ_0	...	φ_{i-1}	φ_i	φ_{i+1}	...	φ_n
s	s_0	...	s_{i-1}	s_i	s_{i+1}	...	s_n

If given the table of values $s_i = s_i(\varphi_i)$, where $\varphi_i = i\Delta\varphi$, then the derivatives can be calculated directly using the method of finite differences

$$\left.\frac{ds}{d\varphi}\right|_{\varphi=\varphi_i} = \frac{s_{i+1} - s_{i-1}}{2\Delta\varphi}, \quad \left.\frac{d^2s}{d\varphi^2}\right|_{\varphi=\varphi_i} = \frac{s_{i+1} - 2s_i + s_{i-1}}{(\Delta\varphi)^2}, \tag{6.26}$$

where the angles φ_i and $\Delta\varphi$ are expressed in radians. The precision of (6.25) and (6.26) is ensured by taking large values for n. If $\varphi \in [0°, 360°]$, then $n = 360$; in this situation $\Delta\varphi = \frac{\pi}{180}$.

Knowing the velocity v and the acceleration a of the follower, it is only necessary to determine the relative velocities and accelerations at the cam–follower contact points. If the follower is in rotational motion, $\varphi_2 = \varphi_2(\varphi_1)$, then a similar approach is used, the function s being replaced by the function φ_2.

Cam mechanism with axial translational plunger tip and rotational cam

The velocity $\mathbf{v}_{A_2}$ is situated in the direction of the plunger tip (see Figure 6.30) and it is calculated using the first relation in (6.24), $v_{A_2} = \omega\frac{ds}{d\varphi}$. From the representation of the vector relation (Figure 6.30):

$$\mathbf{v}_{A_2} = \mathbf{v}_{A_1} + \mathbf{v}_{A_2A_1}, \tag{6.27}$$

Denoting $OA = \rho$ and knowing that $v_{A_1} = \rho\omega$, $\mathbf{v}_{A_1} \perp OA$ and the relative velocity is on the tangent to the profile of the cam, gives

$$v_{A_2A_1} = \sqrt{v_{A_2}^2 + v_{A_1}^2} = \omega\sqrt{\rho^2 + \left(\frac{ds}{d\varphi}\right)^2}. \tag{6.28}$$

For the calculation of the accelerations we make the hypothesis that $\varepsilon = 0$, which gives

$$\underset{\substack{\| AB}}{\mathbf{a}_{A_2}} = \underset{\substack{A \to O}}{\mathbf{a}_{A_1}} + \underset{\substack{\perp AT}}{\mathbf{a}_{A_2A_1}^C} + \underset{\substack{\perp AT}}{\mathbf{a}_{A_2A_1}^\nu} + \underset{\substack{\| AT}}{\mathbf{a}_{A_2A_1}^\tau}, \tag{6.29}$$

as shown in Figure 6.31.

Figure 6.30 The velocities at point A.

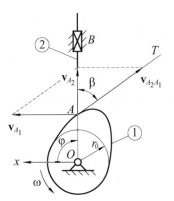

Figure 6.31 Representation of the accelerations.

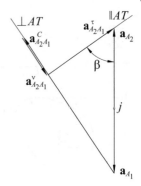

Taking into account the expressions:

$$a_{A_2} = \omega^2 \frac{d^2 s}{d\varphi^2}, \ a_{A_1} = \rho\omega^2, \ a^C_{A_2A_1} = 2\omega v_{A_2A_1} = 2\omega^2 \sqrt{\rho^2 + \left(\frac{ds}{d\varphi}\right)^2},$$

$$\cos\beta = \frac{v_{A_1}}{v_{A_2A_1}} = \frac{\dfrac{ds}{d\varphi}}{\sqrt{\rho^2 + \left(\dfrac{ds}{d\varphi}\right)^2}}, \ \sin\beta = \frac{\rho}{\sqrt{\rho^2 + \left(\dfrac{ds}{d\varphi}\right)^2}},$$

gives, from Figure 6.31:

$$a^v_{A_2A_1} = a^C_{A_2A_1} - \left(a_{A_1} + a_{A_2}\right)\sin\beta = \frac{\omega^2 \left[\rho^2 + 2\left(\dfrac{ds}{d\varphi}\right)^2 - \rho\dfrac{d^2 s}{d\varphi^2}\right]}{\sqrt{\rho^2 + \left(\dfrac{ds}{d\varphi}\right)^2}};$$

since $s = \rho - r_0$ and $a^v_{A_2A_1} = \dfrac{v^2_{A_2A_1}}{R}$, where R is the radius of curvature, this gives (as a verification) the expression for the radius of curvature

$$R = \frac{\left[\rho^2 + \left(\dfrac{d\rho}{d\varphi}\right)^2\right]^{\frac{3}{2}}}{\rho^2 + 2\left(\dfrac{d\rho}{d\varphi}\right)^2 \rho\dfrac{d^2\rho}{d\varphi^2}}. \tag{6.30}$$

The expression of the relative tangential acceleration (Figure 6.31) is given by $a^\tau_{A_2A_1} = \left(a_{A_1} + a_{A_2}\right)\cos\beta$; that is,

$$a^\tau_{A_2A_1} = \omega^2 \frac{\left(\rho + \dfrac{d^2 s}{d\varphi^2}\right)\dfrac{ds}{d\varphi}}{\sqrt{\rho^2 + \left(\dfrac{ds}{d\varphi}\right)^2}}. \tag{6.31}$$

Cam mechanism with flat tappet and rotational cam
In this situation (Figure 6.32), the contact points move on the cam and on the flat tappet. The contact point can be represented as a ring denoted by A_3, which moves on the cam and on the

Figure 6.32 The velocities at point A.

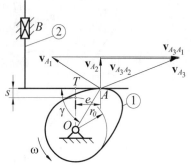

flat tappet. This gives:

$$\mathbf{v}_{A_3} = \underbrace{\mathbf{v}_{A_1}}_{\perp OA} + \underbrace{\mathbf{v}_{A_3A_1}}_{\parallel AT} = \underbrace{\mathbf{v}_{A_2}}_{\parallel OT} + \underbrace{\mathbf{v}_{A_3A_2}}_{\parallel AT},$$

(6.32)

the representation of which is as in Figure 6.32.

Taking into account $v_{A_1} = \rho\omega$ and $v_{A_2} = \omega\dfrac{ds}{d\varphi}$, where $\rho = OA$, gives, from Figure 6.32, $v_{A_1}\cos\delta = v_{A_2}$ or $\rho\cos\delta = \dfrac{ds}{d\varphi}$ and we deduce that:

$$AT = e = \frac{ds}{d\varphi}.$$

(6.33)

This means that $v_{A_3A_2} = \dfrac{de}{dt} = \omega\dfrac{de}{d\varphi}$, so the velocity of the contact point on the flat tappet has the expression

$$v_{A_3A_2} = \omega\frac{d^2s}{d\varphi^2}.$$

(6.34)

Figure 6.32 also means that the velocity $v_{A_3A_1}$ of the contact point on the cam is given by:

$$v_{A_3A_1} = v_{A_1}\sin\gamma + v_{A_3A_2} = \omega\left(\rho\sin\gamma + \frac{d^2s}{d\varphi^2}\right) = \omega\left(OB + \frac{d^2s}{d\varphi^2}\right),$$

so

$$v_{A_3A_1} = \omega\left(r_0 + s + \frac{d^2s}{d\varphi^2}\right).$$

(6.35)

For the calculation of the accelerations we assume that ω is a constant and get the vector relations

$$\mathbf{a}_{A_3} = \underbrace{\mathbf{a}_{A_1}}_{A \to O} + \underbrace{\mathbf{a}_{A_3A_1}^C}_{\perp AT} + \underbrace{\mathbf{a}_{A_3A_1}^\tau}_{\perp AT} + \underbrace{\mathbf{a}_{A_3A_1}^\nu}_{\parallel AT}$$

$$= \underbrace{\mathbf{a}_{A_2}}_{\perp AT} + \underbrace{\mathbf{a}_{A_3A_2}}_{\parallel AT}$$

(6.36)

as shown in Figure 6.33. The relative accelerations of the contact point on the cam $a_{A_3A_1}^\tau$ and on the flat tappet $a_{A_3A_2}$ are calculated using:

$$a_{A_3A_1}^\tau = \frac{dv_{A_3A_1}}{dt}, \quad a_{A_3A_2} = \frac{dv_{A_3A_2}}{dt};$$

Taking into account (6.34) and (6.35) gives

$$a_{A_3A_1}^\tau = \omega^2\left(\frac{ds}{d\varphi} + \frac{d^2s}{d\varphi^2}\right), \quad a_{A_3A_2} = \omega^2\frac{d^2s}{d\varphi^2}.$$

(6.37)

Figure 6.33 Representation of the accelerations.

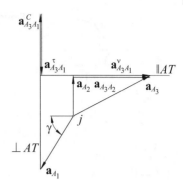

Then, considering:

$$a_{A_1} = \rho\omega^2, \quad a^C_{A_3A_1} = 2\omega^2\left(r_0 + s + \frac{d^2s}{d\varphi^2}\right), \quad a^v_{A_3A_1} = \frac{v^2_{A_3A_1}}{R},$$

where R is the radius of curvature, and the equality (Figure 6.33)

$$a^v_{A_3A_1} = a^C_{A_3A_1} - a_{A_2} - a_{A_3A_2} - a_{A_1}\sin\gamma,$$

gives

$$\frac{\omega^2\left(r_0 + s + \dfrac{d^2s}{d\varphi^2}\right)}{R} = 2\omega^2\left(r_0 + s + \frac{d^2s}{d\varphi^2}\right) - \omega^2\frac{d^2s}{d\varphi^2} - \omega^2\rho\sin\gamma;$$

Since $\rho\sin\gamma = r_0 + s$, this gives the expression for the radius of curvature:

$$R = r_0 + s + \frac{d^2s}{d\varphi^2}. \tag{6.38}$$

Equation (6.37) shows that, on the portion of the cam's profile on which the radius of curvature R is constant, the law of motion is of the form:

$$s = R - r_0 + C_1\cos\varphi + C_2\sin\varphi, \tag{6.39}$$

where C_1 and C_2 are constants of integration.

In the portion of the cam's profile in which the contact point does not move on the flat tappet, $v_{A_3A_2} = 0$; that is, $e = AT = $ const (Figure 6.33), and the law of motion is of the form of an involute cam:

$$s = e\varphi + C_1, \tag{6.40}$$

where C_1 is a constant of integration. In this case, the normal to the curve at the contact point is tangent to the circle centred at point O and of radius e; in the portion in which $e = $ const the cam's profile is an involute arc of the circle.

6.3.2 Graphical Method: Graphical Derivation

For the determination of the reduced velocity $\dfrac{ds}{d\varphi}$ and reduced acceleration $\dfrac{d^2s}{d\varphi^2}$, we proceed to the graphical derivation of the functions s and $\dfrac{ds}{d\varphi}$. For that purpose, scales k_s and k_φ for the parameters s and φ, respectively, mean that the derivative at a point φ is

$$\frac{ds}{d\varphi} = \frac{d[k_s(s)]}{d[k_\varphi(\varphi)]} = \frac{k_s}{k_\varphi}\frac{d(s)}{d(\varphi)}.$$

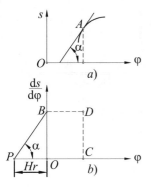

Figure 6.34 The graphical derivation.

At this point, if we construct the tangent to the curve and denote by α the angle between this tangent and axis Ox (Figure 6.34a), we obtain $\dfrac{d(s)}{d(\varphi)} = \tan \alpha$; hence

$$\frac{ds}{d\varphi} = \frac{k_s}{k_\varphi} \tan \alpha. \tag{6.41}$$

Choosing point P on axis $O\varphi$ (Figure 6.34b), and denoting $H_v = PO$, then constructing at point P the parallel to the tangent that intersects ordinate axis at point B, we obtain $OB = H_v \tan \alpha$; for the same φ and taking the segment $CD = OB$, mean that the segment $CD = H_v \tan \alpha$ represents, at the scale k_v, the value of the derivative at point A (Figure 6.34a). The scale k_v is given by:

$$k_v = \frac{\dfrac{ds}{d\varphi}}{\left(\dfrac{ds}{d\varphi}\right)} = \frac{\dfrac{k_s}{k_\varphi} \tan \varphi}{H_v \tan \alpha}$$

or

$$k_v = \frac{k_v}{H_v k_\varphi}. \tag{6.42}$$

For the determination of the reduced acceleration $\left(\dfrac{d^2 s}{d\varphi^2}\right)$ we proceed in a similar way and obtain the scale $k_a = \dfrac{k_v}{H_a k_\varphi}$. If the derivative is performed repeatedly, then $H_a = H_v$.

As an example, consider Figure 6.35a, where $k_v = 0.3 \left[\dfrac{m/s}{mm}\right]$ and $k_\varphi = \dfrac{\pi}{21} \left[\dfrac{rad}{mm}\right]$.

Choosing $PO = H_v = 15\,mm$ gives $k_v = \dfrac{21}{50\pi} \left[\dfrac{mm/rad}{mm}\right]$, $\left(\dfrac{ds}{d\varphi}\right)\Big|_{max} = 18\,mm$, and $\left(\dfrac{ds}{d\varphi}\right)\Big|_{min} = -18\,mm$. The maximum value for the reduced velocity is $18\dfrac{21}{50\pi} = 2.40 \left[\dfrac{mm}{rad}\right]$, while the minimum value of the reduced velocity is $-18\dfrac{21}{50\pi} = -2.40 \left[\dfrac{mm}{rad}\right]$.

The working procedure was as follows:

1) Divide the interval $[0°, 360°]$ into 12 equal parts obtaining points $A_1, A_2,..., A_{12}$.
2) At points $A_1, A_2,..., A_{12}$ construct the tangents to the curve.
3) From point P, construct parallels to these tangents obtaining, on the axis of the ordinates, points $1, 2,..., 12$. A these points, construct parallels to axis $O\varphi$, obtaining the corresponding points $B_1, B_2,..., B_{12}$.

Figure 6.35 The graphical derivation: (a) law of motion; (b) reduced velocity.

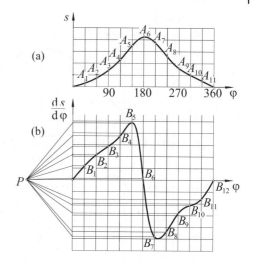

4) Unite points B_1, B_2,..., B_{12}, obtaining the curve of function $\dfrac{ds}{d\varphi}$ at the scale $\dfrac{k_s}{H_v k_\varphi} = 0.1337 \left[\dfrac{mm/rad}{mm}\right]$.

As another example, we consider the cam of a real engine. The law of displacement $s = s(\varphi)$ is given in Figure 6.36a. Our goal is to determine, by graphical derivation in AutoCAD, the reduced velocity. We proceed as follows:

1) Draw the graph $s = s(\varphi)$, based on the table of values given by the manufacturer ($\varphi \in [90°, \dots, 275°]$).
2) Using the command **Divide**, divide axis $O\varphi$ into 37 equal parts using a straight line parallel to axis Os of length 6;
3) To construct tangents to the graph at the points of intersection between the parallel in step 2 and the graph of $s = s(\varphi)$, construct circles of small diameter (the radii can be 0.01 mm) at these points of intersection. Draw a line between the points of intersection between the circle and the curve and save the values in a file (*Tan095,..., Tang270*);
4) Choose a point P (Figure 6.35b) at which to insert the saved blocks (*Tan095,..., Tang270*);
5) Use the procedure used for Figure 6.35b, constructing straight lines parallel to axis $O\varphi$ to obtain points B_1, B_2,..., B_{37}.
6) Draw a poly-line (**Pline**) through points B_1, B_2,..., B_{37}; this poly-line can be modified using spline functions (command **Pedit**).

We thus obtain Figure 6.36b. To obtain the values for the points, one draws the dimensions on the obtained figure, calculates the scale with (6.42), and then performs the calculation as in the previous case. For the reduced acceleration, the scale is determined with $k_a = \dfrac{k_v}{H_a k_\varphi}$; proceeding in a similar way to the determination of the reduced velocity, gives the reduced acceleration as shown in Figure 6.36c.

6.4 Dynamical Analysis

6.4.1 Pre-load in the Spring

Considers the simple cam mechanism with flat tappet in Figure 6.37a, in which the cam is in rotational motion with a constant angular velocity. The forces that act upon the flat tappet

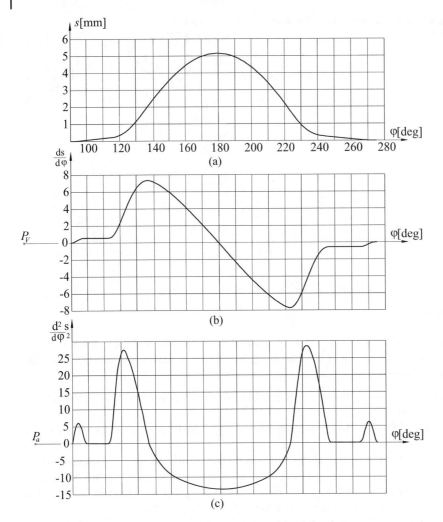

Figure 6.36 The displacement, reduced velocity, and reduced acceleration of the cam of a real engine: (a) displacement; (b) reduced velocity; (c) reduced acceleration.

are represented in Figure 6.37b, where Q is the technological force, while F_s is the force in the spring.

If the pre-load force of the spring produces a deformation s_0, then the force in the spring is $F_s = k\,(s + s_0)$, where k is the stiffness of the spring and s is the displacement of the flat tappet. Denoting by m the mass of the flat tappet gives its equation of motion $m\ddot{s} = N - F_s - Q$, from which:

$$N = m\frac{d^2 s}{d\varphi^2}\omega^2 + k\,(s + s_0) + Q. \tag{6.43}$$

The condition of permanent contact between the cam and the flat tappet may be transcribed as $N > 0$. From this, and neglecting force Q, the pre-load deformation of the spring becomes:

$$s_0 > -\frac{m\omega^2}{k}\frac{d^2 s}{d\varphi^2} - s. \tag{6.44}$$

As a numerical example, consider $m = 0.1$ kg, $\omega = 300$ rad/s, $k = 9000$ N/m, and $s = h_0\left[1 - \cos\left(2\varphi\right)\right]$, for $\varphi \in [0, \pi]$, $s = 0$, for $\varphi \in (\pi, 2\pi)$, and $h_0 = 0.005$ mm. This gives

Figure 6.37 The forces in a cam mechanism with flat tapet: (a) cam mechanism with rotational cam and flat tappet; (b) forces which act upon the flat tappet.

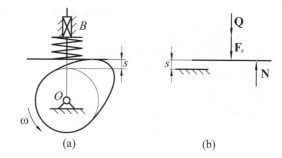

(a)   (b)

$\dfrac{m\omega^2}{k} = 1$ and thus $s_0 \geq \left(-\dfrac{d^2 s}{d\varphi^2} - s\right) = 0.01$ m. Therefore, the minimum pre-load deformation of the spring is 10 mm.

6.4.2 The Work of Friction

The elementary work of friction is $dW_f = \mu N v_r dt$, where μ is the coefficient of sliding friction and v_r is the relative velocity at the contact point:

$$v_r = (r_0 + s)\,\omega.$$

This gives the work of friction in a cycle as:

$$W_f = \mu \int_0^{2\pi} N\,(r_0 + s)\,d\varphi. \tag{6.45}$$

For the example considered in Section 6.4.1, we obtain, for $\varphi \in [0, 2\pi]$, the expression

$$N\,(r_0 + s) = \left[4m\omega^2 h_0 \cos^2 (2\varphi) + k s_0 + k h_0 - k h_0 \cos (2\varphi)\right]$$
$$\cdot \left[r_0 + h_0 - h_0 \cos (2\varphi)\right] = k\left[(s_0 + h_0)\,(r_0 + h_0) + \frac{h_0^2}{2}\right]$$
$$-2m\omega^2 h_0^2 + \left[h_0 \left(4m\omega^2 - k\right)(r_0 + h_0) - k h_0 \left(s_0 + h_0\right)\right] \cos (2\varphi)$$
$$-\frac{1}{2} h_0^2 \left(4m\omega^2 - k\right) \cos (4\varphi);$$

Since $\int_0^\pi \cos(2\varphi)\,d\varphi = \int_0^\pi \cos(4\varphi)\,d\varphi = 0$, this gives

$$W_f = \pi\mu \left[k\left(s_0 r_0 + s_0 h_0 + h_0 r_0 + 1.5 h_0^2\right) - 2m\omega^2 h_0^2\right] + k\mu s_0 r_0 \pi.$$

If $r_0 = 0.02$ m, $s_0 = 0.01$ m, and $\mu = 0.1$, then we obtain $W_f = 1.51 \left[\dfrac{J}{\text{cycle}}\right]$, while the power consumed by friction is $P_f = W_f \dfrac{\omega}{2\pi} = 72.1$ W.

6.4.3 Pressure Angle, Transmission Angle

Consider rigid bodies 1 and 2, which are in contact at point A (Figure 6.38), and let tt' and nn' be the common tangent and the common normal respectively. If the motion is transmitted from rigid body 1 to rigid body 2, and we denotes by $\mathbf{v}_{A_2}$ the velocity of point A on rigid body 2, then the *pressure angle* α_{12} is defined as the angle between the velocity $\mathbf{v}_{A_2}$ and the common normal, while the *transmission angle* γ_{12} is defined as the angle between the velocity $\mathbf{v}_{A_2}$ and the common tangent. It can be seen that $\gamma_{12} = 90° - \alpha_{12}$. The pressure and transmission angles for the cam–follower contact are shown in Figure 6.39.

Good transmission of the motion demands that the pressure angle α is small; that is, the transmission angle is large. This condition is met by the cam mechanism with flat tappet (Figure 6.39b). The maximum value of angle α, denoted by α_{cr}, comes from the requirement to avoid self-locking.

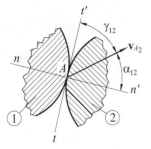

Figure 6.38 The pressure angle and the transmission angle.

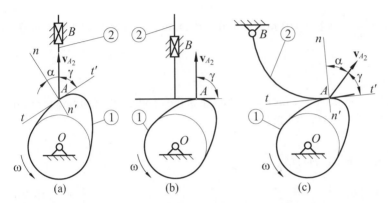

Figure 6.39 The pressure angle and the transmission angle at a cam mechanism: (a) cam mechanism with rotational cam and plunger tip; (b) cam mechanism with rotational cam and translational flat tappet; (c) cam mechanism with rotational cam and rotational follower.

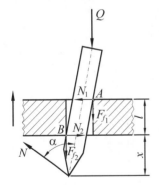

Figure 6.40 The position of the plunger tip in the guide.

For a cam mechanism with a translational plunger tip, the position of the plunger tip in the guides is represented in Figure 6.40. If Q is the technological force, and N is the cam's reaction, then, from the equation of equilibrium relative to the axis of the plunger tip and from the equations of the moments at points A and B, and neglecting the thickness of the plunger tip:

$$Q = F_{f_1} + F_{f_2} + N \cos \alpha, \ N_2 = N \frac{x+l}{l} N \sin \alpha, \ N_1 = \frac{x}{l} N \sin \alpha;$$

Since the friction forces are expressed by $F_{f_1} = \mu N_1$ and $F_{f_2} = \mu N_2$, we get

$$N = \frac{Q}{\cos \alpha - \left(l + \frac{2x}{l}\right) \mu \sin \alpha}. \tag{6.46}$$

Self-locking happens at the angle α at which the denominator vanishes. From this:

$$\tan \alpha = \frac{1}{\mu\left(1 + \frac{2x}{l}\right)}.$$

The maximum value of angle α, α_{cr}, is obtained if $x = x_{min}$; that is,

$$\alpha_{cr} = \arctan\left(\frac{1}{\mu\left(1 + \frac{2x_{min}}{l}\right)}\right).$$

The cam mechanism functions correctly if $\alpha < \alpha_{cr}$.

In a numerical application, for the axial plunger tip, considering $\mu = 0.1$, $l = 20$ mm, $x_{min} = 20$ mm and the displacement of the plunger tip $s_{max} = 10$ mm, $x_{max} = x_{min} + s_{max} = 30$ mm, and we obtain $\alpha_{cr} = 68°$, and $\alpha_{min} = 22°$.

For different cam mechanisms, the value of the pressure (transmission) angle for both the ascending phase and the descending phase of the follower are given. For instance, for the mechanism with a rotational cam and a translational follower, the following values are recommended:

- in the ascending phase $\alpha_{cr} = 35°$, $\gamma_{cr} = 55°$;
- in the descending phase $\alpha_{cr} = 60°$, $\gamma_{cr} = 30°$.

6.4.4 Determination of the Base Circle's Radius

The radius $r_0 = \rho_{min}$ of the circle of the base is determined from the requirement to avoid self-locking ($\gamma > \gamma_{cr}$). For the cam mechanism with plunger tip (Figure 6.41), representing the velocity at the point of contact as

$$\begin{array}{ccc} \mathbf{v}_{A_2} & = & \mathbf{v}_{A_1} & + & \mathbf{v}_{A_2 A_1} \ , \\ \| AG & \perp OA & \| tt' \end{array} \tag{6.47}$$

gives the triangle of velocities ADE. Denoting by G the intersection point between the normal from O to the direction of the plunger tip, and the normal from A to the cam's profile, means that the triangles ADE and OAG are similar because $\widehat{AGO} = \widehat{ADE} = \gamma$ and $\widehat{OAG} = \widehat{DEA} = \beta$ (angles with reciprocal perpendicular edges). Taking into account the relation of similarity $\frac{AE}{OA} = \frac{AD}{OG}$ and the equalities

$$AE = v_{A_1} = OA\omega, \quad AD = v_{a_2} = \omega\frac{ds}{d\varphi},$$

Figure 6.41 Cam mechanism with non-axial plunger tip.

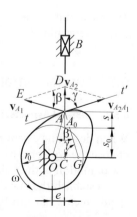

gives $OG = \dfrac{ds}{d\varphi}$. Now taking into account the notation in the figure, from the right-angled triangle GCA:

$$\tan \gamma = \frac{CA}{CG} = \frac{CA_0 + A_0A}{OG - OC} = \frac{s_0 + s}{\dfrac{ds}{d\varphi} - e}.$$

If point C is situated on the left-hand side of point O, then $e < 0$. The requirement to avoid self-locking is met if $|\gamma| > \gamma_{cr}$ or $\left| \dfrac{s_0 - s}{\dfrac{ds}{d\varphi} - e} \right| > \tan \gamma_{cr}$; that is,

$$s_0 = max \left[\left| \frac{ds}{d\varphi} - e \right| \tan \gamma_{cr} - s \right]. \tag{6.48}$$

In practice, s_0 is determined from (6.48), while the radius r_0 of the circle of the base is calculated using

$$r_0 = \sqrt{e^2 + s_0^2}. \tag{6.49}$$

For instance, consider the case in which $e = 0$, $\gamma_{cr} = 56.31°$ ($\tan \gamma_{cr} = 1.5$) and

$$s = \begin{cases} h_0 \left[1 + \cos(2\varphi) \right] & \text{if } \varphi \in \left[90°, 270° \right], \\ 0 & \text{if } \varphi \in \left(0°, 90° \right) \cup \left(270°, 360° \right). \end{cases}$$

Equation (6.48) reduces to the calculation of the maximum of the function $f(\varphi)$, where

$$f(\varphi) = \begin{cases} h_0 \left[-3 \sin(2\varphi) - \cos(2\varphi) - 1 \right] & \text{if } \varphi \in \left[90°, 180° \right], \\ h_0 \left[3 \sin(2\varphi) - \cos(2\varphi) - 1 \right] & \text{if } \varphi \in \left[180°, 270° \right]. \end{cases}$$

By derivation:

$$f'(\varphi) = \begin{cases} 2h_0 \left[-3 \cos(2\varphi) + \sin(2\varphi) \right] & \text{if } \varphi \in \left[90°, 180° \right], \\ 2h_0 \left[3 \cos(2\varphi) + \sin(2\varphi) \right] & \text{if } \varphi \in \left[180°, 270° \right]. \end{cases}$$

The derivative vanishes for $\varphi = 125.78°$ and $\varphi = 234.22°$, which corresponds to $max(f) = 2.16h_0$. This gives $r_0 = s_0 > 2.16h_0$.

If the function $f(\varphi) = \left| \dfrac{ds}{d\varphi} - e \right| \tan \gamma_{cr} - s$ can not be expressed analytically, a table of values can be created:

φ	$\cdots$	φ_{i-1}	φ_i	φ_{i+1}	$\cdots$
$f(\varphi)$	$\cdots$	f_{i-1}	f_i	f_{i+1}	$\cdots$

The maximum value is retained, from which s_0 and then the radius r_0 of the circle of the base can be determined.

6.5 Fundamental Laws of the Follower's Motion

6.5.1 General Aspects: Phases of Motion of the Follower

When designing a cam mechanism, depending on the specific requirements, it is necessary to choose the type of cam mechanism, to determine the radius of the circle of base, establish the law of motion of the follower, and, finally, determine the cam's profile.

To establish the law of motion of the follower the kinematic, dynamical, and technological requirements have to be taken into account. These are set out using the cyclogram, which

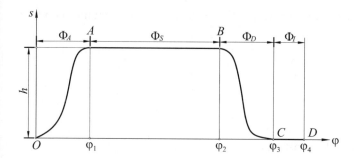

Figure 6.42 The phases of motion of a translational follower for a cam mechanism with a rotational cam.

specifies the phases of motion of the follower. For the rotational cam and translational follower arrangement (Figure 6.42) the following phases can be distinguished:

- the ascending phase OA, in which the cam rotates with angle $\Phi_A = \varphi_1$
- the stationary phase AB in superior position, in which the cam rotates through an angle $\Phi_S = \varphi_2 - \varphi_1$
- the descending phase BC, in which the cam rotates through an angle $\Phi_D = \varphi_3 - \varphi_2$
- the stationary phase CD in the lower position, in which the cam rotates through an angle $\Phi_I = \varphi_4 - \varphi_3$, where, in most cases, $\varphi_4 = 2\pi$.

For the determination of the effective laws of the ascending OA and descending BC phases, parameters h, Φ_A, Φ_S, Φ_D, and Φ_I are assumed as known.

6.5.2 The Linear Law

The linear law of motion is represented in Figure 6.43a. In the interval OA, $s = C_1\varphi + C_2$, where C_1 and C_2 are constants of integration determined from the boundary conditions:

$$\varphi = 0 \rightarrow s = 0,$$
$$\varphi = \Phi_A \rightarrow s = h.$$

This gives $C_2 = 0$ and $C_1 = \dfrac{h}{\Phi_A}$, leading to

$$s = h\frac{\varphi}{\Phi_A}. \tag{6.50}$$

Figure 6.43 The linear law of motion, the reduced velocity, and the reduced acceleration: (a) linear law of motion; (b) reduced velocity; (c) reduced acceleration.

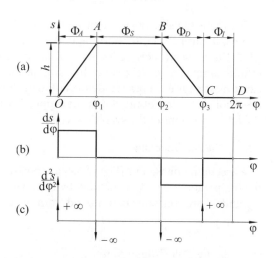

In the interval BC the displacement reads $s = C_3\varphi + C_4$ and, taking into account the boundary conditions,

$$\varphi = \varphi_2 \to s = h,$$
$$\varphi = \varphi_3 \to s = 0,$$

this gives the system

$$\begin{cases} h = C_3\varphi_2 + C_4, \\ 0 = C_3\varphi_3 + C_4, \end{cases}$$

from which it is possible to deduce the constants

$$C_3 = -\frac{h}{\varphi_3 - \varphi_2}, \quad C_4 = -\frac{h\varphi_3}{\varphi_3 - \varphi_2}.$$

This gives the expression

$$s = \frac{\varphi_3 - \varphi}{\varphi_3 - \varphi_2}h. \tag{6.51}$$

For a complete cycle this gives:

$$s = \begin{cases} \dfrac{\varphi}{\varphi_1}h & \text{for } \varphi \in [0, \varphi_1], \\ h & \text{for } \varphi \in (\varphi_1, \varphi_2], \\ \dfrac{\varphi_3 - \varphi}{\varphi_3 - \varphi_2} & \text{for } \varphi \in (\varphi_2, \varphi_3], \\ 0 & \text{for } \varphi \in (\varphi_3, 2\pi], \end{cases} \tag{6.52}$$

where $\varphi_1 = \Phi_A$, $\varphi_2 = \Phi_A + \Phi_S$, $\varphi_2 = \Phi_A + \Phi_S + \Phi_D$, and $2\pi = \Phi_A + \Phi_S + \Phi_D + \Phi_I$.

The reduced velocity and reduced acceleration are obtained by successive differentiation of (6.52)

$$\frac{ds}{d\varphi} = \begin{cases} \dfrac{h}{\varphi_1} & \text{for } \varphi \in [0, \varphi_1], \\ 0 & \text{for } \varphi \in (\varphi_1, \varphi_2], \\ -\dfrac{h}{\varphi_3 - \varphi_2} & \text{for } \varphi \in (\varphi_2, \varphi_3], \\ 0 & \text{for } \varphi \in (\varphi_3, 2\pi], \end{cases} \tag{6.53}$$

$$\frac{d^2s}{d\varphi^2} = \begin{cases} 0 & \text{for } \varphi \neq \varphi_1, \ \varphi \neq \varphi_2 \ \varphi \neq \varphi_3, \\ +\infty & \text{for } \varphi = 0 \text{ or } \varphi = \varphi_3, \\ -\infty & \text{for } \varphi = \varphi_1 \text{ or } \varphi = \varphi_2. \end{cases} \tag{6.54}$$

The reduced velocity and reduced acceleration are shown in Figure 6.43b,c. Examining the graphs, it can be seen that at points 0, φ_1, φ_2, and φ_3 the reduced velocity has finite jumps, while the reduced acceleration has infinite jumps. The infinite jumps for accelerations lead to infinite jumps for the forces, so the cam mechanism will be affected by huge shocks that may cause elastic (or even plastic) deformation of the elements, increased wear, or decreases in the precision of the law of motion of the tappet. To reduce the shocks one may connect by curves the linear portions of the graph $s = s(\varphi)$.

6.5.3 The Parabolic Law

The ascending phase OA (Figure 6.44a) consists of the parabolic arcs OE and EA. In general, for the parabolic arc A_1A_2 defined by the law of motion $s = C_1 + C_2\varphi + C_3\varphi^2$, and where C_1, C_2, and C_3 are constants determined from the conditions:

$$A_1 \to \varphi = \tilde{\varphi}_1, \ s = \tilde{s}_1, \ \frac{ds}{d\varphi},$$
$$A_2 \to \varphi = \tilde{\varphi}_2, \ s = \tilde{s}_2;$$

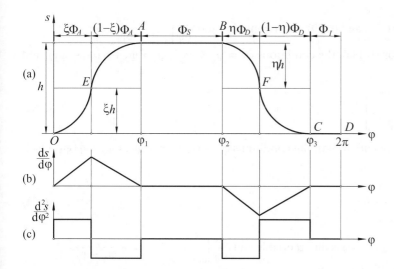

Figure 6.44 The parabolic law of motion: (a) law of motion; (b) reduced velocity; (c) reduced acceleration.

the following system results

$$\begin{cases} C_1 + C_2\varphi + C_3\varphi^2 = s, \\ C_1 + C_2\tilde{\varphi}_1 + C_3\tilde{\varphi}_1^2 = s_1, \\ C_1 + C_2\tilde{\varphi}_2 + C_3\tilde{\varphi}_2^2 = s_2, \\ C_2 + 2C_3\tilde{\varphi}_1 = 0, \end{cases} \tag{6.55}$$

In this system, the unknowns are C_1, C_2, and C_3. The compatibility condition for system (6.55) reads

$$\begin{vmatrix} 1 & \varphi & \varphi^2 & s \\ 1 & \tilde{\varphi}_1 & \tilde{\varphi}_1^2 & \tilde{s}_1 \\ 1 & \tilde{\varphi}_2 & \tilde{\varphi}_2^2 & \tilde{s}_2 \\ 0 & 1 & 2\tilde{\varphi}_1 & 0 \end{vmatrix} \tag{6.56}$$

and this gives

$$\begin{vmatrix} 0 & \varphi - \tilde{\varphi}_1 & \varphi^2 - \tilde{\varphi}_1^2 & s - \tilde{s}_1 \\ 1 & \tilde{\varphi}_1 & \tilde{\varphi}_1^2 & \tilde{s}_1 \\ 0 & \tilde{\varphi}_2 - \tilde{\varphi}_1 & \tilde{\varphi}_2^2 - \tilde{\varphi}_1^2 & \tilde{s}_2 - \tilde{s}_1 \\ 0 & 1 & 2\tilde{\varphi}_1 & 0 \end{vmatrix} = 0, \quad \begin{vmatrix} \varphi - \tilde{\varphi}_1 & \varphi^2 - \tilde{\varphi}_1^2 & s - \tilde{s}_1 \\ \tilde{\varphi}_2 - \tilde{\varphi}_1 & \tilde{\varphi}_2^2 - \tilde{\varphi}_1^2 & \tilde{s}_2 - \tilde{s}_1 \\ 1 & 2\tilde{\varphi}_1 & 0 \end{vmatrix} = 0$$

or

$$s = \tilde{s}_1 + \left(\tilde{s}_2 - \tilde{s}_1\right) \frac{\left(\varphi - \tilde{\varphi}_1\right)^2}{\left(\tilde{\varphi}_2 - \tilde{\varphi}_1\right)^2}. \tag{6.57}$$

On the interval OE, point A_1 becomes point O, while point A_2 becomes point E (Figure 6.44); that is, $\tilde{\varphi} = 0$, $\tilde{s}_1 = 0$, $\tilde{\varphi}_2 = \xi\varphi_1$, and $\tilde{s}_2 = \xi h$, and the law of motion reads

$$s = \frac{\varphi^2}{\xi\varphi_1^2} h. \tag{6.58}$$

Similarly, the upgrading on the interval EA gives: $\tilde{\varphi}_1 = \varphi_1$, $\tilde{s}_1 = h$, $\tilde{\varphi}_2 = \xi\varphi_1$, and $\tilde{s}_2 = \xi h$, and the law of motion (6.57) takes the form

$$s = \left[1 - \frac{\left(\varphi - \varphi_1\right)^2}{\left(1 - \xi\right)\varphi_1^2}\right] h. \tag{6.59}$$

The function defined by (6.58) and (6.59) is differentiable and $\left.\dfrac{ds}{d\varphi}\right|_{\varphi=\xi\varphi_1} = \dfrac{2h}{\varphi_1}$.

In the interval BF (Figure 6.44a), the conditions $\tilde{\varphi}_1 = \varphi_2$, $\tilde{s}_1 = h$, $\tilde{\varphi}_2 = \varphi_2 + \eta\,(\varphi_3 - \varphi_2)$, and $\tilde{s}_2 = (1-\eta)\,h$ give

$$s = \left[1 - \frac{(\varphi - \varphi_2)^2}{\eta\,(\varphi_3 - \varphi_2)^2}\right] h, \tag{6.60}$$

while on the interval FC (Figure 6.44a), the conditions $\tilde{\varphi}_1 = \varphi_3$, $\tilde{s}_1 = 0$, $\tilde{\varphi}_2 = \varphi_2 + \eta\,(\varphi_3 - \varphi_2)$, and $\tilde{s}_2 = (1-\eta)\,h$ give

$$s = \frac{(\varphi - \varphi_3)^2}{(1-\eta)\,(\varphi_3 - \varphi_2)^2} h. \tag{6.61}$$

At point F the function $s = s\,(\varphi)$ is differentiable and $\left.\dfrac{ds}{d\varphi}\right|_{\varphi=\varphi_2+\eta(\varphi_3-\varphi_2)} = -\dfrac{2h}{\varphi_3 - \varphi_2}$.

Finally, on the entire interval, the following law of motion results:

$$s\,(\varphi) = \begin{cases} \dfrac{\varphi_2}{\xi\varphi_1^2}h \text{ if } \varphi \in \left[0, \xi\varphi_1\right], \\[2mm] \left[1 - \dfrac{(\varphi - \varphi_1)^2}{(1-\xi)\,\varphi_1^2}\right] h, \text{ if } \varphi \in \left(\xi\varphi_1, \varphi_1\right], \\[2mm] h \text{ if } \varphi \in \left(\varphi_1, \varphi_2\right], \\[2mm] \left[1 - \dfrac{(\varphi - \varphi_2)^2}{\eta\,(\varphi_3 - \varphi_2)^2}\right] h \text{ if } \varphi \in \left(\varphi_2, \varphi_2 + \eta\,(\varphi_3 - \varphi_2)\right], \\[2mm] \dfrac{(\varphi - \varphi_3)^2}{(1-\eta)\,(\varphi_3 - \varphi_2)^2}h \text{ if } \varphi \in \left(\varphi_2 + \eta\,(\varphi_3 - \varphi_2), \varphi_3\right], \\[2mm] 0 \text{ if } \varphi \in \left(\varphi_3, 2\pi\right]. \end{cases} \tag{6.62}$$

Graphical representations of the displacement s, reduced velocity $\dfrac{ds}{d\varphi}$, and reduced accelera-tion $\dfrac{d^2s}{d\varphi^2}$ are given in Figure 6.44a–c. As a result, the reduced velocity has a continuous varia-tion, while the reduced acceleration presents discontinuities of the first kind (small shocks).

The jumps of the reduced acceleration at points E, and F have the values $\dfrac{2h}{\varphi_1^2\xi\,(1-\xi)}$ and $\dfrac{2h}{(\varphi_3 - \varphi_2)^2\eta\,(1-\eta)}$; they are minimum for $\xi = \eta = \dfrac{1}{2}$.

6.5.4 The Harmonic Law

The harmonic law (Figure 6.45) is characterized by the fact that in an interval $\left[\tilde{\varphi}_1, \tilde{\varphi}_2\right]$ the reduced acceleration is defined by a relation in the form

$$\frac{d^2s}{d\varphi^2} = C_1 \sin\left(k\,(\varphi - \tilde{\varphi}_1)\right), \tag{6.63}$$

where C_1 and k are constant and are determined from the frontier conditions at points A_1 and A_2

$$\varphi = \tilde{\varphi}_1 \rightarrow \frac{ds}{d\varphi} = 0, \; s = \tilde{s}_1, \tag{6.64}$$

Figure 6.45 The harmonic law.

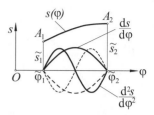

$$\varphi = \tilde{\varphi}_2 \rightarrow \frac{d^2 s}{d\varphi_2} = 0, \ \frac{ds}{d\varphi} = 0, \ s = \tilde{s}_2. \tag{6.65}$$

Integration of (6.63) gives

$$\frac{ds}{d\varphi} = C_2 - \frac{C_1}{k} \cos\left(k\left(\varphi - \tilde{\varphi}_1\right)\right),$$

$$s = C_3 + C_2 \left(\varphi - \tilde{\varphi}_1\right) - \frac{C_1}{k^2} \sin\left(l\left(\varphi - \tilde{\varphi}_1\right)\right);$$

and (6.64) gives

$$C_2 \frac{C_1}{k}, \ C_3 = \tilde{s}_1.$$

Equation (6.65) gives

$$\sin\left(k\left(\tilde{\varphi}_2 - \tilde{\varphi}_1\right)\right) = 0, \ \cos\left(k\left(\tilde{\varphi}_2 - \tilde{\varphi}_1\right)\right) = 1, \tag{6.66}$$

$$\tilde{s}_2 = \tilde{s}_1 + \frac{C_1}{k}\left(\tilde{\varphi}_2 - \tilde{\varphi}_1\right) - \frac{C_1}{k^2} \sin\left(k\left(\tilde{\varphi}_2 - \tilde{\varphi}_1\right)\right). \tag{6.67}$$

The constant k comes from (6.66) in which $\tilde{\varphi}_2 \neq \tilde{\varphi}_1$, while the constant C_1 is obtained from (6.67):

$$k = \frac{2\pi}{\tilde{\varphi}_2 - \tilde{\varphi}_1}, \ C_1 = \frac{2\pi}{\left(\tilde{\varphi}_2 - \tilde{\varphi}_1\right)^2}.$$

Finally, the law of motion in the interval $A_1 A_2$ is defined by the expression

$$\tilde{s}_2 = \tilde{s}_1 + \frac{\tilde{s}_2 - \tilde{s}_1}{\tilde{\varphi}_2 - \tilde{\varphi}_1}\left(\varphi - \tilde{\varphi}_1\right) - \frac{\tilde{s}_2 - \tilde{s}_1}{2\pi} \sin\left(\frac{2\pi\left(\varphi - \tilde{\varphi}_1\right)}{\tilde{\varphi}_2 - \tilde{\varphi}_1}\right). \tag{6.68}$$

If all the phases of motion are considered, the law of motion of the follower is as shown in Figure 6.46a. On the interval OA, identifying points A_1 and A_2 with points O and A, respectively, gives: $\tilde{\varphi}_1 = 0$, $\tilde{s}_1 = 0$, $\tilde{\varphi}_2 = \varphi_1$, and $\tilde{s}_2 = h$, which leads to:

$$s = \left[\frac{\varphi}{\varphi_1} - \frac{1}{2\pi} \sin\left(\frac{2\pi\varphi}{\varphi_1}\right)\right] h;$$

On the interval BC, identification of points A_1, A_2 with points B, C, respectively, gives: $\tilde{\varphi}_1 = \varphi_2$, $\tilde{s}_1 = h$, $\tilde{\varphi}_2 = \varphi_3$, and $\tilde{s}_2 = 0$, and the following law of motion is obtained:

$$s = \left[1 - \frac{\varphi - \varphi_2}{\varphi_3 - \varphi_2} + \frac{1}{2\pi} \sin\left(\frac{2\pi\left(\varphi - \varphi_2\right)}{\varphi_3 - \varphi_2}\right)\right] h.$$

On the interval OD, the following law of motion is obtained:

$$s\left(\varphi\right) = \begin{cases} \left[\dfrac{\varphi}{\varphi_1} - \dfrac{1}{2\pi} \sin\left(\dfrac{2\pi\varphi}{\varphi_1}\right)\right] h \ \text{for } \varphi \in \left[0, \varphi_1\right), \\ h \ \text{for } \varphi \in \left[\varphi_1, \varphi_2\right), \\ s = \left[1 - \dfrac{\varphi - \varphi_2}{\varphi_3 - \varphi_2} + \dfrac{1}{2\pi} \sin\left(\dfrac{2\pi\left(\varphi - \varphi_2\right)}{\varphi_3 - \varphi_2}\right)\right] h \ \text{for } \varphi \in \left[\varphi_2, \varphi_3\right), \\ 0 \ \text{for } \varphi \in \left[\varphi_3, 2\pi\right]. \end{cases} \tag{6.69}$$

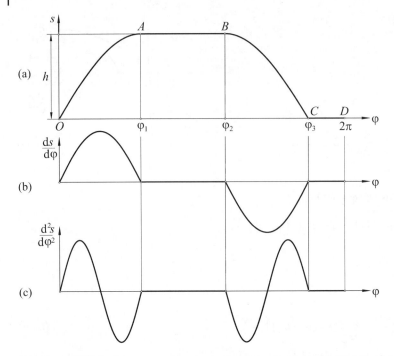

Figure 6.46 The displacement, reduced velocity, and reduced acceleration for a harmonic cam: (a) law of motion; (b) reduced velocity; (c) reduced acceleration.

The reduced velocity $\dfrac{ds}{d\varphi}$ and the reduced acceleration $\dfrac{d^2 s}{d\varphi^2}$ are as shown in Figure 6.46b,c. It can be seen that, in this case, the reduced acceleration is a continuous function.

6.5.5 The Polynomial Law: Polydyne Cams

The law of motion of the follower between points A_1 and A_2 (Figure 6.47) is polynomial if it is expressed by a polynomial function of the form

$$s = \tilde{s}_1 + \tilde{v}_1 \left(\varphi - \tilde{\varphi}_1\right) + \frac{\tilde{a}_1}{2}\left(\varphi - \tilde{\varphi}_1\right)^2$$
$$+ C_1\left(\varphi - \tilde{\varphi}_1\right)^n + C_2\left(\varphi - \tilde{\varphi}_1\right)^{n+1} + C_3\left(\varphi - \tilde{\varphi}_1\right)^{n+2}, \tag{6.70}$$

where $n \geq 3$, $\tilde{v}_1 = \left.\dfrac{ds}{d\varphi}\right|_{\varphi=\tilde{\varphi}_1}$, and $\tilde{a}_1 = \left.\dfrac{d^2 s}{d\varphi^2}\right|_{\varphi=\tilde{\varphi}_1}$. The constants C_1, C_2, and C_3 are determined (at angle $\varphi = \tilde{\varphi}_2$) from the conditions

$$s = \tilde{s}_2, \quad \frac{ds}{d\varphi} = \tilde{v}_2, \quad \frac{d^2 s}{d\varphi_2} = \tilde{a}_2.$$

These conditions give the possibility of adjusting the curves (C_1) and (C_2) at points A_1 and A_2, so that the displacement s, the reduced velocity $\dfrac{ds}{d\varphi}$, and the reduced acceleration $\dfrac{d^2 s}{d\varphi^2}$ are

Figure 6.47 The adjustment between the curves (C_1) and (C_2).

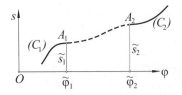

continuous functions. Due to the way in which the law of motion (6.70) is written, the continuity conditions of the functions a, v, and s at point A_1 are assured. The continuity of functions s, v, and a at point A_2 leads to:

$$
\begin{aligned}
&C_1 \left(\tilde{\varphi}_2 - \tilde{\varphi}_1\right)^n + C_2 \left(\tilde{\varphi}_2 - \tilde{\varphi}_1\right)^{n+1} + C_3 \left(\tilde{\varphi}_2 - \tilde{\varphi}_1\right)^{n+2} \\
&\quad = \tilde{s}_2 - \tilde{s}_1 - \tilde{v}_1 \left(\tilde{\varphi}_2 - \tilde{\varphi}_1\right) - \frac{\tilde{a}_1}{2} \left(\tilde{\varphi}_2 - \tilde{\varphi}_1\right)^2, \\
&C_1 n \left(\tilde{\varphi}_2 - \tilde{\varphi}_1\right)^{n-1} + C_2 (n+1) \left(\tilde{\varphi}_2 - \tilde{\varphi}_1\right)^n + C_3 (n+2) \left(\tilde{\varphi}_2 - \tilde{\varphi}_1\right)^{n+1} \\
&\quad = \tilde{v}_2 - \tilde{v}_1 - \tilde{a}_1 \left(\tilde{\varphi}_2 - \tilde{\varphi}_1\right), \\
&C_1 n(n-1) \left(\tilde{\varphi}_2 - \tilde{\varphi}_1\right)^{n-2} + C_2 n(n+1) \left(\tilde{\varphi}_2 - \tilde{\varphi}_1\right)^{n-1} \\
&\quad + C_3 (n+2)(n+1) \left(\tilde{\varphi}_2 - \tilde{\varphi}_1\right)^n = \tilde{a}_2 - \tilde{a}_1.
\end{aligned}
\tag{6.71}
$$

The expressions in (6.70) and (6.71) form a system of four equations with three unknowns (C_1, C_2, and C_3). The compatibility condition of the system requires the vanishing of the determinant; that is

$$
\begin{vmatrix}
\left(\varphi - \tilde{\varphi}_1\right)^n & \left(\varphi - \tilde{\varphi}_1\right)^{n+1} & \left(\varphi - \tilde{\varphi}_1\right)^{n+2} & \begin{matrix} s - \tilde{s}_1 \\ -\tilde{v}_1 \left(\varphi - \tilde{\varphi}_1\right) \\ -\frac{\tilde{a}_1}{2} \left(\varphi - \tilde{\varphi}_1\right)^2 \end{matrix} \\[3ex]
\left(\tilde{\varphi}_2 - \tilde{\varphi}_1\right)^n & \left(\tilde{\varphi}_2 - \tilde{\varphi}_1\right)^{n+1} & \left(\tilde{\varphi}_2 - \tilde{\varphi}_1\right)^{n+2} & \begin{matrix} \tilde{s} - \tilde{s}_1 \\ -\tilde{v}_1 \left(\tilde{\varphi}_2 - \tilde{\varphi}_1\right) \\ -\frac{\tilde{a}_1}{2} \left(\tilde{\varphi}_2 - \tilde{\varphi}_1\right)^2 \end{matrix} \\[3ex]
\begin{matrix} n \\ \left(\tilde{\varphi}_2 - \tilde{\varphi}_1\right)^{n-1} \\ n(n-1) \\ \left(\tilde{\varphi}_2 - \tilde{\varphi}_1\right)^{n-2} \end{matrix} & \begin{matrix} (n+1) \\ \left(\tilde{\varphi}_2 - \tilde{\varphi}_1\right)^n \\ n(n+1) \\ \left(\tilde{\varphi}_2 - \tilde{\varphi}_1\right)^{n-1} \end{matrix} & \begin{matrix} (n+2) \\ \left(\tilde{\varphi}_2 - \tilde{\varphi}_1\right)^{n+1} \\ (n+1)(n+2) \\ \left(\tilde{\varphi}_2 - \tilde{\varphi}_1\right)^n \end{matrix} & \begin{matrix} \tilde{v}_2 - \tilde{v}_1 \\ -\tilde{a}_1 \left(\tilde{\varphi}_2 - \tilde{\varphi}_1\right) \\ \tilde{a}_2 - \tilde{a}_1 \end{matrix}
\end{vmatrix}
\tag{6.72}
$$
$$
= 0.
$$

The following notation is used:

$$
\Phi = \frac{\varphi - \tilde{\varphi}_1}{\tilde{\varphi}_2 - \tilde{\varphi}_1},
$$
$$
P(\Phi) = \frac{1}{2} \left[(n+1)(n+2)\Phi^n - 2n(n+2)\Phi^{n+1} + n(n+1)\Phi^{n+2}\right],
$$
$$
Q(\Phi) = (n+1)\Phi^n - (2n+1)\Phi^{n+1} + n\Phi^{n+2},
$$
$$
R(\Phi) = \frac{1}{2}\left(\Phi^n - 2\Phi^{n+1} + \Phi^{n+2}\right),
$$

where the polynomials P, Q, and R have the properties

$$
\begin{aligned}
&P(0) = 0, \ P'(0) = 0, \ P''(0) = 0, \ P(1) = 1, \ P'(1) = 0, \ P''(1) = 0, \\
&Q(0) = 0, \ Q'(0) = 0, \ Q''(0) = 0, \ Q(1) = 0, \ Q'(1) = -1, \ Q''(1) = 0, \\
&R(0) = 0, \ R'(0) = 0, \ R''(0) = 0, \ R(1) = 0, \ R'(1) = 0, \ R''(1) = 1;
\end{aligned}
\tag{6.73}
$$

Developing the determinant gives

$$
\begin{aligned}
s = &\left[1 - P(\Phi)\right]\tilde{s}_1 + P(\Phi)\tilde{s}_2 + \left[\Phi - P(\Phi) + Q(\Phi)\right]\tilde{v}_1 \left(\tilde{\varphi}_2 - \tilde{\varphi}_1\right) \\
&- Q(\Phi)\left(\tilde{\varphi}_2 - \tilde{\varphi}_1\right)\tilde{v}_2 + \left[\frac{1}{2}\Phi^2 - P(\Phi) + Q(\Phi) - R(\Phi)\right]\tilde{a}_1 \left(\tilde{\varphi}_2 - \tilde{\varphi}_1\right)^2 \\
&+ R(\Phi)\left(\tilde{\varphi}_2 - \tilde{\varphi}_1\right)^2 \tilde{a}_2.
\end{aligned}
\tag{6.74}
$$

Cams that generate polynomial laws of motion at the follower are called *polydyne cams*. If a polynomial function is used to adjust two straight lines parallel to axis $O\varphi$ (Figure 6.48), it gives $\tilde{v}_1 = \tilde{v}_2 = 0$ and $\tilde{a}_1 = \tilde{a}_2 = 0$, while the law of motion between points A_1 and A_2 reads

$$
s = \left[1 - P(\Phi)\right]\tilde{s}_1 + P(\Phi)\tilde{s}_2.
\tag{6.75}
$$

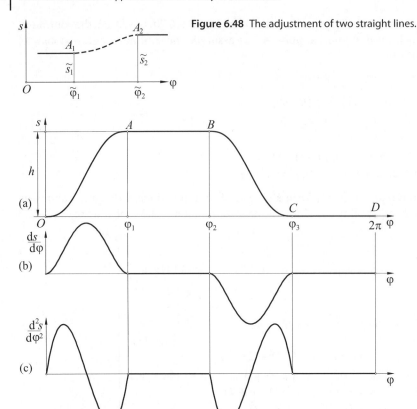

Figure 6.48 The adjustment of two straight lines.

Figure 6.49 The polynomial law: (a) law of motion; (b) reduced velocity; (c) reduced acceleration.

The polynomial law of motion for the cyclogram in Figure 6.49 is

$$
s = \begin{cases}
hP\left(\dfrac{\varphi}{\varphi_1}\right) & \text{if } \varphi \in \left[0, \varphi_1\right], \\[2mm]
h & \text{if } \varphi \in \left[\varphi_1, \varphi_2\right), \\[2mm]
h\left[1 - P\left(\dfrac{\varphi - \varphi_2}{\varphi_3 - \varphi_2}\right)\right] & \text{if } \varphi \in \left[\varphi_2, \varphi_3\right], \\[2mm]
0 & \text{if } \varphi \in \left(\varphi_3, 2\pi\right].
\end{cases}
\tag{6.76}
$$

As a numerical example, consider $n = 3$, $\varphi_1 = \dfrac{\pi}{2}$, $\varphi_2 = \pi$, and $\varphi_3 = \dfrac{3\pi}{2}$. This gives:

$$
s = \begin{cases}
\left[10\left(\dfrac{2\varphi}{\pi}\right)^3 - 15\left(\dfrac{2\varphi}{\pi}\right)^4 + 6\left(\dfrac{2\varphi}{\pi}\right)^5\right] h & \text{if } \varphi \in \left[0, \varphi_1\right], \\[3mm]
h & \text{if } \varphi \in \left[\varphi_1, \varphi_2\right), \\[3mm]
\left[1 - 10\left(\dfrac{2\varphi - 2\pi}{\pi}\right)^3 + 15\left(\dfrac{2\varphi - 2\pi}{\pi}\right)^4 - 6\left(\dfrac{2\varphi - 2\pi}{\pi}\right)^5\right] & \text{if } \varphi \in \left[\varphi_2, \varphi_3\right], \\[3mm]
0 & \text{if } \varphi \in \left(\varphi_3, 2\pi\right].
\end{cases}
\tag{6.77}
$$

6.6 Synthesis of Cam Mechanisms

6.6.1 Formulation of the Problem

For the design of a cam mechanism, the following factors must be considered:

- the type of the cam (rotational, translational)
- the type of the follower (rotational, translational) and the shape of the follower
- the maximum displacement of the follower.

Knowing the maximum displacement, the methods outlined in this chapter can be used to establish the law of motion of the follower and to determine, using the critical angle of transmission, the radius of the circle of base.

Now, knowing the law of the follower's displacement, we can determine the cam's profile.

6.6.2 The Equation of Synthesis

Consider the mechanism with a rotational cam in Figure 6.50, in which a point O_2 of the follower moves on a straight line (D) parallel to axis OY; the law of motion is $s = s(\varphi)$, where φ is the rotational angle of the cam; the angle of the follower relative to the initial position is $\theta = \theta(\varphi)$.

This model permits the establishment of the equation of synthesis both for a follower in rotational motion ($s = 0$) and for a follower in translational motion ($\theta = 0$). The equations of the follower's profile are given in parametric form

$$x_2 = x_2(\lambda), \ y_2 = y_2(\lambda). \tag{6.78}$$

A fixed reference system OXY is chosen, the system Ox_1y_1 rigidly linked to the cam, and the system $O_2x_2y_2$ rigidly linked to the follower. The system OXY is chosen so that the straight line (D) is parallel to axis OY (Figure 6.50b).

With the notation in Figure 6.50, between the coordinates X, Y, x_1, y_1, x_2, y_2 of the contact point A:

$$X = x_1 \cos \varphi - y_1 \sin \varphi = X_0 + x_2 \cos \theta - y_2 \sin \theta, \tag{6.79}$$

$$Y = x_1 \sin \varphi + y_1 \cos \varphi = Y_0 + x_2 \sin \theta + y_2 \cos \theta + s; \tag{6.80}$$

Since $\theta = \theta(\varphi)$ and $s = s(\varphi)$, the following relations hold true:

$$f_1(x_1, y_1, \varphi, \lambda) = x_1 \cos \varphi - y_1 \sin \varphi - X_0 - x_2 \cos \theta + y_2 \sin \theta = 0, \tag{6.81}$$

$$f_2(x_1, y_1, \varphi, \lambda) = x_1 \sin \varphi + y_1 \cos \varphi - Y_0 - x_2 \sin \theta - y_2 \cos \theta - s = 0. \tag{6.82}$$

Equations (6.81) and (6.82) define in the reference system Ox_1y_1 a family of curves (the successive positions of the follower), the envelope of which is the cam. The equation of the envelope reads

$$\begin{vmatrix} \dfrac{\partial f_1}{\partial \varphi} & \dfrac{\partial f_1}{\partial \lambda} \\ \dfrac{\partial f_2}{\partial \varphi} & \dfrac{\partial f_2}{\partial \lambda} \end{vmatrix} - 0 \tag{6.83}$$

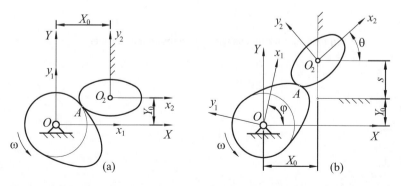

Figure 6.50 Cam mechanism with rotational cam: (a) initial position; (b) arbitrary position.

or

$$\left| \begin{array}{cc} -x_1 \sin\varphi - y_1 \cos\varphi + \dfrac{d\theta}{d\varphi}\left(x_2 \sin\theta + y_2 \cos\theta\right) & -\dfrac{dx_2}{d\lambda}\cos\theta + \dfrac{dy_2}{d\lambda}\sin\theta \\[3mm] \begin{array}{l} x_1 \cos\varphi - y_1 \sin\varphi \\[1mm] +\dfrac{d\theta}{d\varphi}\left(-x_2 \cos\theta + y_2 \sin\theta\right) - \dfrac{ds}{d\theta} \end{array} & -\dfrac{dx_2}{d\lambda}\sin\theta - \dfrac{dy_2}{d\lambda}\cos\theta \end{array} \right| = 0,$$

which leads to the equation of synthesis

$$\begin{aligned} &\frac{dx_2}{d\lambda}\left[\left(Y_0 + s\right)\sin\theta + \left(X_0 - \frac{ds}{d\varphi}\right)\cos\theta - x_2\left(\frac{d\theta}{d\varphi} - 1\right)\right] \\ &+ \frac{dy_2}{d\lambda}\left[\left(Y_0 + s\right)\cos\theta - \left(X_0 - \frac{ds}{d\varphi}\right)\sin\theta - y_2\left(\frac{d\theta}{d\varphi} - 1\right)\right] = 0. \end{aligned} \tag{6.84}$$

6.6.3 Synthesis of Mechanism with Rotational Cam and Translational Follower

The general case

In the general case $\theta = 0$ (Figure 6.51) and the equations of synthesis (6.79), (6.80) and (6.84) read

$$\begin{aligned} x_1 \cos\varphi - y_1 \sin\varphi &= x_2\left(\lambda\right) + X_0, \\ x_1 \sin\varphi + y_1 \cos\varphi &= y_2\left(\lambda\right) + s + y_0, \end{aligned} \tag{6.85}$$

$$\frac{dx_2}{d\lambda}\left[x_2\left(\lambda\right) + X_0 - \frac{ds}{d\varphi}\right] + \frac{dy_2}{d\lambda}\left[y_2\left(\lambda\right) + Y_0 + s\left(\varphi\right)\right] = 0. \tag{6.86}$$

From (6.85) we get:

$$\begin{aligned} x_1 &= \left[x_2\left(\lambda\right) + X_0\right]\cos\varphi + \left[y_2\left(\lambda\right) + Y_0 + s\left(\varphi\right)\right]\sin\varphi, \\ y_1 &= -\left[x_2\left(\lambda\right) + X_0\right]\sin\varphi + \left[y_2\left(\lambda\right) + Y_0 + s\left(\varphi\right)\right]\cos\varphi. \end{aligned} \tag{6.87}$$

Adding condition (6.86) to (6.87) allow the equations of the cam's profile to be defined. For a numerical calculation, values are assigned to angle φ. Equation (6.86) gives the parameter λ, and the coordinates x_1 and y_1 of the current point of the cam can be calculated from (6.87).

Plunger tip

From (6.87), and making $x_2 = y_2 = 0$, the parametric equations of the cam's profile can be obtained:

$$x_1 = X_0 \cos\varphi + \left[Y_0 + s\left(\varphi\right)\right]\sin\varphi, \quad y_1 = -X_0 \sin\varphi + \left[Y_0 + s\left(\varphi\right)\right]\cos\varphi. \tag{6.88}$$

Flat tappet

In the reference system $O_2 x_2 y_2$, the flat tappet (Figure 6.51) has the equations $x_2 = \lambda$ and $y_2 = 0$; the condition (6.86) becomes

$$X_0 + \lambda = \frac{ds}{d\varphi}.$$

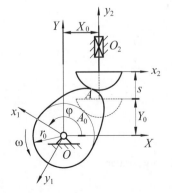

Figure 6.51 Cam mechanism with translational follower.

Figure 6.52 Cam mechanism with roller tappet.

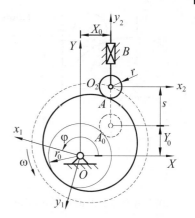

Replacing this expression in t (6.87) and taking into account that in this case $Y_0 = r_0$ (where r_0 is the radius of the circle of base), gives the parametric equations of the cam

$$x_1 = \frac{ds}{d\varphi} \cos \varphi + (r_0 + s) \sin \varphi, \quad y_1 = -\frac{ds}{d\varphi} \sin \varphi + (r_0 + s) \cos \varphi. \tag{6.89}$$

Roller tappet

Denoting by r the radius of the roller (Figure 6.52), its parametric equations read $x_2 = r \cos \lambda$ and $y_2 = r \sin \lambda$, and the equation of synthesis (6.86) becomes

$$\tan \lambda = \frac{s(\varphi) + Y_0}{X_0 - \dfrac{ds}{d\varphi}}. \tag{6.90}$$

This gives the parametric equations of the cam (C) as:

$$\begin{aligned}
x_1 &= r \cos(\varphi - \lambda) + X_0 \cos \varphi + Y_0 \sin \varphi + s(\varphi) \sin \varphi, \\
y_1 &= -r \sin(\varphi - \lambda) - X_0 \sin \varphi + Y_0 \cos \varphi + s(\varphi) \cos \varphi,
\end{aligned} \tag{6.91}$$

where λ is expressed as a function of φ using (6.90). For numerical applications, values are given to angle φ, then λ is calculated from (6.90), and x_1 and y_1 are determined from (6.91).

Another way in which we can approach this problem involves, firstly, the determination of the theoretical profile of the cam (the dotted line in Figure 6.52), denoted by (C_0). The parametric equations of this curve are obtained by considering the flat tappet as a plunger tip; hence we may write

$$x_{10} = X_0 \cos \varphi + [Y_0 + s(\varphi)] \sin \varphi, \quad y_{10} = -X_0 \sin \varphi + [Y_0 + s(\varphi)] \cos \varphi. \tag{6.92}$$

Then (Figure 6.53), the coordinates of point A on the real cam are calculated using:

$$x_1 = x_{10} - r \cos \alpha, \quad y_1 = y_{10} - r \sin \alpha. \tag{6.93}$$

where α is the angle between the normal at the curve (C_0), at point O_2 and axis Ox_1.

Figure 6.53 The theoretical cam.

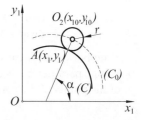

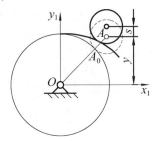

Figure 6.54 Roller tappet.

Angle α is calculated from the expression $\tan \alpha = -\dfrac{\dfrac{dx_{10}}{d\varphi}}{\dfrac{dy_{10}}{d\varphi}}$; that is,

$$\tan \alpha = -\frac{-X_0 \sin \varphi + Y_0 \cos \varphi + s' \sin \varphi + s \cos \varphi}{-X_0 \cos \varphi - Y_0 \sin \varphi + s' \cos \varphi - s \sin \varphi}. \tag{6.94}$$

In numerical applications, values are given to angle φ, the coordinates x_{10}, y_{10} are calculated using (6.92), then α is determined form (6.94), and the coordinates x_1, y_1 of point A on the real cam (C) are found from (6.93).

Numerical example

Problem: Consider (Figure 6.54) values of $\varphi \in [0, 2\pi]$, $r_0 = 15$ mm, $r = 5$ mm, $X_0 = 12$ mm, and $s = 6 \sin^2 \varphi$. For $\varphi = \frac{\pi}{4}$ these give results of

$$Y_0 = \sqrt{(r_0 + r)^2 - x_0^2} = 16\,\text{mm},$$

$$\tan \lambda = \frac{6 \sin^2 \varphi + 16}{12 - 6 \sin(2\varphi)} - 2.792, \ \lambda = 250.29°,$$

$x_1 = 17.4$ mm, $y_1 = 2.81$ mm, $x_{10} = 21.92$ mm, $y_{10} = 4.94$ mm, and $\alpha = 25.29°$. The values obtained for x_1 and y_1 are identical in both in both cases (roller tappet and plunger tip).

6.6.4 Synthesis of Mechanism with Rotational Cam and Rotational Follower

The general case

In this case $s = 0$, and the equation of synthesis (6.84) becomes

$$\frac{dx_2}{d\lambda}\left[Y_0 \sin \theta + X_0 \cos \theta - x_2 \left(\frac{d\theta}{d\varphi} - 1 \right) \right]$$
$$+ \frac{dy_2}{d\lambda}\left[Y_0 \cos \theta - X_0 \sin \theta - y_2 \left(\frac{d\theta}{d\varphi} - 1 \right) \right] = 0; \tag{6.95}$$

Then (6.79) and (6.80) read

$$\begin{cases} x_1 \cos \varphi - y_1 \sin \varphi = X_0 + x_2 \cos \theta - y_2 \sin \theta, \\ x_1 \sin \varphi + y_1 \cos \varphi = Y_0 + x_2 \sin \theta + y_2 \cos \theta \end{cases}$$

or

$$\begin{aligned} x_1 &= X_0 \cos \varphi + Y_0 \sin \varphi + x_2 \cos (\varphi - \theta) + y_2 \sin (\varphi - \theta), \\ y_1 &= -X_0 \sin \varphi + Y_0 \cos \varphi - x_2 \sin (\varphi - \theta) + y_2 \cos (\varphi - \theta). \end{aligned} \tag{6.96}$$

In numerical applications, values are given to angle φ, angle θ is calculated from the law $\theta = \theta(\varphi)$, parameter λ is determined from (6.95), and then, using (6.96) the coordinates x_1 and y_1 of the points of the cam's profile are deduced.

Figure 6.55 Cam mechanism with rotational plunger tip.

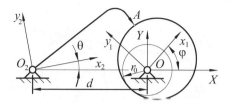

Plunger tip

In this case the coordinates x_1 and y_1 of point A relative to the reference system $O_2x_2y_2$ are constant. In addition (Figure 6.55), $X_0 = -d$, $Y_0 = 0$.

In the initial position ($\varphi = 0$, $\theta = 0$) point A is situated on the circle of base so

$$(x_2 - d)^2 + y_2^2 = r_0^2.$$

Knowing the function $\theta = \theta(\varphi)$, the parametric equations of the cam can be obtained, angle φ being the parameter

$$\begin{aligned} x_1 &= -d \cos \varphi + x_2 \cos(\varphi - \theta) + y_2 \sin(\varphi - \theta), \\ y_1 &= d \sin \varphi - x_2 \sin(\varphi - \theta) + y_2 \cos(\varphi - \theta). \end{aligned} \tag{6.97}$$

Roller tappet

Denoting with x_0 and y_0 the coordinates of the centre of roller relative to the reference system $O_2x_2y_2$, means that the parametric equations of the roll in the system $O_2x_2y_2$ are

$$x_2 = x_0 + r \cos \lambda, \ y_2 = y_0 + r \sin \lambda, \tag{6.98}$$

where r is the radius of the roll.

Since $X_0 = -d$ and $Y_0 = 0$ (Figure 6.56), the equation of synthesis (6.95) becomes

$$\tan \lambda = -\dfrac{-d \sin \theta + y_0 \left(\dfrac{d\theta}{d\varphi} - 1 \right)}{d \cos \theta + x_0 \left(\dfrac{d\theta}{d\varphi} - 1 \right)}. \tag{6.99}$$

Angle φ is chosen and the parameter λ is calculated from (6.99). The values of x_2 and y_2 (the coordinates of point A in the system $O_2x_2y_2$) are calculated using (6.98), and then the coordinates of point A in the system Ox_1y_1 are determined using; that is

$$\begin{aligned} x_1 &= -d \cos \varphi + x_2 \cos(\varphi - \theta) + y_2 \sin(\varphi - \theta), \\ y_1 &= d \sin \varphi - x_2 \sin(\varphi - \theta) + y_2 \cos(\varphi - \theta). \end{aligned} \tag{6.100}$$

Another possibility to obtain the cam's profile is to first determine the theoretical cam (the dotted curve in Figure 6.56) with:

$$\begin{aligned} x_{10} &= -d \cos \varphi + x_0 \cos(\varphi - \theta) + y_0 \sin(\varphi - \theta), \\ y_{10} &= d \sin \varphi - x_0 \sin(\varphi - \theta) + y_0 \cos(\varphi - \theta), \end{aligned} \tag{6.101}$$

Figure 6.56 Cam mechanism with rotational roller tappet.

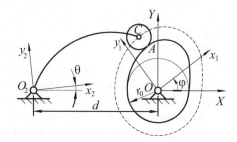

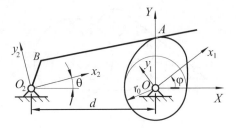

Figure 6.57 Cam mechanism with rotational flat tappet.

and then to calculate the real cam using the expressions

$$x_1 = x_{10} - r \cos \alpha, \quad y_1 = y_{10} - r \sin \alpha, \tag{6.102}$$

where $\tan \alpha = -\dfrac{-\dfrac{dx_{10}}{d\varphi}}{\dfrac{dy_{10}}{d\varphi}}$ or

$$\tan \alpha = \frac{d \sin \varphi + \left[-x_0 \sin \dfrac{dx_{10}}{d\varphi} + y_0 \cos \dfrac{dx_{10}}{d\varphi} \right] \left(1 - \dfrac{d\theta}{d\varphi} \right)}{-d \cos \varphi + \left[x_0 \cos \dfrac{dx_{10}}{d\varphi} + y_0 \sin \dfrac{dx_{10}}{d\varphi} \right] \left(1 - \dfrac{d\theta}{d\varphi} \right)}. \tag{6.103}$$

Flat tappet

If the parametric equations of the flat tappet (straight line AB in Figure 6.57) are

$$x_2 = x_0 + \lambda \cos \alpha, \quad y_2 = y_0 + \lambda \sin \alpha, \tag{6.104}$$

then the equation of synthesis (6.95) becomes

$$\lambda = \frac{d \cos (\theta + \alpha)}{1 - \dfrac{d\theta}{d\varphi}} - x_0 \cos \alpha - y_0 \sin \alpha. \tag{6.105}$$

This means that to determine the coordinates of a point of the cam (the point of contact), the value of angle φ must be chosen and angle θ calculated from the law $\theta = \theta(\varphi)$.

Now the parameter λ is calculated from (6.105), x_2 and y_2 from (6.104) and, finally, x_1 and y_1 from:

$$\begin{aligned} x_1 &= -d \cos \varphi + x_2 \cos (\varphi - \theta) + y_2 \sin (\varphi - \theta), \\ y_1 &= d \sin \varphi - x_2 \sin (\varphi - \theta) + y_2 \cos (\varphi - \theta). \end{aligned} \tag{6.106}$$

6.6.5 Cam Synthesis using AutoLisp Functions

The methods of synthesis shown in this section are inspired from practical applications. One way to obtain a cam is to manufacture it on a copying machine. On a traditional machine tool the work piece from which the cam is manufactured has a geometric shape that covers its final shape. Following the exterior profile of a mother-cam, the stock of material is cut, obtaining the final cam. Below, we present a method to obtain cams using AutoCAD. It is similar to a cutting process and involves positioning the follower in successive positions while rotating the cam and removing excess material, as is normal in the traditional manufacturing process of cams. The method is called *by cutting* and uses the following algorithm:

1) Place a solid model (the raw material) that approximates the cam in the position called 0°. The raw material that represents the cam is obtained from a solid with AutoCAD commands; this solid includes the exterior profile of the cam. For instance, to obtain a cam of the radius of the circle of base r_0, the maximum height of the follower's lift h, and thickness equal to g, a cylinder can be used, with radius at least of $R = r_0 + h$.

2) Add a solid that represents the follower. This will later be eliminated from the cam's material. The exterior surface of the solid will have to coincide with the shape of the follower. For instance, the solid for the follower of the mechanism in Figure 6.56 is a cylinder of length L (at least equal to the thickness of the cam) and radius r. The AutoCAD command for obtaining the solid is **Cylinder**.

3) From the solid that represents the cam, exclude the solid that represents the follower, using the AutoCAD command **Subtract**.

4) Rotate the cam to a new position and repeat steps 1–3.

To obtain a cam using this algorithm, at least 360 such operations (i.e. an angular step of one degree) have to be performed. For this purpose we have created an AutoLisp function for the solid that represents the cam of a mechanism with a rotational cam and a translational follower with a curvilinear boundary, such as the one in Figure 6.51. The law of displacement of the follower is the polynomial one in (6.77).

The data for the mechanism in Figure 6.51 are: $r_0 = 20\,\text{mm}$, $r = 15\,\text{mm}$, and $d = 20\,\text{mm}$. In (6.77) the following values are known: $h = 10\,\text{mm}$, $\varphi_1 = \frac{\pi}{2}$, $\varphi_2 = \pi$, and $\varphi_3 = \frac{3\pi}{2}$. The thickness of the cam and follower is $g = 10\,\text{mm}$. If we denote by x_C, y_C the coordinates of the centre C of the follower, then these coordinates have the expressions $x_C = d$ and $y_C = \left(r_0 + r\right)\sin\alpha_0 + s$, where α_0 stands for the value of angle α corresponding to the $0°$ position, $\alpha_0 = \arccos\left(\dfrac{d}{r_0 + r}\right)$.

The code for the AutoLisp function can be found in the companion website of the book. The name of the function is *Polydyne_Cam.lsp* and it was written for the numerical data used in the example just outlined. The function has no parameters and variables and follows the algorithm just presented. The function:

- creates (with the AutoCAD command **Cylinder**) a cylinder *cam* at point $O\,(0, 0, 0)$;
- in a repetitive *while* loop in the interval $[0, 2\pi]$:
  - calculates (degree by degree) the displacement x of the follower, and the coordinates of point C, which permits the placing of the cylinder that represents the follower (this cylinder is constructed with the AutoCAD command **Cylinder** and it is named *follower*);
  - extracts (with the AutoCAD command **Subtract**) the *follower* from the *cam*; the new entity is also named *cam*;
  - rotates (the AutoCAD command **Rotate**) the solid obtained through one degree, passing to a new position and then repeating the operations of placing of the follower and extracting it from the solid obtained.

Figure 6.58 shows the resulting solid for five angular positions ($0°$, $90°$, $180°$, $270°$, and $360°$); the last position is for the final cam.

Spatial cams are obtained using the same algorithm.

6.6.6 Examples

In this sction we will present some examples.

Figure 6.58 The solid of the cam at intermediate positions of the cutting process.

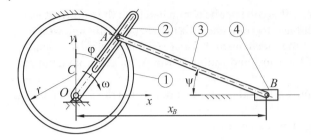

Figure 6.59 Cam mechanism with fixed cam.

Example 1

The fixed cam of the mechanism in Figure 6.59 is a circle of radius r with its centre at point C. Inside the guide manufactured in the crank 1, there is a slider 2, to which is jointed a coupler 3. The slider has a bolt that moves on the fixed cam. At point B of coupler 3, slider 4 is jointed, the latter being in translational motion.

Knowing the angular velocity ω of the crank, the length l of the coupler, and the eccentricity $e = OC$ of the cam, we want to determine the law of motion $x(\varphi)$ of slider 4. The values used are $r = 100\,\text{mm}$, $e = 40\,\text{mm}$, and $l = 300\,\text{mm}$.

Solution: Denoting $\rho = OA$ and applying the theorem of cosine in the triangle CAO, gives $r^2 = e^2 + \rho^2 - 2e\rho\cos\varphi$, from which:

$$\rho = e\cos\varphi + \sqrt{r^2 - e^2\sin^2\varphi}. \tag{6.107}$$

Projecting the vector relation $\mathbf{OB} = \mathbf{OA} + \mathbf{AB}$ onto axis Oy gives $\rho\cos\varphi = l\sin\psi$ or

$$\psi = \arcsin\left(\frac{\rho}{l}\cos\varphi\right); \tag{6.108}$$

Projecting it onto axis Ox, on the other hand, gives

$$x_B = \rho\sin\varphi + l\cos\psi. \tag{6.109}$$

Equations (6.107)–(6.109) give the values in Table 6.6 and the graph of function $x_B(\varphi)$ in Figure 6.60.

Example 2

Consider the mechanism with a translational cam and a translational roller tappet in Figure 6.61, for which the following are known: $OA = e$, r (the radius of the roller), and the equation of the cam $y_1 = b + Kx_1^2$. In addition, when $\rho = 0$, $s_2 = 0$ and $s_1 = 0$. We want to determine the equations of the real cam and the law of motion of the follower. The values to be used are $e = 10\,\text{mm}$, $b = 20\,\text{mm}$, $K = 0.02\,\text{mm}^{-1}$, and $r = 10\,\text{mm}$.

Table 6.6 The values of the functions ρ, ψ, and x_B.

φ	0°	30°	60°	90°	120°	150°	180°	210°	240°	270°	300°	330°	360°
ρ	140	132.6	113.8	91.7	73.8	63.3	60	63.3	73.8	91.7	113.8	132.6	140
ψ	27.8	22.5	10.9	0	−7.1	−10.5	−11.5	−10.5	−7.1	0	10.9	22.5	27.8
x_B	265.3	343.5	393.1	391.7	361.6	326.6	293.9	263.3	233.8	208.3	196.3	210.9	265.3

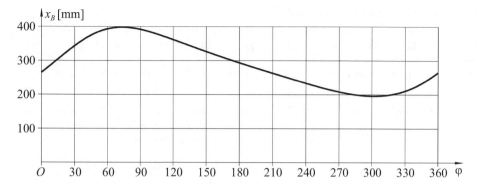

Figure 6.60 The slider's displacement.

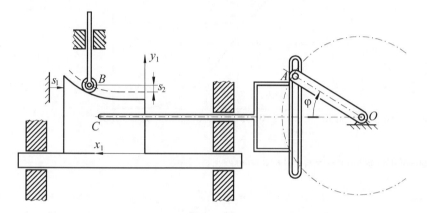

Figure 6.61 Cam mechanism with translational cam and roller tappet.

Figure 6.62 Determination of the equations of the real cam.

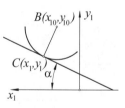

Solution: From Figure 6.62, $\tan \alpha = \dfrac{dy_1}{dx_1} = 2Kx_1$, $x_{10} = x_1 - r \sin \alpha$, and $y_{10} = y_1 + r \cos \alpha$ or

$$x_{10} = x_1 - \frac{2Kx_1 r}{\sqrt{1 + 4K^2 x_1^2}}, \quad y_{10} = y_1 + \frac{r}{\sqrt{1 + 4K^2 x_1^2}}. \tag{6.110}$$

The expressions in (6.110) are the equations of the real cam.

The displacement of the follower is expressed in:

$$s_2 = y_{10} - y_{10}\big|_{x_1 = 0}.$$

Replacing the value of y_{10}, we obtain the expression

$$s_2 = y_{10} - b - r. \tag{6.111}$$

For the determination of the laws of motion, $s_2 = s_2\left(s_1\right)$ and $s_2 = s_2\left(s_1\right)$, values are given to the parameter x_1, then x_{10} and y_{10} are determined from (6.110), and x_2 from (6.111), since

Table 6.7 The values of the functions s_1, φ, y_{10}, and x_2.

x_1	0	5	10	15	20	25	27	20	15	10	5	0
s_1	0	3.04	6.29	9.86	13.75	17.93	19.66	13.75	9.86	6.29	3.04	0
$\varphi\,[^\circ]$	0	45.88	68.20	89.17	111.0	142.5	165.1	245.4	270.8	291.8	314.1	360
y_{10}	30	30.31	31.28	33.07	35.81	39.57	41.37	35.81	33.07	31.28	30.31	30
s_2	0	0.31	1.28	3.07	5.81	9.57	11.37	5.81	3.07	1.28	0.31	0

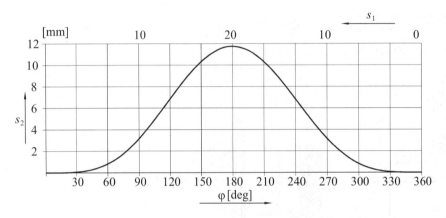

Figure 6.63 The graphs of the functions $s_2 = s_2\left(s_1\right)$ and $s_2\left(\varphi\right)$.

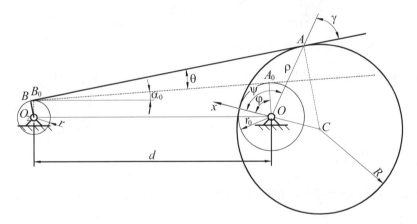

Figure 6.64 Cam mechanism with rotational circular cam and flat tappet.

$s_1 = x_{10} = e\left(1 - \cos\varphi\right)$. A table of values for the follower's displacement is obtained. For the numbers given above, this is Table 6.7 and the graphs $s_2 = s_2\left(s_1\right)$ and $s_2\left(\varphi\right)$ are as shown in Figure 6.63.

Example 3

Consider the mechanism with a rotational circular cam and a rotational (oscillatory) flat tappet in Figure 6.64. Using the derivation by finite differences, we want to determine the law of motion and the reduced angular velocity for a degree-by-degree variation of the parameter ψ, in an interval of $10°$. The following values are known: $R = 100\,\text{mm}$, $r_0 = 40\,\text{mm}$, $r = O_2B = 20\,\text{mm}$, and $d = 250\,\text{mm}$.

Solution: Denoting $OA = \rho$, from the triangle OCA:

$$R^2 = (R - r_0)^2 + \rho^2 + 2\rho (R - r_0) \cos \psi$$

or

$$\rho = - (R - r_0) \cos \psi + \sqrt{R^2 - (R - r_0)^2 \sin^2 \psi}. \tag{6.112}$$

The angle α_0 between the straight lines B_0A_0 and O_2O is given by $\alpha_0 = \arcsin \left(\frac{r_0 - r}{d} \right)$. In the numerical case we get $\alpha_0 = 4.558°$.

The derivative $\dfrac{d\rho}{d\psi}$ is expressed using finite differences

$$\left. \frac{d\rho}{d\psi} \right|_{\psi = \psi_1} = (\rho_{i+1} - \rho_{i-1}) \frac{90}{\pi}; \tag{6.113}$$

At point $\psi = 0$ the derivative has the expression

$$\left. \frac{d\rho}{d\psi} \right|_{\psi = 0} = (\rho_1 - \rho_0) \frac{180}{\pi}. \tag{6.114}$$

In the interval $[0°, 10°]$, this gives

$$\gamma = \arctan \left(\frac{\rho}{\dfrac{d\rho}{d\psi}} \right). \tag{6.115}$$

The rotational angle of the follower is given by (6.113); that is,

$$\theta = \arcsin \left(\frac{\rho \sin \gamma - r}{d} \right) - \alpha_0, \tag{6.116}$$

and the rotational angle of the cam is expressed by:

$$\varphi = \psi + \gamma + \theta - \frac{\pi}{2}. \tag{6.117}$$

Due to the degree-by-degree variation of angle ψ, the reduced angular velocity may be calculated with $\dfrac{d\theta}{d\varphi} = \dfrac{\dfrac{d\theta}{d\psi}}{\dfrac{d\varphi}{d\psi}}$. Since

$$\left. \frac{d\theta}{d\psi} \right|_{\psi = \psi_i} = \frac{\theta_{i+1} - \theta_{i-1}}{2\Delta\psi}, \quad \left. \frac{d\varphi}{d\psi} \right|_{\psi = \psi_i} = \frac{\varphi_{i+1} - \varphi_{i-1}}{2\Delta\psi}$$

then

$$\left. \frac{d\theta}{d\varphi} \right|_{\varphi = \varphi_i} = \frac{\theta_{i+1} - \theta_{i-1}}{\varphi_{i+1} - \varphi_{i-1}}. \tag{6.118}$$

Since the ratio $\dfrac{d\theta}{d\varphi}$ is non-dimensional, angles θ and φ can also be given in degrees.

To calculate the derivatives at point $\psi = 10°$ the necessary values calculated for $\psi = 11°$ are called, while for the calculation of the derivative at point $\psi = 0°$ the following expression is used:

$$\frac{d\theta}{d\varphi} = \frac{\theta_1 - \theta_0}{\varphi_1 - \varphi_0}. \tag{6.119}$$

Equations (6.112)–(6.119) give the values in Table 6.8 and the graphs in Figure 6.65.

Table 6.8 The use of the finite differences method.

$\psi\,[^\circ]$	0	1	2	3	4	5	6	7	8	9	10	11
$\rho\,[\text{mm}]$	40.000	40.004	40.015	40.033	40.059	40.091	40.132	40.180	40.235	40.297	40.367	40.445
$\dfrac{d\rho}{d\psi}$	0.209	0.419	0.838	1.258	1.678	2.100	2.522	2.946	3.372	3.799	4.229	4.661
$\gamma\,[^\circ]$	89.70	89.40	88.80	88.20	87.60	87.00	86.40	85.81	85.21	84.61	84.02	83.43
$\gamma\,[^\circ]$	0.000	0.000	0.001	0.003	0.005	0.008	0.012	0.017	0.022	0.027	0.034	0.041
$\varphi\,[^\circ]$	0.000	0.400	0.801	1.203	1.606	2.011	2.416	2.823	3.231	3.641	4.053	4.467
$\dfrac{d\theta}{d\psi}$	0.0007	0.0013	0.0034	0.0050	0.0067	0.0083	0.0100	0.0117	0.0133	0.0150	0.0166	0.0084

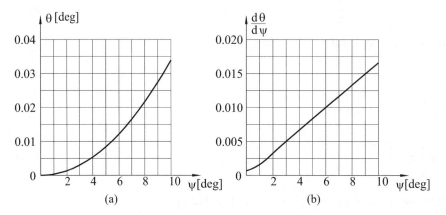

Figure 6.65 The variations of the parameters θ and $\dfrac{d\theta}{d\psi}$ as functions of the angle ψ: (a) $\theta = \theta\,(\psi)$;
(b) $\dfrac{d\theta}{d\psi} = \dfrac{d\theta}{d\psi}\,(\psi)$.

Example 4

point A of the robot mechanism in Figure 6.66 describes an imposed closed curve (Γ), in a period T, the equations of the coupler curve being $X_A = f\,(t)$ and $Y_A = g\,(t)$, where t is the time parameter. Knowing the lengths l_1, l_2, the coordinates X_B, Y_B, the coordinates x_0, y_0 of point D relative to the system Cx_1y_1, the radius r of the roller centred at point D, we want to determine the constant angular velocity ω and the parametric equations in the reference system Ox_1y_1 of the cams that border the roller.

The values used are $l = 0.35\,\text{m}$, $l_1 = l_2 = l$, $X_B = -l$, $Y_B = 2.1l$, $x_0 = -\dfrac{l}{2}$, $y_0 = -l\dfrac{\sqrt{3}}{2}$, $X_A = \dfrac{l}{4}\sin\left(\dfrac{\pi t}{2}\right)$, $Y_A = Y_B - \dfrac{l}{10}\cos\left(\dfrac{\pi t}{2}\right)$, and $r = 0.25\,\text{m}$.

Solution: The period T in which the curve (Γ) is covered is

$$T = \frac{2\pi}{\dfrac{\pi}{2}} = 4\,\text{s}, \tag{6.120}$$

while the constant angular velocity of the cam is

$$\omega = \frac{2\pi}{T} = \frac{\pi}{2}\,\text{s}^{-1}. \tag{6.121}$$

Given the relation

$$\left(X_C - X_B\right)^2 + \left(Y_C - Y_B\right)^2 = l_2^2, \tag{6.122}$$

Figure 6.66 Robot mechanism with cams.

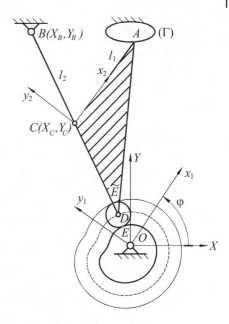

and taking into account the equalities

$$X_C = X_A - l_1 \cos\theta, \ Y_C = Y_A - l_1 \sin\theta \tag{6.123}$$

and the notation

$$A = X_A - X_B, \ B = Y_A - Y_B, \ C = \frac{l_2^2 - l_1^2 - A^2 - B^2}{2l_1}, \tag{6.124}$$

becomes

$$A \cos\theta + B \sin\theta + C = 0; \tag{6.125}$$

The solution of (6.125) is

$$\theta = 2\arctan\left(\frac{-B \pm \sqrt{A^2 + B^2 - C^2}}{C - A}\right). \tag{6.126}$$

In practice, the sign in front of the radical is defined using the initial position; in the present case, the sign is a minus. The coordinates x_1, y_1 of point D, situated on the theoretical cam, satisfy:

$$\begin{cases} X_D = x_1 \cos\varphi - y_1 \sin\varphi = X_C + x_0 \cos\theta - y_0 \sin\theta, \\ Y_D = x_1 \sin\varphi + y_1 \cos\varphi = Y_C + x_0 \sin\theta + y_0 \cos\theta \end{cases}$$

or

$$\begin{aligned} x_1 &= X_C \cos\varphi + Y_1 \sin\varphi + x_0 \cos(\varphi - \theta) + y_0 \sin(\varphi - \theta), \\ y_1 &= -X_C \sin\varphi + Y_1 \cos\varphi - x_0 \sin(\varphi - \theta) + y_0 \cos(\varphi - \theta). \end{aligned} \tag{6.127}$$

Equations (6.127) represent the parametric coordinates (as functions of time) of the theoretical cam. For the determination of the parametric equations of the cams that border the roller it is necessary (Figure 6.67) to determine angle α. For that purpose, by differentiating (6.123)–(6.125) with respect to time, we obtain

$$\frac{dX_C}{dt} = \frac{dX_A}{dt} + h\frac{d\theta}{dt}\sin\varphi, \ \frac{dY_C}{dt} = \frac{dY_A}{dt} - h\frac{d\theta}{dt}\cos\varphi, \tag{6.128}$$

Figure 6.67 The real cam and the theoretical cam.

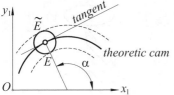

$$\frac{dA}{dt} = \frac{dX_A}{dt}, \quad \frac{dB}{dt} = \frac{dY_A}{dt}, \quad \frac{dC}{dt} = -\frac{1}{l_1}\left(A\frac{dA}{dt} + B\frac{dB}{dt}\right),$$

(6.129)

$$\frac{d\theta}{dt} = \frac{\dfrac{dA}{dt}\cos\theta + \dfrac{dB}{dt}\sin\theta + \dfrac{dC}{dt}}{A\sin\theta - B\cos\theta};$$

(6.130)

Now differentiating (6.127) with respect to time, gives

$$\begin{aligned}
\frac{dx_1}{dt} &= \frac{dX_C}{dt}\cos\varphi + \frac{dY_C}{dt}\sin\varphi + \omega\left(-X_C\sin\varphi + Y_C\cos\varphi\right) \\
&\quad + \left(\omega - \frac{d\theta}{dt}\right)\left[-x_0\sin(\varphi - \theta) + y_0\cos(\varphi - \theta)\right], \\
\frac{dy_1}{dt} &= -\frac{dX_C}{dt}\sin\varphi + \frac{dY_C}{dt}\cos\varphi - \omega\left(X_C\cos\varphi + Y_C\sin\varphi\right) \\
&\quad - \left(\omega - \frac{d\theta}{dt}\right)\left[x_0\cos(\varphi - \theta) + y_0\sin(\varphi - \theta)\right],
\end{aligned}$$

(6.131)

from which

$$\tan\alpha = -\frac{-\dfrac{dx_1}{dt}}{\dfrac{dy_1}{dt}}.$$

(6.132)

The parametric equations of the cams that border the roller are

$$\begin{aligned}
x_E &= x_1 - r\cos\alpha, \quad y_E = y_1 - r\sin\alpha, \\
x_{\tilde{E}} &= x_1 + r\cos\alpha, \quad y_{\tilde{E}} = y_1 + r\sin\alpha.
\end{aligned}$$

(6.133)

For the determination of the real cam described by point E, the following algorithm is used

1) Calculate the angular velocity from (6.121).
2) Give values to parameter t, calculate X_A and Y_A using the numerical data, and angle φ from $\varphi = \omega t$.
3) Calculate the parameters A, B, C, θ, X_C, and Y_C using (6.124), (6.126), and (6.123).
4) Calculate x_1 and y_1 from (6.127).
5) Using (6.128)–(6.132), calculate α and, using (6.133), calculate the coordinates x_E, y_E and $x_{\tilde{E}}$, $y_{\tilde{E}}$.

We have created a Turbo Pascal calculation program based on this algorithm. The code can be found in the companion website of this book. In a repetitive loop, the program determines the numerical values for (6.123)–(6.133), in 400 steps, in the time interval $[0, 4]$ s. We thus obtain, point-by-point, the profiles of the two cams in the general reference system.

For the creation of tables of values and graphs, the calculation program writes the parameters t, θ, α, x_D, y_D, x_E, y_E, $x_{\tilde{E}}$ and $y_{\tilde{E}}$, to a text file called *values.txt*.

For the graphical representations, the program creates four script files, from which AutoCAD can create:

- the drawing of curve (Γ) (the file *curve.scr*);
- the drawing of the exterior profile of the cam described by point E (the file *cam1.scr*;)
- the drawing of the exterior profile of the cam described by point $\tilde{E}$ (the file *cam2.scr*;)
- the animation of the mechanism $BCDA$ (the file *animation.scr*);

Figure 6.68 The cams of the robot mechanism.

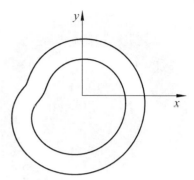

Figure 6.68 is based on the script files *cam1.scr* and *cam2.scr* and shows the shapes of the two cams.

The cams of the robot mechanism can also be obtained in AutoCAD, with solids, using an AutoLisp function (the code for which can be found in the companion website of the book). This is based on the algorithm in Section 6.5.5. The AutoLisp function is called *Robot_Cam*. Values are first assigned to the constant parameters l, l_1, l_2, x_B, y_B, r, and ω; a constant step is chosen for the time, *step* = 0.01 s (consequently, in the interval $[0, 4]$ s, there will be 400 iterations).

The solid which represents the raw material of the cam is a cylinder placed at point $O\,(0, 0, 0)$ with radius $R = 0.22$ m, and thickness $g = 0.05$ m. The radius of the raw material is sufficient to create, after removal of the stock material, the two cams. The solid obtained is named *cam*.

In a *while* loop, during which the time t takes values in the interval $[0, 4]$ s, the following operations are undetaken at each step:

1) Determine angle φ with the relation $\varphi = \omega t = \frac{\pi}{2} t$.
2) Calculate the coordinates X_A, Y_A of points A on the curve (Γ)
3) Determine angle θ using (6.126), and then the parameters A, B, C using (6.124).
4) Determine the coordinates X_C, Y_C of points C using (6.123), and the coordinates X_D, Y_D of point S with:

$$X_D = X_C + CE \cos (\theta - \beta), \ \ Y_D = X_C + CE \sin (\theta - \beta),$$

where angle β is angle $\widehat{ACD}$, having the value $\beta = 120°$, while $CE = l_1$.
5) At point D, construct a cylinder of radius r and height g; this cylinder is called *tappet*.
6) Extract the tappet from the cam by the Boolean operation of subtracting (*Subtract*); the resulting solid is called *cam*.
7) Rotate the solid obtained by one step and repeat the algorithm (up to 400 times) until the time reaches a value of $t = 4$ s.

Ultimately, the solid in Figure 6.69 is obtained. This contains both cams. If the curves in Figure 6.68 are overlaid on the solid, it can be seen that that the results of the two methods are identical.

To separate the two cams from the solid obtained, the AutoCAD command **Solidedit** is used with the options *Body* and *Separate*. After separating and moving to another position, the representation in Figure 6.70 is obtained.

Another possibility to compare the results of the analytical method with those obtained using solids is the transformation of the curves in Figure 6.68 into solids using the AutoCAD command **Extrude**, after which the results can be compared with the command **Massprop**. The **Massprop** command gives information about the mass, volume, position of the centre of mass, and moments of inertia, and so on.

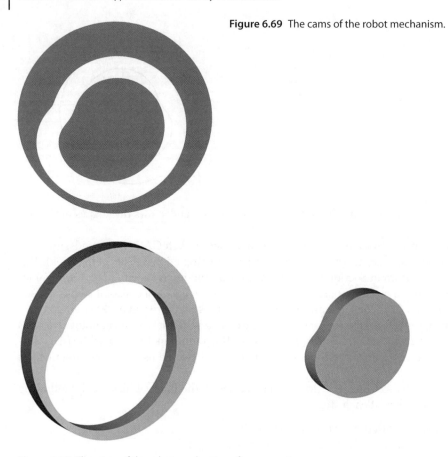

Figure 6.69 The cams of the robot mechanism.

Figure 6.70 The cams of the robot mechanism after separation.

7

Gear Mechanisms

Gear mechanisms are formed by two gears or geared sectors that rotate around two axes with invariant relative positions. They are frequently encountered in design problems, such as reductors, gearboxes of motor cars, in machine-tools, and in industrial robots, and so on.

7.1 General Aspects: Classifications

A pair of gears defines a gear train. The gear train transmits motion between two axes of invariant relative positions. One of the gears (the driving gear) sets the other (the driven gear) in motion by the action of teeth, which are successively and continuously situated in contact with each other, on both gears.

The teeth make contact with each other on their lateral surfaces. These surfaces are called *flanks*. The two flanks of a tooth are symmetrical (see Figure 7.1). The teeth are upper bounded by *head surfaces*, and lower bounded by *foot surfaces*. The section through the flanks determines the *profile of the teeth*. The intersections between the flanks and the head surfaces determine the *head lines* of the flanks, while the intersections between the flanks and the foot surfaces determine the *foot lines* of the flanks.

If the head (foot) lines are straight and parallel to the axis of the gear, then the tooth is called a *spur* (Figure 7.1), while if they are are curvilinear, the tooth is called *helical* (Figure 7.2).

The gear is called *cylindrical* (*conical*) if the head (foot) surfaces are cylinders (cones) coaxial to the axis of the gear.

The conventional representations in Figure 7.3b,c are used for cylindrical gears with spur teeth (Figure 7.3a), and the conventional representations in Figure 7.4b,c for conical gears with spur teeth (Figure 7.4a).

When the radius of the cylinder becomes infinite, a gear with spur teeth becomes a rack with spur teeth (Figure 7.5).

The classification of gear trains is made according to the criteria in Table 7.1.

7.2 Relative Motion of Gears: Rolling Surfaces

Generally for a gear train, motion is transmitted by the contact between the flanks of the teeth from one rotational axis Δ_1 to another Δ_2 that crosses it (see Figure 7.6a). Let us denote by O_1O_2 the common perpendicular, with a its length, Σ the crossing angle (the angle between the rotational axes), ω_1 and ω_2 the angular velocities, and i_{12} the transmission ratio,

$$i_{12} = \frac{\omega_1}{\omega_2};$$

(7.1)

Classical and Modern Approaches in the Theory of Mechanisms, First Edition.
Nicolae Pandrea, Dinel Popa and Nicolae-Doru Stănescu.
© 2017 John Wiley & Sons Ltd. Published 2017 by John Wiley & Sons Ltd.
Companion Website: www.wiley.com/go/pandmech17

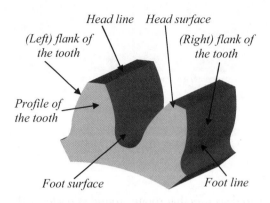

Figure 7.1 Gear teeth.

Figure 7.2 Cylindrical gear with helical teeth.

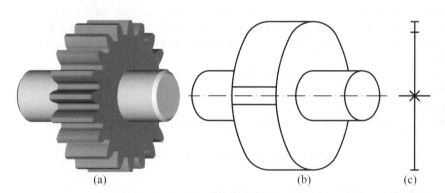

(a) (b) (c)

Figure 7.3 Cylindrical gear with spur teeth: (a) photographic representation; (b) and (c) conventional representations.

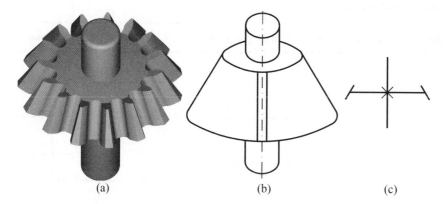

Figure 7.4 Conical gear with spur teeth: (a) photographic representation; (b) and (c) conventional representations.

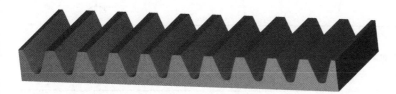

Figure 7.5 Rack with spur teeth.

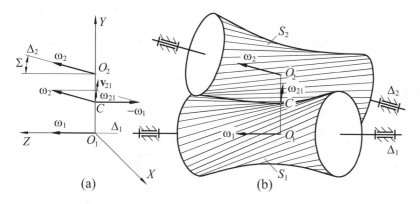

Figure 7.6 Rolling surfaces: (a) reference systems; (b) rolling surfaces.

Table 7.1 The classification of gear trains.

Nr.	Criterion of classification	The name of the gear train	Conventional representation
1	Position of the axes	With parallel axes	
		With concurrent axes	
		With cross axes	
2	The form of the teeth	Spur	
		Inclined	
		In V	
3	Profile of the teeth	Involute Cycloidal Special profile	
4	Motion of the axes	With fixed axes	
		With mobile axes (planetary)	

(Continued)

Table 7.1 (Continued)

Nr.	Criterion of classification	The name of the gear train	Conventional representation
5	Mobility degree	Mono-mobile	
		Differential	
6	Transmission ratio $i_{12} = \dfrac{\omega_1}{\omega_2}$	With constant transmission ratio	
		With variable transmission ratio	

The transmission ratio is, obviously, a constant for a given gear train.

The relative motion of a body rigidly linked to axis Δ_2 relative to a body rigidly linked to axis Δ_1 is defined by the kinematic torsor, which, at an arbitrary point M in space, has the components $(\boldsymbol{\omega}_{21}, \mathbf{v}_{21M})$, where

$$\boldsymbol{\omega}_{21} = \boldsymbol{\omega}_2 - \boldsymbol{\omega}_1, \quad \mathbf{v}_{21M} = \boldsymbol{\omega}_2 \times \mathbf{O}_2\mathbf{M} - \boldsymbol{\omega}_1 \times \mathbf{O}_1\mathbf{M}. \tag{7.2}$$

Choosing the reference system O_1XYZ so that axis O_1Z coincides with axis Δ_1, and axis O_1Y coincides with the common perpendicular, means that the kinematic torsor reduced at point O_1, that is $(\boldsymbol{\omega}_{21}, \mathbf{v}_{12O_1})$, has the components

$$\boldsymbol{\omega}_{21} = (\omega_2 \cos \Sigma - \omega_1) \mathbf{k} + \omega_2 \sin \Sigma \mathbf{i} = \omega_2 \left[(\cos \Sigma - i_{12}) \mathbf{k} + \sin\Sigma \mathbf{i} \right], \tag{7.3}$$

$$\mathbf{v}_{21O_1} = \boldsymbol{\omega}_2 \times \mathbf{O}_2\mathbf{O}_1 = \omega_2 a (- \sin \Sigma \mathbf{k} + \cos \Sigma \mathbf{i}). \tag{7.4}$$

The theory of the reduction of sliding vectors means that the instantaneous axis Δ_{21} of the helical relative motion passes through a point C defined by the vector

$$\mathbf{O}_1\mathbf{C} = \frac{\boldsymbol{\omega}_{21} \times \mathbf{v}_{21O_1}}{\omega_{21}^2}, \tag{7.5}$$

and it has the direction of the vector $\boldsymbol{\omega}_{21}$. Performing the calculations and taking into account (7.1), (7.3), and (7.4), gives point C (Figure 7.6a) situated on axis O_1O_2, at a distance $O_1C = r_{w_1}$ from point O_1,

$$r_{w_1} = \frac{a (1 - i_{12} \cos \Sigma)}{1 + i_{12}^2 - 2i_{12} \cos \Sigma}, \tag{7.6}$$

In other words, point C is situated at a distance $O_2C = r_{w_2}$ from point O_2,

$$r_{w_2} = \frac{ai_{12} (i_{12} - \cos \Sigma)}{1 + i_{12}^2 - 2i_{12} \cos \Sigma}. \tag{7.7}$$

The minimum relative velocity $\mathbf{v}_{21min}$ is given by:

$$\mathbf{v}_{21min} = p\boldsymbol{\omega}_{21}, \tag{7.8}$$

where p is the *parameter* or the *step* of the kinematic torsor; it is given by:

$$p = \frac{\boldsymbol{\omega}_{21} \cdot \mathbf{v}_{21O_1}}{\omega_{21}^2}. \tag{7.9}$$

Performing the calculations using (7.1), (7.3), and (7.4), we obtain

$$p = a \frac{i_{12} \sin \Sigma}{1 + i_{12}^2 - 2 i_{12} \cos \Sigma}. \tag{7.10}$$

The vector equation of axis Δ_{21} of the instantaneous relative motion is

$$\mathbf{r} = r_{w_1} \mathbf{j} + \lambda \left[\left(\cos \Sigma - i_{12} \right) \mathbf{k} + \sin \Sigma \mathbf{i} \right], \tag{7.11}$$

where λ is a real parameter. Dividing (7.7) by (7.6), we obtain:

$$\frac{r_{w_2}}{r_{w_1}} = \frac{i_{12} \left(i_{12} - \cos \Sigma \right)}{1 - i_{12} \cos \Sigma}; \tag{7.12}$$

From the sine theorem in the triangle of the composition of the angular velocities, this gives at point C (Figure 7.6a):

$$i_{12} = \frac{\omega_1}{\omega_2} = \frac{\sin \delta_2}{\sin \delta_1}. \tag{7.13}$$

If the axes are parallel ($\Sigma = 0$), then from (7.12) this gives $i_{12} = -\dfrac{r_{w_2}}{r_{w_1}}$; if the axes are concurrent, then the transmission ratio is calculated using (7.13). In its motion relative to the axes Δ_1 and Δ_2, axis Δ_{21} generates two rolling surfaces S_1 and S_2, which represents two one-wind hyperboloids of rotation (Figure 7.6b). These hyperboloids are tangential to one another along the axis Δ_{21}. The hyperboloid S_2 rolls with sliding (with velocity $\mathbf{v}_{21min}$ along axis Δ_{21}) over the hyperboloid S_1.

When the axes Δ_1 and Δ_2 are parallel, the rolling surfaces are two cylinders of rotation, with radii r_{w_1} and r_{w_2}, respectively; when the axes are concurrent, the rolling surfaces are two cones of rotation. In the latter cases, the relative motions of the rolling surfaces are rolling without sliding along axis Δ_{21}.

7.3 Reciprocal Wrapped Surfaces

As we have already shown, the transmission of motion from the driving axis Δ_1 to the driven axis Δ_2 is made by the contact of the flanks of the teeth. The flanks of the contact teeth represent two surfaces that remain tangential to one another during the motion. These surfaces must be created so that they transmit the motion with a constant defined transmission ratio i_{12}. For instance, if the parameters a, Σ, i_{12}, and the surface S_2 (Figure 7.7) are known, then the determination of the conjugate surface S_1 is based on the fact that the common normal $\mathbf{N}$ to the two surfaces, at the contact point M, is perpendicular to the relative motion $\mathbf{v}_{21M}$; that is,

$$\mathbf{v}_{21M} \cdot \mathbf{N} = 0. \tag{7.14}$$

Let the parametric equations of the surface S_2 be

$$x = x(u, v), \; y = y(u, v), \; z = z(u, v), \tag{7.15}$$

and let us denote by φ_1 and φ_2 the rotation angles of the mobile systems. In these conditions, the normal $\mathbf{N}$ has, in the system $O_2 x_2 y_2 z_2$, the projections A, B, and C, given by:

$$
\begin{aligned}
A &= \frac{\partial y_2}{\partial u} \frac{\partial z_2}{\partial v} - \frac{\partial y_2}{\partial v} \frac{\partial z_2}{\partial u}, \\
B &= \frac{\partial z_2}{\partial u} \frac{\partial x_2}{\partial v} - \frac{\partial z_2}{\partial v} \frac{\partial x_2}{\partial u}, \\
C &= \frac{\partial x_2}{\partial u} \frac{\partial y_2}{\partial v} - \frac{\partial x_2}{\partial v} \frac{\partial y_2}{\partial u}.
\end{aligned} \tag{7.16}
$$

Figure 7.7 Reciprocal wrapped surfaces.

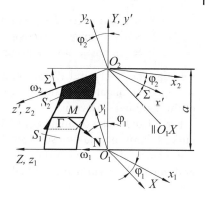

Between the unit vectors $\mathbf{i}$, $\mathbf{j}$, $\mathbf{k}$, $\mathbf{i}'$, $\mathbf{j}'$, $\mathbf{k}'$, $\mathbf{i}_1$, $\mathbf{j}_1$, $\mathbf{k}_1$, and $\mathbf{i}_2$, $\mathbf{j}_2$, $\mathbf{k}_2$ of the axes O_1X_1, O_1Y_1, O_1Z_1, O_1X', O_1Y', O_1Z', O_1x_1, O_1y_1, O_1z_1, and O_2x_2, O_2y_2, O_2z_2, respectively, there are the relations

$$
\begin{aligned}
\mathbf{i}_1 &= \mathbf{i}\cos\varphi_1 + \mathbf{j}\sin\varphi_1,\ \mathbf{j}_1 = -\mathbf{i}\sin\varphi_1 + \mathbf{j}\cos\varphi_1\ \mathbf{k}_1 = \mathbf{k},\\
\mathbf{i} &= \mathbf{i}'\cos\Sigma + \mathbf{k}'\sin\Sigma,\ \mathbf{j} = \mathbf{j}',\ \mathbf{k} = -\mathbf{i}'\sin\Sigma + \mathbf{k}'\cos\Sigma,\\
\mathbf{i}' &= \mathbf{i}_2\cos\varphi_2 - \mathbf{j}_2\sin\varphi_2,\ \mathbf{j}' = \mathbf{i}_2\sin\varphi_2 + \mathbf{j}_2\cos\varphi_2,\ \mathbf{k}' = \mathbf{k}_2,
\end{aligned}
\tag{7.17}
$$

from which

$$
\begin{aligned}
\mathbf{i}_1 &= \left(\cos\varphi_1\cos\Sigma\cos\varphi_2 + \sin\varphi_1\sin\varphi_2\right)\mathbf{i}_2\\
&\quad + \left(-\cos\varphi_1\cos\Sigma\sin\varphi_2 + \sin\varphi_1\cos\varphi_2\right)\mathbf{j}_2 + \cos\varphi_1\sin\Sigma\mathbf{k}_2,\\
\mathbf{j}_1 &= \left(-\sin\varphi_1\cos\Sigma\cos\varphi_2 + \cos\varphi_1\sin\varphi_2\right)\mathbf{i}_2\\
&\quad + \left(\sin\varphi_1\cos\Sigma\sin\varphi_2 + \cos\varphi_1\cos\varphi_2\right)\mathbf{j}_2 + \sin\varphi_1\sin\Sigma\mathbf{k}_2,\\
\mathbf{k}_1 &= -\sin\Sigma\cos\varphi_2\mathbf{i}_2 + \sin\Sigma\sin\varphi_2\mathbf{j}_2 + \cos\Sigma\mathbf{k}_2.
\end{aligned}
\tag{7.18}
$$

Expressing the velocity $\mathbf{v}_{21M}$ by:

$$
\mathbf{v}_{21M} = \boldsymbol{\omega}_2 \times O_2M - \boldsymbol{\omega}_1 \times O_1M = \left(\boldsymbol{\omega}_2 - \boldsymbol{\omega}_1\right) \times O_2M - \boldsymbol{\omega}_1 \times O_1O_2,
\tag{7.19}
$$

taking into account the equalities

$$
\boldsymbol{\omega}_2 = \omega_2\mathbf{k}_2,\ \boldsymbol{\omega}_1 = \omega_2\mathbf{k}_1,\ \omega_1 = i_{12}\omega_2,\ O_1O_2 = a\mathbf{k}
\tag{7.20}
$$

and recalling (7.17), then from (7.14) we deduce the fundamental equation of toothing

$$
\begin{aligned}
&\left[y_2\left(\cos\Sigma - \frac{1}{i_{12}}\right) - z_2\sin\Sigma\sin\varphi_2 + a\cos\Sigma\cos\varphi_2\right]A\\
&+ \left[-x_2\left(\cos\Sigma - \frac{1}{i_{12}}\right) - z_2\sin\Sigma\cos\varphi_2 - a\cos\Sigma\sin\varphi_2\right]B\\
&+ \sin\Sigma\left(-x_2\sin\varphi_2 - y_2\cos\varphi_2\right)C = 0.
\end{aligned}
\tag{7.21}
$$

Taking into account that angle φ_2 can be written in the form

$$
\varphi_2 = \varphi_{20} + \frac{\varphi_1 - \varphi_{10}}{i_{12}},
\tag{7.22}
$$

where φ_{10} and φ_{20} are the initial values of the angles φ_1 and φ_2, respectively, (7.21) is a function that depends on the parameters φ_1, u, and v.

The coordinates of point M in the reference systems $O_1x_1y_1z_1$ and $OXYZ$ read

$$
\begin{aligned}
x_1 &= a\sin\varphi_1 + x_2\left(\cos\varphi_1\cos\Sigma\cos\varphi_2 + \sin\varphi_1\sin\varphi_2\right)\\
&\quad y_2\left(-\cos\varphi_1\cos\Sigma\sin\varphi_2 + \sin\varphi_1\cos\varphi_2\right) + z_2\cos\varphi_1\sin\Sigma,\\
y_1 &= a\cos\varphi_1 + x_2\left(-\sin\varphi_1\cos\Sigma\cos\varphi_2 + \sin\varphi_1\sin\varphi_2\right)\\
&\quad + y_2\left(\sin\varphi_1\cos\Sigma\sin\varphi_2 + \cos\varphi_1\cos\varphi_2\right) - z_2\sin\varphi_1\sin\Sigma,\\
z_1 &= -x_2\sin\Sigma\cos\varphi_2 + y_2\sin\Sigma\sin\varphi_2 + z_2\cos\Sigma,
\end{aligned}
\tag{7.23}
$$

$$X = x_2 \cos \varphi_2 \cos \Sigma - y_2 \sin \varphi_2 \cos \Sigma + z_2 \sin \Sigma,$$
$$Y = a + x_2 \sin \varphi_2 + y_2 \cos \varphi_2, \tag{7.24}$$
$$Z = -x_2 \cos \varphi_2 \sin \Sigma + y_2 \sin \varphi_2 \sin \Sigma + z_2 \cos \Sigma.$$

Equations (7.21) and (7.23) define the parametric equations of the surface S_1 reciprocal wrapped (conjugated) to the surface S_2, while (7.21) and (7.24) define the parametric equations of the toothing surface S.

For a fixed value of angle φ_1, (7.21)–(7.23) define the parametric equations of the contact curve Γ between the two surfaces. When angle φ varies, the characteristic curve Γ describes the reciprocal wrapped surface S_1 of (7.23) relative to the reference mobile system $O_1 x_1 y_1 z_1$, and the toothing surface S of (7.24) relative to the fixed reference system $O_1 XYZ$.

7.4 Fundamental Law of Toothing

Let us consider an arbitrary point M of the characteristic curve Γ (Figure 7.8) and let $\mathbf{v}_{21}$ be the relative velocity at this point, $\mathbf{N}$ the common normal to the two conjugate surfaces, Δ_{21} the helical axis of the instantaneous relative motion, $\boldsymbol{\omega}_{21}$ the relative angular velocity, α the angle between the normal $\mathbf{N}$ and axis Δ_{21}, DE the common perpendicular to axis Δ_{21} and the support of the normal $\mathbf{N}$, and d the length of this common perpendicular.

The relative velocity $\mathbf{v}_{21}$ at point M is perpendicular to the normal $\mathbf{N}$; this gives the expression

$$\mathbf{v}_{21} \cdot \mathbf{N} = 0. \tag{7.25}$$

Writing the relative velocity at point M in the form

$$\mathbf{v}_{21} = p\boldsymbol{\omega}_{21} + \boldsymbol{\omega}_{21} \times \mathbf{CM}, \tag{7.26}$$

and using the unit vector $\mathbf{u}$ of axis Δ_{21}, from (7.25), we obtain the equality

$$p\mathbf{N} \cdot \mathbf{u} + \mathbf{u} \cdot (\mathbf{CM} \times \mathbf{N}) = 0. \tag{7.27}$$

Then, taking into account the obvious relations

$$\mathbf{N} \cdot \mathbf{u} = N \cos \alpha, \quad \mathbf{u} \cdot (\mathbf{CM} \times \mathbf{N}) = -dN \sin \alpha, \tag{7.28}$$

from (7.27), we deduce the expression

$$p \cos \alpha - d \sin \alpha = 0, \tag{7.29}$$

which represents an expression of the law of toothing.

This relation can be interpreted in the following way: to create a constant transmission ratio between two arbitrary crossed axes in space, it is necessary that normals at the contact points are inclined relative to the axis of the relative motion Δ_{21} so that they fulfil the condition

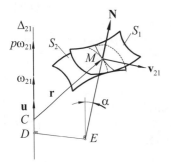

Figure 7.8 The position of the common normal relative to axis Δ_{21}.

$\tan \alpha = \frac{p}{d}$, where p is the step of the kinematic torsor of the relative motion and d is the distance between the support of the common normal and axis Δ_{21} of the helical relative motion.

When the rotational axes are parallel, step p is null and from (7.29) we obtain $d = 0$. In other words we rediscover the well known observation that for a parallel gear train the normal at the contact of the conjugate profiles passes through a fixed point C, referred to as the *pole of the toothing*. This property holds true for conical gears, for which, from the condition $p = 0$, we deduce that $d = 0$; in other words, the normals at the contact points of the conjugate flanks intersect axis Δ_{21} of the relative motion.

7.5 Parallel Gears with Spur Teeth

7.5.1 Generalities. Notations

We have already seen that, for parallel gears, the fundamental law of toothing reduces to the property that the normal at the contact point M (Figure 7.9) of the conjugate profiles passes through a point C, referred to as the *pole of the toothing*. For exterior toothing, the distances $O_1 C = r_{w_1}$ and $O_2 C = r_{w_2}$ satisfy:

$$i_{12} = \frac{\omega_1}{\omega_2} = -\frac{r_{w_1}}{r_{w_2}}, \tag{7.30}$$

resulting in the equalities

$$r_{w_1} = \frac{a}{1 - i_{12}}, \quad r_{w_2} = -\frac{a i_{12}}{1 - i_{12}}, \tag{7.31}$$

where a is the distance between the axes of the gears.

The circles with their centres at points O_1 and O_2 that pass through point C and have radii equal to r_{w_1} and r_{w_2}, respectively, are referred to as the *rolling circles*. The geometric locus described by the contact point M of the conjugate profiles relative to the fixed reference system is called the *toothing curve*.

7.5.2 Determination of the Conjugate Profile and Toothing Curve

Consider the profile (C_2) in Figure 7.10, given by the parametric equations

$$x_2 = x_2(u), \quad y_2 = y_2(u), \tag{7.32}$$

Figure 7.9 The geometry of parallel gear trains.

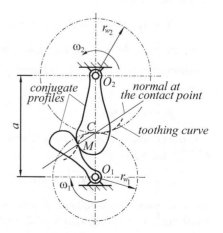

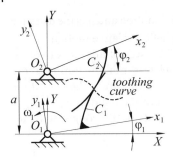

Figure 7.10 Conjugate profiles and toothing curves.

The distance a between the axes, and the transmission ratio i_{12} are also known. We want to find the equations of the conjugate profile (C_1) and the toothing curve.

For the case represented in Figure 7.10, (7.30) becomes

$$\varphi_2 = \frac{\varphi_1}{i_{12}}. \tag{7.33}$$

To apply the general theory of reciprocal wrapped surfaces, we consider the toothing between the cylindrical surface S_2, the section of which in the plane $z_2 = 0$ is the curve (C_2) given by the parametric equations (7.32) and the surface S_1. In these conditions, surface S_2 has the parametric equations

$$x_2 = x_2(u), \; y_2 = y_2(u), \; z_2 = v, \tag{7.34}$$

while the projections of the normal N onto the axes of the system $O_2 x_2 y_2 z_2$ are

$$A = y_{2u}, \; B = -x_{2u}, \; C = 0, \tag{7.35}$$

where x_{2u} and y_{2u} are the derivatives of the functions x_2 and y_2 with respect to u.

As a consequence, and making $\Sigma = 0$, from (7.21) and (7.23) we obtain the expressions

$$\left[x_2 \left(1 - \frac{1}{i_{12}} \right) + a \sin \varphi_2 \right] x_{2u} + \left[y_2 \left(1 - \frac{1}{i_{12}} \right) + a \cos \varphi_2 \right] y_{2u} = 0, \tag{7.36}$$

$$
\begin{aligned}
x_1 &= a \sin \varphi_1 + x_2 \cos(\varphi_1 - \varphi_2) + y_2 \sin(\varphi_1 - \varphi_2), \\
y_1 &= a \cos \varphi_1 - x_2 \sin(\varphi_1 - \varphi_2) + y_2 \cos(\varphi_1 - \varphi_2),
\end{aligned} \tag{7.37}
$$

which define the parametric equations of the conjugate profile (C_1).

Determination of the parametric equations of the toothing curve uses (7.24), in which we set $\Sigma = 0$; this gives

$$X = x_2 \cos \varphi_2 - y_2 \sin \varphi_2, \; Y = a + x_2 \sin \varphi_2 + y_2 \cos \varphi_2. \tag{7.38}$$

As an example, we consider the case in which the profile (C_2) is axis $O_2 y_2$ and the transmission ratio is $i_{12} = \frac{1}{2}$. The relation $\varphi_2 = 2\varphi_1$ holds true, and the parametric equations of the curve (C_2) are

$$x_2 = 0, \; y_2 = u. \tag{7.39}$$

From (7.36), $y_2 = a \cos(2\varphi_1)$. The parametric equations of the conjugate profile are given by (7.37), and read

$$x_1 = 2a \sin^3 \varphi_1, \; y_1 = 2a \cos^3 \varphi_1; \tag{7.40}$$

In conclusion, the conjugate profile is an astroid. The toothing is shown in Figure 7.11. The toothing line is the semicircle of diameter $O_2 C$.

Figure 7.11 Toothing between a segment of a straight line and an astroid.

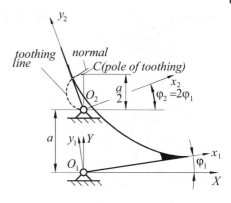

Figure 7.12 The involute of a circle.

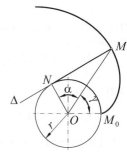

7.5.3 The Involute of a Circle

The involute of a circle (Figure 7.12) represents the most commonly used curve for gear teeth profiles. The involute of a circle is the geometric locus of a point M on a straight line that rolls without sliding on a circle.

From the condition of rolling without sliding, $NM = \text{arc}\,(NM_0)$, from which, with the notation in Figure 7.12, we deduce:

$$r \tan \alpha = r(\alpha + \gamma). \tag{7.41}$$

Using the *involute function*

$$\text{inv}\,\alpha = \tan \alpha - \alpha, \tag{7.42}$$

we obtain the polar angle

$$\gamma = \text{inv}\,\alpha. \tag{7.43}$$

The polar radius $\rho = OM$ is given by:

$$\rho = \frac{r}{\cos \gamma}. \tag{7.44}$$

The straight line MN is the normal to the involute at point M, while point N is the centre of the curvature of the involute at the same point M.

7.5.4 Involute Conjugate Profile and Toothing Line

Consider as known: the distance between the centres O_1 and O_2 of two bodies, the transmission ratio i_{12}, and an involute (C_1) that belongs to the first body (see Figure 7.13). We calculate the distances $O_1 C = r_{w_1}$ and $r_{w_1} = \dfrac{a}{1 - i_{12}}$, mark point C, at which we construct the normal to the given involute profile (C_1). This normal must be a tangent to the generator circle. For

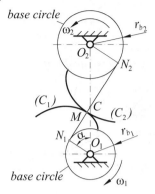

base circle

base circle

Figure 7.13 The conjugate profile of the involute.

that purpose, at point O_1 we construct the perpendicular O_1N_1 to the normal from point C to the involute.

The distance O_1N_1 represents exactly the radius of the generator circle, referred to as the *base circle*. Denoting with α the angle between the directions O_1O_2 and O_1N_1, means that the radius of the base circle r_{b_1} centred at point O_1 reads

$$r_{b_1} = r_{w_1} \cos \alpha \tag{7.45}$$

Let us consider now an involute (C_2), tangential to the involute (C_1) at point M, and let us construct from point O_2 the perpendicular O_2N_2 to the normal CN_1. The straight line N_2M being normal to the involute (C_2) means that the circle centred at point O_2, having the radius

$$r_{b_2} = r_{w_2} \cos \alpha \tag{7.46}$$

and passing through point N_2 is the generator circle for the involute (C_2). This is the second base circle.

When the base circle centred at O_1 rotates, the involute (C_1) moves, remaining normal to the straight line N_1N_2. Similarly, the involute (C_2) moves too, remaining normal to the straight line N_1N_2. The point M moves on the straight line N_1N_2, which is the toothing line. Consequently, the conjugate profile of the involute is also an involute, while the toothing line is the straight line that is tangent to the two base circles.

7.5.5 The Main Dimensions of Involute Gears

The geometric dimensions of a gear are defined with respect to the *reference rack*. The reference rack is shown in Figure 7.14. Its complementary is called the *generator rack*. The essential elements of the reference rack are the pressure angle α_0 and the modulus m. In Romania, the

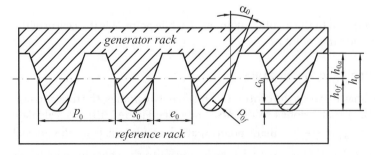

Figure 7.14 The reference rack and the generator rack.

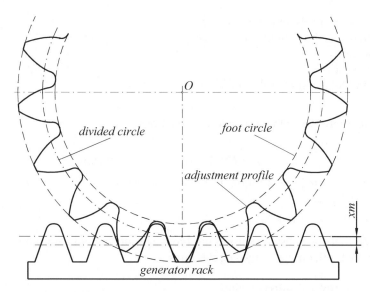

Figure 7.15 The generation of the toothing.

pressure angle α_0 is equal to $20°$ and the modulus may have the values (in mm): 0.15, 0.2, 0.25, 0.3, 0.4, 0.5, 0.6, 0.8, 1, 1.25, 1.5, 2, 2.5, 3, 4, 5, 6, 8, 10, 12 etc.[1]

The geometric elements in Figure 7.14, which define the reference rack, are given as functions of the modulus m and have the following dimensions and magnitudes:

- p_0, the step of the rack: $p_0 = \pi m$
- s_0, the chord of the tooth: $s_0 = \dfrac{p_0}{2}$
- e_0, the chord of the blank: $e_0 = \dfrac{p_0}{2}$
- h_0, the height of the tooth: $h_0 = 2.25m$
- h_{0a}, the height of the head of the tooth: $h_{0a} = m$
- h_{0f}, the height of the foot of the tooth: $h_{0f} = 1.25m$
- c_0, the clearance: $c_0 = 0.25m$
- ρ_{0f}, the radius of the adjustment circle: $\rho_{0f} = 0.38m$.

To establish the main dimensions of a gear we will consider its generation procedure from the generator rack. If the generator rack has step m and we want to generate a gear with z teeth, then the circle of diameter $d = mz$ is called the *divided circle*, while the distance between the teeth, measure on the divided circle, is equal to the step $p_0 = \pi m$ of the reference rack.

If the reference line of the rack is displaced in the exterior (interior) of the divided circle, then we say that we have a *positive (negative) displaced toothing*, while if the reference line of the generator rack is tangential to the divided rack, then we say that we have a *zero displaced (non-displaced) toothing*. The displacement xm is expressed using the displacement coefficient x, which is positive in positive-displaced toothing, negative in negative-displaced toothing, and null in zero-displaced toothing.

In the generation process, the flanks of the rack generate (see Figure 7.15) the involute flanks of the teeth of the gear, while the head circular profile of the rack generates the adjustment profile. Referring the toothing to the reference rack, gives the representation in Figure 7.16, in which the following notation is used:

- r, d, radius and diameter, respectively, of the divided circle
- r_b, d_b, radius and diameter, respectively, of the base circle

---

1 These values and those in the list below are valid for Romania. For other countries the values may differ.

- r_f, d_f, radius and diameter, respectively, of the foot circle
- r_a, d_a, radius and diameter, respectively, of the head circle
- s, chord of the tooth of the gear
- g, chord of the blank of the gear
- h_a, h_f, height of the head and foot, respectively, of the tooth
- h, height of the tooth.

On strict geometrical considerations, we obtain the following relations

$$d = 2r = mz,$$
$$h_a = xm + h_{0a} = (1+x)m, \; h_f = h_{0f} - xm = (1.25 - x)m, \; h = h_0 = 2.25m,$$
$$s = m\left(\frac{\pi}{2} + 2x \tan\alpha_0\right), \; e = m\left(\frac{\pi}{2} - 2x \tan\alpha_0\right), \; p = s + e = m\pi = p_0, \tag{7.47}$$
$$r_b = r\cos\alpha_0, \; r_f = r - h_f, \; r_a = r + h_a,$$
$$d_b = d\cos\alpha_0, \; d_f = d - 2h_f, \; d_a = d + 2h_a.$$

The equations in (7.47) mean that for a displaced toothing the height of the tooth does not change, but the thickness of the tooth on the divided circles does.

7.5.6 Thickness of a Tooth on a Circle of Arbitrary Radius

Let us consider a circle of arbitrary radius r_x, and let us denote by X, X' and T, T' the points at which this circle and the divided circle, respectively, intersect the involutes concurrent at point V; the involutes bordering the tooth (Figure 7.17). Using the notation in Figure 7.17, we may write:

$$\gamma = \text{inv}\alpha, \tag{7.48}$$

$$\gamma_0 = \text{inv}\alpha_0, \tag{7.49}$$

$$r_x \cos\alpha = r \cos\alpha_0 = r_b. \tag{7.50}$$

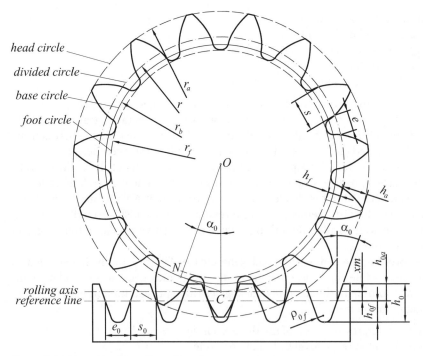

Figure 7.16 The main dimensions of a rack.

Figure 7.17 The thickness of a tooth on a circle of arbitrary radius.

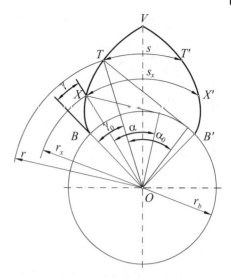

From symmetry, the thickness $s_x = \text{arc}(XX')$ of a tooth on the circle of radius r_x may be written as

$$s_x = 2r_x \widehat{XOV} = 2r_x \left(\widehat{BOV} - \widehat{BOX} \right)$$
$$= 2r_x \left(\widehat{TOV} + \widehat{BOT} - \widehat{BOX} \right) = 2r_x \left(\widehat{TOV} + \gamma_0 - \gamma \right).$$

Then, expressing angle $\widehat{TOV}$ in the form $\widehat{TOV} = \dfrac{TT'}{2r} = \dfrac{s}{2r}$, and expressing r_x from (7.48), we obtain

$$s_x = 2r \frac{\cos \alpha_0}{\cos \alpha} \left(\frac{s}{2r} + \text{inv}\alpha_0 - \text{inv}\alpha \right). \tag{7.51}$$

Now using:

$$2r = mz, \quad s = m\frac{\pi}{2} + 2xm \tan \alpha_0, \tag{7.52}$$

we finally obtain the expression

$$s_x = m\frac{\cos \alpha_0}{\cos \alpha} \left[\frac{\pi}{2} + 2x \tan \alpha_0 + z \left(\text{inv}\alpha_0 - \text{inv}\alpha \right) \right]. \tag{7.53}$$

The circular arc e which measures the blank between the teeth is given by $e = p_x - s_x$; since $p_x = \dfrac{2\pi r_x}{z} = \dfrac{2\pi \cos \alpha_0}{z \cos \alpha}$:

$$e_x = m\frac{\cos \alpha_0}{\cos \alpha} \left[\frac{\pi}{2} - 2x \tan \alpha_0 - z \left(\text{inv}\alpha_0 - \text{inv}\alpha \right) \right]. \tag{7.54}$$

7.5.7 Building-up of Gear Trains

Let us consider two gears (Figure 7.18) generated with the same rack and forming a gear train. In order to do this, it is necessary that the thickness s_{x_1} of the teeth of the first gear on the rolling circle of radius r_{w_1} is equal to the blank e_{x_2} of the second gear on the rolling circle of radius r_{w_2}; that is,

$$m\frac{\cos \alpha_0}{\cos \alpha_w} \left[\frac{\pi}{2} + 2x_1 \tan \alpha_0 + z_1 \left(\text{inv}\alpha_0 - \text{inv}\alpha_w \right) \right]$$
$$= m\frac{\cos \alpha_0}{\cos \alpha_w} \left[\frac{\pi}{2} - 2x_2 \tan \alpha_0 - z_2 \left(\text{inv}\alpha_0 - \text{inv}\alpha_w \right) \right];$$

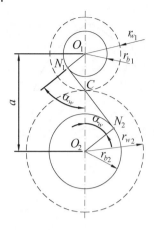

Figure 7.18 The building-up of gear trains.

This gives:

$$\mathrm{inv}\alpha_w = \mathrm{inv}\alpha_0 + 2\frac{x_1 + x_2}{z_1 + z_2}\tan\alpha_0, \tag{7.55}$$

from which the toothing angle α_w can be determined.

The radii of the rolling circles are deduced from $r_w = \dfrac{r_b}{\cos\alpha_w}$. which gives

$$r_{w_1} = r_1\frac{\cos\alpha_0}{\cos\alpha_w}, \tag{7.56}$$

$$r_{w_2} = r_2\frac{\cos\alpha_0}{\cos\alpha_w}, \tag{7.57}$$

If we denote with $a_0 = r_1 + r_2$ the distance a between the axes for non-displaced toothing, then we deduce the distance between the axes O_1 and O_2 as

$$a = a_0\frac{\cos\alpha_w}{\cos\alpha_0}. \tag{7.58}$$

Since the function $\mathrm{inv}\alpha_w$ is an increasing one, if $x_1 + x_2 > 0$, then $\alpha_w > \alpha_0$ and, in this case, the distance between the axes of the gears verifies the inequality

$$a > a_0. \tag{7.59}$$

7.5.8 The Contact Ratio

We consider a gear train (Figure 7.19) and let $T_1 T_2$ be the toothing line. Let us denote with A and E the intersection points between the toothing line and the head circles of radii r_{a_2} and r_{a_1}, respectively. Points A and E determine the *toothing segment*. In other words, the toothing starts at point A with the contact between the foot of the tooth of the first gear and the head of the second, and ends at point E with the contact between the head of the first gear and the foot of the second.

If we wrap the straight line $T_1 T_2$ around the first circle, we obtain points A_1 and E_1, which overlap on the circle coinciding with points A and E; if we wrap the straight line $N_1 N_2$ on the second circle, we obtain points A_2 and E_2, which overlap on the circle coinciding with points A and E. This gives the equalities

$$\mathrm{arc}\left(A_1 E_1\right) = \mathrm{arc}\left(A_2 E_2\right) = AE. \tag{7.60}$$

The step of the teeth on the base circle is the same, namely

$$p_0 = \frac{2\pi r_{b_1}}{z_1} = \frac{2\pi r_1 \cos\alpha_0}{z_1} = \pi m \cos\alpha_0. \tag{7.61}$$

Figure 7.19 The toothing segment.

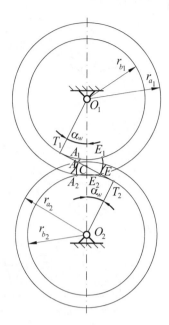

The condition of continuous motion requires that the second group of teeth enters the toothing before the first group of teeth ends its toothing. This means that the step of the teeth (for instance, on the first circle) is less than the circular arc $A_1 E_1$; that is

$$p_0 < AE. \tag{7.62}$$

The contact ratio ε is defined by:

$$\varepsilon = \frac{AE}{p_0}; \tag{7.63}$$

and must satisfy the condition $1 < \varepsilon < 2$. This means that

$$AE = AC + CE = AT_2 - CT_2 + ET_1 - CT_1$$
$$= \sqrt{r_{a_2}^2 - r_{b_2}^2} + \sqrt{r_{a_1}^2 - r_{b_1}^2} - \left(r_{w_1} + r_{w_2} \right) \sin \alpha_w;$$

Finally, we obtain the expression for the contact ratio

$$\varepsilon = \frac{\sqrt{r_{a_1}^2 - r_{b_1}^2} + \sqrt{r_{a_2}^2 - r_{b_2}^2} - a \sin \alpha_w}{\pi m \cos \alpha_0}. \tag{7.64}$$

7.5.9 Interference of Generation

In some geometric conditions, during the generation of toothing using a generator rack, it is possible that the head of the generator tooth cuts from the base of the generated tooth. This phenomenon is called *undercut* or *interference of generation*. To avoid this phenomenon it is necessary that the head line d of the generator rack (Figure 7.20) intersects the toothing line CT at a point K situated between points C and T; that is, the following inequality must hold true:

$$CK < CT \tag{7.65}$$

or

$$\frac{h_{0a} - xm}{\sin \alpha_0} < r \sin \alpha_0. \tag{7.66}$$

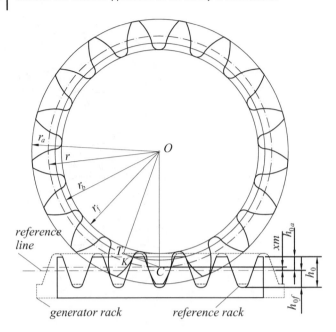

Figure 7.20 Generator and reference racks.

Replacing $r = \frac{mz}{2}$, $h_{0a} = m$ and using:

$$z_0 = \frac{2}{\sin^2 \alpha_0} \approx 17, \; z_{min} = z_0 (1 - x),$$ (7.67)

means that the interference is avoided if

$$z > z_{min}.$$ (7.68)

Equation (7.68) shows that, for non-displaced toothing, interference of generation is avoided if the number of teeth is greater than 17, while for the displaced toothing it is avoided if the number of teeth is greater than $17 (1 - x)$.

7.6 Parallel Gears with Inclined Teeth

7.6.1 Generation of the Flanks

Figure 7.21 shows a parallel gear train with spur teeth, in which $O_1 O_1'$, $O_2 O_2'$ are the axes of rotation, P is the toothing plan, AB is the toothing axis, S_1 and S_2 are the flanks of the teeth, while C_1 and C_2 are the base cylinders, of radii r_{b_1} and r_{b_2}, respectively.

The flanks S_1, and S_2 of the teeth, which, obviously, have involute normal sections, may be generated by the straight line AB when the plane P rolls without sliding successively on the cylinders C_1 and C_2, respectively. The intersections between the flanks S_1 and S_2, and the two cylinders C_1 and C_2 are the generator straight lines $A_{01} B_{01}$ and $A_{02} B_{02}$, respectively, of the two cylinders.

Let us consider now the case in which the straight line AB in plane P (Figure 7.22) is inclined at an angle β_b relative to the straight line AA' of intersection between plane P and the plane formed by the rotation axes $O_1 O_1'$ and $O_2 O_2'$. The geometric loci described by the straight line AB when plane P rolls without sliding on the two base cylinders C_1, and C_2 are the two flanks

Figure 7.21 Parallel gears with spur teeth.

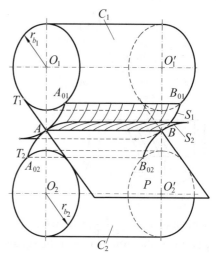

Figure 7.22 Parallel gears with inclined teeth.

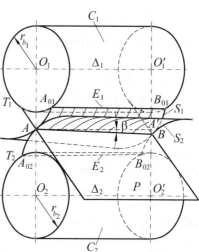

S_1 and S_2, respectively, tangential to one another along the straight line AB. The intersections between two flanks and the cylinders C_1 and C_2 are the helices E_1 and E_2, respectively. During motion, the flanks remain tangential along a straight line situated in the toothing plane P.

The intersections between the flanks and the planes normal to the rotation axes are involutes, so the flanks may be considered to be the geometric locus described by an involute in translational motion, which moves so that one of its points describes a helix, the plane of the involute remaining perpendicular to the rotational axes Δ_1 and Δ_2.

The step of the helix E_1 (or E_2) is given by:

$$p_z = 2\pi r_b \cot \beta_b. \tag{7.69}$$

By limiting the flanks with head and foot cylinders, we obtain the inclined toothing of the parallel gear train. The axoids of the motion are the cylinders of radii r_{w_1} and r_{w_2}.

7.6.2 The Equivalent Planar Gear

Sectioning the gear with inclined teeth (Figure 7.22) with a plane N–N normal to the median helix of the blank on the divided cylinder results in the *equivalent planar gear*. The plane intersects the divided cylinder of radius r along an ellipse, which has semi-minor axis r, and

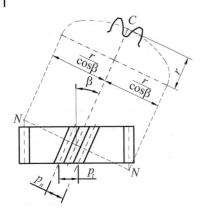

Figure 7.23 The equivalent planar gear.

semi-major axis $\dfrac{r}{\cos \beta}$, where β is the inclination angle of the teeth. The curvature radius ρ_c at point C is the ratio between the square of the semi-major and semi-minor axes; that is,

$$\rho_c = \frac{r}{\cos^2 \beta}. \tag{7.70}$$

The divided circle of the equivalent planar gear has radius r_n equal to the curvature radius

$$r_n = \frac{r}{\cos^2 \beta}. \tag{7.71}$$

The equivalent gear will have an equivalent number of teeth denoted by z_e; hence

$$r_n = \frac{m_n z_e}{2}, \tag{7.72}$$

where m_n is the modulus of the equivalent gear. Expressing the radius r using the step $p_t = \pi m_t$ and of the real number of teeth,

$$r = \frac{m_t z}{2}, \tag{7.73}$$

then taking into account $p_n = p_t \cos \beta$, obtained from the Figure 7.23 and leading to $m_n = m = m_t \cos \beta$, we obtain

$$r = \frac{mz}{2 \cos \beta}. \tag{7.74}$$

Recalling (7.71), in the conditions of (7.72) and (7.74), we obtain the number of teeth of the equivalent gear

$$z_e = \frac{z}{\cos^3 \beta}. \tag{7.75}$$

In the conditions of cutting without interference $z_e = z_{e_{min}} = 17$ teeth. This means that the minimum number of teeth z_{min} of the gear with inclined teeth is

$$z_{min} = 17 \cos^3 \beta. \tag{7.76}$$

This discussion highlights the advantage of gears with inclined teeth, which can operate without the appearance of the phenomenon of interference, with a smaller minimum number of teeth than gears with spur teeth. Other advantages of inclined toothing are the larger contact ratio and reduced noise levels compared to spur gears.

The transmission ratio for exterior parallel toothing with inclined teeth is given by:

$$i_{12} = -\frac{z_2}{z_1}. \tag{7.77}$$

7.7 Conical Concurrent Gears with Spur Teeth

As we have shown in Section 7.2, in gear trains with concurrent axes, the instantaneous relative motion is one of rolling with sliding, and the instantaneous axis of relative motion Δ_{21} is situated in the plane of the rotation axes Δ_1 and Δ_2, the transmission ratio being given by:

$$i_{12} = \frac{\sin \delta_2}{\sin \delta_1}, \tag{7.78}$$

where δ_1, and δ_2 are the angles between axis Δ_{21} and axes Δ_1 and Δ_2, respectively. Axis Δ_{21}, in motion relative to axes Δ_1 and Δ_2, generates the *rolling cones*, which are tangential to one another along axis Δ_{21}. For the case in which $\Sigma = 90°$, the rolling cones c_{w_1} and c_{w_2} are as shown in Figure 7.24.

The flanks of conical gears are reciprocal wrapped surfaces of different shapes. Using the base cones C_{b_1} and C_{b_2} (Figure 7.25), and the plane (P) tangent interior to the two cones, then, by successively rolling this plane on the two base cones, the straight line Δ_{21} generates the involute flanks F_1 and F_2 rigidly linked to the cones C_{b_1} and C_{b_2}, respectively; these two flanks are reciprocal wrapped surfaces. Using these surfaces it is possible to construct the flanks of the teeth of the conical gears. The teeth will be exterior bordered by head cones defined by angles δ_{a_i}, and interior bordered by foot cones defined by angles δ_{f_i}, $i = 1, 2$.

The elements of the reference profile are defined on the exterior frontal cone, relative to standard reference planar gear[2]. The exterior frontal cones have their centres at points O_1 and O_2. Unfolding these cones gives the substitute cylindrical toothing with the centres at points O_{v_1} and O_{v_2}, and the rolling radii (for conical toothing they are also the divided radii)

$$r_{v_i} = \frac{d_i}{2 \cos \delta_i}, \quad i = 1, 2, \tag{7.79}$$

where d_i, $i = 1, 2$, are the divided diameters (Figure 7.26).

Figure 7.24 Rolling cones.

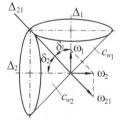

Figure 7.25 The base cones C_{b_1} and C_{b_2}.

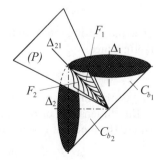

---

2 The standards are those required by Romanian legislation. The standard elements may vary in other countries.

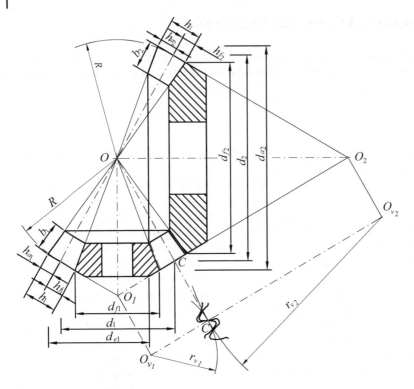

Figure 7.26 Conical toothing.

Denoting by z_{v_i}, $i = 1, 2$ the numbers of teeth of the replacing gears, and by z_i, $i = 1, 2$ the numbers of teeth of the conical gears, we get:

$$r_{v_i} = \frac{m_v z_{v_i}}{2}, \quad r_i = \frac{m z_i}{2}, \tag{7.80}$$

where m_v and m are the moduli of the gears.

We take as the standard modulus the modulus m_v that is equal to m and then, if we use:

$$r_{v_i} = \frac{r_i}{\cos \delta_i}, \tag{7.81}$$

we obtain the numbers of teeth of the replacing gears

$$z_{v_i} = \frac{z_i}{\cos \delta_i}, \quad i = 1, 2. \tag{7.82}$$

To avoid cutting interference, the minimum number of teeth of the replacing gear is $z_{v_{min}} = 17$ teeth; hence

$$z_{min} = 17 \cos \delta_i, \tag{7.83}$$

that is, the minimum number of teeth for conical toothing is 17.

The dimensions of the elements of the conical gears in the exterior section (Figure 7.26), where the modulus is standardized, are given by:

$$\begin{aligned}
h_{a_i} &= m, \ h_{f_i} = 1.25m, \ h_i = 2.25m, \\
d_{a_i} &= 2 \left(r_i + m \cos \delta_i \right) = m \left(z_i + 2 \cos \delta_i \right), \\
d_{f_i} &= 2 \left(r_i - 1.25m \cos \delta_i \right) = m \left(z_i - 2.5 \cos \delta_i \right).
\end{aligned} \tag{7.84}$$

Using the notation in Figure 7.26:

$$\sin \delta_1 = \frac{d_1}{2OC}, \quad \sin \delta_2 = \frac{d_2}{2OC} \tag{7.85}$$

and the second relation in (7.87), the transmission ratio (7.78) becomes

$$i_{12} = \frac{z_2}{z_1}. \tag{7.86}$$

7.8 Crossing Gears

7.8.1 Helical Gears

The helical gears consist of two gears with helical-involute toothing, inclined on divided cylinders with the angles β_1 and β_2, respectively (Figure 7.27). The generation of the flanks of the teeth of the two gears is performed with a rack with inclined teeth. The toothing between the teeth has punctiform contact.

The (relative) sliding velocity $\mathbf{v}_{21}$ is in the direction of the straight line Δ (along the teeth). In the position in which the contact point arrives at point A (Figure 7.27) we can write

$$\mathbf{v}_2 = \mathbf{v}_1 + \mathbf{v}_{21}, \tag{7.87}$$

where the velocities $\mathbf{v}_1$ and $\mathbf{v}_2$ are perpendicular to the rotation axes Δ_1 and Δ_2 and have magnitudes $v_1 = r_1\omega_1$ and $v_2 = r_2\omega_2$, respectively.

For un-displaced toothing, r_1 and r_2 are the radii of the divided cylinders which, in this case, coincide with the radii of the minimum sections of the rolling hyperboloids. Projecting (7.87) onto the direction of the normal to the straight line Δ gives

$$v_2 \cos \beta_2 = v_1 \cos \beta_1, \tag{7.88}$$

from which, using (7.88), we deduce the transmission ratio

$$i_{12} = \frac{\omega_1}{\omega_2} = \frac{r_2 \cos \beta_2}{r_1 \cos \beta_1}. \tag{7.89}$$

Then, denoting by m_{t_1} and m_{t_2} the moduli of the divided circles, gives

$$r_1 = \frac{m_{t_1} z_1}{2}, \tag{7.90}$$

$$r_2 = \frac{m_{t_2} z_2}{2}. \tag{7.91}$$

In the normal section we obtain the same non-standard modulus, which is connected to the moduli m_{t_1} and m_{t_2} by:

$$m = m_{t_1} \cos \beta_1 = m_{t_2} \cos \beta_2. \tag{7.92}$$

Figure 7.27 Helical toothing.

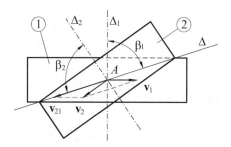

From (7.90)–(7.92)

$$r_1 = \frac{mz_1}{2\cos\beta_1}, \ r_2 = \frac{mz_2}{2\cos\beta_2}, \ \frac{r_2}{r_1} = \frac{z_2\cos\beta_1}{z_1\cos\beta_2}; \tag{7.93}$$

and the transmission ratio (7.89) becomes

$$i_{12} = \frac{z_2}{z_1}. \tag{7.94}$$

According to (7.93), the distance between the axes $a = r_1 + r_2$ is

$$a = \frac{m}{2}\left(\frac{z_1}{\cos\beta_1} + \frac{z_2}{\cos\beta_2}\right). \tag{7.95}$$

7.8.2 Cylindrical Worm and Wheel Toothing

The toothing between a cylindrical worm and a wheel may be considered as helical toothing having an angle between the axes of $\Sigma = \frac{\pi}{2}$; in addition the driving gear (the worm) must have a much smaller diameter than the driven gear (the worm wheel).

The worm may have multiple starts z_1, which represent the number of teeth of the worm, so according to (7.92), the transmission ratio is

$$i_{12} = \frac{z_2}{z_1}. \tag{7.96}$$

The flank of the worm's helix may or may not be a ruled surface. If it is, we obtain the representation in Figure 7.28.

The geometrical elements of the worm and wheel toothing are standardized[3]. The axial modulus m_s is standardized, while angle α_{0x} is equal to 20°.

We will denote by r_i, $i = 1, 2$, the radii of the divided cylinders, and by q the *diametral coefficient*

$$q = \frac{d_1}{m_s}, \tag{7.97}$$

The latter is also standardized at integer values between 7 and 16.

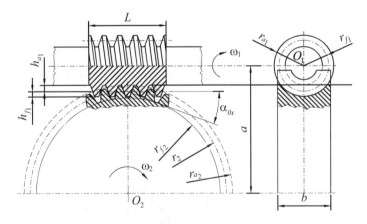

Figure 7.28 Worm and wheel toothing.

---

3 This statement is valid for Romania. For other countries the elements may be different.

With these remarks in mind and denoting by γ the angle of the reference helix, we obtain:

$$\tan \gamma = \frac{\pi m_s z_1}{2\pi r_1} = \frac{z_1}{q}, \tag{7.98}$$

$$h_{a_1} = m_s, \; h_{f_1} = 1.25 m_s, \tag{7.99}$$

$$r_{a_1} = \frac{m_s}{2}(q+2), \; r_{f_1} - \frac{m_s}{2}(q-2.5), \tag{7.100}$$

$$r_2 = \frac{m_s z_2}{2}, \; r_{a_2} = \frac{m_s}{2}(z_2+2), \; r_{f_2} = \frac{m_s}{2}(z_2 - 2.5), \tag{7.101}$$

$$a = 0.5(r_1 + r_2) = 0.5 m_s(q + z_2), \tag{7.102}$$

$$L = \begin{cases} (11 + 0.06 z_2) \, m_s, & \text{if } z_1 = 1 \text{ or } 2, \\ (12.5 + 0.09 z_2) \, m_s, & \text{if } z_1 = 3 \text{ or } 4, \end{cases} \tag{7.103}$$

$$b = \begin{cases} 0.75 d_{a_1} & \text{if } z_1 \leq 3, \\ 0.67 d_{a_1} & \text{if } z_1 = 4. \end{cases} \tag{7.104}$$

7.9 Generation of the Gears using a CAD Soft

7.9.1 Gear Tooth Manufacture

Gears are toothed by copying or by rolling. Copying can involve casting, milling with a disc cutter, cutting, or shaping with a pattern tool. Continuous-tooth formation is based on the toothing between a tool and a raw material which will be toothed. The tool may be a cylindrical gear without displacement, or a rack. The most commonly used profile for the teeth is the involute one. Some gears use teeth profiles of cycloids, circular arcs, or other shapes. The most used teeth are those with involute profiles.

7.9.2 Algorithm and AutoLisp Functions for Creating Gears from Solids

As we have already seen in Sections 6.2.4 and 6.6.5, the availability of Boolean operations permits composite solids to be designed in AutoCAD. We now use these facilities to create the gears from solids.

The procedure to obtain the toothing in AutoCAD is copied from the practical applications. We will use the most common procedure, namely rolling without sliding. This procedure involves toothing between a tool and a raw material that is to be toothed. The profiles in toothing cover cover equal circular arcs or distances. The tool may be a cylindrical gear or a rack.

In practice, the most used profile for the teeth of the gears is the involute one. We now present the five-step algorithm which we will use to obtain gears in AutoCAD:

1) Create the solid that will be the tool.
2) Place another solid (the raw material), which will become the gear, in a convenient position; it must be large enough to allow it to become the gear (there must be a sufficient stock of material).
3) Place the tool solid in contact with the raw material solid.
4) Give a constant rotational motion to the raw material solid; simultaneously ensure the displacement or the rotation of the tool solid.
5) Eliminates the tool solid from the raw material solid.

To obtain a gear using this algorithm we will have to perform minimum of 360 extractions of the tool from the raw material. Unlike the classic manufacturing procedure, in AutoCAD the gear is obtained after a single rotation, the stock being completely eliminated.

For the performing of the operations with solid previously presented, we wrote the following AutoLisp function:

```
(Defun C:Gear ()
(SetVar "Cmdecho" 0)
(Setq din (Getvar "Osmode"))
(Command "-Osnap" "Off" "Ortho" "Off" "Erase" "All" "")
(Data)
(Command "Zoom" "W" W1 W2)
(Generator_Rack)
(Positioning_of_Rack)
(Construction_of_Raw_Material)
(Operation_Elimination)
(While (< phi 360)
(Move_G_R)
(Operation_Elimination)
)
(Setvar "Cmdecho" 1 "Osmode" din)
)
```

The function has no local parameters or variables; it calls other functions to define the geometrical dimensions (*Data*), the construction of the solids for the tool (*Generator_Rack*), and the raw material (*Construction_of_Raw_Material*), the positioning of the tool (*Positioning_of_Rack*), the simultaneous displacement of the gear and rack (*Move_G_R*), and the elimination of the generator tool from the raw material (*Operation_Elimination*).

The function is called *Gear* and it successively:

- sets system variables and calls the function *Data*, which defines the main dimensions of the generator rack and gear
- sets up a visualization space and calls function *Generator_Rack* to construct the solid for the tool; this solid is called *Rack*
- positions *Rack* for toothing by calling function *Positioning_of_Rack*l; this gives, by copying, a new solid called *Extraction*
- constructs the raw material for the gear with the function *Construction_of_Raw_Material*; this new solid will be called *Raw_Material*
- extracts the solid *Extraction* from the solid *Raw_Material* using the function *Operation_Elimination*; the new solid will be called *Gear*
- in a repetitive *while* loop, gives degree-by-degree increments to angle φ in the interval (0, 2π); depending on angle φ, using the AutoLisp function *Move_G_R* it rotates and positions the solid for the gear, positions the solid for the generator rack, extracts the generator rack from the gear using the function *Operation_Elimination*, and finally moves to the next step.

Now we will present the AutoLisp functions used in the previous function, for different gear mechanisms.

7.9.3 Generation of the Cylindrical Gears with Spur and Inclined Teeth

In fabrication, the rack for generating spur or inclined teeth for cylindrical gears is called the *basic rack-type cutter*, the *tooth tool*, or the *Mag tooth*. It has the geometrical dimensions specified in Section 7.5.5.

In Figure 7.29 we show the characteristic points necessary to construct the teeth of the reference rack. The coordinates of the points depend on the modulus *m*; the geometrical elements

Figure 7.29 Characteristic points for the construction of the reference rack.

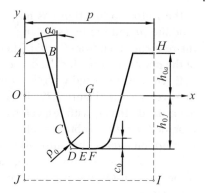

that define the reference rack are given by:

$$p = \pi m, \ \alpha_0 = 20°, \ h_{0a} = m, \ h_{0f} = 1.25m, \ \rho_0 = 0.38m, \ c_0 = 0.25m. \tag{7.105}$$

The coordinates of the points necessary to construct a tooth of the reference rack in Figure 7.29 are

$$x_A = 0, \ y_A = h_{0a}, \ x_B = \frac{p}{4} - h_{0a} \tan \alpha_0, \ y_B = y_{0a},$$

$$x_D = \frac{p}{4} + h_{0a} \tan \alpha_0, \ y_D = -h_{0f}, \ CD = \frac{\rho_0}{\cos \alpha_0} - \rho_0 \tan \alpha_0,$$

$$x_C = x_D - CD \sin \alpha_0, \ y_C = y_D + CD \cos \alpha_0, \tag{7.106}$$

$$x_E = x_C + \rho_0 \cos \alpha_0, \ y_E = y_D, \ x_F = \frac{p}{2}, \ y_F = y_D,$$

$$x_G = x_F, \ y_G = 0, \ x_H = p, \ y_H = h_{0a}, \ x_I = x_H, \ y_I = -2h_{0a}, \ x_J = x_A, \ y_J = y_I.$$

For the generator rack, the coordinates of point I are

$$x_I = x_H, \ y_I = 2h_{0a}. \tag{7.107}$$

For the construction in Figure 7.29 we wrote an AutoLisp function called *Construction_Tooth*, the code for which is as follows:

```
(Defun Construction_Tooth ()
(Setq ha m hf (* 1.25 m) hsmall (+ ha hf) radius (* 0.38 m) alpha0_rad
(* alpha0 (/ pi 180)) sinx (Sin alpha0_rad) cosx (Cos alpha0_rad) tan_alfa0
(/ sinx cosx) step (* Pi m) xA 0 yA ha A (List xA yA) xB (- (/ step 4)
(* tan_alpha0 ha)) yB ha B (List xB yB) xD (+ (/ pas 4) (* tan_alpha0 hf))
yD (* -1 hf) D (List xD yD) CD (- (/ radius (Cos alpha0_rad)) (* radius
tan_alpha0)) xC (- xD (* CD (Sin alpha0_rad))) yC (+ yD (* CD (Cos
alpha0_rad))) C (List xC yC) xE (+ xC (* radius (Cos alpha0_rad))) yE yD
E (List xE yE) xF (/ step 2) yF yD F (List xF yF) xG xF yG 0 G (List xG yG)
xH step yH (* -1 ha) H (List xH yH) xI xH yI (* -2 ha) I (List xI yI) xJ xA
yJ yI J (List xJ yJ) xK 0 yK yH K (List xK yK) )
(Command "Pline" A B C "A" E "L" F"")
(Setq curve1 (EntLast))
(Command "Mirror" curve1 "" F G "")
(Setq curve2 (EntLast))
(Command "Mirror" curve1 curve2 "" "0,0" "10,0" "Y")
(Command "Pline" H I J K "")
(Setq curve3 (EntLast))
(Command "Pedit" "M" curve1 curve2 curve3 "" "J" "" "")
(Setq Tooth2D (EntLast))
)
```

The AutoLisp function has no parameters or variables. In a first stage, depending on the modulus *m* and the pressure angle α_0, values are assigned to the geometrical dimensions of the reference rack from (7.105). Then, based on (7.106), values are assigned to the coordinates of the points required for the construction in Figure 7.29.

Three curves created with the command **Pline** contribute to the construction of the tooth. The first is *ABCF* (Figure 7.29); it is named *curve1*. The second is obtained by mirroring the first (command **Mirror**) with respect to axis *GF*. The resulting curve is called *curve2*. The third curve is named *curve3*. To obtain a solid from the three curves they are transformed in a single entity with option *Joint* of the AutoCAD command **Pedit**. The resulting entity is named *Tooth2D*.

If we wish to use the tooth for a generator rack, then *curve1* and *curve2* must be mirrored with respect to axis *Ox*; the originals are not kept. The generator rack is obtained with the function *Generator_Rack*.

```
(Defun Generator_Rack ()
(Construction_Tooth)
(Command "Extrude" Tooth2D "" (* 3 thickness) "0")
(Setq Tooth3D(EntLast))
(Command "Array" Tooth3D "" "R" "1" teeth step)
(Command "Union" "All" "")
(Setq Rack(Entlast))
)
```

This function calls function *Construction_Tooth* for the 2D construction of a tooth of the generator rack. After this, the tooth is transformed into a solid with the command **Extrude** and is given the name *Tooth3D*; the tooth is multiplied in the horizontal direction with the command **Array**. The resulting solids are united using the command **Union**, and the resulting solid is given the name *Rack*.

To use the generator rack to obtain the inclined toothing too, the extrusion height of the 2D tooth is equal to three times the thickness of the gear tooth. The data required for the previous constructions is specified in AutoLisp function *Data*. For example, for this construction only, it will have the following content:

```
(Defun Data ()
(Setq alpha0 20 m 1.5 teeth 10 thickness 20)
)
```

Calling the function will give the solid in Figure 7.5.

The positioning of the generator rack is a function of the solid for the raw material solid used to generate the gear. The raw material will be placed at point *O* of coordinates 0, 0, 0. According to (7.47), for a cylindrical gear, it will be a cylinder of radius $R_e = \frac{mz}{2} + (1 + x)\,m$ and height *g*. We denote by *z* the number of teeth and by *x* the coefficient of the profile's displacement.

Since the generator rack was constructed with point $O\,(0, 0, 0)$ as reference, it will have to be re-positioned to be used to obtain a gear; this re-positioning is via the AutoLisp function *Positioning_of_Rack*.

```
(Defun Positioning_of_Rack ()
(Setq OCapital(List 0 0 0) xOprime(*Step (/ teeth -2)) yOprime(* -1 (+
xm R)) zOprime(* thickness -1.5)Oprime(List xOprime yOprime zOprime))
(Command "Move" Rack "" OCapital Oprime)
(Setq P(List 0 0 (* thickness -0.5)))
)
```

The point $O'(x_{O'}, y_{O'}, z_{O'})$ is the point at which the rack is re-positioned because point $O(0,0,0)$ is used as the reference point for the positioning of the raw material. The generator rack (*Rack*) has a sufficient number of teeth (*teeth*) for the complete generation of the gear; its thickness is three times the thickness of the gear (*thickness*) to permit the toothing of gears with spur or inclined teeth. It is placed with the reference line tangential to the divided circle for non-displaced toothing, or it is moved with the parameter *xm* for displaced toothing. The coordinates of point O' are

$$x_{O'} = -\pi m\frac{n}{2}, \; y_{O'} = -(R+xm), \; z_{O'} = -\frac{3g}{2}, \tag{7.108}$$

where n stands for the number of teeth of the generator rack and g is the thickness of the gear. In this way, the generator rack is positioned at the middle of the gear. We denote with P the point of coordinates $0, 0, -\frac{g}{2}$.

The construction of the raw material is done using the following function:

```
(Defun Construction_of_Raw_Material ()
(Command "Cylinder" P Re thickness)
(Setq Raw_Material(Entlast))
)
```

The cylinder is placed at point P, and has exterior radius $R_e = \frac{mz}{2} + (1+x)m$ and thickness g. The axes of coordinates OX, OY, and OZ are the central principal axes of inertia of the cylinder.

With the AutoLisp function *Operation_Elimination*, the solid for the generator rack is extracted from the raw material. To preserve the generator rack before the elimination (the AutoCAD command **Subtract**) a copy of the generator rack is made and is named *Extraction*.

For toothing with inclined teeth the copy of the generator rack is rotated with angle β of inclination of the toothing. The resulting solid is given the name *Raw_Material*. The content of the function is:

```
(Defun Operation_Elimination ()
(Setq C(list ssmall 0))
(Command "Copy" Rack "" "0,0" C)
(Setq Extraction(Entlast))
(If (/= beta 0)
(Command "Rotate3D" Extraction "" "Y" "0,0,0" beta)
)
(Command "Subtract" Raw_Material "" Extraction "")
(Setq Raw_Material(Entlast))
)
```

Depending on the rotation angle φ in the repetitive *while* loop, the raw material is positioned with the AutoLisp function *Move_G_R*, the displacement s (*ssmall*) of the generator rack is calculated, before passing to a new angle by adding the angular step $\Delta\varphi$ (*ang_step*) to the current angle φ. The contents of the function are given below:

```
(Defun Move_G_R ()
(Command "Rotate" Raw_Material "" P (* -1 ang_step))
(Setq phi(+ phi ang_step))
(Setq ssmall(- ssmall lin_step))
(If (< ssmall (* -1 pi m))
(Setq ssmall(+ ssmall (* pi m)))
)
)
```

Figure 7.30 The solid for a gear with inclined teeth.

All of these functions called by the function *Gear* permit the construction of the gears with spur and inclined teeth. As an example, consider the construction of a gear with the following data: modulus $m = 3$ mm, the pressure angle of the reference rack $\alpha_0 = 20°$, the number of teeth of the gear $z = 20$, the displacement of profile $x = 0.5$, the thickness of the gear $g = 20$ mm, the angle of inclination of the toothing $\beta = 15°$, and the toothing is performed with an angular step $\Delta\varphi = 1°$ in the interval $\varphi \in [0°, 360°]$. The condition of rolling without sliding means that a rotation of $1°$ of the gear corresponds to a linear displacement $\Delta s = R\frac{\pi}{180}$ of the generator rack. For this gear, the function *Data* is as follows

```
(Defun Data ()
(Setq alpha0 20 z 30 m 3.0 xm(* 0.5 m) teeth 10 thickness 20 beta 15
ang_step 1.0 phi 0.0 ssmall 0.0 R(* m z 0.5) Re(+ R m xm) lin_step(* R
(/ Pi 180)) W1(List (* -2 R) (* -1.5 R)) W2(List (* 2 R) (* 1.5 R)))
)
```

Running the function gives the solid in Figure 7.30. The construction of the solid was completed with:

- a hole of assembly for an axle of diameter $\Phi = 52$ mm
- a notch for a parallel chock $16 \times 10 \times 56$
- 6 equidistant holes with $\Phi = 35$ mm, situated on a diameter $\Phi = 135$ mm
- 2 disks for a smaller weight, symmetrically situated, with interior diameter $\Phi = 92$ mm, exterior diameter $\Phi = 182$ mm, and width 10 mm.

7.9.4 The Generation of the Cylindrical Gears with Curvilinear Teeth

The algorithm in the last subsection for the generation of the gears with spur or inclined teeth can also be used for gears with curvilinear teeth; the only difference is the function used for creation of the generator rack.

For the construction of the solid for a generator rack with curvilinear teeth, the following are known:

- *m*, the modulus of the gear
- the dimensions of the reference rack with spur teeth
- *g*, the thickness of the gear
- *R*, the curvature radius of the teeth.

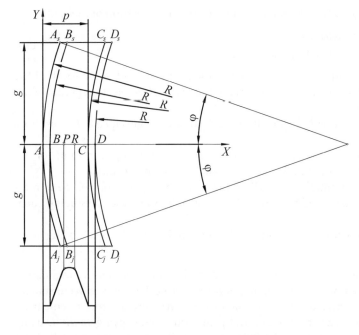

Figure 7.31 The planar construction for a tooth of a generator rack with curvilinear teeth.

The coordinates of the points required for the planar construction of a curvilinear tooth of the reference rack in Figure 7.31 are

$$x_A = 0, \ x_B = \frac{p}{4} - h_{0a}\tan\alpha_0, \ x_C = p - x_B, \ x_D = p, \ \varphi = \arcsin\left(\frac{g}{R}\right),$$

$$A_S\left(x_{A_S}, y_{A_S}\right) : \ x_{A_S} = R - R\cos\varphi, \ y_{A_S} = g,$$

$$A_J\left(x_{A_J}, y_{A_J}\right) : \ x_{A_J} = x_{A_S}, \ y_{A_J} = -g,$$

$$B_S\left(x_{B_S}, y_{B_S}\right) : \ x_{B_S} = x_{A_S} + x_B, \ y_{B_S} = g,$$

$$B_J\left(x_{B_J}, y_{B_J}\right) : \ x_{B_J} = x_{B_S}, \ y_{B_J} = -g, \qquad\qquad (7.109)$$

$$C_S\left(x_{C_S}, y_{C_S}\right) : \ x_{C_S} = x_{A_S} + x_C, \ y_{C_S} = g,$$

$$C_J\left(x_{C_J}, y_{C_J}\right) : \ x_{C_J} = x_{C_S}, \ y_{C_J} = -g,$$

$$D_S\left(x_{D_S}, y_{D_S}\right) : \ x_{D_S} = x_{A_S} + x_D, \ y_{D_S} = g,$$

$$D_J\left(x_{D_J}, y_{D_J}\right) : \ x_{D_J} = x_{D_S}, \ y_{D_J} = -g.$$

Based on the relations (7.109), the construction is made as follows:

1) With the command **Pline**, construct the circular arc A_1A_2 and multiply it at points B, C, and D; A is the reference point.
2) With the command **Pline**, construct the segments of straight lines B_SC_S, B_JC_J, A_SD_S, and A_JD_J.
3) Using the option *Joint* of the command **Pedit**, create the closed loop $B_SB_JC_JC_SB_S$.
4) Using the option *Joint* of the command **Pedit**, create the closed loop $A_SA_JD_JD_SA_S$.
5) With the AutoCAD command **Extrude**, extrude the loop $B_SB_JC_JC_SB_S$, with height h equal to that of the tooth and extrusion angle equal to α_0, givings the raw contour (without nose radius) of the tooth of the generator rack.

6) Calculate the coordinates of points P and R, resulting from the extrusion of the tooth

$$P\left(x_P, y_P\right) : \quad x_P = x_B + h\tan\alpha_0, \; y_P = 0,$$
$$R\left(x_R, y_R\right) : \quad x_R = x_C - h\tan\alpha_0, \; y_R = 0;$$

(7.110)

7) With the command **Fillet** and the radius ρ_{0f}, create the rounded head of the tooth of the future generator rack with curvilinear teeth using points P and R as selection points of the edges that will be rounded.

8) Extrude the loop $A_S A_J D_J D_S A_S$ with the height m and extrusion angle equal to $0°$, resulting in the base of the tooth.

9) With the command **Union**, unite the solids from steps 5 and 8 to give a tooth of the generator rack with curvilinear teeth; the tooth will have the name *Tooth3D*.

10) With the command **Array**, multiply *Tooth3D* 10 times; the multiplication will be in rectangular form, with one column and ten rows, the distance between the rows being equal to the step $p = \pi m$.

11) Using the command **Union**, unite the 10 solids obtained in Step 10, the result being given the name *Rack*.

This gives the solid shown in Figure 7.32. For inclined teeth, if one gear has an inclination angle of β, then the conjugate gear will have an inclination angle of $-\beta$. For gears with curvilinear teeth, we need to construct a generator rack with the inverse curvature in order to construct the conjugate gear. This is done in a very simple way, by mirroring and copying *Tooth3D* using the command **Mirror3D**; the mirror plane is YOZ, situated in the middle of the tooth. After mirroring the original is discarded. The result is the solid in Figure 7.33.

The AutoLisp function for constructing racks with curvilinear teeth using the algorithm outlined in this section can be found in the companion website of the book.

Figure 7.34 shows the toothing for two gears with curvilinear teeth for which: $m = 10\,\text{mm}$, $z_1 = z_2 = 30$ teeth, $\alpha_0 = 20°$, $x_1 m = x_2 m = 0$, $g = 90\,\text{mm}$, and $R = 3g$, obtained using the algorithm just described.

Figure 7.32 Solid for a generator rack with curvilinear teeth.

Figure 7.33 Solid for a generator rack with curvilinear teeth, as required for conjugate gears.

Figure 7.34 The solids for toothing of two gears with curvilinear teeth.

Figure 7.35 The toothing of conical gears: (a) pyramidal; (b) prismatic.

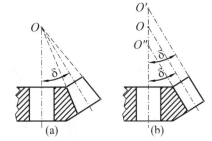

7.9.5 The Generation of Conical Gears with Spur Teeth

The method of rolling without sliding is used in classical toothing procedures for conical gears too. The procedure involves toothing between an imaginary planar gear and a raw material that approximates the gear to be constructed.

The teeth of conical gears may be truncated pyramids (pyramidal toothing, Figure 7.35a) or may have constant height (Figure 7.35b); in the latter case the toothing is referred to as *prismatic* or *of constant height*.

For pyramidal toothing, the clearance between the head of the tooth of one gear and the foot of the blank of the tooth of the complementary gear decreases along the tooth; this may lead to the blocking of the toothing. For prismatic toothing, the clearance at the foot of the blank remains constant along the tooth.

No matter the shape of the tooth, the cutting procedure assumes a planar gear with a number of teeth equal to z_g, which results from the equality

$$\frac{mz_g}{2} = \frac{mz_{1,2}}{2 \sin \delta_{1,2}}, \tag{7.111}$$

This leads to

$$z_g = \frac{z_{1,2}}{\sin \delta_{1,2}}, \tag{7.112}$$

where $z_{1,2}$ stands for the numbers of teeth of the conical gears, and $\delta_{1,2}$ are the angles of the divided cones. In practice, gears with the numbers of teeth z_1 and z_2 are cut using the same planar gear.

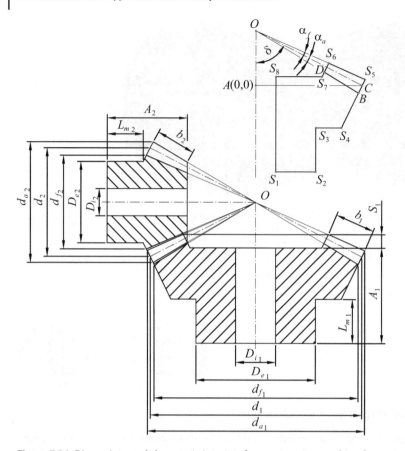

Figure 7.36 Dimensions and characteristic points for constructing toothing for conical gears.

To construct the solids that will be the conical gears, we use the same AutoLisp function *Gear*, which was described in Section 7.9.2. We will also use other AutoLisp functions for the solids for the raw material, the generator rack (in this case a planar gear), the positioning of the rack, and the simultaneous displacements of the raw material and generator rack.

Based on the notation in Figure 7.36, we have written an AutoLisp function for the construction of the raw material. The code is set out in the companion website of the book. This function gives the solids for the raw materials used in the pyramidal toothing. The following are known: the modulus m, the numbers of teeth of the gears z_1, z_2, and the width b of the toothing. For a gear, we calculate (see Figure 7.26)

$$i_{12} = \frac{z_2}{z_1}, \ h_a = m, \ h_f = 1.25m, \ h = h_a + h_f. \tag{7.113}$$

We determine the angle of the divided cone

$$\delta_1 = \arctan\left(\frac{1}{i_{12}}\right), \ \left(i_{12} = \frac{z_2}{z_1} = \frac{\sin\delta_2}{\sin\delta_1}\frac{\sin\delta_2}{\sin\delta_1} = \frac{\cos\delta_1}{\cos\delta_2} = i_{12}\right), \tag{7.114}$$

and the geometric elements of the gear $(\delta = \delta_1)$

$$r = mz, \ r_a = r + h_a\cos\delta, \ r_f = r - h_f\cos\delta, \ L = OC = \frac{r}{\sin\delta},$$

$$\alpha_a = \arctan\left(\frac{h_a}{L}\right), \ \alpha_f = \arctan\left(\frac{h_f}{L}\right) \tag{7.115}$$

and calculate the coordinates of the points

$$x_A = 0, \ y_A = 0, \ x_B = r_f, \ y_B = -h_f \sin\delta, \ x_C = r, \ y_C = 0,$$
$$x_{S_5} = x_C + h_a \cos\delta, \ y_{S_5} = h_a \sin\delta,$$
$$x_{S_6} = x_{S_5} + \frac{b}{\cos\alpha_a}\cos\left(\delta + \frac{\pi}{2} + \alpha_a\right), \ y_{S_6} = y_{S_5} + \frac{b}{\cos\alpha_a}\sin\left(\delta + \frac{\pi}{2} + \alpha_a\right),$$
$$x_D = x_B + \frac{b}{\cos\alpha_f}\cos\left(\delta + \frac{\pi}{2} - \alpha_f\right), \ y_D = y_B + \frac{b}{\cos\alpha_f}\sin\left(\delta + \frac{\pi}{2} - \alpha_f\right),$$
$$DS_7 = \frac{S}{\sin\delta} - h + b\left(\tan\alpha_a + \tan\alpha_f\right), \tag{7.116}$$
$$x_{S_7} = x_D - DS_7\cos\delta, \ y_{S_7} = y_D - DS_7\sin\delta,$$
$$x_{S_8} = \frac{D_i}{2}, \ y_{S_8} = y_{S_7}, \ x_{S_1} = x_{S_8}, \ y_{S_1} = y_{S_8} - A, \ x_{S_2} = \frac{D_e}{2}, \ y_{S_2} = y_{S_1},$$
$$x_{S_3} = x_{S_2}, \ y_{S_3} = y_{S_2} - L_m, \ S_4B = \frac{y_{S_5} - y_{S_3}}{\sin\delta} - h,$$
$$x_{S_4} = x_{S_5} - \left(S_4B + h\right)\cos\delta, \ y_{S_4} = y_{S_3},$$
$$\text{If } x_{S_4} \le x_{S_3} \text{ then } x_{S_4} = x_{S_3}.$$

The AutoLisp function called *Construction_of_Raw_Material*, the code for which is shown in the companion website of the book, uses (7.116) and calculates the coordinates of the points necessary to obtain the closed loop $S_1S_2S_3S_4S_5S_6S_7S_8$, which is given the name *Curve*. The solid is obtained by rotating the curve $S_1S_2S_3S_4S_5S_6S_7S_8$ around axis OY with the command **Revolve**. At the end of the function the obtained solid is given the name *Raw_Material*. For the raw material for prismatic toothing, a process similar to that used for pyramidal toothing is followed; only the coordinates of points S_6 and S_7 change.

Next, we will have to obtain a solid for the planar gear; this is done using the AutoLisp function *Generator_Rack*. In a first stage we create a solid that represents a tooth, and then duplicate it. To create a tooth of the planar gear we use the notation in Figure 7.37. Based on this, the coordinates of the points in Figure 7.37, relative to the reference system with its origin at point $O(0,0)$, are:

$$x_{P_1} = (L - b)\cos\left(\frac{\varphi_C}{2}\right), \ y_{P_1} = (L - b)\sin\left(\frac{\varphi_C}{2}\right),$$
$$x_{P_2} = L\cos\left(\frac{\varphi_C}{2}\right), \ , y_{P_2} = L\sin\left(\frac{\varphi_C}{2}\right), \tag{7.117}$$
$$x_{P_3} = x_{P_2}, \ y_{P_3} = -y_{P_2}, \ x_{P_4} = x_{P_1}, \ y_{P_4} = -y_{P_1},$$

while the angles φ_r, φ_C, α_1, and α_2 have the expressions

$$\varphi_r = \frac{2\pi}{z}, \ \varphi_C = \varphi_r\sin\delta, \ \alpha_1 = \varphi_C\left(\frac{1}{4} - \frac{h_a}{p}\tan\alpha_0\right), \ \alpha_2 = \arctan\left(\frac{h}{L}\right). \tag{7.118}$$

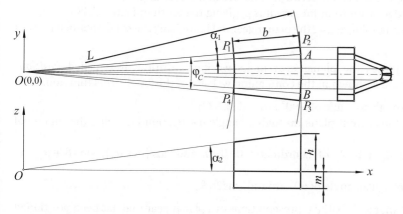

Figure 7.37 Characteristic points of a planar gear.

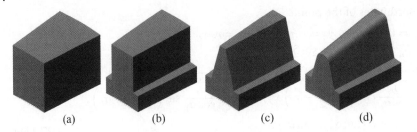

(a)   (b)   (c)   (d)

Figure 7.38 The stages of the construction of a tooth for a planar gear: (a) step 1; (b) step 2; (c) step (4); (d) step 5.

Figure 7.39 Planar gear.

These expressions are encapsulated in the function *Generator_Rack*, the code for which is shown in the companion website of the book. The construction following these steps:

1) Construct the close poly-line $P_1 P_2 O_3 P_4 P_1$. This is extruded to height $h + m$ (h for the height of the tooth, and m for the base of the tooth), giving the solid in Figure 7.38a; the resulting solid is moved a distance $-m$ in the direction of axis OZ.
2) Construct two parallelepiped solids. Using these, eliminate the lateral portions bordering the tooth of the planar gear, giving the solid in Figure 7.38b; at this point, the height of the tooth is constant.
3) Create the lateral flanks of the tooth by eliminating two parallelepiped solids of the same dimensions as the ones in step 2 and inclined at angles α_0 and $-\alpha_0$, respectively.
4) Using a parallelepiped solid inclined at an angle φ_2, create the final shape of the tooth as in Figure 7.38c; this operation is not performed for prismatic toothing.
5) Round the upper edges with roundness radius ρ, giving the tooth in Figure 7.38d.
6) Duplicate the resulting solid, giving the planar gear in Figure 7.39; the solid obtained is given the name *Rack*.

The next function places the planar gear in its working position. It is called *Positioning_of_Rack* and the code is shown in the companion website of this book. This function:

- rotates the solid through an angle of $-90°$ around axis Oy
- rotates the solid in the current plane through an angle $\delta + \alpha_a$, point O being the rotational centre
- defines two points R_1 and R_2 of coordinates 0, 0, 0, and $\cos(\delta + \alpha_a)$, $\sin(\delta + \alpha_a)$, 0, respectively
- rotates the solid through an angle of $90°$ around axis $R_1 R_2$.

Figure 7.40 shows planar and spatial representations of a planar gear that has been positioned for the cutting process; it is in contact with the raw material that will become the new gear.

Figure 7.40 The placement of the planar gear relative to the raw material: (a) planar representation; (b) spatial representation.

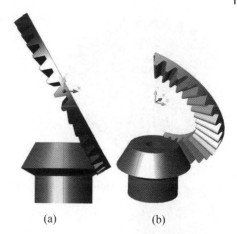

(a)　　　　　　　　　(b)

Figure 7.41 Design of a conical gear with pyramidal toothing.

To eliminate the material of the planar gear from the solid for the raw material, the function *Operation_Elimination* is used. The code is shown in the companion website of the book. The function copies the solid for the gear, assigns the name *Extraction* to the new solid, and eliminates it from the solid *Raw_Material*.

The companion website of this book also includes code for the function *Move_G_R*. This function is used in a *while* loop and it performs, in i ($i \in [1, 360]$) steps, the rotation of the raw material for the gear through an angle $\varphi_i = \varphi_{i-1} + 1°$, around axis Oy, and the rotation of a copy of the planar gear around axis R_1R_2, through an angle $\alpha_i = \varphi_i \sin \delta$.

After 360 steps are complete, the conical gear in Figure 7.41 is obtained. In the figure we have retained the solid *Rack*; in other words, the planar gear.

We construct the solids for two gears with spur teeth, of dimensions (Figure 7.36):

$$m = 2\,\text{mm},\ z_1 = 32\ \text{teeth},\ z_2 = 16\ \text{teeth},\ D_{i_1} = 12\,\text{mm},\ D_{i_2} = 8\,\text{mm},$$
$$D_{m_1} = 36\,\text{mm},\ D_{m_2} = 24\,\text{mm},\ b_1 = b_2 = 8\,\text{mm},\ L_{m_1} = 13\,\text{mm},\ L_{m_2} = 11\,\text{mm},$$
$$A_1 = 28.08\,\text{mm},\ A_2 = 24.56\,\text{mm},\ S_1 = 4\,\text{mm},\ S_2 = 0\,\text{mm}.$$

These values are written, separately for each gear, using the function *Data* (see the companion website of the book); the function also contains the values of the pressure angle (20°) and the angular step *ang_step 1.0*. We call the function twice. For a gear with z_1 teeth, we obtain Figure 7.42, while for a gear with z_2 teeth, we obtain the representation in Figure 7.43. The toothing of the two gears is shown in Figure 7.44.

Figure 7.42 The z_1 gear.

Figure 7.43 The z_2 gear.

Figure 7.44 The conical toothing $z_1 - z_2$.

To obtain a toothing with inclined or curvilinear teeth we only have to modify the AutoLisp function *Planar_Gear* (see the companion website of the book). The modification implies only the inclination of the solid in Figure 7.37 by an angle β, or another shape for the tooth.

In conclusion, we may say that the methods outlined here are identical to classical methods for obtaining gears using mechanical processes. The algorithm to obtain the solids for the gears, accompanied by a programming language is a general one and may be applied to any toothing, no matter the shape of the gears. The principle of the method is based on the ability of CAD software to use Boolean operations (addition, subtraction, intersection etc.) on solids. The conjugate gear can thus be obtained by extracting the solid representing the tool or the generator gear from the solid representing the raw material. Because of the huge volume of operations involved, the operation is performed using the programming language AutoLisp. In fact, with this method, virtual gears are obtained; the process is identical to that used on a traditional machine-tool for manufacture of gears.

The AutoLisp functions we have created and presented in the companion website of the book may be used, with small modifications, to obtain other toothing, such as helical or cylindrical worm and wheel toothing. For instance, for cylindrical worm and wheel, the solid which represents the worm is obtained first. It is then used as the tool for obtaining the wheel. To reduce the time required to obtain the wheel, a single tooth is generated and is then duplicated around a pole.

7.10   Kinematics of Gear Mechanisms with Parallel Axes

7.10.1   Gear Mechanisms with Fixed Parallel Axes

Determination of the angular velocities in gear mechanisms with fixed parallel axes involves the transmission ratio for a single toothing with two parallel axes; this transmission ratio is given, for exterior toothing, by:

$$i_{12} = -\frac{z_2}{z_1} = \frac{\omega_1}{\omega_2} \tag{7.119}$$

and by:

$$i_{12} = \frac{z_2}{z_1} = \frac{\omega_1}{\omega_2} \tag{7.120}$$

for interior toothing. For instance, for the gear mechanism in Figure 7.45 we have four fixed axes and three points of toothing (A, B, and C). Denoting by z_1, and z_4 the numbers of teeth of the gears rigidly linked to axes 1 and 4, respectively, and by z_2', z_2'' and z_3', z_3'' the numbers of teeth of the gears rigidly linked to axes 2 and 3, respectively, at the contact points A, B, and C, we obtain:

$$\frac{\omega_1}{\omega_2} = -\frac{z_1}{z_2}, \; \frac{\omega_3}{\omega_2} = -\frac{z_2''}{z_3'}, \; \frac{\omega_4}{\omega_3} = -\frac{z_3''}{z_4}. \tag{7.121}$$

Multiplying (7.121), we obtain an expression from which the angular velocity ω_4 and its sense (which is identical to that of angular velocity ω_1) can be determined, namely:

$$\frac{\omega_4}{\omega_1} = \frac{z_1 z_2'' z_3''}{z_2' z_3' z_4}. \tag{7.122}$$

7.10.2   The Willis Method

For gear mechanisms with mobile axes (planetary mechanisms), the motion relative to the port-satellite (Figure 7.46) must be examined:

Figure 7.45 Gear mechanism with four fixed axes and three points of toothing.

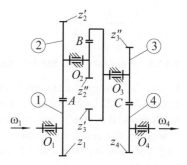

Figure 7.46 Motion relative to to the port-satellite.

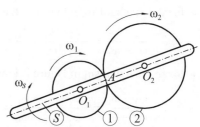

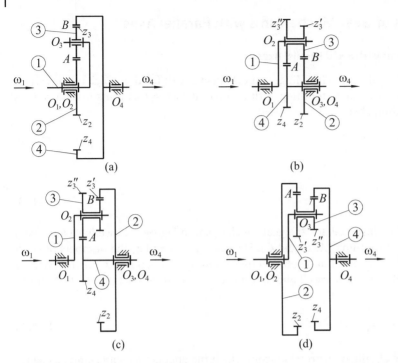

Figure 7.47 Differential planetary mechanisms with four mobile elements: (a) planetary mechanism with simple satellite; (b)–(d) planetary mechanisms with double satellite.

If ω_1 and ω_2 are the absolute angular velocities of the gears and ω_S is the absolute angular velocity of the port-satellite, then the relative angular velocities of the gears with respect to the port-satellite are $\omega_1 - \omega_S$ and $\omega_2 - \omega_S$. Examining the exterior toothing at point A gives

$$\frac{\omega_2 - \omega_S}{\omega_1 - \omega_S} = -\frac{z_1}{z_2}. \tag{7.123}$$

For interior toothing, we have

$$\frac{\omega_2 - \omega_S}{\omega_1 - \omega_S} = \frac{z_1}{z_2}. \tag{7.124}$$

7.10.3 Planetary Gear Mechanisms with Four Elements

Figure 7.47 shows the schemata of various differential planetary gear mechanisms (of mobility degree $M \geq 2$) with four elements. Knowing the angular velocities ω_1, ω_2, and the numbers of teeth marked in the figures, we want to determine the angular velocity ω_4.

Writing the relations between the angular velocities and the numbers of teeth at points A and B of toothing (element 1 being the port-satellite), we successively obtain:

- for the mechanism in Figure 7.47a:

$$\frac{\omega_3 - \omega_1}{\omega_2 - \omega_1} = -\frac{z_2}{z_3}, \quad \frac{\omega_4 - \omega_1}{\omega_3 - \omega_1} = \frac{z_3}{z_4}, \quad \omega_4 = \omega_1 - \frac{z_2}{z_4}(\omega_2 - \omega_1); \tag{7.125}$$

- for the mechanism in Figure 7.47b:

$$\frac{\omega_3 - \omega_1}{\omega_2 - \omega_1} = -\frac{z_2}{z_3'}, \quad \frac{\omega_4 - \omega_1}{\omega_3 - \omega_1} = -\frac{z_3''}{z_4}, \quad \omega_4 = \omega_1 + \frac{z_2 z_3''}{z_1' z_4}(\omega_2 - \omega_1); \tag{7.126}$$

- for the mechanism in Figure 7.47c:

$$\frac{\omega_3 - \omega_1}{\omega_2 - \omega_1} = \frac{z_2}{z_3'}, \quad \frac{\omega_4 - \omega_1}{\omega_3 - \omega_1} = -\frac{z_3''}{z_4}, \quad \omega_4 = \omega_1 - \frac{z_2 z_3''}{z_1' z_4}\left(\omega_2 - \omega_1\right); \tag{7.127}$$

- for the mechanism in Figure 7.47d:

$$\frac{\omega_3 - \omega_1}{\omega_2 - \omega_1} = \frac{z_2}{z_3'}, \quad \frac{\omega_4 - \omega_1}{\omega_3 - \omega_1} = \frac{z_3''}{z_4}, \quad \omega_4 = \omega_1 + \frac{z_2 z_3''}{z_1' z_4}\left(\omega_2 - \omega_1\right). \tag{7.128}$$

It is interesting to observe that for the planetary gear mechanism in Figure 7.47b if $\omega_2 = 0$ and the numbers of teeth are $z_2 = 59$, $z_3' = 60$, $z_3'' = 59$, and $z_4 = 60$, then a gear mechanism is obtained, the transmission ratio for which is

$$i_{14} = \frac{\omega_1}{\omega_4} = 3600. \tag{7.129}$$

7.10.4 Planetary Gear Mechanisms with Six Mobile Elements

Figure 7.48 shows the kinematic schema of a mono-mobile gear mechanism with six mobile elements. We want to determine the angular velocity ω_6 as a function of the angular velocity ω_1 and the numbers of teeth.

At points A and B we have toothing with fixed axes, so

$$\frac{\omega_6}{\omega_5} = -\frac{z_5}{z_6'}, \quad \frac{\omega_4}{\omega_6} = -\frac{z_6''}{z_4'} \Rightarrow \omega_4 = -\frac{z_6''}{z_4'}\omega_6, \quad \omega_5 = -\frac{z_6'}{z_5}\omega_6. \tag{7.130}$$

At point E there is another toothing with fixed axes, so

$$\frac{\omega_1}{\omega_2} = -\frac{z_2''}{z_1} \Rightarrow \omega_2 = -\frac{z_1}{z_2''}\omega_1.$$

For the toothing with mobile axes at points C and D, element 5 being the port-satellite, this gives

$$\frac{\omega_4 - \omega_5}{\omega_3 - \omega_5} = -\frac{z_3'}{z_4''}, \quad \frac{\omega_3 - \omega_5}{\omega_2 - \omega_5} = -\frac{z_2'}{z_3''} \Rightarrow \frac{\omega_4 - \omega_5}{\omega_2 - \omega_5} = \frac{z_2' z_3'}{z_4'' z_3''}. \tag{7.131}$$

Thenn, taking into account (7.130) and (7.131), we obtain

$$\omega_6 = \frac{\dfrac{z_1}{z_2''} \cdot \dfrac{z_2'}{z_3''} \cdot \dfrac{z_3'}{z_4''}}{\dfrac{z_6''}{z_4'} - \dfrac{z_6'}{z_5} + \dfrac{z_2' z_3' z_6'}{z_3'' z_4'' z_5}}\omega_1. \tag{7.132}$$

Figure 7.48 Planetary gear mechanism with six elements.

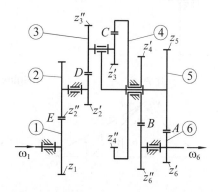

7.11 Kinematics of Mechanisms with Conical Gears

7.11.1 Planetary Transmission with Three Elements

Figure 7.49 shows the kinematic schema of a conical gear mechanism for which the angular velocities of elements 1 and 2 are known. We want to determine the angular velocity of element 3. Expressing the angular velocity of element 3 as

$$\underline{\underline{\omega_3}} = \underline{\underline{\omega_1}} + \underbrace{\underline{\omega_{31}}}_{\parallel\, OA} = \underline{\underline{\omega_2}} + \underbrace{\underline{\omega_{32}}}_{\parallel\, OB} \tag{7.133}$$

and representing the vector relation (7.133), we obtain the vector polygon in Figure 7.50.

The expression for the relative angular velocity is given by:

$$\omega_{32} = \frac{\omega_2 - \omega_1}{\sin \beta}\sin(\alpha + \beta), \tag{7.134}$$

and the angular velocity ω_3 is

$$\omega_3 = \sqrt{\omega_2^2 + \omega_{32}^2 - 2\omega_2\omega_{32}\cos\alpha}. \tag{7.135}$$

7.11.2 Planetary Transmission with Four Elements

If an element is added to the planetary gear mechanism in Figure 7.49, a planetary gear mechanism with *four* elements is obtained. Its kinematic schema is shown in Figure 7.51. The angular velocities of elements 1 and 2 are known. We want to determine the angular velocity ω_4 of element 4.

The angular velocity of element 3 is determined using (7.133), while the angular velocity of element 4 is given by:

$$\underbrace{\underline{\omega_4}}_{\parallel\, OA} = \underline{\underline{\omega_3}} + \underbrace{\underline{\omega_{43}}}_{\parallel\, OD}, \tag{7.136}$$

This is shown in Figure 7.52.

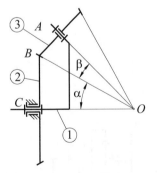

Figure 7.49 Conical gear mechanism with three elements.

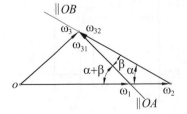

Figure 7.50 The polygon of angular velocities.

Figure 7.51 Conical planetary gear mechanism with four elements.

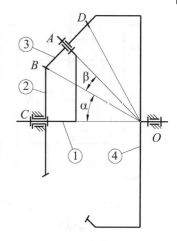

Figure 7.52 The polygon of angular velocities.

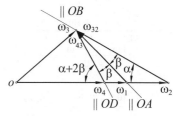

We successively obtain

$$\omega_{31} = \frac{\omega_2 - \omega_1}{\sin \beta} \sin \alpha, \qquad (7.137)$$

$$\omega_1 - \omega_4 = \frac{\sin \beta}{\sin (\alpha + 2\beta)} \omega_{32}. \qquad (7.138)$$

Then, knowing that

$$\frac{\sin \alpha}{\sin \beta} = \frac{z_2}{z_3}, \quad \frac{\sin \beta}{\sin (\alpha + 2\beta)} = \frac{z_3}{z_4}, \qquad (7.139)$$

we obtain the expression for the relative angular velocity ω_{31},

$$\omega_{31} = \frac{z_2}{z_3} \left(\omega_2 - \omega_1 \right); \qquad (7.140)$$

Finally, the expression for the angular velocity of element 4 is

$$\omega_4 = \omega_1 - \frac{z_2}{z_4} \left(\omega_2 - \omega_1 \right). \qquad (7.141)$$

7.11.3 Automotive Differentials

A differential with the kinematic schema shown in Figure 7.53 consists of planetary gear pinions 1 and 2, satellite gear pinions 3 and 3′, and the box of the differential 4. It is assembled between the main transmission (entry gear pinion 6 and the gear crown of differential 5) and the wheels 7 and 8 of the motor car; the gear differential allows there to be different angular velocities in the driven wheels.

Performing the kinematic analysis gives expressions for the angular velocities of elements 3 and 4:

$$\underline{\underline{\omega_3}} = \underline{\underline{\omega_1}} + \underline{\underline{\omega_{31}}} = \underline{\underline{\omega_2}} + \underline{\underline{\omega_{32}}},$$
$$ \| OA \| OB \qquad (7.142)$$

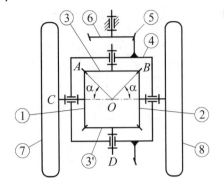

Figure 7.53 Motor car differential.

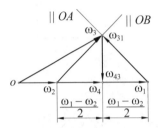

Figure 7.54 The polygon of angular velocities.

$$\underset{\parallel OC}{\underline{\omega_4}} = \underset{}{\underline{\underline{\omega_3}}} + \underset{\parallel OD}{\underline{\omega_{43}}} \,, \tag{7.143}$$

This is shown in Figure 7.54; in the figure we have assumed that the differential is symmetric ($z_1 = z_2, z_3 = z_{3'}$). Taking into account the notation in Figure 7.54, we obtain the expression for the relative angular velocity ω_{43},

$$\omega_{43} = \frac{\omega_1 - \omega_2}{2} \tan \alpha \tag{7.144}$$

and the expression for the angular velocity of element 4

$$\omega_4 = \frac{\omega_2 + \omega_1}{2}. \tag{7.145}$$

Using (7.144) and (7.145) we may examine the functioning of a differential gear mechanism in a motor car. For instance, when the car is travelling in a straight line, the angular velocities of planetary gear pinions 1 and 2 are equal ($\omega_1 = \omega_2$) and we get

$$\omega_{43} = 0, \ \omega_1 = \omega_2 = \omega_4. \tag{7.146}$$

It can be seen that there is no relative movement between the satellite pinions and the box of the gear differential ($\omega_{43} = 0$) and that the angular velocities of the driving wheels, which are equal to one another, are also equal to the angular velocity of the box of the gear differential.

When one wheel has null angular velocity ($\omega_1 = 0$), we get

$$\omega_2 = 2\omega_4; \tag{7.147}$$

so the angular velocity of the wheel that slides is twice the angular velocity of the box of the gear differential.

When the box of the gear differential is blocked ($\omega_4 = 0$), we obtain

$$\omega_1 = -\omega_2, \tag{7.148}$$

In other words, the angular velocities of the wheels have opposite sense and equal moduli.

8

Spatial Mechanisms

In this chapter we will examine some common spatial mechanisms and their kinematic analysis using some of their simple geometric properties.

8.1 Kinematics of Spatial Mechanisms: Generalities

The creation of modern machines that can replace the work of people or perform movements of great complexity implies the need for the widespread use of spatial mechanisms. With theoretical developments and the spread of numerical methods giving rise to new and efficient methods of analysis and synthesis, spatial mechanisms are nowadays common both in hi-tech applications such as manipulators, industrial robots, aircrafts, and nuclear installations, and also in common ones such as motor cars, combine harvesters, hydrostatic pumps, and rolling mills.

As for planar mechanisms, it is the kinematics, dynamics and synthesis of spatial mechanism that are examined most closely. For the positional and kinematic analyses of the complex movements of the elements of spatial mechanism, general methods have been developed. These are based on either matrix calculation or mathematical hyper-complex calculations such as complex numbers, dual numbers, dual vectors, and quaternions.

8.1.1 Kinematics of the RSSR Mechanism

In a constructive form, the RSSR mechanism is shown in Figure 8.1a, while its kinematic schema is shown in Figure 8.1b.

The number of elements is $n = 3$, the number of fifth-class kinematic pairs is $c_5 = 2\,(O_1, O_2)$, and the number of third-class kinematic pairs is $c_3 = 2\,(B, C)$. This gives a mobility degree $M = 6n - 5c_5 - 3c_3 = 2$; the rotation of the coupler BC around its own axis represents a redundant mobility degree.

In Figure 8.2 we have represented the suspension and steering mechanisms of an automobile. The suspension mechanism $ABCD$ is an RSSR spatial one, the revolute kinematic joints being situated at points A and D, while the spherical kinematic joints are situated at points B and C. To perform the positional analysis, we consider the kinematic schema in Figure 8.1b, and the common perpendicular OE of the axes of the rotational joints at points O_1 and O_2. We denote by α the angle between these two axes, as well as the dimensions $AO = a$, $OE = b$, $ED = c$, $AB = l_1$, $BC = l_2$, and $CD = l_3$. We choose the fixed reference system $Oxyz$, so that axis Ox coincides with OO_1 and axis Oz coincides with OE (Figure 8.1b). We construct the straight line OD' parallel to ED. The unit vector $\mathbf{v}$ of this axis has the expression

$$\mathbf{v} = \mathbf{i}\cos\alpha + \mathbf{j}\sin\alpha, \tag{8.1}$$

Classical and Modern Approaches in the Theory of Mechanisms, First Edition.
Nicolae Pandrea, Dinel Popa and Nicolae-Doru Stănescu.
© 2017 John Wiley & Sons Ltd. Published 2017 by John Wiley & Sons Ltd.
Companion Website: www.wiley.com/go/pandmech17

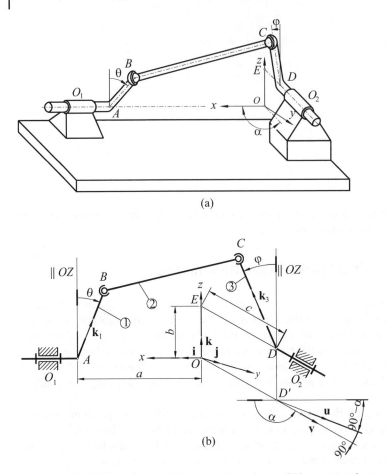

Figure 8.1 The RSSR mechanism: (a) constructive schema; (b) kinematic schema.

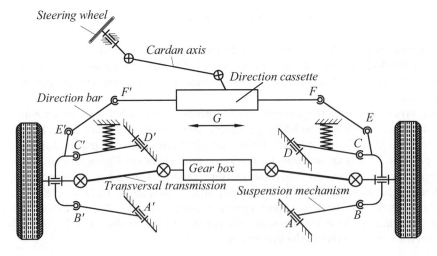

Figure 8.2 The suspension and steering mechanisms of an automobile.

where $\mathbf{i}$, $\mathbf{j}$, and $\mathbf{k}$ are the unit vectors of the axes Ox, Oy, and Oz, respectively. In addition, we denote by $\mathbf{u}$ the unit vector of the straight line situated in the plane Oxy and perpendicular to ED, and by $\mathbf{k}_1$ and $\mathbf{k}_3$ the unit vectors of the straight lines AB and DC, respectively. Since the unit vector $\mathbf{u}$ is perpendicular to Ox, it is also perpendicular to OD', so it reads

$$\mathbf{u} = -\mathbf{i} \sin \alpha + \mathbf{j} \cos \alpha. \tag{8.2}$$

The straight line AB is perpendicular to Ox, so the unit vector $\mathbf{k}_1$ has the expression

$$\mathbf{k}_1 = \mathbf{j} \sin \theta + \mathbf{k} \cos \theta, \tag{8.3}$$

where θ is the rotation angle of element 1. In addition, straight line DC is perpendicular to ED and therefore the unit vector $\mathbf{k}_3$ has the expression

$$\mathbf{k}_3 = -\mathbf{u} \sin \varphi + \mathbf{k} \cos \varphi, \tag{8.4}$$

in which φ is the rotation angle of element 3. Taking into account (8.2), (8.4) becomes

$$\mathbf{k}_3 = \mathbf{i} \sin \varphi \sin \alpha - \mathbf{j} \sin \varphi \cos \alpha + \mathbf{k} \cos \varphi. \tag{8.5}$$

Writing the vectors $\mathbf{AB}$ and $\mathbf{DC}$ as

$$\mathbf{AB} = l_1 \mathbf{k}_1, \quad \mathbf{DC} = l_3 \mathbf{k}_3 \tag{8.6}$$

and taking into account:

$$\mathbf{OB} = \mathbf{OA} + \mathbf{AB}, \quad \mathbf{OC} = \mathbf{OE} + \mathbf{ED} + \mathbf{DC},$$

gives

$$\begin{aligned}
\mathbf{OB} &= a\mathbf{i} + l_1 \mathbf{k}_1 = a\mathbf{i} + l_1 \sin \theta \mathbf{j} + l_1 \cos \theta \mathbf{k}, \\
\mathbf{OC} &= b\mathbf{k} + c\mathbf{v} + l_3 \mathbf{k}_3 \\
&= b\mathbf{k} + c(\mathbf{i} \cos \alpha + \mathbf{j} \sin \alpha) + l_3(\mathbf{i} \sin \varphi \sin \alpha - \mathbf{j} \sin \varphi \cos \alpha + \mathbf{k} \cos \varphi)
\end{aligned}$$

or

$$\mathbf{OB} = a\mathbf{i} + l_1 \sin \theta \mathbf{j} + l_1 \cos \theta \mathbf{k}, \tag{8.7}$$

$$\mathbf{OC} = (c \cos \alpha + l_3 \sin \varphi \sin \alpha)\mathbf{i} + (c \sin \alpha - l_3 \sin \varphi \cos \alpha)\mathbf{j} + (b + l_3 \cos \varphi)\mathbf{k}. \tag{8.8}$$

Since $|\mathbf{BC}| = l_2$, this gives $|\mathbf{OC} - \mathbf{OB}|^2 = l_2^2$, or

$$\begin{aligned}
(c \cos \alpha + l_3 \sin \varphi \sin \alpha - a)^2 + (c \sin \alpha - l_3 \sin \varphi \cos \alpha - l_1 \sin \theta)^2 \\
+ (b + l_3 \cos \varphi - l_1 \cos \theta)^2 = l_2^2.
\end{aligned}$$

By squaring, we get

$$\begin{aligned}
l_3^2 \sin^2 \varphi \sin^2 \alpha + 2l_3 \sin \varphi \sin \alpha (c \cos \alpha - a) + (c \cos \alpha - a)^2 + l_3^2 \sin^2 \varphi \cos^2 \alpha \\
- 2l_3 \sin\varphi \cos \alpha (c \cos \alpha - l_1 \sin \theta) + (c \sin \alpha - l_1 \sin \theta)^2 + l_3^2 \cos^2 \varphi \\
+ 2l_3 \cos \varphi (b - l_1 \cos \theta) + (b - l_1 \cos \theta)^2 - l_2^2 = 0.
\end{aligned}$$

Then denoting

$$A = 2l_3 (b - l_1 \cos \theta), \tag{8.9}$$

$$B = 2l_3 \left[\sin \alpha (c \cos \alpha - a) - \cos \alpha (c \sin \alpha - l_1 \sin \theta)\right], \tag{8.10}$$

$$C = l_3^2 + (c \cos \alpha - a)^2 + (c \sin \alpha - l_1 \sin \theta)^2 + (b - l_1 \cos \theta)^2 - l_2^2, \tag{8.11}$$

we obtain the equation

$$A \cos \varphi + B \sin \varphi + C = 0,$$

with the solution

$$\varphi = 2\arctan\left(\frac{-B \pm \sqrt{A^2 + B^2 - C^2}}{C - A}\right).$$ (8.12)

Equation (8.12) allows us to numerically deduce the input–output function $\varphi = \varphi(\theta)$ by giving values to angle θ and calculating the corresponding values for angle φ.

Then, in the kinematic calculation, (numerical) differentiation of the input–output function gives the expression for ω_3,

$$\omega_3 = \frac{d\varphi}{d\theta}\omega_1,$$ (8.13)

where $\omega_1 = \dot{\theta}$ is the angular velocity of element AB (Figure 8.1b). For the calculation of the derivative $\frac{d\varphi}{d\theta}$ we use the finite differences method

$$\left.\frac{d\varphi}{d\theta}\right|_{\theta=\theta_i} = \frac{\varphi\left(\theta_{i+1}\right) - \varphi\left(\theta_{i-1}\right)}{2\Delta\theta},$$ (8.14)

$\Delta\theta$ being the variation step of angle θ (it can be set to $\Delta\theta = 1°$ or $\Delta\theta = \frac{\pi}{180}$).

In a numerical example, we considered the following dimensions for the RSSR mechanism in Figure 8.1: $a = -0.0345\,\mathrm{m}$, $b = 0.173\,\mathrm{m}$, $c = 0.301\,\mathrm{m}$, $\alpha = \frac{2\pi}{3}$, $l_1 = 0.05\,\mathrm{m}$, $l_2 = 0.334\,\mathrm{m}$, $l_3 = 0.100\,\mathrm{m}$, and $\omega = 15\,\mathrm{rad/s}$. Based on the equations above, the angle is varied with a step of $1°$ from $0°$ to $360°$, giving the corresponding numerical values for angle φ, as shown in Figure 8.3, and the angular velocity ω_3, shown in Figure 8.4.

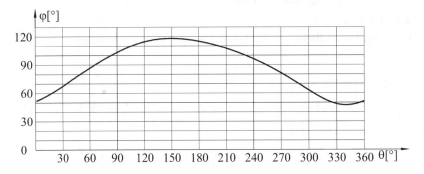

Figure 8.3 The function $\varphi = \varphi(\theta)$.

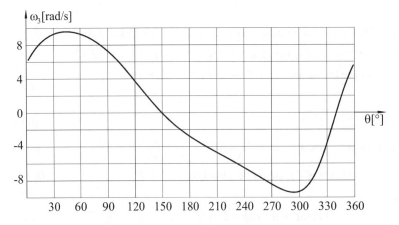

Figure 8.4 The function $\omega_3 = \omega_3(\theta)$.

8.1.2 Kinematics of the RSST Mechanism

The constructive schema of an RSST mechanism is shown in Figure 8.5a, while the kinematic one is in Figure 8.5b. The mobility of this mechanism is also $M = 2$, the rotation of the coupler BC around its proper axis being again a redundant mobility degree.

An application of this mechanism is the steering mechanism of a car (Figure 8.2), separately represented in Figure 8.6. From the representation in Figure 8.6, spherical joints B and C permit the rotation of body 1 around axis BC. In addition, the displacement of element 3 produced by the steering cassette (kinematic pair G) is transformed, using coupler 2, into a rotation of body 1 (the wheel of the car).

To determine the input–output equation, recalls the kinematic schema in Figure 8.5b. We denote:

- OE, the common perpendicular to the axes of the revolute kinematic joint at point O_1 and the translational kinematic joint at point D
- $Oxyz$, the reference system in which axis Ox coincides with the axis of the revolute kinematic joint OA and axis Oz coincides with OE
- **i**, **j**, and **k**, the unit vectors of the axes Ox, Oy, and Oz, respectively

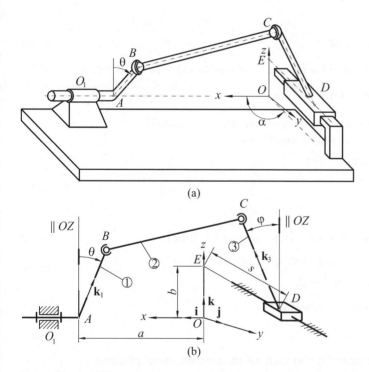

(a)

(b)

Figure 8.5 The RSST mechanism: (a) constructive schema; (b) kinematic schema.

Figure 8.6 Steering mechanism.

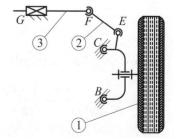

- α, the angle between the rotation axis OA and the axis of the translational kinematic joint ED
- the segments $OA = a$, $OE = b$, $AB = l_1$, $BC = l_2$, and $CD = l_3$
- φ, the constant angle between the axes CD and Ox
- θ, the rotation angle of the bar AB
- $s = ED$, the displacement parameter of the translational kinematic pair
- $\mathbf{k}_1$ and $\mathbf{k}_3$, the unit vectors of the axes AB and DC, respectively
- $\mathbf{v}$, the unit vector of axis ED
- $\mathbf{u}$, the unit vector situated in the plane Oxy and perpendicular to $\mathbf{v}$.

Similar to the RSSR mechanism, we get the expressions:

$$\mathbf{v} = \mathbf{i}\cos\alpha + \mathbf{j}\sin\alpha, \quad \mathbf{u} = -\mathbf{i}\sin\alpha + \mathbf{j}\cos\alpha, \tag{8.15}$$

$$\mathbf{k}_1 = \mathbf{j}\sin\theta + \mathbf{k}\cos\theta, \tag{8.16}$$

$$\mathbf{k}_3 = -\mathbf{u}\sin\varphi + \mathbf{k}\cos\varphi = \mathbf{i}\sin\varphi\sin\alpha - \mathbf{j}\sin\varphi\cos\alpha + \mathbf{k}\cos\varphi,$$

$$\mathbf{OB} = \mathbf{OA} + \mathbf{AB} = a\mathbf{i} + l_1\mathbf{k}_1 = a\mathbf{i} + l_1\sin\theta\mathbf{j} + L_1\cos\theta\mathbf{k},$$

$$\mathbf{OC} = b\mathbf{k} + c\mathbf{v} + l_3\mathbf{k}_3 = b\mathbf{k} + s(\mathbf{i}\cos\alpha + \mathbf{j}\sin\alpha) \tag{8.17}$$

$$+ l_3(\mathbf{i}\sin\varphi\sin\alpha - \mathbf{j}\sin\varphi\cos\alpha + \mathbf{k}\cos\varphi)$$

or

$$\mathbf{OB} = a\mathbf{i} + l_1\sin\theta\mathbf{j} + l_1\cos\theta\mathbf{k}, \tag{8.18}$$

$$\mathbf{OC} = (s\cos\alpha + l_3\sin\varphi\sin\alpha)\mathbf{i} = (s\sin\alpha - l_3\sin\varphi\cos\alpha)\mathbf{j} + (b + l_3\cos\varphi)\mathbf{k}. \tag{8.19}$$

Since $|\mathbf{BC}| = l_2$, this gives $|\mathbf{OC} - \mathbf{OB}|^2 = l_2^2$, or

$$(s\cos\alpha + l_3\sin\varphi\sin\alpha - a)^2 + (s\sin\alpha - l_3\sin\varphi\cos\alpha - l_1\sin\theta)^2$$
$$+ (b + l_3\cos\varphi - l_1\cos\theta)^2 = l_2^2. \tag{8.20}$$

By squaring, we get the equation

$$s^2 + 2As + B = 0, \tag{8.21}$$

where

$$A = \cos\alpha(l_3\sin\varphi\sin\alpha - a) - \sin\alpha(l_3\sin\varphi\cos\alpha + l_1\sin\varphi), \tag{8.22}$$

$$B = (l_3\sin\alpha\sin\varphi - a)^2 + (l_3\cos\alpha\sin\varphi + l_1\sin\theta)^2$$
$$+ (b + l_3\cos\varphi - l_1\cos\theta)^2 - l_2^2. \tag{8.23}$$

The function $s = s(\theta)$ is given by the expression

$$s = -A \pm \sqrt{A^2 - B}, \tag{8.24}$$

where the sign in front of the radical is chosen depending on the initial position.

By numerical derivation of the expression $s(\theta)$ we obtain the velocity in the translational kinematic joint

$$v = \omega_1 \frac{ds}{d\theta} \tag{8.25}$$

where $\omega_1 = \dot{\theta}$. For the calculation of the derivative $\dfrac{ds}{d\theta}$ we use the finite differences method:

$$\left.\frac{ds}{d\theta}\right|_{\theta=\theta_i} = \frac{s(\theta_{i+1}) - s(\theta_{i-1})}{2\Delta\theta}. \tag{8.26}$$

8.1.3 Spatial Mechanism Generating Oscillatory Motion

Spatial mechanisms for generating oscillatory motion are show in two variants in Figures 8.7 and 8.8.

The mechanism in Figure 8.7 has chassis 0, axle 1 bent at angle α, an element with fixed point 2, and a roller 3, which moves in guide 4. When bent axle 1 rotates, it moves element 2, which, using roller 3, which moves in guide 4, gives an oscillatory rotating motion to segment OB, the amplitude of the rotation angle θ being equal to angle α.

Similarly, the mechanism with an oscillatory fork shown in Figure 8.8 consists of a chassis 0, an axle 1 bent at angle α, an element with fixed point 2, and the oscillatory fork 3, which maintains segment OB in the plane perpendicular to axis OC. When bent axle 1 rotates, segment OB, being kept in the plane perpendicular to axis OC by fork 3, is given an oscillatory motion with an amplitude equal to angle α.

In Figure 8.9 we show a hydrostatic pump driven by a mechanism that generates oscillatory motion, of the type described in Figure 8.7.

For the kinematic analysis we use the kinematic schema in Figure 8.10, in which bent axle 1, at angle α, rotates in the revolute joint D putting into motion the element with fixed point 2, using the revolute kinematic joint O. Element 2, via the revolute kinematic joint B, puts into motion body 3 which gains an oscillatory rotational motion.

The three elements have (instantaneous) rotational motion and therefore the family of the mechanism is $f = 3$; since the mechanism has three elements and four kinematic pairs, it has mobility $M = 3 \cdot 3 - 2 \cdot 4 = 1$.

The reference system $Oxyz$ is chosen so that the axes Ox and Oy are situated on the axes of the fixed revolute kinematic joints at point C and D, respectively. A point A situated on the

Figure 8.7 Mechanism with oscillatory element.

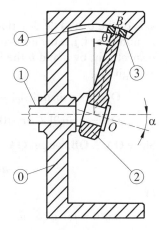

Figure 8.8 Mechanism with oscillatory fork.

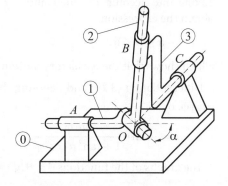

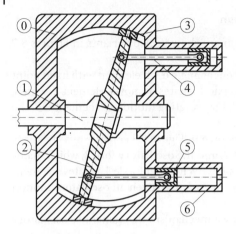

Figure 8.9 Hydrostatic pump.

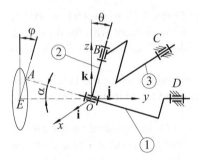

Figure 8.10 Kinematic schema of mechanism with oscillatory fork.

axis of the bent axle describes a circle centred at point E, situated on axis Oy (Figure 8.10). Denoting by $\mathbf{i}, \mathbf{j}, \mathbf{k}$ the unit vectors of the axes, by φ and θ the rotation angles of bodies 1 and 3, respectively, by a and b the lengths OA and OB, respectively, and by α the angle of the bent axle, gives

$$\mathbf{OA} = -a \sin \alpha \sin \varphi \mathbf{i} - a \cos \alpha \mathbf{j} + a \sin \alpha \cos \varphi \mathbf{k}, \tag{8.27}$$

$$\mathbf{OB} = b \sin \theta \mathbf{j} + b \cos \theta \mathbf{k}; \tag{8.28}$$

Since $\mathbf{OA} \perp \mathbf{OB}$, we get $\mathbf{OA} \cdot \mathbf{OB} = 0$, from which

$$-ab \cos \alpha \sin \theta + abd \sin \alpha \cos \varphi \cos \theta = 0,$$

or

$$\tan \theta = \tan \alpha \cos \varphi. \tag{8.29}$$

Taking into account the inequality $-1 \le \cos \varphi \le 1$, (8.29) gives $-\tan \alpha \le \tan \theta \le \alpha$ and we obtain the expression

$$-\alpha \le \theta \le \alpha, \tag{8.30}$$

which shows that the oscillatory motion of element 3 has an amplitude equal to α.

Differentiating (8.29) and denoting the transmission ratio by i, $i = \dfrac{\dot{\theta}}{\dot{\varphi}}$, gives $\dot{\theta} - cos^2\theta \tan \alpha \sin \varphi \dot{\varphi}$, or

$$i = -\tan \alpha \frac{\sin \varphi}{1 + \tan^2 \alpha \cos^2 \varphi}. \tag{8.31}$$

The graphs of the functions $\theta = \theta(\varphi)$ and $i = i(\varphi)$ are shown in Figure 8.11.

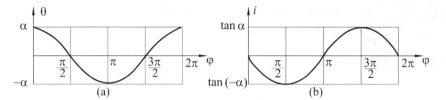

Figure 8.11 Graphs of the functions: (a) $\theta = \theta(\varphi)$; (b) $i = i(\varphi)$.

8.2 Hydrostatic Pumps with Axial Pistons

Rotational hydrostatic pumps (or engines) with axial pistons, due to their unique qualities (small gauge, high pressure, large range of power, large range of angular velocities), are very common nowadays. In older structural variants (Figure 8.12), the actuation of the cylinder block is via a double cardan joint 3; in this way, the equality of the angular velocities of the driving element 1 and the body of the pump 2 is assured.

To these elements, is added the assembly consisting of coupler 4 and piston 5, linked to one another and to element 1 by spherical joints, and to block 2 by a cylindrical joint. In this way, the supplementary kinematic chain has mobility degree $M = 6 \cdot 2 - 2 \cdot 3 - 1 \cdot 4 = 2$, and there are no constraints supplementary to those in the initial mechanism. Due to the inclination of element 1 relative to element 2, when moving, the pistons displace in the cylinders, thus bringing about the the absorption and evacuation phases of the liquid.

Modern pumps with axial pistons have no double cardan joints, the actuation of the body of pump coming about by direct contact between the coupler and the interior surface of the piston (Figure 8.13). Obviously, in this case, the interior surface of the piston has a conical shape, and contact between the coupler and the pistons is continuously assured; during rotation of body 1, at certain positions, each coupler participating in the actuation of the cylinder block 2. Because the linkage between the active coupler and the corresponding piston is created by spherical contact and by supporting on the generatrix of the conical surface, it is a fourth-class kinematic pair (Figure 8.14). Indeed, if the contact between the coupler and the piston is maintained, then

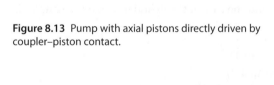

Figure 8.12 Pump with axial pistons actuated by Cardan transmission.

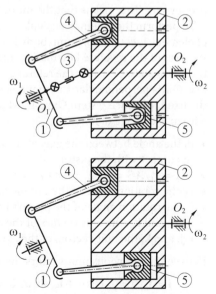

Figure 8.13 Pump with axial pistons directly driven by coupler–piston contact.

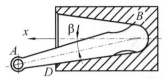

Figure 8.14 The coupler–piston contact.

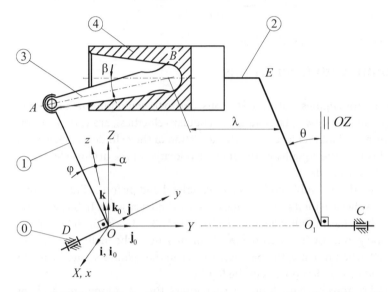

Figure 8.15 The kinematic chain driving axle 1, coupler 3, piston 4, and block of cylinders 2.

the coupler has two possible motions: a rotation around axis Ax and a rotation around its own axis AB.

Now consider the mechanism in Figure 8.15. This consists of driving element 1, the active coupler 3, piston 4, the body of the pump 2 with two revolute kinematic joints $(0, 1)$ and $(0, 2)$, a spherical joint $(1, 3)$ and two fourth-class kinematic pairs $(3, 4)$ and $(4, 2)$. We obtain the mobility $M = 6 \cdot 4 - 5 \cdot 2 - 4 \cdot 2 - 3 = 3$ (the family is $f = 0$). There are two passive mobilities: rotation of coupler 3 around its own axis and the rotation of the piston about its proper axis. This means that the motion of the pump's body 2 is uniquely determined.

To perform the kinematic calculation, consider the kinematic chain in Figure 8.15, consisting of element 1, coupler 3, pump body 2, and piston 4. We denote by O the point of intersection of the axes of the revolute kinematic joints D and C, $(0, 1)$ and $(0, 2)$, and by α the angle between these two axes. Consider the reference system $OXYZ$ in which axis OY coincides with OC, and axis OX is perpendicular to the plane formed by the rotation axes OD and OC. Also consider the fixed reference system $Oxyz$, in which axis Oy coincides with axis OD and axis Ox coincides with axis OX. We denote:

- β, the angle between the axis of the coupler and the axis of the cylinder, when the coupler is in contact with the piston
- $OA = r_1$, $AB = l$, $BE = \lambda$, $O_1E = r_2$, $OO_1 = d$
- φ, the rotation angle of the driving axle (element 1)
- θ, the rotation angle of the pump's body (element 2)
- $\mathbf{i_0}$, $\mathbf{j_0}$, and $\mathbf{k_0}$, the unit vectors of the axes OX, OY, and OZ, respectively
- $\mathbf{i}$, $\mathbf{j}$, and $\mathbf{k}$, the unit vectors of the axes Ox, Oy, and Oz, respectively.

Between the previous unit vectors there are obvious relations

$$\mathbf{i} = \mathbf{i_0}, \ \mathbf{j} = \mathbf{j_0} \cos \alpha + \mathbf{k_0} \sin \alpha, \ \mathbf{k} = -\mathbf{j_0} \sin \alpha + \mathbf{k_0} \cos \alpha. \tag{8.32}$$

The vector **OA** reads

$$\mathbf{OA} = r_1 (\sin \varphi \mathbf{i} + \cos \varphi \mathbf{k}) = r_1 \left(\sin \varphi \mathbf{i}_0 - \sin \alpha \cos \varphi \mathbf{j}_0 + \cos \alpha \cos \varphi \mathbf{k}_0 \right). \tag{8.33}$$

The vector **OB** has the expression

$$\mathbf{OB} = \mathbf{OO}_1 + \mathbf{O}_1\mathbf{E} + \mathbf{EB} = d\mathbf{j}_0 + r_2 \left(\sin \theta \mathbf{i}_0 + \cos \theta \mathbf{k}_0 \right) - \lambda \mathbf{j}_0$$

or

$$\mathbf{OB} = r_2 \sin \theta \mathbf{i}_0 + (d - \lambda) \mathbf{j}_0 + r_2 \cos \theta \mathbf{k}_0. \tag{8.34}$$

From the expression $\mathbf{AB} \cdot \mathbf{j}_0 = |\mathbf{AB}| \cos \beta$, $Y_B - Y_A = l \cos \beta$ or

$$d - \lambda + r_1 \sin \alpha \cos \varphi = l \cos \beta. \tag{8.35}$$

From (8.33) and (8.34), the expression $|\mathbf{AB}| = l$, in other words, $|\mathbf{OB} - \mathbf{OA}|^2 = l^2$, becomes

$$\left(r_2 \sin \theta - r_1 \sin \varphi \right)^2 + \left(d - \lambda + r_1 \sin \alpha \cos \varphi \right)^2 + \left(r_2 \cos \theta - r_1 \cos \alpha \cos \varphi \right)^2 = l^2;$$

Taking into account (8.35), we get

$$r_2^2 - 2r_1 r_2 \sin \theta \sin \varphi - 2r_1 r_2 \cos \theta \cos \alpha \cos \varphi + r_1^2 \sin^2 \varphi$$
$$+ r_1^2 \cos^2 \alpha \cos^2 \varphi - l^2 \sin^2 \beta = 0.$$

Denoting

$$A = 2r_1 r_2 \cos \alpha \cos \varphi, \ B = 2r_1 r_2 \sin \varphi,$$
$$C = l^2 \sin^2 \beta - r_2^2 - r_1^2 \sin^2 \varphi - r_1^2 \cos^2 \alpha \cos^2 \varphi, \tag{8.36}$$

gives the equation

$$A \cos \theta + B \sin \theta + C = 0, \tag{8.37}$$

with the solution

$$\tan \left(\frac{\theta}{2} \right) = \frac{-B \pm \sqrt{A^2 + B^2 - C^2}}{C - A}. \tag{8.38}$$

The graphical representation of the function $\theta = \theta (\varphi)$ is a periodical curve of period 2π. If the pump has three pistons, we obtain three curves denoted by 1, 2, and 3, staggered at $\frac{2\pi}{3}$ to one another.

The interval in which the ith coupler becomes active is that in which the corresponding angle θ_i is greater than angle θ_j corresponding to another coupler. Taking into account this criterion, we obtain the continuous curve $\theta = \theta (\varphi)$ in Figure 8.16.

If the number of pistons increases, then motion with little variation of angular velocity $\dot{\theta}$ is assured.

Figure 8.16 The function $\theta = \theta (\varphi)$.

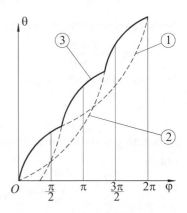

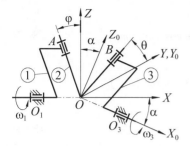

Figure 8.17 Simple mechanism with Cardan joint.

8.3 Cardan Transmissions

To transmit motion between two axles, the Cardan joint is used. Its kinematic schema is drawn in Figure 8.17.

Motion is transmitted from axle 1 to axle 3 by the intermediary element 2, AOB, with the fixed point O and $AO \perp OB$. The axes of the revolute kinematic joints O_1, A, B, and O_3 being concurrent, means that the elements may have only rotational motion and therefore the family f is equal to 3. The mobility is therefore $M = 3n - 2c_5 = 3 \cdot 3 - 2 \cdot 4 = 1$.

To determine the law of motion, the fixed reference system $OXYZ$ is chosen, in which axis OX coincides with the rotation axis O_1O and axis OY is perpendicular to the plane defined by the rotation axes OO_1 and OO_3. A system of axes $OX_0Y_0Z_0$ is also chosen such that axis OX_0 coincides with the straight line OO_3 and axis OY_0 coincides with axis OY.

The notation used is as follows:

- α, the angle between the axes O_1O and OO_3
- φ, the rotation angle of the primary axle 1
- θ, the rotation angle of the secondary axle 3
- $OA = OB = l$
- $\mathbf{i}$, $\mathbf{j}$, and $\mathbf{k}$, the unit vectors of the axes OX, OY, and OZ, respectively
- $\mathbf{i}_0$, $\mathbf{j}_0$, and $\mathbf{k}_0$, the unit vectors of the axes OX_0, OY_0, and OZ_0, respectively

This gives:

$$\mathbf{i}_0 = \mathbf{i}\cos\alpha - \mathbf{k}\sin\alpha, \ \mathbf{j}_0 = \mathbf{j}, \ \mathbf{k}_0 = \mathbf{i}\sin\alpha + \mathbf{k}\cos\alpha, \tag{8.39}$$

$$\mathbf{OA} = -l\sin\varphi\mathbf{j} + l\cos\varphi\mathbf{k}, \tag{8.40}$$

$$\mathbf{OB} = l\cos\theta\mathbf{j} + l\sin\theta\mathbf{k}_0 = l\sin\alpha\sin\theta\mathbf{i} + l\cos\theta\mathbf{j} + l\cos\alpha\sin\theta\mathbf{k}; \tag{8.41}$$

Since $OA \perp OB$, we get $\mathbf{OA} \cdot \mathbf{OB} = 0$ or

$$-\sin\varphi\cos\theta + \cos\varphi\cos\alpha\sin\theta = 0,$$

from which we obtain the expression

$$\tan\theta = \frac{\tan\varphi}{\cos\alpha}. \tag{8.42}$$

Differentiating (8.42), we get

$$\frac{\dot{\theta}}{\cos^2\theta} = \frac{\dot{\varphi}}{\cos\alpha\cos^2\varphi}.$$

Denoting by i the transmission ratio $i = \frac{\dot{\theta}}{\dot{\varphi}}$ and taking into account (8.42), we successively obtain

$$i = \frac{\cos^2\theta}{\cos\alpha\cos^2\varphi} = \frac{1}{\cos\alpha\cos^2\varphi\left(1 + \frac{\tan^2\varphi}{\cos^2\alpha}\right)} = \frac{\cos\alpha}{\cos^2\varphi\cos^2\alpha + \sin^2\varphi};$$

Figure 8.18 The function $\theta = \theta(\varphi)$.

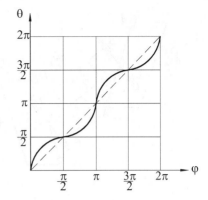

Figure 8.19 The function $i = i(\varphi)$.

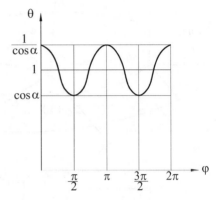

Figure 8.20 Linkage with Cardan cross.

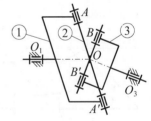

hence

$$i = \frac{\cos\alpha}{\cos^2\varphi\cos^2\alpha + \sin^2\varphi}. \tag{8.43}$$

The graphs of the functions $\theta = \theta(\varphi)$ and $i = i(\varphi)$ are shown in Figures 8.18 and 8.19, respectively.

The smaller the angle α, the closer to unity is the transmission ratio. For good dynamic behavior, coupler 2 is replaced by a symmetric component with a cross shape (Figure 8.20). In this way the rigidity increases. Since the linkages between the kinematic pairs A' and B' are passive, they do not influence the motion of the mechanism.

To create a homokinetic transmission ($i = 1$) with Cardan linkages, these linkages must be put in series using an intermediary axle (Figure 8.21). With the notation in Figure 8.21, denoting by φ_1, φ_2, and φ_3 the rotation angles, and taking into account (8.42), we get the expressions:

$$\tan\varphi_2 = \frac{\tan\varphi_1}{\cos\alpha}, \quad \tan\varphi_2 = \frac{\tan\varphi_3}{\cos\alpha}, \tag{8.44}$$

from which we obtain the relation for the homokinetic transmission $\varphi_1 = \varphi_3$.

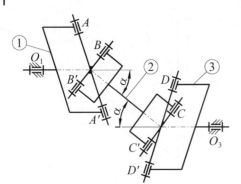

Figure 8.21 Homokinetic Cardan transmission.

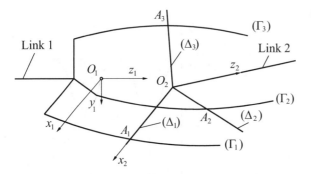

Figure 8.22 The tripod C–C kinematic pair.

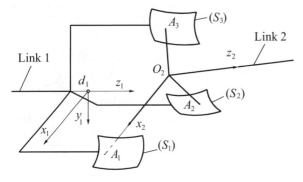

Figure 8.23 The tripod S–P kinematic pair.

8.4 Tripod Transmissions

8.4.1 General Aspects

The main structural variants of the tripod joints are those based on:

- a tripod kinematic pair with curve–curve contact type (C–C; Figure 8.22)
- a tripod kinematic pair with surface–point contact type (S–P; Figure 8.23).

In C–C tripod kinematic pairs, one of the joints is made up of the rigid curves (Γ_1), (Γ_2), and (Γ_3), rigidly linked to one another, while the other joint is made up of the coplanar half-lines (Δ_1), (Δ_2), and (Δ_3), concurrent at point O_2, rigidly linked to one another, and in permanent contact with the curves (Γ_1), (Γ_2), and (Γ_3) (Figure 8.22) at points A_1, A_2, A_3, respectively.

Similarly, in the vase of the P–S kinematic pair, one of the joints is made up of the rigid surfaces (S_1), (S_2), (S_3), rigidly linked to one another, while the other joint is made up of the coplanar linear segments O_2A_1, O_2A_2, O_2A_3, concurrent at point O_2 and having the A_1, A_2, and A_3 extremities permanently situated on (S_1), (S_2), and (S_3), respectively.

With the tripod kinematic pairs, which correspond to the tripod joints used for technical purposes, the (Γ_j) and the (Δ_j) half-lines, and respectively the (S_1) surfaces and the linear segments O_2A_j, with $j = 1, 2, 3$, are meridional symmetric at an angle of 120° to one another.

In the rest of this section we will only analyze these kinds of kinematic pairs, and we will denote by $C1$–C and $C2$–C the C–C tripod kinematic pairs for which the curves (Γ_j), $j = 1, 2, 3$, are identical curvilinear lines and parallel straight lines situated in the meridional plans, concurrent on one axis and placed at angles equal to 120° to one another, respectively. We will also denote by $S1$–P and $S2$–P the S–P tripod kinematic pairs for which the surfaces (S_j), $j = 1, 2, 3$, are meridional planar surfaces and spherical surfaces of equal radii and placed at angles equal to 120° to one another, respectively.

The corresponding structural solutions for the $S2$–P, $S1$–P, and $C1$–C kinematic pairs are shown in Figures 8.24–8.26, respectively.

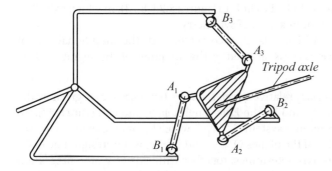

Figure 8.24 The $S2$–P tripod joint.

Figure 8.25 The $S1$–P tripod joint.

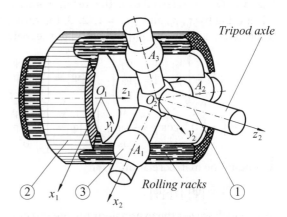

Figure 8.26 The tripod $C1$–C joint.

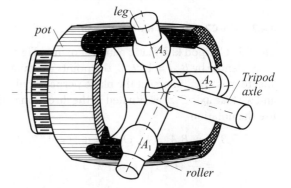

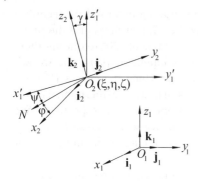

Figure 8.27 Defining the position of reference system $O_2x_2y_2z_2$ using Euler angles.

To study the kinematic and geometric properties of tripod kinematic pairs, we define

- the $O_1x_1y_1z_1$ reference system, rigidly linked to link 1 (Figures 8.22, 8.23), in which axis O_1z_1 corresponds to the meridional longitudinal axis of symmetry
- the $O_2x_2y_2z_2$ reference system, in which axis O_2z_2 coincides with the meridional axis of symmetry of the link 2, while axis O_2x_2 is situated in the direction of the segment O_2A_1 (Figures 8.22, 8.23).

The position of reference system $O_2x_2y_2z_2$ relative to reference system $O_1x_1y_1z_1$ is given by the coordinates ξ, η, and ζ of the origin O_2, and by the Euler angles ψ, γ, and φ (Figure 8.27). To define the Euler angle, consider the reference system $O_2x_1'y_1'z_1'$, which has its axes parallel to axes O_1x_1, O_1y_1, and O_1z_1. The intersection of the planes $O_2x_1'y_1'$ and $O_2x_2y_2$ is the straight line O_2N, referred to as the *nutation axis*. This axis determines, together with axis O_2x_1', the precession angle ψ. With axis O_2x_2, it determines the self-rotation angle φ. The angle γ defined by axes O_2z_2 and O_2z_1' is referred to as the *nutation angle*.

Denoting by α_i, β_i, and γ_i, $i = 1, 2, 3$, the director cosines of the axes O_2x_2, O_2y_2 and O_2z_2, respectively, and by $[T]$, $[\psi]$, $[\gamma]$, and $[\varphi]$ the matrices

$$[T] = \begin{bmatrix} \alpha_1 & \alpha_2 & \alpha_3 \\ \beta_1 & \beta_2 & \beta_3 \\ \nu_1 & \nu_2 & \nu_3 \end{bmatrix}, \quad [\psi] = \begin{bmatrix} \cos\psi & -\sin\psi & 0 \\ \sin\psi & \cos\psi & 0 \\ 0 & 0 & 1 \end{bmatrix},$$

$$[\gamma] = \begin{bmatrix} 1 & 0 & 0 \\ 0 & \cos\gamma & -\sin\gamma \\ 0 & \sin\gamma & \cos\gamma \end{bmatrix}, \quad [\varphi] = \begin{bmatrix} \cos\varphi & -\sin\varphi & 0 \\ \sin\varphi & \cos\varphi & 0 \\ 0 & 0 & 1 \end{bmatrix},$$

$$(8.45)$$

we obtain the matrix relation

$$[T] = [\psi]\,[\gamma]\,[\varphi], \tag{8.46}$$

from which we get the expressions for the director cosines as functions of the Euler angles.

8.4.2 The C2–C Tripod Kinematic Pair

The *C2–C* tripod kinematic pair is shown in Figure 8.28.

We denote by B_j, $j = 1, 2, 3$, the points at which the straight lines (Γ_j), equidistant and parallel to axis O_1z_1, intersect the plane perpendicular to axis O_1z_1 and passing through point O_1, and we take axis O_1z_1 as situated in the direction of the straight line O_1B_1 (Figure 8.28). We also use the notation

$$O_jB_j = r, \ B_jA_j = \lambda_j, \ O_2A_j = \mu_j, \ j = 1, 2, 3, \tag{8.47}$$

$$\delta_j = \frac{2\pi}{3}\,(j - 1), \ j = 1, 2, 3. \tag{8.48}$$

Figure 8.28 The *C2–C* tripod kinematic pair.

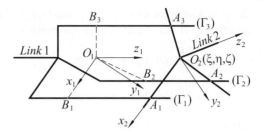

Let x_{1j}, y_{1j}, z_{1j} and x_{2j}, y_{2j}, z_{2j} be the coordinates of points A_j, $j = 1,\ 2,\ 3$ relative to the reference systems $O_1x_1y_1z_1$ and $O_2x_2y_2z_2$, respectively. Taking into account the expressions

$$x_{1j} = r\cos\delta_j,\ y_{1j} = r\sin\delta_j,\ z_{1j} = \lambda_j,\tag{8.49}$$

$$x_{2j} = \mu_j\cos\delta_j,\ y_{1j} = \mu_j\sin\delta_j,\ z_{1j} = 0,\tag{8.50}$$

and the matrix relation

$$\begin{bmatrix} x_{1j} \\ y_{1j} \\ z_{1j} \end{bmatrix} = \begin{bmatrix} \xi \\ \eta \\ \zeta \end{bmatrix} + \begin{bmatrix} \alpha_1 & \alpha_2 & \alpha_3 \\ \beta_1 & \beta_2 & \beta_3 \\ \gamma_1 & \gamma_2 & \gamma_3 \end{bmatrix} \begin{bmatrix} x_{2j} \\ y_{2j} \\ z_{2j} \end{bmatrix},\tag{8.51}$$

by eliminating the parameter μ_j, we obtain the equalities

$$\xi\left(\beta_1\cos\delta_j + \beta_2\sin\delta_j\right) - \eta\left(\alpha_1\cos\delta_j + \alpha_2\sin\delta_j\right)$$
$$-r\left[\beta_1\cos^2\delta_j + \left(\beta_2 - \alpha_1\right)\sin\delta_j\cos\delta_j - \alpha_2\sin\delta_j\right],\ j = 1,\ 2,\ 3.\tag{8.52}$$

Adding (8.52) and taking into account the well known trigonometric relations

$$\sum_{i=1}^{3}\cos\delta_j = \sum_{i=1}^{3}\sin\delta_j = \sum_{i=1}^{3}\sin\delta_j\cos\delta_j = 0,\tag{8.53}$$

$$\sum_{i=1}^{3}\cos^2\delta_j = \sum_{i=1}^{3}\sin^2\delta_j = \frac{3}{2},\tag{8.54}$$

we obtain the *essential condition* of the *C2–C* tripod kinematic pair

$$\beta_1 = \alpha_2;\tag{8.55}$$

Using the Euler angles, (8.55) takes the form

$$\psi = -\varphi.\tag{8.56}$$

Using (8.55), (8.56), (8.45), (8.46), (8.51), and (8.52), we can also obtain the expressions

$$\xi = \frac{r\left(1 - \cos\gamma\right)}{2\cos\gamma}\left[\cos\left(3\varphi\right)\cos\varphi + \cos\gamma\sin\left(3\varphi\right)\sin\varphi\right],\tag{8.57}$$

$$\eta = \frac{r\left(1 - \cos\gamma\right)}{2\cos\gamma}\left[-\cos\left(3\varphi\right)\sin\varphi + \cos\gamma\sin\left(3\varphi\right)\cos\varphi\right],\tag{8.58}$$

$$\mu_j = \frac{r\cos\delta_j - \xi}{\alpha_1\cos\delta_j + \alpha_2\sin\delta_j},\tag{8.59}$$

$$\lambda_j = \zeta + \left(\gamma_1\cos\delta_j + \gamma_2\sin\delta_j\right)\mu_j,\tag{8.60}$$

in which ζ, φ, and γ are independent parameters.

In real tripod joints, the variation of angle γ during use is small, so a good approximation can be obtained by assuming that angle γ is constant. In this situation, the projection of the curve described by point O_2 onto the plane $O_1x_1y_1$ (Figure 8.28) is the curve given by equations (8.57) and (8.58). These equations may be written as

$$\xi = \frac{2R}{3} \cos(2\varphi) + h\cos(4\varphi),\tag{8.61}$$

$$\eta = \frac{2R}{3} \sin(2\varphi) - h\sin(4\varphi),\tag{8.62}$$

where

$$R = \frac{3r\left(1 - \cos^2\gamma\right)}{8\cos\gamma}, \; h = \frac{r\left(1 - \cos\gamma\right)^2}{4\cos\gamma};\tag{8.63}$$

They represent a shortened, normal, or elongated hypocycloid (Γ), for which the ratio p between the radius of the fixed circle and the circle which rolls in the interior is 3. Various graphical representations of the curve (Γ) are shown in Table 8.1.

Table 8.1 Different cases of hypocycloids.

γ	Condition	Denomination	Graphical representation
$\cos^{-1}\left(\frac{1}{3}\right) < \gamma < \frac{\pi}{2}$	$\rho_{max} > r$	Elongated hypocycloid ($\rho = 3$)	
$\gamma = \cos^{-1}\left(\frac{1}{3}\right)$	$\rho_{max} = r, \rho \geq \frac{r}{3}$	Steiner's hypocycloid	
$\cos^{-1}\left(\frac{3}{5}\right) < \gamma < \cos^{-1}\left(\frac{1}{3}\right)$	$\frac{r}{6} < \rho_{max} < r,$ $\frac{r}{10} < \rho_{min} < \frac{r}{3},$ $\rho_{min} < e$	Shortened hypocycloid ($\rho = 3$)	
$0 < \gamma \leq \cos^{-1}\left(\frac{3}{5}\right)$	$0 < \rho_{max} \leq \frac{r}{6},$ $0 < \rho_{min} \leq \frac{r}{10}$		

Figure 8.29 The C2–C tripod angular joint.

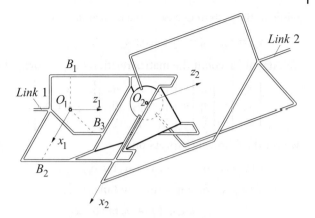

From the analysis of the results presented in Table 8.1, it follows that when $\gamma > \cos^{-1}\left(\frac{1}{3}\right)$, point O_2 becomes exterior to the circle of radius r; since this condition is difficult to meet in practice, a first limit for angle γ is given by:

$$\gamma < \cos^{-1}\left(\frac{1}{3}\right). \tag{8.64}$$

Next, we calculate the distance s between points O_2 and O_1 (Figure 8.28), and we get the following expression

$$s = \frac{r(1 - \cos\gamma)}{2\cos\gamma}\sqrt{\cos^2(3\varphi) + \cos^2\gamma\sin^2(3\varphi) + \zeta^2}, \tag{8.65}$$

the minimum variation of which is when $\zeta = 0$. It follows that for $\gamma \neq 0$, a C2–C tripod joint for which the axes O_2z_2 and O_1z_1 are concurrent cannot be made, but it is possible to construct a pseudo-angular tripod joint for which the distance s between points O_1 and O_2 has minimal variation for $\zeta = 0$ and in addition which verifies:

$$\frac{r(1 - \cos\gamma)}{2\cos\gamma} \geq s \geq \frac{r(1 - \cos\gamma)}{2}. \tag{8.66}$$

From the structural point of view, the condition for an angular joint ($\zeta = 0$) is fulfilled using a bilateral kinematic pair of the sphere–plane type (Figure 8.29)

8.4.3 The C1–C Tripod Kinematic Pair

The C1–C tripod kinematic pair is shown in Figure 8.22. The curves $(\Gamma_j), j = 1, 2, 3$ are symmetrically situated in space, so if the curve (Γ_1) has the equations

$$x = f(z_1), \ y_1 = 0, \tag{8.67}$$

in the system $O_1x_1y_1z_1$, then the curves $(\Gamma_j), j = 1, 2, 3$ will be described by the equations

$$x_1 = f(z_1)\cos\delta_1, \ y_1 = f(z_1)\sin\delta_1, \ j = 1, 2, 3. \tag{8.68}$$

Similarly, for the straight lines $(\Delta_j), j = 1, 2, 3$, we obtain the following equations

$$y_2 = x_2\tan\delta_j, \ z_2 = 0 \tag{8.69}$$

in the system $O_2x_2y_2z_2$. This means that in the system $O_1x_1y_1z_1$, the coordinates of point A_j, $j = 1, 2, 3$ are

$$x_{1j} = f(z_{1j})\cos\delta_j, \ y_{1j} = f(z_{1j})\sin\delta_j, \ z_{1j} = z_{1j}, \tag{8.70}$$

while in the system $O_2 x_2 y_2 z_2$ the coordinates are

$$x_{2j} = x_{2j}, \ y_{2j} = x_{2j} \tan \delta_j, \ z_{2j} = 0; \tag{8.71}$$

Taking into account the matrix relation of transformation

$$\begin{bmatrix} x_{1j} \\ y_{1j} \\ z_{1j} \end{bmatrix} = \begin{bmatrix} \xi \\ \eta \\ \zeta \end{bmatrix} + \begin{bmatrix} \alpha_1 & \alpha_2 & \alpha_3 \\ \beta_1 & \beta_2 & \beta_3 \\ \gamma_1 & \gamma_2 & \gamma_3 \end{bmatrix} \begin{bmatrix} x_{2j} \\ x_{2j} \tan \delta_j \\ 0 \end{bmatrix}, \tag{8.72}$$

we get the following expressions

$$\begin{aligned} f(z_{1j}) \cos \delta_j &= \xi + (\alpha_1 + \alpha_2 \tan \delta_j) x_{2j}, \\ f(z_{1j}) \sin \delta_j &= \eta + (\beta_1 + \beta_2 \tan \delta_j) x_{2j}, \\ z_{1j} &= \zeta + (\gamma_1 + \gamma_2 \tan \delta_j) x_{2j}. \end{aligned} \tag{8.73}$$

Eliminating x_{2j} and x_{1j} from (8.73) and denoting

$$U_j = \xi (\beta_1 \cos \delta_j + \beta_2 \sin \delta_j) - \eta (\alpha_1 \cos \delta_j + \alpha_2 \sin \delta_j), \tag{8.74}$$

$$V_j = (\beta_2 - \alpha_1) \sin \delta_j \cos \delta_j + \beta_1 \cos^2 \delta_j - \alpha_2 \sin^2 \delta_j, \tag{8.75}$$

$$W_j = (\xi \sin \delta_j - \eta \cos \delta_j)(\gamma_1 \cos \delta_j + \gamma_2 \sin \delta_j), \tag{8.76}$$

we obtain:

$$V_j f \left(\xi + \frac{W_j}{V_j} \right) - U_j = 0, \ j = 1, \ 2, \ 3. \tag{8.77}$$

The three relations in (8.77) represent the equations of the $C1$–C tripod kinematic pairs. If the curves $(\Gamma_j), j = 1, \ 2, \ 3$ are straight lines, then $f(z_1) = r$, and (8.77) may be written as

$$r V_j - U_j = 0, \ j = 1, \ 2, \ 3; \tag{8.78}$$

Addition of these relations gives once again condition (8.55).

If the joint is an *angular* one, then setting $\zeta = 0$ and using Euler angles, for set values of angle γ, from (8.77) it is possible to determine the parameters $\psi, \ \xi$, and η as functions of angle φ. In practice, the function f may be defined as a third-degree polynomial function of the form

$$f(z_1) = r + \lambda_1^* z_1 + \lambda_2^* z_1^2 + \lambda_3^* z_1^3 \tag{8.79}$$

and (8.77) becomes

$$- U_j V_j^2 + r V_j^3 + \lambda_1^* V_j^2 W_j + \lambda_2^* V_j W_j^2 + \lambda_3^* W_j^3 = 0. \tag{8.80}$$

The system of equations (8.80) can be solved using the Newton–Raphson method and a calculation program. Considering the numerical data

$$\gamma = 46°, \ r = 22.85, \ \lambda_1^* = -0.0124, \ \lambda_2^* = -0.00083, \ \lambda_3^* = -0.000018,$$

the parameters $\xi, \ \eta$, and ψ are determined as functions of angle φ, as well as the coordinates of point O_2 in the two reference systems. These results are compared with those obtained for the $C2$–C tripod kinematic pair for which $\gamma = 46°$ and $r = 22.85$. The results are shown in Figures 8.30–8.32. We find that that the curve (Γ_1) of the $C1$–C tripod kinematic pair (Figure 8.30) corresponds the curve (Γ) in Figure 8.31, for which the distance

$$\rho = \sqrt{\xi^2 + \eta^2}$$

has the property that the distance between its extreme values is shorter than in the case of the curve (Γ) corresponding to the $C2$–C tripod kinematic pair.

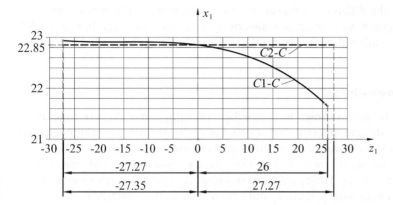

Figure 8.30 The function $x_1 = x_1\left(z_1\right)$ for the C1–C and C2–C tripod kinematic pairs.

Figure 8.31 The function $\eta = \eta\left(\xi\right)$ for the C1–C and C2–C tripod kinematic pairs.

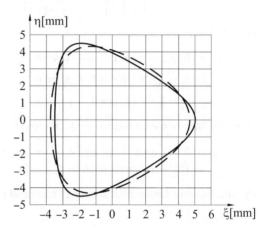

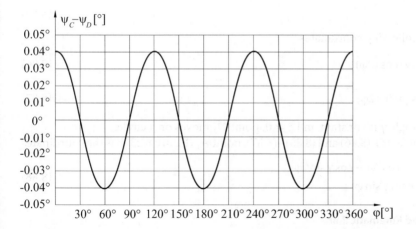

Figure 8.32 The function $\psi_C - \psi_D = \left(\psi_C - \psi_D\right)\left(\varphi\right)$ for the C1–C and C2–C tripod kinematic pairs.

Denoting by $\psi_C - \psi_D$ the difference between the values of angle ψ which correspond to the $C1-C$ and $C2-C$ tripod kinematic pairs, respectively, we obtain the graph in Figure 8.32, from which it follows that this difference is periodic and, in addition, it has a small maximum value.

8.4.4 The S1–P Tripod Kinematic Pair

The $S1-P$ tripod kinematic pair is shown in Figure 8.23. (S_1), (S_2), and (S_3) are concurrent planar surfaces along the straight line O_1z_1 and situated at angles of $120°$ to one another. Axis O_1x_1 is situated in plane (S_1), and we denote by r the equal distances $O_2A_1 = O_2A_2 = O_2A_3 = r$, by λ_j the distances from points A_j, $j = 1$, 2, 3 to plane $O_1x_1y_1$, and by μ_j the distances from points A_j, $j = 1$, 2, 3 to axis O_1z_1. The coordinates of points A_j, in the two reference systems $O_1x_1y_1z_1$, and $O_2x_2y_2z_2$, are $\mu_j \cos \delta_j$, $\mu_j \sin \delta_j$, λ_j and $r \cos \delta_j$, $r \sin \delta_j$, 0, respectively.

The matrix relation

$$\begin{bmatrix} \mu_j \cos \delta_j \\ \mu_j \sin \delta_j \\ \lambda_j \end{bmatrix} = \begin{bmatrix} \xi \\ \eta \\ \zeta \end{bmatrix} + \begin{bmatrix} \alpha_1 & \alpha_2 & \alpha_3 \\ \beta_1 & \beta_2 & \beta_3 \\ \gamma_1 & \gamma_2 & \gamma_3 \end{bmatrix} \begin{bmatrix} r \cos \delta_j \\ r \sin \delta_j \\ 0 \end{bmatrix} \tag{8.81}$$

leads to the expressions

$$\mu_j \cos \delta_j = \xi + r \left(\alpha_1 \cos \delta_j + \alpha_2 \sin \delta_j \right),$$
$$\mu_j \sin \delta_j = \eta + r \left(\beta_1 \cos \delta_j + \beta_2 \sin \delta_j \right), \tag{8.82}$$
$$\lambda_j = \zeta + r \left(\gamma_1 \cos \delta_j + \gamma_2 \sin \delta_j \right).$$

From the first two expressions in the system (8.82), eliminating the parameter μ_j, we get

$$\eta \cos \delta_j - \xi \sin \delta_j + r\beta_1 \cos^2 \delta_j - r\alpha_2 \sin^2 \delta_j$$
$$+ r \left(\beta_2 - \alpha_2 \right) \cos \delta_j \sin \delta_j = 0, \ j = 1, \ 2, \ 3; \tag{8.83}$$

By addition and taking into account (8.53) and (8.54), we obtain the condition

$$\alpha_2 = \beta_1, \tag{8.84}$$

which, using Euler angles, becomes

$$\psi = -\varphi. \tag{8.85}$$

Then, we obtain the following coordinates

$$\xi = \frac{r}{2} (1 - \cos \gamma) \cos (2\varphi), \tag{8.86}$$

$$\eta = \frac{r}{2} (1 - \cos \gamma) \sin (2\varphi), \tag{8.87}$$

which shows that if angle γ is constant and $\xi = 0$, point O_2 moves on a circle.

Taking into account (8.45), (8.46) and (8.85)–(8.87), point A_1 (Figure 8.23) has the coordinates

$$x_1 = \mu_1 = \xi + r \left(\cos^2 \varphi + \cos \gamma \sin^2 \varphi \right),$$
$$z_1 = \lambda_1 = \zeta + r \sin \gamma \sin \varphi. \tag{8.88}$$

8.4.5 The S2–P Tripod Kinematic Pair

The $S2-P$ tripod kinematic pairis shown in Figure 8.23. Here, (S_i), $i = 1$, 2, 3 are equal spheres of radius l, centred at points (B_i), $i = 1$, 2, 3 and situated at distances r_1 from axis O_1z_1. Point B_1 is situated on axis O_1x_1, while the lengths of the distances O_2A_j are assumed to be equal to r_2.

In these conditions, we may write the coordinates x_{1j}, y_{1j}, and z_{1j} of point A_j in the reference system $O_1 x_1 y_1 z_1$, using the matrix relation

$$\begin{bmatrix} x_{1j} \\ y_{1j} \\ z_{1j} \end{bmatrix} = \begin{bmatrix} \xi \\ \eta \\ \zeta \end{bmatrix} + \begin{bmatrix} \alpha_1 & \alpha_2 & \alpha_3 \\ \beta_1 & \beta_2 & \beta_3 \\ \gamma_1 & \gamma_2 & \gamma_3 \end{bmatrix} \begin{bmatrix} r_2 \cos \delta_j \\ r_2 \sin \delta_2 \\ 0 \end{bmatrix} ; \tag{8.89}$$

Taking into account that points B_j have the coordinates $r_2 \cos \delta_j$, $r_2 \sin \delta_j$, and 0 in the reference system $O_1 x_1 y_1 z_1$, from:

$$\left(B_j A_j \right)^2 = l^2, \tag{8.90}$$

we get the expressions

$$\begin{aligned} \xi^2 + \eta^2 + \zeta^2 + r_1^2 + r_2^2 - l^2 + 2\xi \left[r_2 \left(\alpha_1 \cos \delta_j + \alpha_2 \sin \delta_j \right) - r_1 \cos \delta_j \right] \\ + 2\eta \left[r_2 \left(\beta_1 \cos \delta_j + \beta_2 \sin \delta_j \right) - r_1 \sin \delta_1 \right] + 2\zeta r_2 \left(\gamma_1 \cos \delta_j + \gamma_2 \sin \delta_j \right) \\ - 2 r_1 r_2 \left[\alpha_1 \cos^2 \delta_j + \left(\alpha_2 + \beta_1 \right) \cos \delta_j \sin \delta_j + \beta_2 \sin^2 \delta_j \right] = 0, \end{aligned} \tag{8.91}$$

where $j = 1, 2, 3$.

When $r_1 = r_2 = r$, if we denote

$$\cos \tilde{\beta} = \frac{2 r^2 - l^2}{2 r^2} \tag{8.92}$$

then, from (8.91), the equations of the $S2$–P tripod kinematic pair are of the form

$$\xi^2 + \eta^2 + \zeta^2 + 2 r^2 \cos \tilde{\beta} - r^2 \left(\alpha_1 + \beta_2 \right) = 0. \tag{8.93}$$

$$2\xi \alpha_2 + 2\eta \left(\beta_2 - 1 \right) + 2\zeta \gamma_2 = -r \left(\alpha_2 + \beta_1 \right) , \tag{8.94}$$

$$2\xi \left(\alpha_1 - 1 \right) + 2\eta \beta_1 + 2\zeta \gamma_1 = r \left(\alpha_1 - \beta_2 \right) . \tag{8.95}$$

As an approximation, it is assumed that $\gamma = \text{const.}$ and the expression $\xi^2 + \eta^2 + \zeta^2$ is negligible relative to r^2. Denoting

$$\beta_0 = \arccos \left(\frac{2 \cos \tilde{\beta}}{1 + \cos \gamma} \right) , \tag{8.96}$$

from (8.93)–(8.95) we get

$$\psi = \beta_0 - \varphi, \tag{8.97}$$

$$\xi = r^* \left[\sin \left(\varphi - \beta_0 \right) \sin \left(3\varphi - \beta_0 \right) + \cos \gamma \cos \left(\varphi - \beta_0 \right) \cos \left(3\varphi - \beta_0 \right) - \cos \left(2\varphi - \beta_0 \right) \right] , \tag{8.98}$$

$$\eta = r^* \left[\cos \left(\varphi - \beta_0 \right) \sin \left(3\varphi - \beta_0 \right) - \cos \gamma \sin \left(\varphi - \beta_0 \right) \cos \left(3\varphi - \beta_0 \right) - \sin \left(2\varphi - \beta_0 \right) \right] , \tag{8.99}$$

where

$$r^* = r \frac{1 - \cos \gamma}{2 \left(1 + \cos \gamma \right) \left(1 - \cos \beta_0 \right)} . \tag{8.100}$$

The curve (Γ) described by point O_2, when $\beta_0 = \frac{\pi}{2}$, $\cos \gamma = 0.8$, and $r = 1$, is shown in Figure 8.33. It can be seen that the shape is similar to those of the curves (Γ) obtained for C–C tripod joints, but the extreme values of the polar radius $\rho = \sqrt{\xi^2 + \eta^2}$ are smaller.

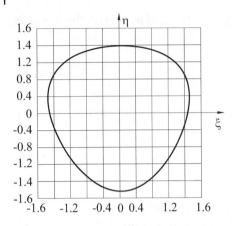

8.4.6 Simple Mechanisms with Tripod Joints

The transmissions of some cars are use an axial-angular tripod joint situated next to the gear box, and an angular joint of Rzeppa, Weis, or Cardan type situated near the wheel. In such cases, the study of the transmissions is reduced to the study of the movement of the simple axial-angular tripod joint mechanism, an example of which is shown in Figure 8.34. Here, links 1 and 2 are connected by a revolute pair, and a spherical kinematic pair, respectively.

The analysis of such mechanisms is performed on the basis of the vector relation

$$\mathbf{O_1O_3} = \mathbf{O_1O_2} + \mathbf{O_2O_3} \tag{8.101}$$

written in the reference system $O_1x_1y_1z_1$ (Figure 8.34). To this is added the equations of the tripod kinematic pairs, namely

- (8.56)–(8.58) for C2–C tripod kinematic pairs
- (8.74)–(8.77) for C1–C tripod kinematic pairs
- (8.85)–(8.88) for S1–P tripod kinematic pairs
- (8.93)–(8.95) for S2–P tripod kinematic pairs.

We denote by l the distance O_2O_3, by O_0 the point situated on the axis of link 1 so that $O_0O_3 = O_2O_3$, by $O_0x_0y_0z_0$ the fixed reference system in which axis O_0z_0 coincides with the axis of link 1, by $O_1x_1y_1z_1$ the system rigidly linked to link 1 so that O_1 coincides with O_0 and axis O_1z_1 coincides with axis O_0z_0, by $O_2x_2y_2z_2$ the mobile reference system rigidly linked to the tripod spider, by α the angle between axes O_0z_0 and O_0O_3, by β the angle between the projections of segment O_0O_3 onto the plane $O_0x_0y_0z_0$ and axis O_0x_0, and by θ the rotation angle of link 1.

This all means that point O_3 has the coordinates $i\sin\alpha\cos\beta$, $l\sin\alpha\sin\beta$, and $l\cos\alpha$ in the $O_0x_0y_0z_0$ system. Denoting now

$$[\boldsymbol{\theta}] = \begin{bmatrix} \cos\theta & -\sin\theta & 0 \\ \sin\theta & \cos\theta & 0 \\ 0 & 0 & 1 \end{bmatrix} \tag{8.102}$$

and taking into account (8.45) and (8.46), equation (8.101) becomes

$$[\boldsymbol{\theta}]^{-1} \begin{bmatrix} l\sin\alpha\cos\beta \\ l\sin\alpha\sin\beta \\ l\cos\alpha \end{bmatrix} = \begin{bmatrix} \xi \\ \eta \\ \zeta \end{bmatrix} + [\boldsymbol{\psi}]\,[\boldsymbol{\gamma}]\,[\boldsymbol{\phi}] \begin{bmatrix} 0 \\ 0 \\ l \end{bmatrix}; \tag{8.103}$$

hence

$$\begin{aligned} \xi &= l\sin\alpha\cos(\theta-\beta) - l\sin\gamma\sin\psi, \\ \eta &= -l\sin\alpha\sin(\theta-\beta) + l\sin\gamma\cos\psi, \\ \zeta &= l\cos\alpha - l\cos\gamma. \end{aligned} \tag{8.104}$$

The C2–C tripod kinematic pair

In this situation, (8.56)–(8.58) have to be added to (8.104). Denoting:

$$\tilde{\lambda} = \frac{r}{2l} \tag{8.105}$$

we obtain (for $\tilde{\lambda} \ll 1$):

$$\sin(3\varphi) = \frac{-\cos^2\gamma + \sqrt{\cos^2\alpha\cos^2\gamma + \tilde{\lambda}^2(1-\cos\gamma)^2}}{\tilde{\lambda}\sin\gamma(1-\cos\gamma)}, \tag{8.106}$$

where

$$\gamma \in \left[\gamma_{min}, \gamma_{max}\right]. \tag{8.107}$$

The values γ_{min} and γ_{max} of angle γ are obtained from the inequality

$$|\sin(3\varphi)| \le 1; \tag{8.108}$$

and we get

$$\gamma_{min} = \arcsin\left(\frac{\sin\alpha - \tilde{\lambda}}{\sqrt{1+\tilde{\lambda}^2}}\right) + \arctan\tilde{\lambda}, \tag{8.109}$$

$$\gamma_{max} = \arcsin\left(\frac{\sin\alpha + \tilde{\lambda}}{\sqrt{1+\tilde{\lambda}^2}}\right) - \arctan\tilde{\lambda}. \tag{8.110}$$

First the calculation on the basis of (8.104) and (8.106) has to be carried out to determinedthe parameters φ, ξ, η, ζ, and θ depending on angle γ; in a second stage, by rearrangement of the results, the parameters φ, γ, ξ, η, and ζ are obtained as functions of angle θ.

As an approximation, the parameters λ, $\lambda\dfrac{1-\cos\gamma}{\cos\gamma}$, $\gamma - \alpha$, and $\varphi - \theta$ can be assumed as small, and we obtain the following approximate expressions

$$\varphi \approx \theta - \frac{\lambda(1-\cos\alpha)}{\cos\alpha\sin\alpha}\sin(3(\theta-\beta)),$$

$$\gamma \approx \alpha + \frac{\lambda(1-\cos\alpha)}{\cos\alpha}\cos(3(\theta-\beta)), \tag{8.111}$$

$$\frac{\zeta}{l} \approx \lambda(1-\cos\alpha)\tan\alpha\cos(3(\theta-\beta)).$$

Then, from (8.104) we can determine the parameters ζ and η.

In the system of coordinates O_0XYZ (Figure 8.34), point O_2 has the coordinates X, Y, and Z, approximated by:

$$\frac{X}{r} \approx \frac{1-\cos\alpha}{2\cos\alpha}\cos(3\theta), \quad \frac{Y}{r} \approx \frac{1-\cos\alpha}{2\cos\alpha}\sin(3\theta), \quad \frac{Z}{r} \approx (1-\cos\alpha)\tan\alpha\sin(3\theta). \tag{8.112}$$

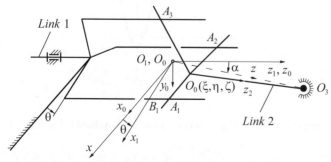

Figure 8.34 Simple mechanism with tripod joint.

This means that for a change of angle θ equal to 2π, axis O_3O_2 performs (in a first approximation) three complete rotations around the fixed axis O_3O_0.

The C1–C tripod kinematic pair

In this situation, the parameters of position φ, γ, ψ, ξ, η, and ζ are obtained from the equations (8.104) to which are added (8.74)–(8.77). The system of equations is solved numerically using the Newton–Raphson method, taking as initial values for parameters ψ, φ, and γ the following values

$$\psi_0 = -\varphi_0 = -\theta, \quad \gamma_0 = \alpha;$$

The initial values for parameters ξ, η, and ζ are the corresponding values obtained from (8.104). As an example, consider the following data

$$l = 100, \ \alpha = 30°, \beta = \frac{3\pi}{2}$$

and the second-degree polynomial

$$x_1 = f(z_1) = r + \lambda_1^* z_1 + \lambda_2^* z_1^2. \tag{8.113}$$

The different values of parameters r, λ_1^*, and λ_2^* are used to obtain the values of the unknown quantities γ, φ, ξ, and η, with which we have drawn the diagrams in Figures 8.35–8.37.

From the diagrams presented in Figure 8.35, the reader can easily see the influence of the curves (Γ_j) on the variation of angle γ. It follows that for concave curves the maximum value

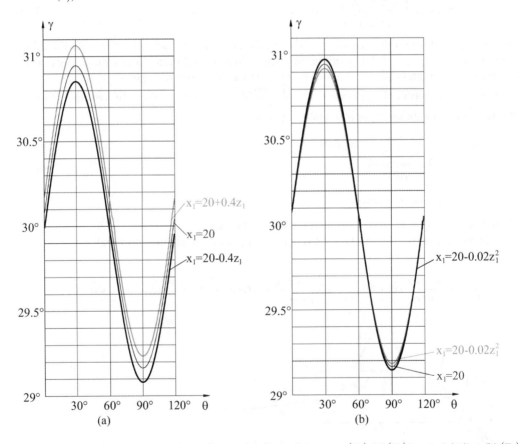

Figure 8.35 The function $\gamma = \gamma(\theta)$ depending on the shape of the curve (Γ_1): (a) (Γ_1) is a straight line; (b) (Γ_1) is a parabola.

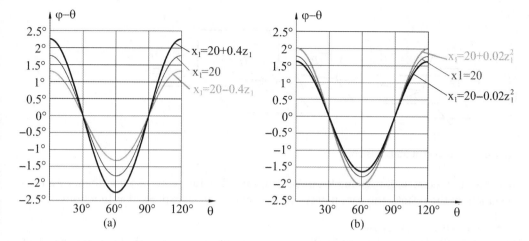

Figure 8.36 The function $\varphi - \theta = (\varphi - \theta)(\theta)$ depending on the shape of the curve (Γ_1): (a) (Γ_1) is a straight line; (b) (Γ_1) is a parabola.

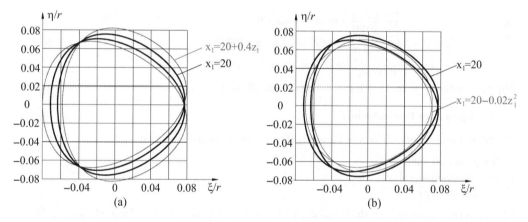

Figure 8.37 The curves described by point O_2 (the variation $\frac{\eta}{r} = \frac{\eta}{r}\left(\frac{\xi}{r}\right)$) depending on the shape of the curves (Γ_1): (a) (Γ_1) is a straight line; (b) (Γ_1) is a parabola.

of angle γ is smaller than the values for straight lines or convex curves. The same conclusion holds true for the variation of the difference $\varphi - \theta$.

From the diagrams in Figure 8.37 it follows that the curve described by point O_2 for concave curves is in a more restricted zone than for the other cases.

The S1–P tripod kinematic pair

Here, the parameters of position are obtained from the (8.104) and (8.85)–(8.87). Eliminating angle θ from these relations, we get

$$\sin(3\varphi) = \frac{\lambda^2 (1 - \cos\gamma)^2 + \sin^2\gamma - \sin^2\alpha}{2\lambda(1 - \cos\gamma)\sin\gamma}, \tag{8.114}$$

where

$$\gamma \in [\gamma_{min}, \gamma_{max}];$$

the values γ_{min} and γ_{max} are deduced from the inequality

$$|\sin(3\varphi)| \leq 1.$$

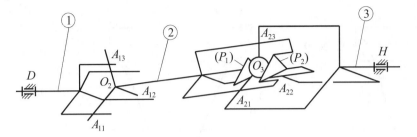

Figure 8.38 Tripod joint transmission.

First the parameters φ, ξ, η, ζ, and θ can be determined as functions of γ and then, by a rearrangement of the results, the parameters φ, γ, ξ, η, and ζ can be expressed as functions of θ.

The S2–P tripod kinematic pair

Here, (8.93)–(8.95) are added to (8.104) to give a non-linear system with the unknowns given by the parameters ξ, η, ζ, ξ, γ, and φ. These parameters are determined as functions of θ.

As an approximation, with $\beta = \frac{3\pi}{2}$ and $\beta_0 = \frac{\pi}{2}$, we get

$$\gamma = \alpha - \lambda \frac{1 - \cos\alpha}{1 + \cos\alpha} \left[\cos(3\theta) + \sin(3\theta)\right],$$
$$\varphi = \frac{\pi}{2} + \theta - \lambda \frac{1 - \cos\alpha}{1 + \cos\alpha} \left[\cos(3\theta) - \cos\alpha\sin(3\theta)\right], \tag{8.115}$$
$$\psi = -\theta + \lambda \frac{1 - \cos\alpha}{1 + \cos\alpha} \left[\cos(3\theta) - \cos\alpha\sin(3\theta)\right];$$

The parameters ξ, η, and ζ are determined from (8.104).

8.4.7 Tripod Joint Transmissions

The tripod joint transmission in Figure 8.38 is made up of three shafts:

1) the principal shaft
2) the intermediary shaft
3) the secondary shaft.

Motion is transmitted from shaft 1 to shaft 2 by means of an angular-axial tripod joint, and from intermediary shaft 2 to secondary shaft 3 by means of an *angular* tripod joint (Figure 8.38) This angular joint comprises the sphere centred at point O_2, rigidly linked to shaft 3, tangential to the planes (P_1) and (P_2), which are parallel to one another, and rigidly linked to shaft 2. To carry out the calculations, we use the following notation (Figures 8.38 and 8.39):

- D and H, the revolute kinematic pairs of shafts 1 and 3, respectively (Figure 8.18)
- O_2, the point of concurrence for the segments O_2A_{1j}, $j = 1, 2, 3$ (Figure 8.38)
- O_2'', the intersection point between the axis of shaft 2 and the plane (P_1) (Figure 8.38)
- R_S, the radius of the sphere (Figure 8.38) centred at point O_2
- l, the length $O_2O_2'' + R_S$ (Figure 8.39)
- O_2', the point situated on the axis of shaft 2 (Figure 8.39) so that $O_2O_2' = l$
- O_0, the point of intersection between the axis of the revolute joint situated at point D and the sphere centred at point O_2 and radius l (Figure 8.39a)
- $A_{1j}, A_{2j}, j = 1, 2, 3$, the representative points of the tripod kinematic pairs
- $O_0x_0y_0z_0$, the fixed reference system (Figure 8.39a) for which axis O_0z_0 coincides with the axis of the revolute kinematic pair situated at point D
- α, the angle between the straight line O_0O_3 and axis O_0z_0 (Figure 8.39a)
- θ_1 and θ_2, the angles of rotation (Figure 8.39a) of shafts 1 and 3, respectively
- $O_1x_1y_1z_1$, the mobile reference system rigidly linked to shaft 1 so that point O_1 coincides with point O_0 (Figure 8.39a)

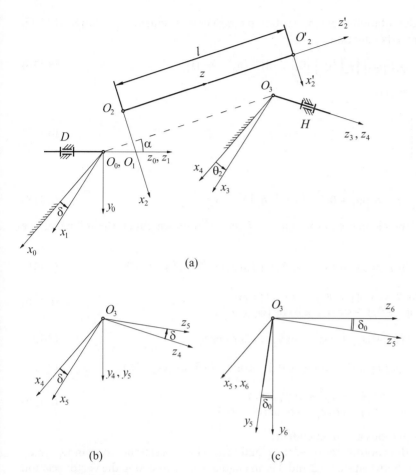

(a)

(b)   (c)

Figure 8.39 Reference systems.

- $O_2x_2y_2z_2$, the mobile reference system rigidly linked to shaft 2 (Figure 8.39a)
- $O_2'x_2'y_2'z_2'$, the reference system having the origin at point O_2' (Figure 8.39a) rigidly linked to shaft 2 and with the axes parallel to the axes of the reference system $O_2x_2y_2z_2$
- $O_3x_3y_3z_3$, the mobile reference system rigidly linked to shaft 3 (Figure 8.39a) so that axis O_3z_3 coincides with the axis of the revolute kinematic pair situated at point H and axis O_3x_3 coincides with the axis of the O_3A_{21} leg
- $O_3x_4y_4z_4$, the fixed reference system in which axis O_3z_4 coincides with axis O_3z_3 and axis O_3x_4 is inclined with angle θ_2 with respect to axis O_3x_3 (Figure 8.39a)
- $O_3x_5y_5z_5$, the fixed reference system in which axis O_3y_5 coincides with axis O_3y_4 (Figure 8.39b)
- δ, the angle between axes O_3x_4 and O_3x_5 (Figure 8.39b)
- $O_3x_5y_6z_6$, the fixed reference system having the axes parallel to the axes of the system $O_0x_0y_0z_0$ (Figure 8.39c) and axis O_3x_6 coincides with axis O_3x_5
- δ_0, the angle between axes O_3y_6 and O_3y_5 (Figure 8.39c)
- $\psi_i, \gamma_i, \varphi_i, i = 1, 2$, the Euler angles that define the positions of the reference systems $O_2x_2y_2z_2$ and $O_3x_3y_3z_3$, respectively towards the reference system $O_1x_1y_1z_1$
- $\xi_i, \eta_i, \zeta_i, i = 1, 2$, the coordinates of points O_2 and O_3 relative to the reference systems $O_1x_1y_1z_1$ and $O_2'x_2'y_2'z_2'$, respectively.

Taking into account the rotation angles and rotation matrices mentioned in Section 8.4.1 (8.45, 8.46), we obtain the matrix equation

$$[\theta_1] [\psi_1] [\gamma_1] [\varphi_1] [\psi_2] [\gamma_2] [\varphi_2] [\theta_2]^{-1} [\delta]^{-1} [\delta_0]^{-1} = [\mathbf{I}], \tag{8.116}$$

where [**I**] is the unity matrix

$$[\mathbf{I}] = \begin{bmatrix} 1 & 0 & 0 \\ 0 & 1 & 0 \\ 0 & 0 & 1 \end{bmatrix}. \tag{8.117}$$

Equation (8.116) reduces to

$$[\gamma_1] [\varphi_1 + \psi_2] [\gamma_2] = [\psi_1 + \theta_1]^{-1} [\delta_0]^{-1} [\delta] [\theta_2 - \varphi_2] ; \tag{8.118}$$

From the equality of the elements in the third row and column, we obtain the following five scalar equations

$$\sin (\varphi_1 + \psi_2) \sin \gamma_3 = \sin \delta \cos (\psi_1 + \theta_1) + \sin \delta_0 \cos \delta \sin (\psi_1 + \theta_1), \tag{8.119}$$

$$\begin{aligned} - \cos \gamma_1 \sin \gamma_2 \cos (\varphi_1 + \psi_2) - \sin \gamma_1 \cos \gamma_2 \\ = - \sin \delta \sin (\psi_1 + \theta_1) + \sin \delta_0 \cos \delta \cos (\varphi_1 + \theta_1), \end{aligned} \tag{8.120}$$

$$\sin \gamma_1 \sin \gamma_2 \cos (\varphi_1 + \psi_2) + \cos \gamma_1 \cos \gamma_2 = \cos \delta \cos \delta_0, \tag{8.121}$$

$$\sin \gamma_1 \sin (\varphi_1 + \psi_2) = \sin \delta_0 \sin (\varphi_2 - \theta_2) - \cos \delta_0 \sin \delta \cos (\varphi_2 - \theta_2), \tag{8.122}$$

$$\begin{aligned} \sin \gamma_1 \cos \gamma_2 \cos (\varphi_1 + \psi_2) + \cos \gamma_1 \sin \gamma_2 \\ = - \sin \delta_0 \cos (\varphi_2 - \theta_2) - \cos \delta_0 \sin \delta \sin (\varphi_2 - \theta_2). \end{aligned} \tag{8.123}$$

Only three of these equations are independent.

Since point O_2 has the coordinates 0, $-l \sin \alpha$, and $l \cos \alpha$ in the reference system $O_0 x_0 y_0 z_0$, and the kinematic pair between shafts 2 and 3 is an angular one, expressing the vector relation

$$\mathbf{O_1 O_2} + \mathbf{O_2 O'_2} + \mathbf{O'_2 O_3} = \mathbf{O_1 O_3} \tag{8.124}$$

in the reference frame $O_1 x_1 y_1 z_1$, we obtain the matrix relation

$$\begin{bmatrix} \xi_1 \\ \eta_1 \\ \zeta_1 \end{bmatrix} + [\psi_1] [\gamma_1] [\varphi_1] \begin{bmatrix} \xi_2 \\ \eta_2 \\ l \end{bmatrix} = [\theta_1]^{-1} \begin{bmatrix} 0 \\ -l \sin \alpha \\ l \cos \alpha \end{bmatrix} ; \tag{8.125}$$

From (8.125) we get three scalar equations of translation. If we denote by α_{1i}, β_{1i}, and γ_{1i}, $i = 1, 2, 3$, the elements of the columns of the matrix $[\psi_1] [\gamma_1] [\varphi_1]$, the three equations of translation may be put in the form

$$\begin{aligned} \xi_1 + \alpha_{11} \xi_2 + \alpha_{12} \eta_{12} + \alpha_{13} l = -l \sin \alpha \sin \theta_1, \\ \eta_1 + \beta_{11} \xi_2 + \beta_{12} \eta_2 + \beta_{13} l = - \sin \alpha \cos \theta_1, \\ \zeta_1 + \gamma_{11} \xi_2 + \gamma_{12} \eta_2 + \gamma_{13} l = l \cos \alpha. \end{aligned} \tag{8.126}$$

To the three equations of rotation (8.119)–(8.123) and the three equations of translation (8.126) are added the equations of the two tripod kinematic pairs:

- the equations of the form (8.56)–(8.58) if the tripod kinematic pairs are of the type $C2-C$
- the equations of the form (8.74)–(8.77) if the tripod kinematic pairs are of the type $C1-C$
- the equations of the form (8.85)–(8.88) if the tripod kinematic pairs are of the type $S1-P$
- the equations of the form (8.93)–(8.95) if the tripod kinematic pairs are of the type $S2-P$.

In this way, a system of 12 non-linear equations is obtained, from which the unknown parameters $\xi_1, \eta_1, \zeta_1, \xi_2, \eta_2, \psi_1, \gamma_1, \varphi_1, \psi_2, \gamma_2, \varphi_2$, and θ_2 are determined as functions of angle θ_1. Thus, in the case when the two tripod kinematic pairs are of $C2$–C type, taking into account (8.56)–(8.58) and denoting

$$\lambda_1 = \frac{r_1 (1 - \cos \gamma_1)}{2l \cos \gamma_1}, \quad \lambda_2 = \frac{r_2 (1 - \cos \gamma_2)}{2l \cos \gamma_2}, \tag{8.127}$$

the equations (8.126) may be written in the form

$$\lambda_1 \cos (3\varphi_1) + \lambda_2 \left[\cos (3\varphi_2) \cos (\varphi_1 - \varphi_2) - \cos \gamma_2 \sin (3\varphi_2) \sin (\varphi_1 - \varphi_2) \right] \\ - \sin \alpha \sin (\varphi_1 - \theta_1) = 0, \tag{8.128}$$

$$\lambda_1 \sin (3\varphi_1) + \lambda_2 \left[\cos (3\varphi_2) \sin (\varphi_1 - \varphi_2) + \cos \gamma_2 \sin (3\varphi_2) \cos (\varphi_1 - \varphi_2) \right] \\ - \sin \gamma_1 + \sin \alpha \cos (\varphi_1 - \theta_1) = 0, \tag{8.129}$$

$$\frac{\zeta_1}{l} + \lambda_2 \left[\cos (3\varphi_2) \sin (\varphi_1 - \varphi_2) + \cos \gamma_2 \sin (3\varphi_2) \cos (\varphi_1 - \varphi_2) \right] \\ + \cos \gamma_1 - \cos \alpha = 0. \tag{8.130}$$

In this case, from (8.119) and (8.120), in which $\psi_1 = -\varphi_1$ and $\psi_2 = -\varphi_2$, and from the (8.128) and (8.129), we can determine the parameters $\varphi_1, \varphi_2, \gamma_1$, and γ_2 as functions of angle θ_1; angle φ_2 is determined from (8.122) and (8.123), while the displacement ζ_1 is determined from (8.130). If the parameters λ_1 and λ_2 fulfill the conditions

$$\lambda_1 < 0.1, \ \lambda_2 < 0.1, \tag{8.131}$$

then it is possible to obtain, with sufficient accuracy, good results using an approximate method as well. Thus, setting as approximations

$$\varphi_1 \approx \theta_1, \ \gamma_1 \approx \alpha, \ \gamma_2 \approx \beta, \tag{8.132}$$

equations (8.119)–(8.123) lead to

$$\cos \beta = \cos \delta \cos (\alpha + \delta_0), \\ \varphi_2 \approx \widetilde{\psi}_1 + \theta_1 - \pi, \\ \theta_2 \approx \theta_1 + \widetilde{\psi}_1 - \widetilde{\psi}_2, \tag{8.133}$$

where the angles $\widetilde{\psi}_1$ and $\widetilde{\psi}_2$ are given by the following expressions

$$\sin \widetilde{\psi}_1 = \frac{\sin \delta}{\sin \beta}, \quad \cos \widetilde{\psi}_1 = \frac{\cos \delta \sin (\alpha + \delta_0)}{\sin \beta}, \tag{8.134}$$

$$\sin \widetilde{\psi}_2 = \frac{\sin \delta \cos (\alpha + \delta_0)}{\sin \beta}, \quad \cos \widetilde{\psi}_2 = \frac{\sin (\alpha + \delta_0)}{\sin \beta}. \tag{8.135}$$

Then we denote

$$\Delta \gamma_1 = \gamma_1 - \theta_1, \ \Delta \gamma_2 = \gamma_2 - \beta, \\ \Delta \varphi_1 = \varphi_1 - \theta_1, \ \Delta \varphi_2 = \varphi_2 - (\theta_1 + \widetilde{\psi}_1 - \pi), \\ \Delta \theta_2 = \theta_2 - (\theta_1 + \widetilde{\psi}_1 - \widetilde{\psi}_2) \tag{8.136}$$

and we assume that these parameters have small values. Developing (8.119), (8.120), (8.128)–(8.130), and (8.56)–(8.58) into a series and keeping only the linear terms, we obtain

$$\Delta \varphi_1 = A_1 \cos (3\theta_1) + B_1 \sin (3\theta_1), \ \Delta \varphi_2 = A_2 \cos (3\theta_1) + B_1 \sin (3\theta_1), \\ \Delta \gamma_1 = A_3 \cos (3\theta_1) + B_3 \sin (3\theta_1), \ \Delta \gamma_2 = A_4 \cos (3\theta_1) + B_4 \sin (3\theta_1), \\ \Delta \theta_2 = A_5 \cos (3\theta_1) + B_5 \sin (3\theta_1), \tag{8.137}$$

in which the coefficients $A_j, B_j, j = 1, 2, \ldots, 5$ are expressed as functions of the parameters r_1, $r_2, l, \alpha, \beta, \tilde{\psi}_1$, and $\tilde{\psi}_2$. For example, the coefficients A_1 and B_1 are given by the following relations

$$
\begin{aligned}
A_1 &= \frac{r_1 (1 - \cos \alpha)}{2l \cos \alpha \sin \alpha} + \frac{r_2 (1 - \cos \beta)}{2l \sin \alpha \cos \beta} \left[\cos \tilde{\psi}_1 \sin (3\psi_1) + \cos \beta \sin \tilde{\psi}_1 \sin (3\tilde{\psi}_1) \right], \\
B_1 &= \frac{r_2 (1 - \cos \beta)}{2l \sin \alpha \cos \beta} \left[-\cos \tilde{\psi}_1 \sin (3\tilde{\psi}_1) + \cos \beta \sin \tilde{\psi}_1 \cos (3\tilde{\psi}_1) \right].
\end{aligned}
\tag{8.138}
$$

For the following numerical data

$$
\alpha = 10°, \ \delta = 45°, \ \delta_0 = 10°, \ \frac{r_1}{2l} = 0.0261, \ \frac{r_2}{2l} = 0.0237,
$$

we determined the geometric parameters for the transmission, both as an exact calculation using the Newton–Raphson method and as an approximation. The results of the two approaches vary by less than 1%, and the deviations defined by (8.136) are periodical, of period $T = \frac{2\pi}{3}$, and have small variations, namely: $|\Delta\varphi_1| < 3°$, $|\Delta\gamma_1| < 0.7°$, $|\Delta\gamma_2| < 0.5°$, $|\Delta\varphi_2| < 1.2°$, and $|\Delta\theta_2| < 0.3°$.

The variation of the deviations $\Delta\gamma_1$, $\Delta\gamma_2$, and $\Delta\theta_2$ with angle θ_1 are shown in Figure 8.40.

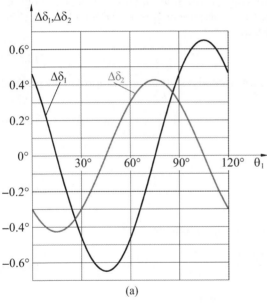

Figure 8.40 The variations (a) $\Delta\gamma_1 = \Delta\gamma_1 (\theta_1)$, $\Delta\gamma_2 = \Delta\gamma_2 (\theta_1)$, and (b) $\Delta\theta_2 = \Delta\theta_2 (\theta_1)$.

8.5   Animation of the Mechanisms

8.5.1   The Need for an Animation

To determine the functional possibilities for a mechanism, the linear dimensions of the elements are not sufficient. When a mechanism functions, the elements and linkages occupy different positions and may sometimes interfere either with other elements, or neighbouring mechanisms, leading to misfunctions even if the kinematic conditions are fulfilled.

For these reasons, it is necessary to model the mechanism's elements and kinematic pairs and to spatially visualize the mechanism obtained for different positions of the driving element: in other words, to create an animation of the mechanism.

8.5.2   The Animation Algorithm

To perform the animation of an arbitrary mechanism it is necessary to cover the following steps:

- the positional analysis of the mechanism as a function of the driving element, with a constant angular step, as well as the determination of the position of each element of the mechanism
- the modelling of the mechanism's elements and kinematic pairs using AutoCAD
- the creation of the animation's frames using script files, which implies the repetition of the following steps for each position of the driving element
  - the positioning of the elements as functions of the kinematic analysis
  - the visualization of the mechanism from a convenient point of view
  - the saving of the obtained image
- performing the animation using specialized software.

In the most cases these steps are repeated until the desired results are obtained. The animation is more realistic if:

- a minimum of 36 positions of the mechanism are used in a complete revolution of the driving element, which implies a kinematic analysis with angular step of 10°
- the elements of the mechanism and the kinematic pairs are modelled as solids
- each image of the mechanism is viewed from the same point and in the same window of visualization
- the images obtained are processed using the AutoCAD command **Render**, applying materials and sources of light only once.

In the next section, we outline the steps in the creation of the animation of a mechanism as well as describing an example.

8.5.3   Positional Analysis

Positional analysis is the first step in the animation of a mechanism. The literature has a great diversity of methods of positional analysis, the particular one used usually depending on the type of mechanism. The general conclusions highlighted are:

- the methods of positional analysis are very different; there is not an universal method;
- for spatial mechanisms, there is a wider range of methods used because many methods cannot deal with the kinematics of spatial mechanisms with complex configuration;
- the methods for positional analysis of spatial mechanisms are complicated; generally, they are different from the methods used for planar mechanisms, which are usually simpler;
- because the positional analysis of spatial mechanisms is laborious, designers avoid it in practice, preferring planar analysis, even if they are dealing with complicated mechanisms, preferring an easy procedure for kinematic analysis;

- the methods of positional analysis of the spatial mechanisms, which include:
  - the method of loops
  - the matrix tensorial method
  - the linear coordinates method
  - the method based on algebraic properties of the screws
  - the method based on the theory of screws and dual matrices
  - the approximate numerical method
  - the matrix iterations method,

  demand a knowledge of certain theoretical notions; the application of these notions implies the determination of certain complicated analytical expressions and, consequently, a great volume of calculations;
- no matter what method is applied for positional analysis of a spatial mechanism, the result is a system of equations that has to be solved.

Next, we will present a method of positional analysis that can be applied to complicated spatial mechanisms, such as mechanisms with six mobile elements and seven revolute joints, known as *7R mechanisms*.

The method of positional analysis requires some simple theoretical knowledge. Starting from the geometric conditions fulfilled by the elements, relative to one another, we obtain simple functions which, finally, lead to a system of non-linear equations, which are solved using the Newton–Raphson method. The approximate position required to initiate the iteration process is obtained using AutoCAD.

Definition of the position functions based on the geometric conditions

The number of functions that define the relative position of an element with respect to a neighbouring one is equal to the number of the geometric constraints introduced by the kinematic pair that links the two elements. These functions can also define the positions of elements linked by kinematic pairs of different classes.

A system of non-linear equations is obtained by taking into account the mobile elements and a simple algorithm to describe how they are assembled assembling. Solving this system of equations does not necessarily demand the writing of analytic expressions for the elements of the Jacobi matrix; the elements of this matrix can be obtained using numerical methods, which considerably reduces the volume of work and, consequently, simplifies the process for reaching the solution of the positional analysis problem.

Let us consider the spatial mono-loop mechanism with n elements, and revolute kinematic joints in a fixed general reference system $OXYZ$, as shown in Figure 8.41. For the ith element of

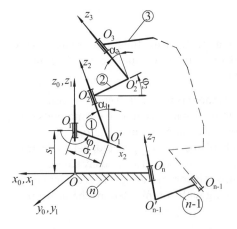

Figure 8.41 Spatial mono-loop mechanism.

the mechanism the following are known:

- the distance O_iO_i' denoted by σ_i
- the distance $O_i'O_{i+1}$ denoted by s_i
- angle $\widehat{O_iO_i'O_{i+1}}$, equal to $\frac{\pi}{2}$
- the angle between axis z_i and the straight line $O_i'O_i$, equal to $\frac{\pi}{2}$,
- the angle between axes z_i and z_{i+1} is denoted by α_i.

In the general reference system, the kinematic pairs at points O_i, and O_{i+1} have the coordinates $X_{O_i}, Y_{O_i}, Z_{O_i}$ and $X_{O_{i+1}}, Y_{O_{i+1}}, Z_{O_{i+1}}$, respectively, while point O_i' has the coordinates $X_{O'}, Y_{O'}, Z_{O'}$. Since the distances σ_i and s_i remain constant no matter the positions of the kinematic pairs at points O_i and O_{i+1}, we may write the functions:

$$F_{\sigma_i} = \sigma_i^2 - \left[\left(X_{O_i} - X_{O_i'}\right)^2 + \left(Y_{O_i} - Y_{O_i'}\right)^2 + \left(Z_{O_i} - Z_{O_i'}\right)^2\right] = 0, \tag{8.139}$$

$$F_{s_i} = s_i^2 - \left[\left(X_{O_{i-1}'} - X_{O_i}\right)^2 + \left(Y_{O_{i-1}'} - Y_{O_i}\right)^2 + \left(Z_{O_{i-1}'} - Z_{O_i}\right)^2\right] = 0. \tag{8.140}$$

From the condition that σ_i is perpendicular to s_{i+1} ($\mathbf{O'}_i\mathbf{O}_i \perp \mathbf{O'}_i\mathbf{O}_{i+1}$), we obtain the function

$$F_{\sigma_i s_{i+1}} = \left(X_{O_i} - X_{O_i'}\right)\left(X_{O_{i+1}} - X_{O_i'}\right) + \left(Y_{O_i} - Y_{O_i'}\right)\left(Y_{O_{i+1}} - Y_{O_i'}\right)$$
$$+ \left(Z_{O_i} - Z_{O_i'}\right)\left(Z_{O_{i+1}} - Z_{O_i'}\right) = 0. \tag{8.141}$$

From the condition that σ_i is perpendicular to s_i ($\mathbf{O'}_i\mathbf{O}_i \perp \mathbf{O'}_{i-1}\mathbf{O}_i$), we get the function

$$F_{\sigma_i s_i} = \left(X_{O_i} - X_{O_{i-1}'}\right)\left(X_{O_i} - X_{O_{i-1}'}\right) + \left(Y_{O_i} - Y_{O_{i-1}'}\right)\left(Y_{O_i} - Y_{O_{i-1}'}\right)$$
$$+ \left(Z_{O_i} - Z_{O_{i-1}'}\right)\left(Z_{O_i} - Z_{O_{i-1}'}\right) = 0. \tag{8.142}$$

The straight lines $\mathbf{O'}_{i-1}\mathbf{O}_i$, and $\mathbf{O'}_i\mathbf{O}_{i+1}$ have the same unit vectors as axes z_i and z_{i+1}, respectively, which means that $\dfrac{\mathbf{O'}_i\mathbf{O}_{i+1}}{|s_{i+1}|} \cdot \dfrac{\mathbf{O'}_{i+1}\mathbf{O}_i}{|s_i|} = \cos\alpha_i$, which is equivalent to

$$F_{\alpha_i} = \left(X_{O_{i+1}} - X_{O_i'}\right)\left(X_{O_i} - X_{O_{i-1}'}\right) + \left(Y_{O_{i+1}} - Y_{O_i'}\right)\left(Y_{O_i} - Y_{O_{i-1}'}\right)$$
$$+ \left(Z_{O_{i+1}} - Z_{O_i'}\right)\left(Z_{O_i} - Z_{O_{i-1}'}\right) - s_i s_{i+1}\cos\alpha_i = 0. \tag{8.143}$$

The functions F_{σ_i}, F_{s_i}, $F_{\sigma_i s_i}$, and F_{α_i} are sufficient to define the position, in the general reference system, of an element i linked by fifth-class kinematic pairs to element $i-1$. By definition, the class of the kinematic pair is given by the number of constraints (not the number of possible motions) of an element in motion relative to another element that is part of the kinematic pair. We may thus state that the number of functions that define the position of an element is equal to the number of constraints introduced by the kinematic pair.

For the case of the fourth-class kinematic pair at point O_i in Figure 8.42, the functions F_{σ_i}, $F_{\sigma_i s_{i+1}}$, $F_{\sigma_i s_i}$, and F_{α_i} are sufficient to define the position, in the general reference system, of element i linked by the fourth-class kinematic pair to element $i-1$. The position angle φ_{i+1} (Figure 8.43) is also determined from geometric considerations, establishing the position of element $i+1$ relative to element i with:

$$\mathbf{O}_i\mathbf{O'}_i \cdot \mathbf{O}_{i+1}\mathbf{O''}_{i+1} = \sigma_i\sigma_{i+1}\cos\varphi_{i+1}, \tag{8.144}$$

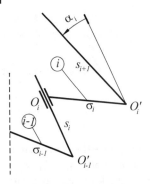

Figure 8.42 Element *i* linked by fourth-class kinematic pairs to element *i* − 1.

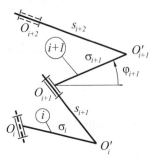

Figure 8.43 Position of element *i* relative to element *i* − 1.

from which we get

$$\cos \varphi_{i+1} = \frac{\left(X_{O_i} - X_{O'_i}\right)\left(X_{O_{i+1}} - X_{O'_{i+1}}\right)}{\sigma_i \sigma_{i+1}}$$
$$+ \frac{\left(Y_{O_i} - Y_{O'_i}\right)\left(Y_{O_{i+1}} - Y_{O'_{i+1}}\right)}{\sigma_i \sigma_{i+1}} + \frac{\left(Z_{O_i} - Z_{O'_i}\right)\left(Z_{O_{i+1}} - Z_{O'_{i+1}}\right)}{\sigma_i \sigma_{i+1}}. \tag{8.145}$$

The condition fulfilled by the straight lines $\mathbf{O}_i\mathbf{O}'_i$, $\mathbf{O}_{i+1}\mathbf{O}'_{i+1}$ and $\mathbf{O}'_i\mathbf{O}_{i+1}$ gives

$$\frac{\mathbf{0}_i\mathbf{O}'_i}{\sigma_i} \cdot \frac{\mathbf{O}_{i+1}\mathbf{O}'_{i+1}}{\sigma_{i+1}} = \frac{\mathbf{O}'_i\mathbf{O}_{i+1}}{s_{i+1}} \sin \varphi_{i+1} \tag{8.146}$$

and from here we obtain

$$\sin \varphi_{i+1} = \frac{\left(X_{O'_i} - X_{O_{i+1}}\right)\left(X_{O_i} - X_{O'_i}\right)\left(X_{O_{i+1}} - X'_{O_{i+1}}\right)}{\sigma_i \sigma_{i+1} s_{i+1}}$$
$$+ \frac{\left(Y_{O'_i} - Y_{O_{i+1}}\right)\left(Y_{O_i} - Y_{O'_i}\right)\left(Y_{O_{i+1}} - Y'_{O_{i+1}}\right)}{\sigma_i \sigma_{i+1} s_{i+1}} \tag{8.147}$$
$$+ \frac{\left(Z_{O'_i} - Z_{O_{i+1}}\right)\left(Z_{O_i} - Z_{O'_i}\right)\left(Z_{O_{i+1}} - Z'_{O_{i+1}}\right)}{\sigma_i \sigma_{i+1} s_{i+1}}.$$

Equations (8.145) and (8.147) are sufficient to define the value of angle φ_{i+1} and its quadrant as well.

Obtaining the system of position equations from geometric constraints

The kinematic analysis of the mechanism in Figure 8.28 leads ultimately to the determination of the positions of the kinematic pairs at points O_i, $i = 1, 2, \ldots, n$. For that purpose, we consider as known the structural dimensions of the elements σ_i, s_i, and α_i, $i = 1, 2, \ldots, n$. We make the following remarks:

- The position of the fixed kinematic pairs O_1 and O_n, linked to the base, can be easily determined in the general reference system, as functions of the structural dimensions of elements 1 and n.
- The position of point O'_{n+1} in the general reference system is known.
- For an arbitrary value of the input angle φ_1, we can determine the position of the kinematic pair at point O_2.

With these remarks in mind, we have to determine the positions of the kinematic pairs at points O_i, $i = 3, 4, \ldots, n - 1$, and of points O'_i, $i = 2, 3, \ldots, n - 2$. Therefore, for the mono-loop mechanism with n elements and revolute kinematic joints in Figure 8.41, we have to determine $N_n = 6\,(3n - 3)$ unknowns ($3\,(n - 3)$ unknowns resulting from the positions of the kinematic pairs at points O_i, and $3\,(n - 3)$ unknowns resulting from the positions of points O'_i). The functions, based on the geometric conditions, are

$$
\begin{cases}
F_{\sigma_i} = 0 \text{ for } i = 2, 3, \ldots, n - 1, \\
F_{s_i} = 0 \text{ for } i = 3, 4, \ldots, n - 1, \\
F_{\sigma_i s_{i+1}} = 0 \text{ for } i = 2, 3, \ldots, n - 1, \\
F_{\sigma_i s_i} = 0 \text{ for } i = 2, 3, \ldots, n - 1, \\
F_{\alpha_i} = 0 \text{ for } i = 2, 3, \ldots, n - 1;
\end{cases}
\tag{8.148}
$$

Therefore, the total number of functions N_f will be $N_f = 5\,(n - 2) - 1$.

For a mono-loop mechanism with one driving element, the number of equations must be equal to the number of unknowns. If we denote by $\{\mathbf{F}\}$ the vector containing the functions based on geometric constraints, then we obtain the system of non-linear equations

$$
\{\mathbf{F}\} = \{\mathbf{0}\}\,,
\tag{8.149}
$$

where function f_j has the expression

$$
F_j = \begin{cases}
F_{\sigma_{j-1}} & \text{for } 1 \leq j < n - 1, \\
F_{s_{j-(n+2)+2}} & \text{for } n - 1 \leq j < 2\,(n - 2), \\
F_{\sigma_{j-2(n-2)+2} s_{j-2(n-2)+3}} & \text{for } 2\,(n - 2) \leq j < 3\,(n - 2), \\
F_{\sigma_{j-3(n-2)+2} s_{j-3(n-2)+3}} & \text{for } 3\,(n - 2) \leq j < 4\,(n - 2), \\
F_{\alpha_{j-4(n-2)+2}} & \text{for } 4\,(n - 2) \leq j < 5\,(n - 2).
\end{cases}
\tag{8.150}
$$

Numerical solution of the system of position equations

To solve the system of position equations we will use the Newton–Raphson method, an iterative approach of successive approximations. We denote by $\{\mathbf{XYZ}\}$ the vector of the unknowns, which will be determined,

$$
\{\mathbf{XYZ}\} = \begin{bmatrix} X_1 & X_2 & X_3 & \ldots & X_{k-1} & X_k & X_{k+1} & \ldots & X_{5(n-2)-1} \end{bmatrix}^{\mathrm{T}}.
\tag{8.151}
$$

For the initiation of the iterative process it is necessary to have approximate solution of system (8.149). Let $\{\mathbf{XYZ}\}_0$ be this approximate solution,

$$
\{\mathbf{XYZ}\}_0 = \begin{bmatrix} X^0_{O'_2} & Y^0_{O'_2} & Z^0_{O'_2} & X^0_{O'_3} & \ldots & X^0_{O'_{n-1}} & Y^0_{O'_{n-1}} & Z^0_{O'_{n-1}} \end{bmatrix}^{\mathrm{T}}.
\tag{8.152}
$$

After the first iteration, the solution $\{\mathbf{XYZ}\}$ reads

$$
\{\mathbf{XYZ}\}_1 = \{\mathbf{XYZ}\}_0 - [\mathbf{W}]_0^{-1} \{\mathbf{F}\}_0\,,
\tag{8.153}
$$

where $[\mathbf{W}]_0^{-1}$ and $\{\mathbf{F}\}_0$ are the inverse of the Jacobi matrix at the first iteration, and the value of the functions $\{\mathbf{F}\}$ at the first iteration, respectively. After m iterations the solution will be

$$
\{\mathbf{XYZ}\}_m = \{\mathbf{XYZ}\}_{m-1} - [\mathbf{W}]_{m-1}^{-1} \{\mathbf{F}\}_{m-1}\,.
\tag{8.154}
$$

Writing analytical expressions for the elements of the Jacobi matrix can be avoided by using numerical methods. Thus the expression for an element W_{jk} of this matrix is given by

$$W_{jk} = \frac{F_j\left(\ldots, X_{k-1}, X_k + \Delta x, X_{k+1}, \ldots\right) - F_j\left(\ldots, X_{k-1}, X_k - \Delta x, X_{k+1}, \ldots\right)}{2\Delta x}, \tag{8.155}$$

where Δx is an arbitrary small value.

The iterative process repeats until the condition

$$\left|\left(X_k\right)_{m+1} - \left(X_k\right)_m\right| \le \varepsilon_k, \; k = 1, 2, 5\,(n-2) - 1 \tag{8.156}$$

is fulfilled, where ε_k is the admissible error for the solution X_k.

For convergence of the iterative process, it is necessary to obtain an approximate solution for the first iteration. For that purpose, AutoCAD can be used. The estimation of this approximate position is not a characteristic of this method; it is found in all methods of positional analysis.

If the positional analysis of the mechanism is performed over a kinematic cycle by changing angle φ_1 (that is, $\varphi_1 = \varphi_1 + \Delta\varphi$, where $\Delta\varphi$ may be equal to $\frac{\pi}{18}$ rad), then the exact solution from one step can be used as the approximate starting point for the next one.

Numerical example in 7R mechanisms

For the 7R mechanism in Figure 8.44, the kinematic schema of which is given in Figure 8.45, the linear dimensions of the elements and the angles between axes z_i and z_{i+1} (the angles α_i) are known. For a numerical example, we will use the values in Table 8.2 for the parameters. Table 8.3 shows the approximate solution when angle φ_1 takes the value of $0°$.

The kinematic schema in Figure 8.45 immediately gives the positions of the kinematic pairs $O_1, O_1', O_2, O_6',$ and O_7 by:

$$X_{O_1} = 0, \; Y_{O_1} = 0, \; Z_{O_1} = s_1, \tag{8.157}$$

$$X_{O_1'} = X_{O_1} + \sigma_1 \cos\varphi_1, \; Y_{O_1'} = Y_{O_1} + \sigma_1 \sin\varphi_1, \; Z_{O_1'} = Z_{O_1}, \tag{8.158}$$

$$\begin{aligned} X_{O_2} &= X_{O_1'} + s_2 \sin\alpha_1 \sin\varphi_1, \\ Y_{O_2} &= Y_{O_1'} - s_2 \sin\alpha_1 \sin\varphi_1, \\ Z_{O_2} &= Z_{O_1'} + s_2 \cos\alpha_1, \end{aligned} \tag{8.159}$$

Figure 8.44 The 7R mechanism.

Figure 8.45 Kinematic schema for the 7R mechanism.

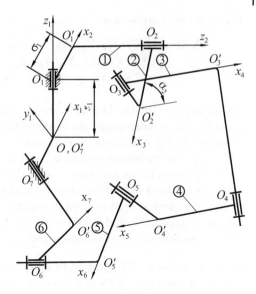

Table 8.2 Linear dimensions of the elements.

Element	1	2	3	4	5	6	7
$\sigma_i = O_iO_i'$ [m]	0.090	0.110	0.150	0.150	0.110	0.090	0.100
$s_i = O_{i-1}'O_i$ [m]	0.100	0.120	0.080	0.200	0.080	0.120	0.100
α_i [^{0}]	80	93	120	120	93	80	35

Table 8.3 Approximate solution for the positions of the kinematic pairs.

Joint	O_2'	O_3	O_3'	O_4	O_4'	O_5	O_5'	O_6
X [m]	−0.0057	−0.0451	−0.0323	−0.1333	−0.1015	−0.1393	−0.2282	−0.1832
Y [m]	−0.1276	−0.1114	−0.2549	−0.3113	−0.1799	−0.1415	−0.1307	−0.0293
Z [m]	0.0675	0.1353	0.1770	0.0138	−0.0513	0.0079	−0.0560	−0.1016

$$X_{O_6'} = -\sigma_7, \ Y_{O_6'} = -s_7 \sin\alpha_7, \ Z_{O_6'} = -s_7 \cos\alpha_7, \tag{8.160}$$

$$X_{O_7} = -\sigma_7, \ Y_{O_7} = 0, \ Z_{O_7} = 0. \tag{8.161}$$

The positional analysis is performed with angular step $\Delta\varphi = \frac{\pi}{18}$ rad, while angle φ_1 varies from 0 to 2π radians, following $\varphi_1 = \varphi_1 + \Delta\varphi$.

We will have to determine the positions of the kinematic pairs O_i, $i = 3, 4, 5, 6$, and of points O_i', $i = 2, 3, 4, 5$. The position functions are given by (8.63)–(8.67), while the vector $\{F\}$, according to (8.74), reads

$$F_j = \begin{cases} F_{\sigma_{j+1}} & \text{for } 1 \leq j < 6, \\ F_{s_{j-3}} & \text{for } 6 \leq j < 10, \\ F_{\sigma_{j-8}s_{j-7}} & \text{for } 10 \leq j < 15, \\ F_{\sigma_{j-13}s_{j-13}} & \text{for } 15 \leq j < 20, \\ F_{\sigma_{j-18}} & \text{for } 20 \leq j < 25. \end{cases} \tag{8.162}$$

The vector of unknowns is

$$\{\mathbf{XYZ}\} = \begin{bmatrix} X_{O'_2} \ Y_{O'_2} \ Z_{O'_2} \ X_{O_3} \ Y_{O_3} \ Z_{O_3} \ X_{O'_3} \ Y_{O'_3} \\ Z_{O'_3} \ X_{O_4} \ Y_{O_4} \ Z_{O_4} \ X_{O'_4} \ Y_{O'_4} \ Z_{O'_4} \ X_{O_5} \\ Y_{O_5} \ Z_{O_5} \ X_{O'_5} \ Y_{O'_5} \ Z_{O'_5} \ X_{O_6} \ Y_{O_6} \ X_{O_6} \end{bmatrix}^{\mathrm{T}} . \tag{8.163}$$

The elements of the Jacobi matrix are obtained using (8.79), where $\Delta x = 10^{-6}$. In this way, we avoid having to write analytical expressions for the elements of the Jacobi matrix (with 576 components, 166 of them being non-zero).

The maximum admissible error at the determination of the coordinates is $\varepsilon = 10^{-8}$.

Using a Turbo Pascal program to solve the problem of the kinematic analysis of a 7R mechanism, we obtained a set of 36 values for each of the 24 unknowns. The code for the program is given in companion website of this book. The program is structured so that it is easily modified for analyzing other spatial mechanisms. To that end it has, in its composition, nine procedures that can be grouped into two categories:

- four procedures – *FCapital12*, *FCapital3*, *FCapital4*, *FCapital5* – which define, for a certain element, the constraints introduced by the element and the related kinematic pairs; these constraints are given by (8.139)–(8.143);
- five procedures – *Fsmalli*, *Fsigmai*, *Fsigmaissmalli*, *Fsigmaissmalliplus1*, *Falphai* – used to define the system of position equations (8.73), given by (8.162).

Thus, according to (8.140) and (8.150), the procedure *Fsmalli* permits the calculation of the values of the function F_{s_i}, for $i = 1, 2, 3, 4$. For that purpose, the procedure also calls the procedure *FCapital12* four times, in order to determine the first four values of the vector *FCapitalVector*. The procedure *FCapital12* has as input data the positions of the neighbouring kinematic pairs at points O'_{i-1} and O_i (six values) and, using the variable j, it transfers, in the vector *XYZ*, the positions of the kinematic pairs situated in the vector *VectPos* (a vector with 42 positions containing the positions of the kinematic pairs from O_1 to O_7).

For instance, for $i = 1$, the components of the vector *XYZ* are assigned from the vector *VectPos* by the correspondence:

- the first component of the vector *XYZ* is the tenth component of the vector *VectPos;*, that is, $X_{O'_2}$
- the second component of the vector *XYZ* is the eleventh component of the vector *VectPos*; that is, $Y_{O'_2}$
- the third component of the vector *XYZ* is the twelfth component of the vector *VectPos*; that is, $Z_{O'_2}$
- the fourth component of the vector *XYZ* is the thirteenth component of the vector *VectPos*; that is, X_{O_3}
- the fifth component of the vector *XYZ* is the fourteenth component of the vector *VectPos*; that is, Y_{O_3}
- the sixth component of the vector *XYZ* is the fifteenth component of the vector *VectPos*; that is, Z_{O_3}.

The sequence of indices from 10 to 15 was came from incrementing j after each assignment; that is, $j = j + 1$.

For $i = 1$, the input data of the procedure *FCapital12* are (in order): s_3, $X_{O'_2}$, $Y_{O'_2}$, $Z_{O'_2}$, X_{O_3}, Y_{O_3}, Z_{O_3}, while the output variable is the value of the function in (8.140), symbolically denoted by F in the program. We thus determine the first component of the vector *FCapitalVector*. In a similar way, we can find the next three values using the procedure *FCapital12*.

The other four procedures are used in a similar way, entering in the vector *FCapitalVector* the values of the functions, as follows:

- the procedure *Fsigmai* for the components with indices from 5 to 9,
- the procedure *Fsigmaissmalli* for the components with indices from 10 to 14,
- the procedure *Fsigmaissmalliplus1* for the components with indices from 15 to 19,
- the procedure *Falphai* for the components with indices from 20 to 24.

The 24 values of the vector *FCapitalVector* are determined in the main program by calling these five procedures. The way in which the nine procedures were written (the functions in (8.139) and (8.140) have the same form; so only one procedure, *Fcapital12*, was needed) permits them to be adapted for any type of mechanism or system of equations obtained by index generation. The main program begins with the definition of the approximate solution (the 42 components of the vector *Positions* corresponding to angle $\varphi_1 = 0$). Then, in a *For* loop, values are given to angle φ_1 in the interval $(0, 2\pi)$, the angular step being $1°$. The positions of the points which remain fixed (O_1, O'_6 and O_7) are determined using (8.157), (8.160) and (8.161) outside the *For* loop, while the positions of the points that can be exactly determined in an analytical way (O'_1 and O_2) as functions of angle φ_1, are determined from (8.158) and (8.159) inside the loop. In a *Repeat - Until* loop, the 24 values of the unknown are determined with a precision of 10^{-8}, as follows:

1) Construct the vectors *VectPos* and *VectP* containing the 42 values of the vector *Positions*.
2) Calls the five procedures *Fssmalli*, *Fsigmai*, *Fsigmaissmalli*, *Fsigmaissmalliplus1*, and *Falphai* to determine the components of the vector *FCapital* (the approximate values for the positions of the kinematic pairs situated at points O'_2, O_2, O'_3, O_4, O'_4, O_5, O'_5, O_6).
3) In two *For* loops, numerically determine the Jacobi matrix $[\mathbf{W}]$, denoted by A in the program, and the components of which are given in (8.155).
4) With (8.154) determine the exact solution for the current step, the inverse of the matrix $[\mathbf{W}]$ being calculated with the procedure *MatInv*, which performs the following correspondences of the input data
   - $[\mathbf{W}] \rightarrow A$,
   - $\{\mathbf{F}\} \rightarrow B$
   and the following correspondences of the output data
   - $A \rightarrow [\mathbf{W}]^{-1}$,
   - $B \rightarrow [\mathbf{W}]^{-1}\{\mathbf{F}\}$;
5) For the current step, determine the error and compare it to the required level. If (8.156) does not hold true, then the exact solution determined for the current step becomes the approximate one for the next iteration.

The exact solution determined in the *Repeat - Until* loop for a value of angle φ_1 is put into the vector *VectPos* and it becomes approximate solution for the next value of angle φ_1. Then, using the components of the vector *VectPos*, the data (from $10°$ to $10°$) is written to a file called *PosKinPairs*.

Finally, in the file *PosKinPairs* we obtain the positions of the kinematic pairs situated at the points from O_1 to O_7 and of the points from O'_1 to O'_6; these values are determined with the required precision for all 36 equidistant positions of the driving element (in total, we have 1512 values). Based on these data, diagrams of variation of the coordinates can be drawn. For example, Figures 8.46 and 8.47 show graphs of variation of the coordinates of points O_3 and O_4.

8.5.4 Modelling the Elements of a Mechanism

Figure 8.48 shows a bar-type element and fifth-class kinematic pairs, both modelled with solids. We choose a circular section for the element and joint, all the elements having diameter equal to

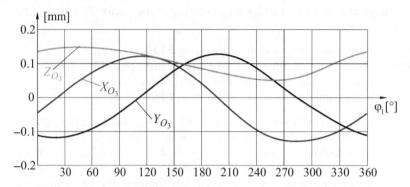

Figure 8.46 Variations of the coordinates of point O_3 as functions of angle φ_1.

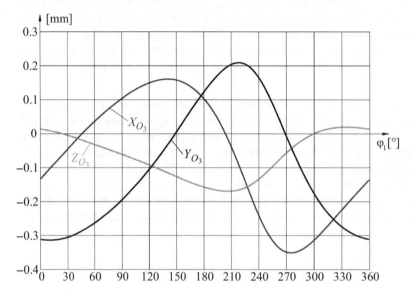

Figure 8.47 Variations of the coordinates of point O_3 as a function of angle φ_1.

Figure 8.48 Element modelled with solids.

Figure 8.49 Dimensions of the elements and kinematic pairs.

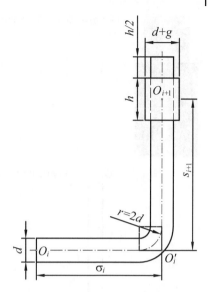

d, while the kinematic pair has diameter equal to $d + g$. For the kinematic pair, we also define the height h, while for an element we define the distance $\frac{h}{2}$ by which it is longer than the kinematic pair, as in Figure 8.49.

The procedure for an element i is as follows:

1) Construct the kinematic pair at point O_{i+1} using the AutoCAD command **Cylinder**, the cylinder having its centre at $(0, 0, -\frac{h}{2})$, base diameter $d = g$, and height h.
2) Construct the cylindrical part that exceeds the kinematic pair, again using the command **Cylinder**; the new cylinder has its centre at $(0, 0, \frac{h}{2})$, base diameter of d and height of $\frac{h}{2}$.
3) Rotate the elements constructed around axis OY by an angle of $-90°$, giving the construction shown in Figure 8.50.
4) Declare as block (**Block**) the construction created, the identification point being $(0,0,0)$ and the name being *Joint*.
5) Draw, with the command **Pline**, the ith element of the mechanism, as shown in Figure 8.51, point O_i having the coordinates $(0,0)$.
6) Fastens, with the command **Fillet**, the right angle at point O_i' with a radius $r = 2d$, obtaining the construction in Figure 8.52.
7) At point O_i, construct a circle of diameter d with the command **Circle** (Figure 8.53); this will be rotated by $90°$ around axis OY (Figure 8.54).
8) With the AutoCAD command **Extrude**, create the solid element by extruding the circle along the poly-line of the symmetry axis of the element, giving the representation in Figure 8.55.
9) At point O_{i+1}, insert the kinematic pair using the command **Insert**, giving the representation in Figure 8.48.
10) Declare as block (AutoCAD command **Block**) the construction created, thus saving it with the name *Bar* followed by the number of the element (e.g. for element number 1, the name is *Bar1*).

Repeating the previous steps, we obtain more blocks saved in the memory with the name of the elements.

If the kinematic pairs have the same diameter and the same height, then it is not necessary to construct them at each step, but it *is* necessary to recall the block *Joint* and to insert it. For a good

Figure 8.50 The joint and the cylindrical part at point O_{i+1}.

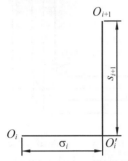

Figure 8.51 Construction of the element.

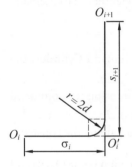

Figure 8.52 Fastening with the command **Fillet.**

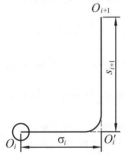

Figure 8.53 Construction of the circle.

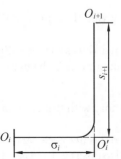

Figure 8.54 Rotation of the circle around axis *OY*.

Figure 8.55 The extruded element *i*.

quality photographic representation it is necessary to apply materials to the elements' surfaces using the command **Rmat**. The properties of the materials – colour, roughness, degree of reflection and transparency – are chosen from the dialog boxes of the command **Rmat** (*Materials* ≫ *Materials Library* ≫ *Modify Standard Material*).

8.5.5 Creation of the Animation Frames

The linear dimensions of the mechanism's elements are given in Table 8.2. From the structural point of view, the elements have a circular shape, all of them with the same diameter $\varphi = 15$ mm. The kinematic pairs are also cylindrical, with interior diameter $\varphi_i = 15$ mm, exterior diameter $\varphi_e = 21$ mm, and height $h = 30$ mm.

It is necessary to model only the elements and the kinematic pairs once, using either a calculation program that will generate a script file, or by modelling the elements and kinematic pairs individually, saving them with the **Block** command.

To position the elements and kinematic pairs, it is necessary to use a programming language that generates a script file, because obtaining the 36 positions of the mechanism needs 2268 values representing the coordinates of the points (nine values for each element).

Next, we will describe automated modelling with solids. This is done using a script file to generate the elements and the kinematic pairs of the mechanism. The kinematic pairs and the elements modelled with solids have the shapes shown in Figure 8.48. There will be seven elements and their corresponding kinematic pairs, saved as blocks, each element having its own name. The construction of the mechanism is the next step. For that purpose, the following steps are required:

1) Successively inserts each element in the plane (**Inset**), using a known point as the insertion point, usually the origin $O\,(0, 0, 0)$.
2) Using the command **Align**, successively align each element, establishing the following correspondences (see Figure 8.32):

$$(0, 0, 0) \rightarrow \left(x_{O_i}, y_{O_i}, z_{O_i}\right),$$
$$(\sigma_i, 0, 0) \rightarrow \left(x_{O_i'}, y_{O_i'}, z_{O_i'}\right),$$
$$(\sigma_i, s_{i+1}, 0) \rightarrow \left(x_{O_{i+1}}, y_{O_{i+1}}, z_{O_{i+1}}\right).$$

3) With the command **Vpoint**, view the mechanism from a convenient point, previously established and which remains unchanged for all of the mechanism's positions.
4) With the command **Zoom**, establish the visualization window, which is the same for all of the mechanism's positions.
5) Create the photographic image with the **Render** command, using, for a good quality image, the rendering mode *Photo Raytrace*; the image is a file having the destination (*Destination*) *File*, the type being selectable using the options *More Option* ≫ *File Output Configuration*.

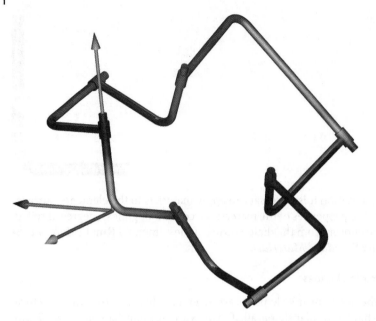

Figure 8.56 The image obtained using the **Render** command for the 0° position of the driving element.

For example, the image might be chosen to be a .bmp file of resolution 1024×768. Images will be saved for each position of the mechanism. W recommend that the the position of the mechanism is incorporated in the file name, for example *Fig000*, *Fig010*, ..., *Fig360*.

Figure 8.56 shows the 7R mechanism at the 0° position of the driving element ($\varphi_1 = 0$), after taking all of the steps in this algorithm. For clarity, the axes of coordinates OX, OY, and OZ have also been drawn, according to the kinematic schema in Figure 8.45. The realization of 36 good-quality photographic images requires a knowledge of photographic representation of objects: the configuration of the photograph, the creation of a stage with optimum placement of the light sources, and the creation of a background. These will remain in place for all of the mechanism's positions.

The script file which gives the images of the mechanism in each of the 36 positions, using the algorithm described above, was created using a Turbo Pascal program, the code for which is found in the companion website of the book. The program has as its input data:

- the structural dimensions of the mechanism (σ_i, s_{i+1})
- the diameter of the bars
- the gauge dimensions of a kinematic pair and the positions of points O_i, O'_i, and O_{i+1} (Figure 8.49) for all 36 positions of the mechanism.

The positions of points O_i, O'_i and O_{i+1} are obtained using the kinematic analysis program presented in the companion website of the book (*Kin_7R.Pas*). As discussed in Section 8.5.3, at the end of the kinematic analysis, a file called *PosKinPairs* is obtained. In this are written (in order) the 42 positions of points O_i, O'_i and O_{i+1}, for each of the 36 positions of the mechanism. From this file we read the corresponding values and construct the matrix *VectPos*, which has 36 rows and 42 columns. The script file from which we will create the animation is called *Animate.scr*. We successively write in it:

1) *The dimensions of the visualization window for the construction of the elements and kinematic pair*: To use the animation program for other mechanisms too, a visualization window is determined by finding the largest linear dimensions of the mechanism (σ_i, s_{i+1}). In the dialog box of the command **Zoom**, when *Specify first corner* is requested, the responses *xbottomw*,

ybottomw are given. When *Specify the opposite corner* is requested, responses *xupw, yupw* are given, where the four values of the visualization window depend on the diameter of the element and the greatest values of the lengths σ_i and s_{i+1}.

2) *The instructions for the construction of the kinematic pair and of the portion of element that exceeds the kinematic pair*: The AutoCAD instructions give the representation in Figure 8.50.

3) *The saving of the construction*: as *Block*, with the name *Joint*.

4) *The construction of the seven elements:* There are six mobile elements and th fixed element. Construction is in a *For* loop, according to the previously described algorithm and the succession of AutoCAD commands that leads to the representations in Figures 8.50–8.55. Finally, the kinematic pair situated at point O_{i+1} is inserted and saved as *Block*; the construction is named *Bar* followed by the number of the element (*Bar1, Bar2, ..., Bar7*).

5) *The construction of the mechanism*: in the 36 positions using two *For* loops. In the first *For* loop is embedded a second *For* loop, in which is written, in the script file, the AutoCAD commands for the successive insertion and positioning (with the AutoCAD command **Align**) of the seven elements. At the end of the second loop the representation of the mechanism for one position of the driving element is obtained. After this, the AutoCAD commands for visualization and settings of various system variables (**Facetres** and **Viewres**) are written to the script file; these commands are required for a good quality representation. Then the command that generates the photographic representation (**Render**) is given, and the image is saved. At the end of the first loop the obtained construction is erased and the instructions to pass from space to plane are used, thus preparing for the next position of the mechanism and for the generation of a new image.

At the calling of the script file *Animate.scr*, AutoCAD will construct the seven elements of the mechanism, and will generate their positioning in the 36 steps of the driving element.

The only intervention required is in the configuration of the image files and entering the name of the image obtained. By calling the command **Render**, the dialog box in Figure 8.57 appears. The mode *Foto Raytrace* is used for the photographic representation and the destination option *File*. For the minimization of the aliasing phenomenon, option *Height* (Figure 8.58) is used in the *Render Option* box.

In the *File Output Configuration* dialog box (Figure 8.59) the type of image and the resolution are chosen. When the configuration is over, the button *Render* (Figure 8.57) is pressed, and after the rendering process and the cassette in Figure 8.60 has appeared, the name of the image is entered. The configuration is set up once, for the first position; for subsequent positions, the *Render* button in the cassette dialogue box in Figure 8.57 is pressed and the the name of the new image is entered in the cassette in Figure 8.60.

Figure 8.57 The **Render** dialog box.

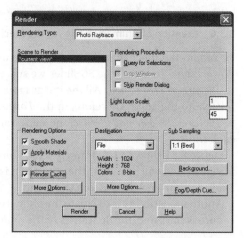

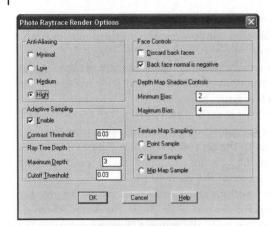

Figure 8.58 The *Render Option* dialog box.

Figure 8.59 The *File Output Configuration* dialog box.

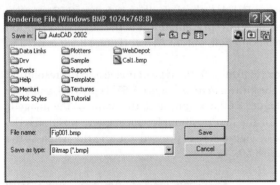

Figure 8.60 The *Rendering File* dialog box.

8.5.6   Creation of Animation File for the Mechanism

To obtain the mechanism's animation we need specialized software, or failing this Microsoft PowerPoint. To create the 36 slides we successively insert the 36 images using the menu *Insert* ≫ *Picture* ≫ *From File...* All the images must have the same dimensions and the same position relative to the page. Therefore, in the *Format Picture* dialog, obtained from the menu *Format* ≫ *Picture...* ≫ *Format Picture*, the same values must be entered for *Size*, *Position*, and *Picture* each time.

Figure 8.61 shows a few frames of the animation. After the inserting of the all 36 images, the file generated will be saved. The animation is obtained by pressing the F5 key or by calling *Slide Show*, option *View Show* from the menu.

Another possibility is to create the animation using Windows Movie Maker. The working procedure is the same: the 36 frames of the animation (*Edit Movie*) are inserted and the animation

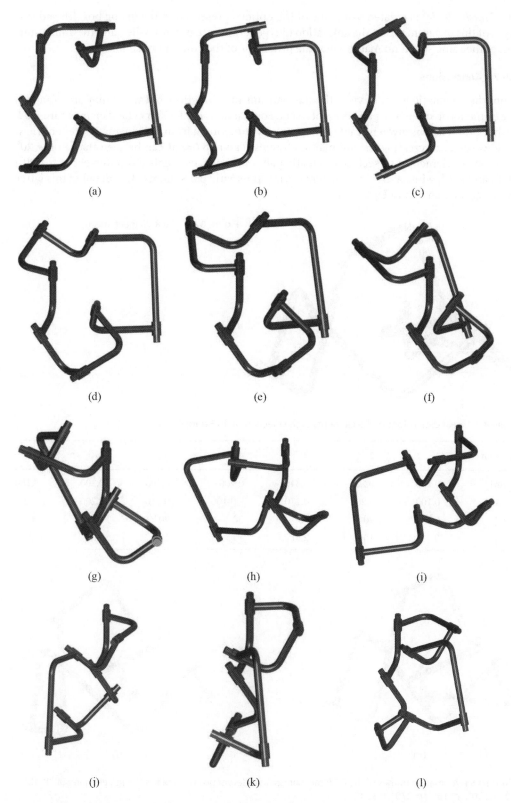

Figure 8.61 Animation frames of the 7R mechanism at different positions of the driving element: (a) 0°; (b) 30°; (c) 60°; (d) 90°; e) 120°; f) 150°; g) 180°; h) 210°; i) 240°; j) 270°; k) 300°; l) 330°.

file (*Finish Movie*) is created using one of the options presented in the dialog box. The advantages of this program is that the animation files generated are much smaller, and the animations run at the same speed no matter what the processor of the computer used.

8.5.7 Conclusions

From the kinematic point of view, the mechanisms may function, but while they are doing so, the elements or kinematic pairs may intersect. For instance, the R-7R mechanism in Figure 8.62, with the dimensions given in Table 8.4, is perfectly functional from the kinematic point of view.

However, after creating the animation's frames (Figure 8.63), it can be seen that, in the 20° position, an element intersects a kinematic pair. Moreover, elements also intersect at the 90°, 160°, and 270°, while at 320° two neighbouring kinematic pairs intersect. A detail of this latter intersection is shown in Figure 8.64.

Figure 8.62 The R-7R mechanism.

Table 8.4 Linear dimensions of the elements of the mechanism in Figure 8.62.

Element	1	2	3	4	5	6	7
σ_i [m]	0.100	0.200	0.340	0.200	0.300	0.200	0.100
s_i [m]	0.100	0.200	0.210	−0.100	0.110	0.200	0.200
α_i [°]	60	90	90	30	90	120	90
θ_i [°]	0	100	220	190	130	340	225

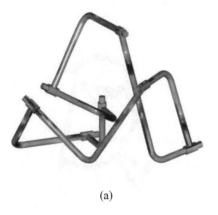

(a)

(b)

Figure 8.63 Animation frames of the R-7R mechanism at different positions of the driving element: (a) 0°; (b) 20°; (c) 90°; (d) 160°; e) 270°; f) 320°.

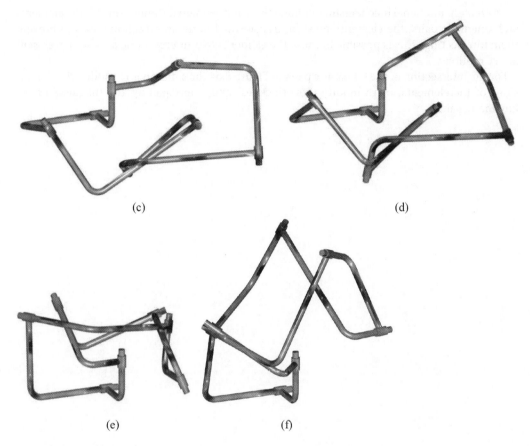

(c)   (d)

(e)   (f)

Figure 8.63 *Continued*

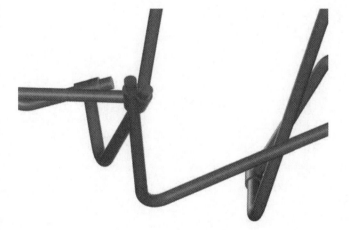

Figure 8.64 Intersection of two kinematic pairs of the R-7R mechanism (at 320°).

To create a functional mechanism, retaining the same geometric dimensions for the elements and kinematic pairs, the elements must be transformed in the intersection zones to prevent them intersecting. This is possible because the elements that intersect one another have oscillatory motions.

For the intersection of the kinematic pairs in Figure 8.64 the solution is to modify the dimensions of the elements, or to modify their thickness and, consequently, the thickness of the kinematic pairs.

9

Industrial Robots

In this chapter we first classify industrial robots, and then we describe the mechanical systems used and the components of their actuating and control systems. We outline general aspects of their mechanical systems, with particular regard to the kinematics of their path-generating mechanisms, their orientation mechanisms and their apprehensiveness (grip mechanism).

9.1 General Aspects

Industrial robots are automatic and autonomous machines, which have universal executive machine parts, such as mechanical arms, the motion of which is controlled by computer programs. Some authors define the industrial robots as 'programable automatic manipulators'. A *manipulator* is a mechanism (such as an artificial arm) with many degrees of freedom.

Depending on the complexity of the functions involved, they can be:

- *simple manipulators*, with working sequences limited to simple successive operations, the control systems being assured by limit switches, cam devices, programable automation and so on
- *industrial robots* (programable manipulators), which have control systems that can call internally saved programs in order to make certain movements
- *intelligent industrial robots*, with different types of sensors (tactile, of force, of proximity, of shape, of presence) and control systems capable of understanding the signals received from the sensors and to adaptively act upon the active elements in them in order to reach a preestablished goal.

In industrial robots we find:

- *The mechanical system*, which has elements linked to one another by kinematic pairs (the kinematic chain) and which creates the motion of the elements in space.
- *The actuating system*, which transforms (hydraulic, pneumatic, or electrical) energy into mechanical energy for the actuation of the mechanical system. It consists of rotary and linear (hydraulic, pneumatic, or electric) motors, and the mechanisms necessary for the transmission of the motion.
- *The control system*, which initiates the operations, and defines the the duration of the motion of the kinematic chain and the velocity of execution. It contains devices for calling a program, and for saving, reproducing, and transmitting it to the actuating system.

Classical and Modern Approaches in the Theory of Mechanisms, First Edition.
Nicolae Pandrea, Dinel Popa and Nicolae-Doru Stănescu.
© 2017 John Wiley & Sons Ltd. Published 2017 by John Wiley & Sons Ltd.
Companion Website: www.wiley.com/go/pandmech17

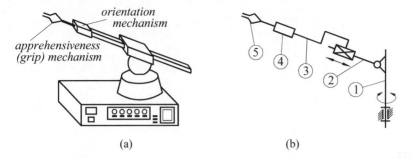

Figure 9.1 Industrial robot in spherical coordinates: (a) constructive schema; (b) kinematic schema.

The criteria for the classification of the industrial robots are:

1) The programming mode:
   - sequential robots, which execute simple successive motions based on a simple program
   - teach robots (repeater), which record a procedure and repeat it
   - robots with numerical control, which receive the passing positions and stop times by using some programs
   - intelligent robots, which have (tactile or visual) sensors to recognize external environmental inputs and which can adapt to them
2) The system of coordinates in which the mechanical system works:
   - Cartesian coordinates
   - cylindrical coordinates
   - spherical coordinates
3) The number of degrees of mobility:
   - with two up to seven mobility degrees
4) The weight of the working task:
   - from hundreds of grams to a few tonnes
5) The use of the robots:
   - painting
   - welding
   - manipulation
   - transport
   - and so on.

Figure 9.1 shows an industrial robot for manipulation and its kinematic schema.

9.2 Mechanical Systems of Industrial Robots

9.2.1 Structure

The mechanical system of an industrial robot consists of:

- the guide mechanism
- the grip mechanism.

In most cases, the guide mechanism consists of:

- a path-generating mechanism
- the orientation mechanism.

In Figure 9.1b the path-generating mechanism is formed by elements 1, 2, and 3, with three degrees of mobility. The orientation mechanism may have 1, 2, or 3 degrees of mobility, so that, in total, the robot in Figure 9.1a can have 4, 5, or 6 degrees of mobility.

If the kinematic chain of the robot is open and the kinematic pairs are of the fifth class, then, in order to assure a desmodromic mechanism, the kinematic pairs must be *driving kinematic pairs*.

9.2.2 The Path-generating Mechanism

Kinematic schemata: working space

To describe a spatial trajectory, this mechanism consists of a kinematic chain formed by three elements linked to one another by fifth-class kinematic pairs. Depending on the reference system associated with these kinematic chains, the robot might work in:

- Cartesian coordinates
- cylindrical coordinates
- spherical coordinates.

Figure 9.2 shows the constructive schema of a path-generating mechanism in Cartesian coordinates (x, y, z), and its kinematic schema. The elements of the mechanism are denoted by 1, 2, and 3.

point A of element 3 (in which the orientation mechanism may be situated) is called the *characteristic point*. The set of positions that can be occupied by the characteristic point is called the *working space*.

For a path-generating mechanism in Cartesian coordinates, the working space is obtained by limiting the displacements along the axes Ox, Oy, and Oz; that is, by writing:

$$x \in [x_1, x_2] , \; y \in [y_1, y_2] , \; z \in [z_1, z_2] . \tag{9.1}$$

This means that the working space is a cuboid having edges equal to $x_2 - x_1, y_2 - y_1$, and $z_2 - z_1$.

Figure 9.3 shows the structural and kinematic schemata of a path-generating mechanism in cylindrical coordinates. Here, the working space is obtained by limiting the cylindrical coordinates θ, r, and z; that is, using the conditions

$$\theta \in [\theta_1, \theta_2] , \; r \in [r_1, r_2] , \; z \in [z_1, z_2] . \tag{9.2}$$

This gives the working space shown in Figure 9.3c. Technical solutions for existing industrial robots permit the use of angle $\alpha = \theta_2 - \theta_1$ up to 270°.

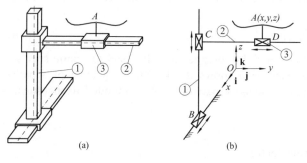

Figure 9.2 Path-generating mechanism in Cartesian coordinates: (a) constructive schema; (b) kinematic schema.

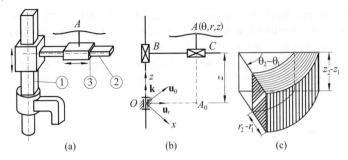

Figure 9.3 Path-generating mechanism in cylindrical coordinates: (a) constructive schema; (b) kinematic schema; (c) working space.

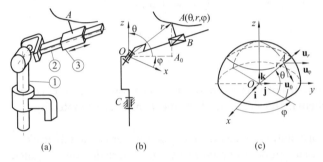

Figure 9.4 Path-generating mechanism in spherical coordinates: (a) constructive schema; (b) kinematic schema; (c) working space.

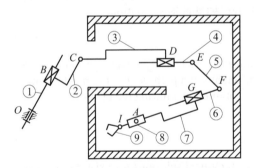

Figure 9.5 Path-generating mechanism with seven mobility degrees.

Figure 9.4 shows the structural and kinematic schemata of a path-generating mechanism in spherical coordinates. The working space is obtained by limiting the spherical coordinates φ, θ, and r (Figure 9.4c); that is, by writing:

$$\varphi \in \left[\varphi_1, \varphi_2\right], \ \theta \in \left[\theta_1, \theta_2\right], \ r \in \left[r_1, r_2\right]. \tag{9.3}$$

Kinematic chains with more than three mobility degrees may be used for some special-purpose path-generating mechanisms. Figure 9.5 shows the kinematic schema of a path-generating mechanism with seven mobility degrees, part of a robot intended to investigate climbs.

Kinematic analysis

The point A (Figure 9.2b) has the position vector

$$\mathbf{r}_A = x\mathbf{i} + y\mathbf{j} + z\mathbf{k}, \tag{9.4}$$

where $x = x(t)$, $y = y(t)$, and $z = z(t)$ represent the coordinates of point A.

For a path-generating mechanism in Cartesian coordinates, the velocity and acceleration of point A are expressed with:

$$\mathbf{v}_A = \dot{x}\mathbf{i} + \dot{y}\mathbf{j} + \dot{x}\mathbf{k}, \tag{9.5}$$

$$\mathbf{a}_A = \ddot{x}\mathbf{i} + \ddot{y}\mathbf{j} + \ddot{x}\mathbf{k}. \tag{9.6}$$

For a path-generating mechanism in cylindrical coordinates (Figure 9.3b) we get

$$\mathbf{r}_A = r\mathbf{u}_r + z\mathbf{k}, \tag{9.7}$$

where $\mathbf{u}_r$ and $\mathbf{k}$ are the unit vectors corresponding to the axes OA_0 and Oz, respectively. This gives the expressions for the velocity and acceleration in cylindrical coordinates

$$\mathbf{v}_A = \dot{r}\mathbf{u}_r + r\dot{\theta}\mathbf{u}_\theta + \dot{z}\mathbf{k}, \tag{9.8}$$

$$\mathbf{a}_A = \left(\ddot{r} - r\dot{\theta}^2\right)\mathbf{u}_r + \left(2\dot{r}\dot{\theta} + r\ddot{\theta}\right)\mathbf{u}_\theta + \ddot{z}\mathbf{k}, \tag{9.9}$$

in which $\mathbf{u}_\theta$ is the unit vector perpendicular to $\mathbf{u}_r$ and $\mathbf{k}$ (Figure 9.3b).

For a path-generating mechanism in spherical coordinates (Figure 9.4b), denoting by $\mathbf{u}_r$, $\mathbf{u}_\theta$, and $\mathbf{u}_\varphi$ the corresponding unit vectors, gives

$$
\begin{aligned}
\mathbf{r}_A &= r\mathbf{u}_r, \\
\mathbf{u}_r &= \sin\theta\cos\varphi\mathbf{i} + \sin\theta\sin\varphi\mathbf{j} + \cos\theta\mathbf{k}, \\
\mathbf{u}_\theta &= \cos\theta\cos\varphi\mathbf{i} + \cos\theta\sin\varphi\mathbf{j} - \sin\theta\mathbf{k}, \\
\mathbf{u}_\varphi &= -\sin\varphi\mathbf{i} + \cos\varphi\mathbf{j}, \\
\dot{\mathbf{u}}_r &= \dot{\theta}\mathbf{u}_\theta - \dot{\varphi}\sin\theta\mathbf{u}_\varphi, \\
\dot{\mathbf{u}}_\theta &= -\dot{\theta}\mathbf{u}_r + \dot{\varphi}\cos\theta\mathbf{u}_\varphi, \\
\dot{\mathbf{u}}_\varphi &= -\dot{\varphi}\left(\cos\theta\mathbf{u}_\theta + \sin\theta\mathbf{u}_r\right), \\
\mathbf{v}_A &= \dot{r}\mathbf{u}_r + r\dot{\theta}\mathbf{u}_\theta + r\dot{\varphi}\sin\theta\mathbf{u}_\varphi, \\
\mathbf{a}_A &= \left(\ddot{r} - r\dot{\theta}^2 + r\dot{\varphi}^2\sin^2\theta\right)\mathbf{u}_r + \left(2\dot{r}\dot{\theta} + r\ddot{\theta} - r\dot{\varphi}^2\sin\theta\cos\theta\right)\mathbf{u}_\theta \\
&\quad + \left(2\dot{r}\dot{\varphi}\sin\theta + r\ddot{\varphi}\sin\theta + 2r\dot{\theta}\dot{\varphi}\cos\theta\right)\mathbf{u}_\varphi.
\end{aligned}
\tag{9.10}
$$

9.2.3 The Orientation Mechanism

Kinematic schemata

The orientation mechanism rotates the transported object relative to the reference system rigidly linked to the last element of the path-generating mechanism; the origin is at the characteristic point A (Figure 9.6).

The orientation mechanism may have 1, 2, or 3 degrees of mobility, depending on the rotations it is to bring about.

Figure 9.7 shows the kinematic schema of an orientation mechanism with two degrees of mobility, highlighted by the angular velocities ω_I and ω_{II}. Axle I, rigidly linked to conical gear 1, sets in motion gear 3, which is rigidly linked to box CA, the angular velocity being ω_3.

Then, axle II, rigidly linked to gear 2, sets in motion gear 4, rigidly linked with gear 4'. The latter sets in motion gear 5, which is rigidly linked to the frame CB of the grip mechanism formed

Figure 9.6 Rotations of the orientation mechanism.

the last element of the
path generating mechanism

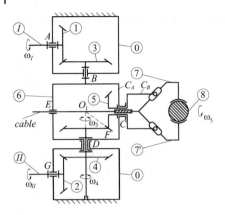

Figure 9.7 Orientation mechanism with two mobility degrees.

by arms 7 and 7' and driven by cable 6. The frame of the grip mechanism rotates, together with the manipulated object 8, the angular velocity being ω_5. If axles I and II are simultaneously in rotational motion, then object 8 also has a rotational motion, the result of the combined rotational motions of the two axles.

Figure 9.8 shows the kinematic schema of a orientation mechanism with three mobility degrees, driven by axles with independent motions I, II, and III. The box of this mechanism, denoted by O, is rigidly linked to the last element of the path-generating mechanism. The motion is transmitted from axle I, using gear 1, to gears 2, 2', and 3. Gear 3 is rigidly linked to box C_1, which has a rotational motion around the axis that passes through bearings A and B.

From axle II, by gears 4, 5, 6, 6', 7, 7', 8, and 8', the motion is transmitted to gear 9, rigidly linked to box C_2; the box now has a rotational motion around the axes that pass through bearings C and D. From axle III, by gears 10, 11, 11', 12, 12', 13, 13', 14, and 14', the motion is transmitted to gear 15, which is rigidly linked to the axle with the grip mechanism situated at point F; gear 15 and the axle rigidly linked to it now have rotational motion around axis EF. If axles I, II, III are simultaneously in rotational motion, the object in the grip mechanism has rotational motion that is a combination of the motions of the three axles.

Figure 9.9 shows a trunk-type orientation mechanism (designed by Renault). The mechanism consists of elements 1, 2, 3, and 4, linked to one another by spherical kinematic pairs situated at points A, B, and C, and by four cables 5–8 (cables 7 and 8 are situated in a plane perpendicular to the figure); the elements are driven by four linear hydro-motors C_1, C_2, C_3, and C_4 (again, C_3 and C_4 are in a plane perpendicular to the figure).

Element 4 may be driven separately too, to give motion around the longitudinal axis CD.

Kinematic analysis

Considers the orientation mechanism with two degrees of mobility in Figure 9.7. point O_1, situated at the intersection of the axes BD and CE, remains fixed.

For the analysis of the positions of object 8, which is(rigidly linked to the frame C_A, we use the reference system of fixed axes $O_1x_0y_0z_0$, in which the axes O_1z_0 and O_1y_0 coincide with O_1B and O_1C, respectively. If axle I rotates through angle α, then it produces a rotation of box C_A (and, consequently, of object 8), the rotation angle being

$$\varphi_1 = \frac{z_1}{z_3}\alpha; \tag{9.11}$$

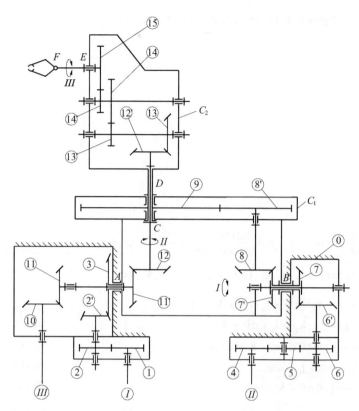

Figure 9.8 Orientation mechanism with three mobility degrees.

Figure 9.9 Trunk-type orientation mechanism.

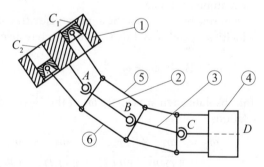

The rotation takes place around axis O_1z_0. There is also an induced rotation of frame C_B around axis O_1y_1, of angle

$$\theta_1 = \varphi_1 \frac{z_4'}{z_5}, \tag{9.12}$$

where z_1, z_3, z_4', and z_5 represent the numbers of teeth of gears 1, 2, 4′, and 5, respectively. If we rotate axle *II* through angle β, then frame C_B rotates through an angle

$$\theta_2 = \beta \frac{z_2 z_4'}{z_4 z_5}. \tag{9.13}$$

If we simultaneously set both axles in motion, then we obtain rotations of angle $\varphi = \varphi_1$ around axis O_1z_0 (Figure 9.10) and of angle $\theta = \theta_1 + \theta_2$ around axis O_1y_1. This gives the following

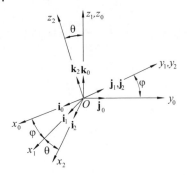

Figure 9.10 The positions of the mobile system.

schema of successive rotations

$$O_1x_0y_0z_0 \xrightarrow[\text{around axis } O_1z_0]{\text{rotation of angle } \varphi} \Rightarrow O_1x_1y_1z_0 \xrightarrow[\text{around axis } O_1z_1]{\text{rotation of angle } \theta} \Rightarrow O_1x_2y_2z_2.$$

Denoting the unit vectors of the axes by $\mathbf{i}_0, \mathbf{j}_0, \mathbf{k}_0, \mathbf{i}_1, \mathbf{j}_1, \mathbf{k}_1, \mathbf{i}_2, \mathbf{j}_2$, and $\mathbf{k}_2$, gives

$$\mathbf{i}_1 = \mathbf{i}_0 \cos \varphi + \mathbf{j}_0 \sin \varphi, \ \mathbf{j}_1 = -\mathbf{i}_0 \sin \varphi + \mathbf{j}_0 \cos \varphi, \ \mathbf{k}_1 = \mathbf{k}_0,$$
$$\mathbf{i}_2 = \mathbf{i}_1 \cos \theta - \mathbf{k}_1 \sin \theta, \ \mathbf{j}_2 = \mathbf{j}_1, \ \mathbf{k}_2 = \mathbf{i}_1 \sin \theta - \mathbf{k}_1 \cos \theta. \tag{9.14}$$

$$\mathbf{i}_2 = \mathbf{i}_0 \cos \theta \cos \varphi + \mathbf{j}_0 \cos \theta \sin \varphi - \mathbf{k}_0 \sin \theta, \tag{9.15}$$

$$\mathbf{j}_2 = -\mathbf{i}_0 \sin \varphi + \mathbf{j}_0 \cos \varphi, \tag{9.16}$$

$$\mathbf{k}_2 = \mathbf{i}_0 \sin \theta \cos \varphi + \mathbf{j}_0 \sin \theta + \mathbf{k}_0 \cos \theta. \tag{9.17}$$

Based on these relations, we can determine the position of frame C_B, and hence the position of the manipulated object.

Denoting by ω_I and ω_{II} the angular velocities of axles I and II, respectively, we get the angular velocities ω_3 and ω_4 of box C_4 and gear 4, respectively

$$\omega_3 = \omega_I \frac{z_1}{z_3}, \ \omega_4 = \omega_{II} \frac{z_2}{z_4}. \tag{9.18}$$

The absolute angular velocity $\boldsymbol{\omega}_5$ of the C_B (rigidly linked to gear 5) is expressed by the vector relations

$$\underset{\| \ \overline{O_1}B}{\boldsymbol{\omega}_5} = \underset{\| \ O_1C}{\boldsymbol{\omega}_3} + \underset{\| \ \overline{O_1}D}{\boldsymbol{\omega}_{53}} = \underset{\| \ FO_1}{\boldsymbol{\omega}_4} + \boldsymbol{\omega}_{54}; \tag{9.19}$$

We get the vector diagram in Figure 9.11, from which we obtain:

$$\omega_{53} = (\omega_3 + \omega_4) \tan \delta, \ \omega_{54} = \frac{\omega_4 + \omega_3}{\cos \beta}, \ \omega_5 = \sqrt{\omega_3^2 + (\omega_3 + \omega_4)^2 \tan^2 \delta};$$

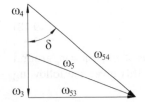

Figure 9.11 The composition of the angular velocities.

Since $\tan \delta = \dfrac{z_4'}{z_5}$, we get

$$\omega_{53} = \left(\omega_I \frac{z_1}{z_3} + \omega_{II} \frac{z_3}{z_4} \right) \frac{z_4'}{z_5},$$

(9.20)

$$\omega_5 = \sqrt{\omega_I^2 \frac{z_1^2}{z_3^2} + \left(\omega_I \frac{z_1}{z_3} + \omega_{II} \frac{z_3}{z_4} \right)^2 \frac{(z_4')^2}{z_5^2}}.$$

(9.21)

9.2.4 The Grip Device

General aspects: structural and kinematic schemata

The grip device fixes manipulated object, maintaining it rigidly linked to the grip elements during transport, and releasing it when it arrives at the destination. The objects manipulated s may be raw materials for working processes, or parts or subassemblies for manufacturing or assembly.

Working objects in fabrication processes serviced by industrial robots are usually cylindrical or prismatic, and have a preset orientation. Modern grip mechanisms have tactile and visual sensors, so that they can grip objects of different shapes and ones that have not not been orientated.

Figure 9.12 shows the structural and kinematic schemata of a grip device with hydraulic (pneumatic) actuation.

In particular,

- the mechanism of the device (here, the mechanism consisting of elements 3, 4, and 6, and its mirror image, the mechanism consisting of elements 3', 4', and 6');
- the fingers of the mechanism (elements 4 and 4');
- pieces 5 and 5', rigidly linked to the fingers, which come into contact with the working object.

The holding of the working object is brought about by displacement of the piston, while the release comes from return spring 7. The drive of the grip device may be electrical, hydraulic or pneumatic.

Figure 9.12 Grip device: (a) constructive schema; (b) kinematic schema.

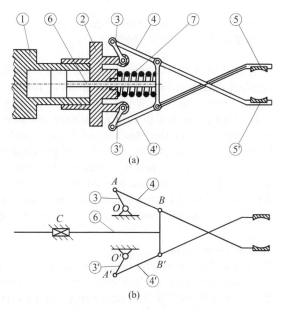

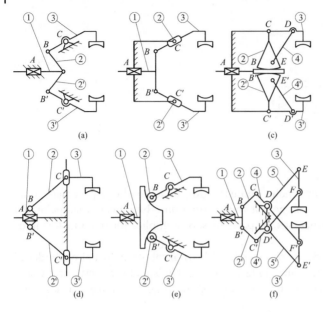

Figure 9.13 Kinematic schemata of some grip devices: (a), (c), (e) fingers 3 and 3′ have rotational motion; (b) fingers 3 and 3′ have plane-parallel motion; (d), (f) fingers 3 and 3′ have translational motion.

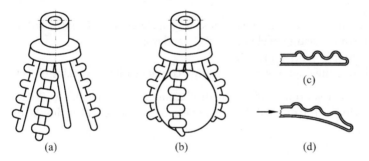

Figure 9.14 Grip devices with pneumatic fingers: (a) grip device in the initial position; (b) grip device in working; (c), (d) shape of the fingers.

Figures 9.13 shows the kinematic schemata for six grip mechanisms. Fingers 3, 3′ in Figure 9.13a,c,e are in rotational motion, fingers 3, 3′ in Figure 9.13b are in plane-parallel motion, while fingers 3, 3′ in Figure 9.13d,f are in translational motion.

Some grip devices (Figure 9.14a have pneumatic fingers. Their functioning is based on the asymmetry of the elasticity of fingers made from an elastic material (Figure 9.14c,d). Figure 9.14b,d shows the curvature of pneumatic fingers produced by the change in fluid pressure. Another model of holding is the grip device with poly-jointed fingers, as shown in Figure 9.15. The fingers, jointed to one another, are rigidly linked to the rollers, the latter driven by cables. By moving the cables, we obtain a relative motion between the fingers; in this way they occupy positions tangential to the working object.

Calculation of forces

Considers the grip mechanism in Figure 9.16, which is driven by the active force F_m. The force N, which grips object 3, is a function of the active force F_m. Decomposing force F_m onto the directions EB and $EB′$, and analyzing (by symmetry) only the equilibrium of finger 2, we obtain

Figure 9.15 Grip device with poly-jointed fingers.

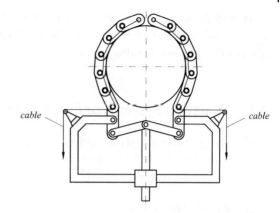

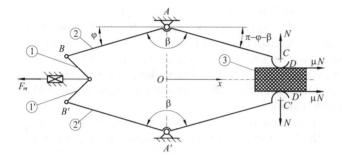

Figure 9.16 The forces on the grip device.

Figure 9.17 The forces on a finger.

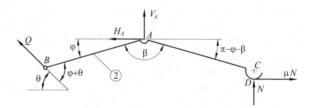

the calculation schema in Figure 9.17, where

$$Q = \frac{F_m}{2\cos\theta}, \tag{9.22}$$

Here, μ is the coefficient of friction between the fingers and the working object. Denoting the distances $AB = b$, $AC = l$, and $CD = r$ and taking into account the notation in Figure 9.17, from the equation of the moments about point A, we get

$$Qb\sin(\varphi + \theta) - Nl\cos(\varphi + \beta) - \mu N\left[l\sin(\varphi + \beta) + r\right] = 0; \tag{9.23}$$

Taking into account (9.22), we get

$$\frac{F_m b}{2}(\sin\varphi + \cos\varphi\tan\theta) - N\left[\cos(\varphi + \beta) - \mu l\sin(\varphi + \beta) - \mu r\right] = 0. \tag{9.24}$$

From here, we obtain

$$\tan\theta = \frac{2Nl}{F_m b\cos\varphi}\left[\cos(\varphi + \beta) + \mu\sin(\varphi + \beta) + \frac{\mu r}{l}\right] - \tan\varphi. \tag{9.25}$$

From (9.25) we can determine the active force F_m as a function of the fixing force N. In the dimensional synthesis we want to maintain constant the value of the mechanical characteristic $\frac{N}{F_m}$. For that purpose we also use (9.25).

9.3 Actuation Systems of Industrial Robots

The mechanical systems of the industrial robots are of mobility degree is $M > 1$. To ensure the desmodromic condition is met, it is necessary that there be M active elements (kinematic pairs). The driving elements are powered by electrical, hydraulic, and pneumatic motors, and so on.

9.3.1 Electrical Actuation

Due to its advantages – the widespread availability of electric power, simple connection to the electric network, high reliability – electrical actuation is very common. The main disadvantage of electrical actuation is the necessity for certain reducers to adjust the velocities and active torques of the motors to those required by the robots. The electrical motors used for actuation of industrial robots are either direct current motors with serial excitation, or alternating current motors. Variation of the angular velocities may be through either a Ward–Leonard installation, or by using electronic variators with thyristors.

Exact placement of the elements of the robot's mechanism may be brought about using step-by-step electrical motors. These have the disadvantage that the active torque diminishes at high velocities.

Figure 9.18 shows the kinematic schema of the mechanical system of an industrial robot with five mobility degrees (three degrees for the path-generating mechanism, and two for the orientation mechanism). Column 1 (Figure 9.19), together with the rotational kinematic pair situated at point A, is driven by the motor and reducer MRA, using gear G_1 and gear sector G_2. Arm 2 is rotated in the kinematic pair situated at point B (Figure 9.19) by the motor and reducer MRB, using gear G_3 and gear sector G_4.

The displacement of element 3 along arm 2 is ensured by the actuation of the motor and reducer MRC, using screw S, which, by nut N_1, rigidly linked to element 3, transmits the translation motion to element 3 using guide G.

The orientation mechanism (Figure 9.20) is driven by the motors and reducers MRD and MRE (fixed on element 2), and by the telescopic axes ATD and ATE. Thus, motor MRD acts the telescopic axis ATD, which drives the gear train consisting of gears G_5 and G_6 and box CA_1, rigidly linked to element 3. In this way, rotation of element 4 in the revolute joint situated at point D (Figure 9.20) is brought about.

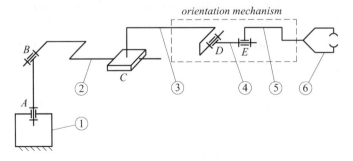

Figure 9.18 The kinematic schema of the mechanical system of an industrial robot with five mobility degrees.

Figure 9.19 Actuation of the path-generating mechanism.

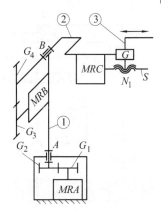

Figure 9.20 The actuation of the orientation mechanism.

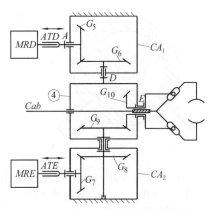

Similarly, motor *MRE*, fixed to element 2 acts on the telescopic axis *ATE*; this axis drives the gear train formed by gears G_7 and G_8 situated in the box CA_2, rigidly linked to element 3. In this way, the motion is transmitted from gear G_9, rigidly linked to gear G_8, to gear G_{10}, producing the rotation in the kinematic pair situated at point E (Figures 9.18 and 9.20). The grip mechanism (Figure 9.20) is driven by the cable *Cab* in the gripping phase, and by a return spring in the releasing phase.

9.3.2 Hydraulic Actuation

Hydraulic actuation is now the most common way of driving robots, because of their ability to apply adjustments. The component elements of a simple hydraulic actuation system are (Figure 9.21): the pump, the pressure adjuster (accumulator) A, the distributor D, the throttle T, the hydraulic motor MH, and the reservoir R.

The pump P delivers, by the pressure adjuster A and the distributor D, the linear hydraulic motor with double effect MH, which is connected, by the distributor D and the throttle T, to the exterior reservoir R. The pumps of actuation systems may be with constant or variable discharge, with radial or axial pistons. The hydraulic motors may be linear (Figure 9.21) or rotary (Figure 9.22).

The pressure adjuster (the accumulator) must accumulate the liquid (energy) when the consumption is small and return it when required; in this way it reduces pressure pulses and the discharges produced by hydraulic shocks in the pipes. The accumulator is similar in its functioning principle to flywheels in mechanical systems and to condensers in electrical systems.

The distributor conducts the fluid in a certain direction, in a way that corresponds functionally to switches in electrical circuits. The distributors used in industrial robots are

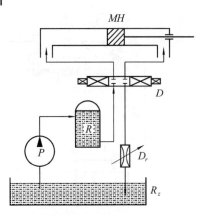

Figure 9.21 Simple hydraulic actuation.

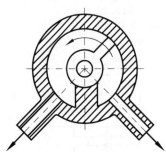

Figure 9.22 Rotary hydraulic motor.

electromagnetically controlled. The throttle dissipates the pressure energy by an adjustable section of it, transforming this energy into heat; this functionally corresponds to a resistor in an electrical circuit.

Figure 9.23a show an industrial robot with five mobility degrees, and hydraulic actuation, in which the rotation around a vertical axis (the axis of the column 1) is driven by a worm and wheel system. The rotation of prop 5, in the revolute joint situated at point A, is driven by linear hydraulic motor 4, while the displacement of the orientation mechanism 6 is driven by linear hydraulic motor 7.

The orientation mechanism, with two mobility degrees, is driven by linear hydraulic motors 8 and 15, which have double effect, and which transmit the motion via chains 9 and 16 from the axle of the pistons to sprockets 10 and 17. From sprocket 10, the motion is transmitted to telescopic axes 11 and 12, which create a rotational motion of angular velocity ω_1 in the orientation mechanism (Figure 9.23b). From sprocket 17 the motion is transmitted to telescopic axes 18 and 19, which create a rotational motion with angular velocity ω_2 in the orientation mechanism.

9.4 Control Systems of Industrial Robots

The control system receives information, processes it, and then transmits commands to the actuation system. The information comes from transducers and sensors.

The control systems can be open-loop (Figure 9.24) or closed-loop (Figure 9.25), the data about the environment and the functioning of the actuation system being delivered by transducers and sensors. The transducers sense the geometrical and kinematic (displacement or linear and angular velocities) parameters, while the sensors detect the grip device's approach to an object (proximity sensors), or contact with it (tactile sensors), or collect visual data about the environment (visual sensors).

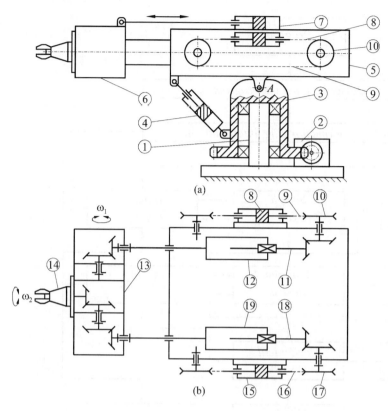

(a)

(b)

Figure 9.23 Robot with five mobility degrees with hydraulic actuation: (a) constructive schema; (b) kinematic schema.

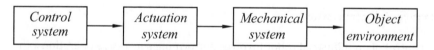

Figure 9.24 Control system with open loop.

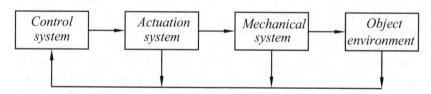

Figure 9.25 Control system with closed loop.

The transducers used for displacement may be:

- *numerical* (incremental): when the (linear angular) displacement generates a number n of impulses, the displacement being equal to $n\Delta$, where Δ is the increment
- *analogical* (potentiometers, resolvers, inductsyn): when the (linear, angular) displacement generates an electrical signal which depends on the displacement.

The sensors used may involve:

- *physical contact* (tactile, sliding, force/torque)
- *no physical contact* (visual, of proximity, auditory).

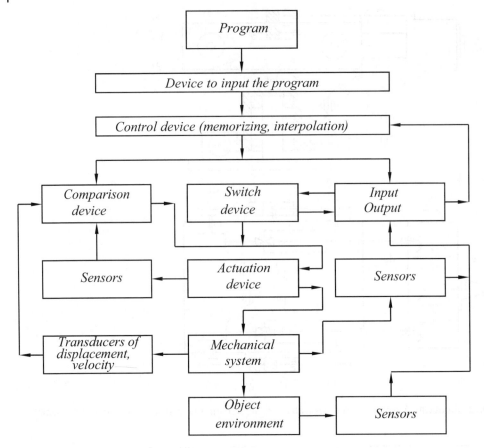

Figure 9.26 The structure of the control system.

The automatic control of an industrial robot involves performing successive operations, the sequence of which is driven by a program. Figure 9.26 shows the structure of a control system. The input to the program may be created by programming or a teach-in. The control device is either a microprocessor or a computer; its task is to create the commands required by the working process and to transmit the necessary commands. In this model, it also saves the program. The comparison device compares the data delivered by the transducers and sensors with programmed values, to establish any differences and to send commands to the actuation system in order to reduce or eliminate them. The switch device sends switch commands in order to change the status of motion of the actuation system. In the input–output block, information about the environment and the functioning of the mechanical system is collected and commands are sent to the switch and control devices.

9.5 Walking Machines

Walking (pedipulator) machines displace themselves by imitating walking. The displacement can be based on different planar or spatial mechanisms. Here we will describe a walking device known as a Tchebychev-type mechanism, which is based on four-bar mechanisms.

Walking mechanisms have interested many scientists. In 1733, the French Academy of Science recorded the patent for a mechanical toy designed by Maillard, which imitated the walk of a horse. The mechanism on which this patent was based was a crank and slotted lever one. In 1878, at the Exposition in Paris, and in 1893 at Chicago, the mathematician P.L. Tchebychev revealed the models of two mechanisms, one of them imitating the walk of a horse. Tchebychev (1821–1894) made some astonishing planar mechanisms using approximate analytical methods, particularly in the case when a point of a certain element must follow a certain curve. For the horse, the foot had to follow approximately a straight line, while the motion of the leg as a whole was a circular arc.

9.5.1   The Mechanical Model of the Walking Mechanism

The basic mechanism that Tchebychev used was a four-bar one, the dimensions of which are given in Figure 9.27; the dimensions were obtained by synthesis, so that a point of the coupler described a segment of straight line.

In the forward movement of the horse, the end of the coupler describes a curvilinear arc and as it does so, the horse lifts up its leg. Figure 9.28 shows the successive positions of a point M on coupler AM as a function of the driving element OA.

The mechanism of the Tchebychev horse, with the kinematic schema in Figure 9.29, consists of four four-bar mechanisms. The first two mechanisms ($O_1A_1B_1C_1$ and $O_2A_2B_2C_2$) which model the legs on one side, are sine-phased, while the other two (modelling the legs on the opposite side ($O_3A_3B_3C_3$ and $O_4A_4B_4C_4$)) are phase-shifted by 180° relative to the first ones. The mechanism also has four vertical legs, rigidly linked to couplers A_iM_i, $i = \overline{1, 4}$, and rigidly linked to one another by horizontal bars.

Figure 9.30 is an AutoCAD representation of the Tchebychev horse mechanism, using solids.

Figure 9.27 The kinematic schema of the walking machine.

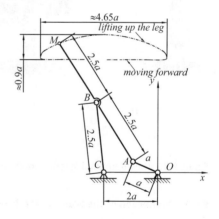

Figure 9.28 The trajectory of the coupler's point.

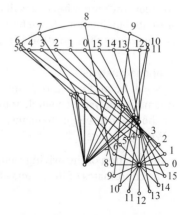

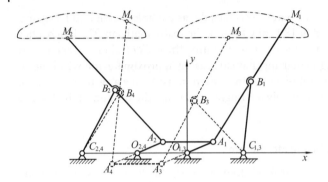

Figure 9.29 The kinematic schema of the *Tchebychev horse*.

Figure 9.30 The photographic representation of *the Tchebychev horse* mechanism.

9.5.2 Animation of the Walking Machine

To create the animation of a walking machine (Tchebychev's horse), according to the algorithm presented in Section 8.5, we first have to perform a positional analysis of the four mechanisms as a function of the position of the driving element.

Positional analysis

This will be performed only for the four-bar mechanisms $O_1A_1B_1C_1$ and $O_2A_2B_2C_2$. In fact, kinematic analysis of a single four-bar mechanism, will give analytical relations that hold true for the other three mechanisms.

We will change:

- the position of the driving element by 180° for the phase-shifted mechanisms
- the coordinates of the points at the base for the sine-phased mechanisms.

We use the kinematic schema of the mechanism in Figure 2.44, for which we already have the kinematic analysis (see Section 2.10.1). For this mechanism we know

- the dimensions of the elements: OA, $AB = 2.5OA$, $BC = 2.5OA$, $BM = 2.5OA$, $\alpha = 180°$
- the positions of the kinematic pairs linked to the base: $O(0, 0)$, $C(2OA, 0)$
- the position angle of element 1.

For mechanism 1, the four-bar mechanism $O_1A_1B_1C_1$ in Figure 9.29, we denote by $\varphi_{11} = \varphi_1$ the position angle of element 1. The kinematic analysis is performed with a constant angular step of 10° and using the analytical method of projections.

Mechanism 3, ($O_3A_3B_3C_3$ in Figure 9.29) has the same kinematic analysis as mechanism 1, with the additional specification that angle φ_{13} of the crank is given by $\varphi_{13} = \varphi_{11} + \pi$. Therefore, the data obtained for this mechanism are staggered by 180° relative to mechanism 1.

For mechanism 2 ($O_2A_2B_2C_2$ in Figure 9.29), the elements have the same lengths, but the positions of the kinematic pairs linked to the base are different: $O(-d, 0)$ and $C(-d - 2OA, 0)$. We denote by $d = O_1O_2$ the distance between mechanisms 1 and 2. The position angle of the crank of mechanism 2 is $\varphi_{12} = \varphi_{11}$.

Mechanism 4 ($O_4A_4B_4C_4$ in Figure 9.29), has the same kinematic analysis as mechanism 2, angle φ_{14} of the crank being $\varphi_{14} = \varphi_{12} + \pi$. Therefore, in this case, the data obtained for this mechanism are staggered by 180° relative to mechanism 2.

Relative to axis OZ, which is perpendicular to the plane of the figure, mechanisms 3 and 2 are displaced by a distance $h = O_1O_2 = O_3O_4$; this aspect does not influence the kinematic analysis of the planar mechanism $OABC$. Distance h is important only for the third coordinate of the points. For the mechanisms situated in the front plane of the figure coordinate $z = 0$, while for the mechanisms situated in the rear plane of the figure (mechanisms 3 and 2) $z = -h$.

To obtain numerical results, we have written a Turbo Pascal program. The code for this can be found in the companion website of the book *(Anim.Pas)*. This program is based on equations (2.219)–(2.224). The program saves the data as matrices. This makes the code and, more importantly, the positioning mode of the elements much simpler. Thus, for instance, the coordinates x_A of the kinematic pair situated at point A is found in a matrix with i rows and j columns, with the elements $x_A[i, j]$, $i = \overline{1, 36}$, $j = \overline{1, 4}$. The rows record the values as a function of the 36 positions of the crank, while the columns contain the values for each of the four mechanisms.

The three loops give:

- matrices *Phi1*, *Phi2*, *Phi3* containing the position angles of the elements of the mechanism
- matrices *xO*, *yO*, *xA*, *yA*, *xB*, *yB*, *xC*, and *yC* containing the coordinates of the points.

One has to consider the connections between the four mechanisms. The values generated by the program can then be used in the animation of the mechanism.

Modelling the elements and kinematic pairs

The models are created using solids in AutoCAD, based on the dimensions and the reference systems shown in Figure 9.31. The calculation program *Anim.Pas* is used for the positional analysis (see the companion of this book). The program is completed with commands to save the elements of the mechanism. These commands are written in a script file, from which we are able to obtain the elements. The program contains a few procedures.

For the bars, there is a single procedure called *Bar*. This is called three times from the main program in order to obtain the three bars with different dimensions. The input data of the procedure are: l, h, g, and d (Figure 9.31), and the number of the bar. The bar is created in a dedicated layer, which is given the name of the bar. After construction, it is given its own name

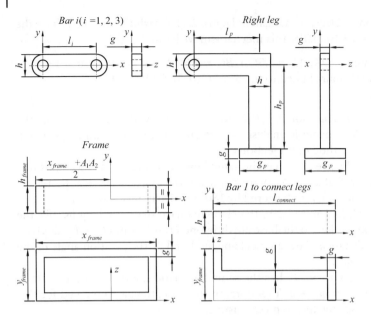

Figure 9.31 The dimensions of the elements of the *Tchebychev horse*.

using an AutoLisp function; then, the layer is closed in order to permit other constructions too. For the connection bars $A_i A_{i+1}$, $i = \overline{1, 3}$, for the front and back mechanisms, procedure *Bar* is called again.

The second procedure, called *Legs*, creates two (left and right) legs, in a dedicated layer called *Legs_L*. Firstly, the right leg is created using the dimensions in Figure 9.31, and this is assigned the name *Leg_1*. The second leg, *Leg_2*, is a mirror-image copy (**Mirror3D**) of the first leg, the plane OXZ being the mirroring plan. The procedure ends by closing the working layer and returning to layer *0*.

The procedure *Frame* generates the script file necessary to construct the frame drawn in Figure 9.31. A new working layer called *Frame_L* is opened and four solids of parallelepiped shape are constructed. These are then united using the command **Union**. The resulting solid is given the name *Frame*. Finally, the working layer is closed and layer *0* becomes active again.

The two connection bars are obtained in a similar way. These are mirror images of each other in plane OXY. The resulting solids are given the names *Union_1* and *Union_2*.

These procedures are called from the main program, as functions of the dimensions of the elements. We also obtain cylinders for the axes of the kinematic pairs $O_i O_{i+2}$, $A_i A_{i+2}$, $i = \overline{1, 3}$, and the solids representing the assembling and safety system of the elements.

Creating the animation frames

This process uses script files, as described in Section 8.5 for the R-7R mechanism. In that application, after construction, the elements of the mechanism were saved with the AutoCAD command **Block** and then inserted at the points having the coordinates determined in the positional analysis. At the insertion of an element with the command **Insert** both the coordinates of the point and the rotation angle relative to axis OX have to be entered. However, here we have obtained the frames of the animation in a different way. Names are assigned to the elements using AutoLisp functions, while the successive positions of the mechanism are saved in 36 layers named after the angle of the crank and which are integral parts of the drawing. To obtain the successive positions of the mechanism, we proceeds as follows:

1) Generates a layer having as its name the position of the mechanism; this layer will be the working layer.
2) For each element:
   - open the layer that contains the element of the mechanism
   - copy the element and assign it a name that contains the prefix *Copy*
   - move the copy to the working layer
   - close the layer that contains the position of the element
   - using the command **Move**, move the copy to the position determined in the positional analysis
   - using the command **Rotate**, rotate the copy through the angle previously determined as a function of the position of the crank.
3) After inserting all the elements, close the layer that has the position of the mechanism as its name and pass to layer *0*.

Although there are many steps involved, they can be automated using the procedures *Opening*, *Closing* and *Positioning*.

The advantage of the present algorithm is that it gives AutoCAD images containing all 36 positions of the mechanism. In the algorithm described in Section 8.5.5, the each position of the model was saved before being erased from the current drawing. In the new algorithm, the positions of the mechanism can be successively viewed in AutoCAD by opening different layers. This can be automated using an AutoLisp function or a script file.

Another advantage of working with layers is that we can assign them materials. It is also simpler to choose the visualization angle of the mechanism and to obtain the frames of the animation.

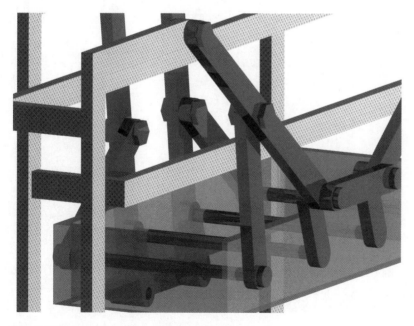

Figure 9.32 Detail from the modeled mechanism.

In the companion website of the book, we present the code for the program that peforms all these operations. In the example used, the dimensions of the mechanism are: $OA = 50\,\mathrm{mm}$, $AB = BC = BM = 125\,\mathrm{mm}$, $OC = 100\,\mathrm{mm}$, $A_1A_2 = A_3A_4 = 90\,\mathrm{mm}$, $g = 10\,\mathrm{mm}$, $h = 20\,\mathrm{mm}$, $d = \dfrac{h}{2}$, $z = 160\,\mathrm{mm}$, $x_{Frame} = 360\,\mathrm{mm}$, $y_{Frame} = z - g$, $h_{Frame} = 60\,\mathrm{mm}$, $l_{Leg} = 145\,\mathrm{mm}$, $h_{Leg} = 145\,\mathrm{mm}$, $h_{Leg} = 350\,\mathrm{mm}$, while the coordinates of the points are: $O_1\,(0,0,0,)$, $O_2\,(-A_1A_2,0,0)$, $O_3\,(0,0,-z)$, $O_4\,(-A_1A_2,0,-z)$.

One detail of the mechanism, after its assembly, is given in Figure 9.32.

Animation of the walking machine.

This can be done in two ways. In the first, the frame is fixed and the legs move (the hanged mechanism), while the second is defined by the motion of the mechanism. The effect of motion is created by displacing the entire assembly, using the comparison of the coordinates corresponding to the front and back legs, the displacement being given by the differences of the abscissae of the same point at two points of time.

10

Variators of Angular Velocity with Bars

The use of the planar mechanisms with bars in variators of angular velocity has been the subject of many patents. Variators of angular velocity transform the continuous rotational motion of a driving element into periodic motion of the driven element, as impulses shape, with the possibility to continuously vary the transmission ratio.

By using a unidirectional clutch at the output axle, the balance motion of the last element of the mechanism that enters in the construction of the variator is transformed (as impulse) into unidirectional motion.

10.1 Generalities

Because they are compact, simple to maintain, and offer good control the machines in which they are used, industrial variators of angular velocity, such as those manufactured by Riedel, Varvel, Gusa, Zero-max, and Morse, are widely used in feed kinematic chains of machine-tools (such as milling machines and grinding machines), saws, conveyer belts, and sewing machines.

The advantages of the use of the variators of angular velocity with bars (also referred to as *impulse variators*) are:

- the possibility of continuously varying the transmission ratio
- the transmission of constant torque
- quiet functioning
- high efficiency.

The main disadvantage of these variators is the irregularity of the angular velocity of the output axle. Keeping the machine operating within acceptable limits is brought about by parallel and equal-phased placement of more identical elementary mechanisms.

10.2 Mono-loop Mechanisms Used in the Construction of the Variators of Angular Velocity with Bars

10.2.1 Kinematic Schemata

The four-bar mechanisms that may be used in the construction of variators of angular velocity with bars are:

- mechanisms with crank and balancer (Figure 10.1)
- mechanisms with oscillatory rocker (Figure 10.2)
- mechanisms with translational rocker (Figure 10.3)

Classical and Modern Approaches in the Theory of Mechanisms, First Edition.
Nicolae Pandrea, Dinel Popa and Nicolae-Doru Stănescu.
© 2017 John Wiley & Sons Ltd. Published 2017 by John Wiley & Sons Ltd.
Companion Website: www.wiley.com/go/pandmech17

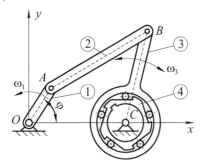

Figure 10.1 Variator of angular velocity with bars with crank and balancer mechanism.

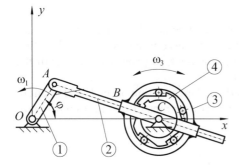

Figure 10.2 Variator of angular velocity with bars with oscillatory rocker.

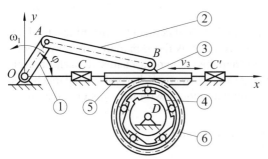

Figure 10.3 Variator of angular velocity with bars with translational rocker.

A common characteristic of these mechanisms is that driving element 1 has continuous rotational motion, elements 2 and 3 have at their ends lower pairs (fifth-class kinematic pairs), while element 3 has periodic motion, which is transformed into impulses by the unidirectional clutch 4.

By using rack 5 and gear 6, the mechanism with translational rocker in Figure 10.3 transforms the translational motion of element 3 into impulses, the motion being transformed by the unidirectional clutch 4, as in the previous cases.

In a complete revolution of crank 1, an impulse is generated and this determines the rotation of output axle 4 of the unidirectional clutch, by an angle ψ, which is less than 2π radians. A large number of impulses is required for a complete revolution of element 1, implying the need for many elementary mechanisms connected in parallel. This arrangement ensures a continuous succession of impulses at the output axle of the unidirectional clutch 4. In industrial variators, up to six elementary mechanisms may be connected in parallel, ensuring that there is little irregularity of the impulses.

Figure 10.4 shows two technical solutions for the parallel connection of four identical crank and balancer mechanisms. In Figure 10.4a, the common crank 1 drives the elementary mechanisms, with are spaced out equally around the input axle. The summing of impulses from the four gears 4, each with a fixed axis and a unidirectional clutch, is made by a central gear placed

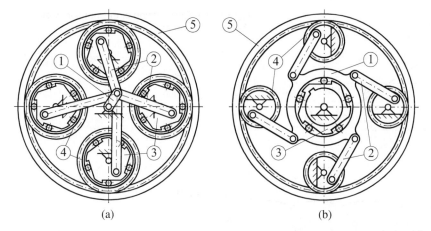

Figure 10.4 The parallel connection of four elementary mechanisms: (a) summing of impulses at central gear 5; (b) summing of impulses at output axle 1.

Figure 10.5 Adjustment of the crank length.

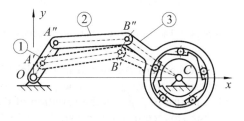

coaxially with the input axle. In Figure 10.4b, the mechanisms are driven by the central gear 5, which is driven by gears 4 having fixed axes. The impulses are created at output axle 1, which has a unidirectional clutch.

The transmission ratio may be changed by modifying the lengths of the mechanism's elements, which reduces the angle of oscillation of the driven element. For example, for the elementary mechanism with crank and balancer in Figure 10.5, the length of crank 1 is adjusted. The oscillation angle of element 3 is a function of the length of the crank 1, and so as the crank length increases the oscillation angle increases, and vice versa.

Usually, for practical variators of angular velocity, we do not modify the dimensions of coupler 2 or balancer 3, the manufacturers choosing to modify the length of the crank. Examples include the variators manufactured by Riedel and Varvel.

10.2.2 Kinematic Aspects

For mono-loop mechanisms used in the construction of variators of angular velocity with bars, the kinematic analysis gives the variation of angular velocity ω_3 of the driven element (Figures 10.1 and 10.2) as a function of the input angle φ and angular velocity ω_1 of the driving element. The *transmission ratio* i_V is defined as :

$$i_V = \frac{\omega_3}{\omega_1},\qquad(10.1)$$

where ω_1 and ω_3 are the angular velocities of elements 1 and 3, respectively. Since angular velocity ω_3 depends on the input angle φ, this means that the transmission ratio i_V depends on the input angle φ too. The shape of the curve of the transmission ratio (Figure 10.6) is identical to that of angular velocity ω_3 because angular velocity ω_1 of the driving element is assumed to be constant.

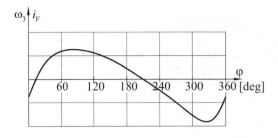

Figure 10.6 The variation of the angular velocity of the output element.

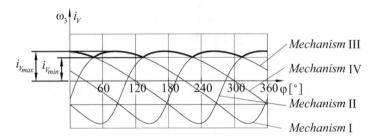

Figure 10.7 The variation of the transmission ratio.

In mechanisms connected in parallel (Figure 10.4), the transmission ratio of the variator is obtained by staggered superposition of the diagrams of variation of the transmission ratio. Figure 10.7 shows the variation of the transmission ratios in four identical elementary mechanisms, connected in parallel as in Figure 10.4; the curve $i_V = i_V(\varphi)$ is drawn with a heavy line.

The maximum and minimum values of the transmission ratio $i_{V_{max}}$ and $i_{V_{min}}$ of the variator are given by:

$$i_{V_{max}} = \frac{\omega_{3_{max}}}{\omega_1}, \; i_{V_{min}} = \frac{\omega_{3_{min}}}{\omega_1}. \tag{10.2}$$

The average value of the transmission ratio $i_{V_{av}}$ is given by the integral

$$i_{V_{av}} = \frac{1}{\varphi_C} \int_0^{\varphi_C} \frac{\omega_3(\varphi)}{\omega_1} d\varphi, \tag{10.3}$$

where φ_C is the length of the kinematic cycle (generally, 2π radians).

The qualitative appreciation of the variator is made using the *degree of irregularity* δ

$$\delta = \frac{\omega_{3_{max}} - \omega_{3_{min}}}{\omega_{3_{max}}} = \frac{i_{V_{max}} - i_{V_{min}}}{i_{V_{av}}}. \tag{10.4}$$

For a variator of angular velocity with bars, we want the smallest possible values for δ, which can be found through a careful choice of the elementary mechanism and the use of a large number of such mechanisms connected in parallel.

10.2.3 Numerical Example

Consider the mechanism with crank and balancer in Figure 10.1, for which the dimensions are $b = AB = 0.07$ m, $c = BC = 0.03$ m, $d = OC = 0.06$ m and the angular velocity of the driving element is $n = 1500$ rev/min. The modification of the transmission ratio of the variator is brought about by modifying the length of the crank. To establish the upper limit of the crank's length $a = OA$, we recall (5.11) – the conditions for the existence of the crank that were established in Section 5.1.1:

$$a \leq d + b - c,$$
$$a \leq d - b + c, \qquad\qquad (10.5)$$
$$a \leq -d + b + c$$

Making the replacements, we get

$$a \leq 0.100,$$
$$a \leq 0.020,$$
$$a \leq 0.040;$$

hence, $a \in [0, 0.020]$ m.

The kinematic analysis of the mechanism $OABC$ of R-3R type in Figure 10.1 is performed using (2.57)–(2.72). These equations were established in Section 2.4.2 and they are incorporated in the procedure *Dyad_RRR*, as described in Section 2.4.3. The code for the Turbo Pascal program which performs the calculations to determine the kinematic parameters is set out in the companion website of the book. The program can deal with any number of elementary mechanisms (*nr_mec*) and any division of the interval $0 - 2\pi$. In the program, the kinematic analysis was performed with an angular step of $1°$, so *Intervals* = 360. To use the procedure *Dyad_RRR*, it is necessary to determine the coordinates, velocities, and accelerations of the input poles A and C. The kinematic analysis is performed with an angular step given by the number of intervals into which the kinematic cycle (2π) was divided. It leads to the positions, angular velocities, and accelerations of elements 2 and 3, and the kinematic parameters of point B too. To create the animation of the mechanism and the graph of the function $\omega_3 = \omega_3 (\varphi)$ simultaneously, the visualization display is divided into two parts: the left-hand side for the animation and the right for the graph.

We use a script file (*animate.scr*) containing the AutoCAD commands for:

- drawing of axes Ox and Oy
- the conventional representation of the kinematic pairs of the four-bar mechanism, situated at points O and C, on the left-hand side of the display
- drawing the system $\omega_3 O'\varphi$ and the graph of $\omega_3 = \omega_3 (\varphi)$ on the right.

Next, we write in the file the commands to create the animation. We use a *For* loop, changing the angle φ in a constant step each time. In the loop the steps are as follows:

- Draw a poly-line representing the bars of the mechanism.
- Place at points A and B two circles with small thickness (**Donut**) representing the revolute joints.
- On the graph $\omega_3 = \omega_3 (\varphi)$, place a circle with thickness in order to localize the point.
- Pause for 100 ms for the visualization of the construction.
- In reverse order, erase (**Erase**) the entities constructed, and pass to the next angle.

Drawing the curve $\omega_3 = \omega_3 (\varphi)$ permits the optimization of the mechanism by modifying the values of the input parameters, and the visualisation of the shape of the angular velocity of the output element.

In order to make it easier to determine the transmission ratio of the variator and the degree of irregularity when more than one elementary mechanism is used, we divided the calculation program into three separate procedures.

The first, called *Separation*, forms from the vector angular velocity $\omega_3 (\varphi)$ (*Omega3mec*) a matrix A, having a number of rows equal to the number of elementary mechanisms, and a

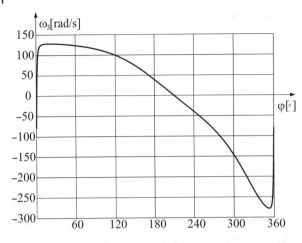

Figure 10.8 The variation of the angular velocity of the output element.

number of columns equal to the ratio between the value of the interval and the number of mechanisms. In other words, the interval $[0, 2\pi]$ will be divided into a number of equal subintervals, this number being equal to the number of elementary mechanisms, while the values in the subintervals are written in separate columns.

The next procedure, called *Forming*, permits the these subintervals to be assembled in the order required to obtain the output angular velocities for each elementary mechanism. These are written to the matrix *Omega3_Mec*, where the number of the row is given by the number of the mechanism, while the columns contain the angular velocities $\omega_3(\varphi)$. For instance, for a variator with four mechanisms, in the procedure *Separation* the interval $[0, 2\pi]$ is divided into four equal parts. If the shape of the graph $\omega_3 = \omega_3(\varphi)$ is that in Figure 10.8, then values are recorded in matrix A as follows: in the first column the values from $0°$ to $89°$, in the second column the values from $90°$ to $179°$, in the third column the values from $180°$ to $274°$, and in the last column the values from $275°$ to $359°$.

In the procedure *Forming* the columns of the matrix A are ordered according to the number of the mechanism:

- for mechanism I, the order is 1, 2, 3, 4
- for mechanism II, the order is 4, 1, 2, 3
- for mechanism III, the order is 3, 4, 1, 2
- for mechanism IV, the order is 2, 3, 4, 1

The vectors obtained are the columns of the matrix *Omega3_Mec*. The representation in Figure 10.9 is obtained.

In the third procedure, *Omega Variator*, we obtain the angular velocity of the variator's output, by comparing the output angular velocities of the elementary mechanism and choosing the greatest value. The values are written in the vector *Omega3_Variator* (the thick line in Figure 10.9).

The procedure *Maximum_Minimum_Average* is used to determined

- the maximum and minimum values of the transmission ratio
- the integral given by (10.3) (using the rectangle formula)
- the average value of the transmission ratio
- the degree of irregularity of the variator, found from (10.4).

Then, with the results given by the program, the kinematic study of the variator can be undertaken. Further values are given to the length OA of the crank. For a certain length of the crank, the number of elementary mechanisms is varied the values for $i_{V_{max}}, i_{V_{min}}, i_{V_{av}}$, and δ are obtained

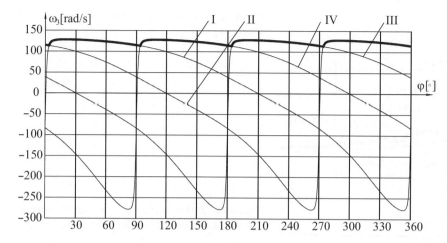

Figure 10.9 The variation of the angular velocities.

Table 10.1 The values obtained for the kinematic analysis.

Nr of mechanisms	a = 0.005			a = 0.010			a = 0.015			a = 0.020		
	3	4	6	3	4	6	3	4	6	3	4	6
$i_{V_{max}}$	0.1891	0.1891	0.1891	0.3718	0.3718	0.3718	0.5581	0.5581	0.5581	0.8221	0.8221	0.8221
$i_{V_{min}}$	0.1025	0.1394	0.1667	0.2173	0.2839	0.3334	0.3535	0.4473	0.5115	0.6346	0.7325	0.7872
$i_{V_{av}}$	0.1595	0.1722	0.1815	0.3196	0.3424	0.3588	0.4919	0.5218	0.5423	0.7674	0.7944	0.8107
δ	0.5433	0.2889	0.1237	0.4836	0.2568	0.1071	0.4160	0.2124	0.0861	0.2444	0.1129	0.0431

Figure 10.10 The variation of $i_{V_{av}}$ as a function of the crank level.

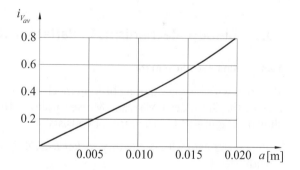

for each case. For the mechanism used in the kinematic study, the values obtained are shown in Table 10.1. These were used to draw Figures 10.10–10.12.

The program may be upgraded by adding two new *For* loops in which the length of the crank *OA* and the number of the elementary mechanisms is varied. This gives the data in Table 10.1 directly, and allows the diagrams to be drawn from within the program. The optimum values for an elementary mechanism, so that it meets the requirements of, say, the largest transmission ratio or the smallest degree of irregularity for a given number of mechanisms, will be determined quickly, and the results can be visualised as modifications are made to the inputs.

If we want to determine the variation of the average transmission ratio $i_{V_{av}}$ as a function of the crank level *OA*, then we can obtain the representation in Figure 10.10. The variation of the degree of irregularity as a function of the number of mechanisms and the crank length is given in Figure 10.11.

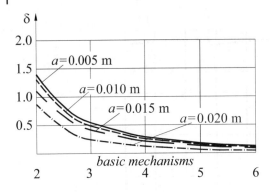

Figure 10.11 The variation of the degree of irregularity.

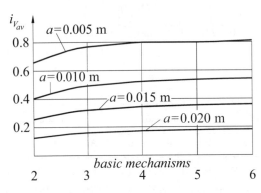

Figure 10.12 The variation of $i_{V_{av}}$ as a function of the crank level and the number of mechanisms.

The variation of the transmission ratio as a function of the number of mechanisms and the crank length is shown in Figure 10.12.

10.3 Bi-loop Mechanisms in Variators of Angular Velocity with Bars

10.3.1 Kinematic Schemata

Variators with bars containing bi-loop mechanisms that are common in Europe and the USA include the Gusa, Zero-Max, and Morse variators. The Gusa variator has the kinematic schema shown in Figure 10.13. The variator consists of:

- the mechanism with oscillatory rocker *R-RTR* (*OABC*), formed by elements 1, 2, 3 together with base 4
- the *TRR* dyad (*DEF*), formed by elements 5 and 6
- the unidirectional clutch 7, which transforms the motion of balancer 6 into impulses.

The transmission ratio of the variator is adjusted by modifying the position of the assembly 3, 4 (that is, the modification of the distance *d*); at a certain value of the transmission ratio, element 4 is fixed. When the assembly 3, 4 moves to the right, the value of the transmission ratio decreases.

In its industrial variant, the Gusa variator has three such mechanisms, connected in parallel and phase-shifted by 120° relative to each other, with the following kinematic characteristics: input angular velocity (of element 1) of 1450 rev/min, output angular velocity (of element 7) 0–300 rev/min. The variator works at output powers between 0.37 and 7.3 kW, transmitted torque between 25 and 400 Nm, and with the mass of the variator between 1.4 and 14.5 kg.

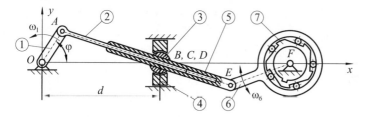

Figure 10.13 The kinematic schema of the Gusa variator.

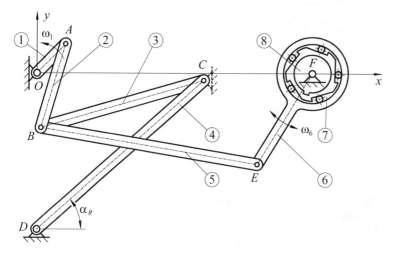

Figure 10.14 The kinematic schema of the Zero-Max variator.

The Zero-Max variator of angular velocity has the kinematic schema in Figure 10.14. It consists of:

- the mechanism with crank and balancer $OABC$, with elements 1, 2, 3 and base 4
- the RRR dyad (BEF), formed by element 5, 6
- the unidirectional clutch 7.

The adjustment of the transmission ratio of the variator is made by modifying angle φ_R (the position of element 4); for a certain value of the transmission ratio element 4 is fixed. The transmission ratio of the variator increases when the value of angle φ_R increases.

In its industrial variant, the Zero-Max variator has four such mechanisms, connected in parallel and phase-shifted by 90°, with the following characteristics: input angular velocity of 1800 rev/min, output angular velocity of 0–400 rev/min, power between 0.12 and 0.55 kW, and maximum transmitted torque between 1.4 and 8.5 Nm.

The mass is equal to 8.8 kg for a variator that permits the transmission of a maximum torque of 8.5 Nm. By putting a special unidirectional clutch at the output, the direction of rotation for axle 8 can be changed.

The kinematic schema of the Morse variator of angular velocity is shown in Figure 10.15. The Morse variator consists of:

- a mechanism with crank and balancer $OABC$, formed by elements 1, 2, 3 and the base 4
- the RRR dyad (BEF) formed by elements 5, 6
- the unidirectional clutch 7.

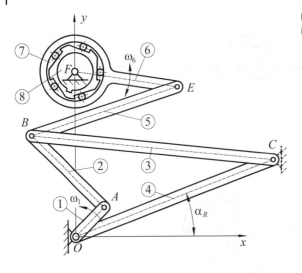

Figure 10.15 The kinematic schema of the Morse variator.

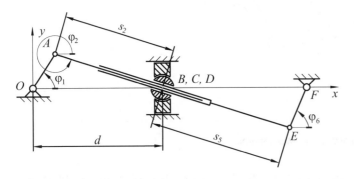

Figure 10.16 The kinematic schema of the Gusa variator.

The adjustment of the transmission ratio is similar to the Zero-Max variator: by modifying angle α_R (the position of the element 4). When $\alpha_R = 0$ the transmission ratio of the variator has its maximum value; it decreases when the value of angle α_R increases.

In its industrial variant, the Morse variator has three such mechanisms, equal-phased and connected in parallel, with the following characteristics: input angular velocity of 250 rev/min, output angular velocity of 1–55 rev/min, maximum transmitted torque of 280 Nm, with an efficiency of 0.95.

10.3.2 Kinematic Analysis

As method for the kinematic analysis we will use the analytical method of projections, as described in Section 2.3.2; the solution of the non-linear system of equations is obtained using the Newton–Raphson method, which was presented in Section 2.3.3.

Gusa variator
For the Gusa variator with the kinematic schema and notation in Figure 10.16, we write the vector relations

$$\begin{cases} \mathbf{OA} + \mathbf{AB} = \mathbf{OB}, \\ \mathbf{OB} + \mathbf{BE} + \mathbf{EF} = \mathbf{OF}, \end{cases} \tag{10.6}$$

This, by projection onto the axes of the reference system Oxy, gives

$$
\begin{aligned}
F_1 &= l_1 \cos \varphi_1 + s_2 \cos \varphi_2 - d = 0, \\
F_2 &= l_1 \sin \varphi_1 + s_2 \sin \varphi_2 = 0, \\
F_3 &= d + s_5 \cos \varphi_2 + l_6 \cos \varphi_6 - l_7 = 0, \\
F_4 &= s_5 \sin \varphi_2 + l_6 \sin \varphi_6 = 0
\end{aligned}
\tag{10.7}
$$

or, in matrix notation,

$$
\{\mathbf{F}\} = \{\mathbf{0}\},
\tag{10.8}
$$

where the vector $\{\mathbf{F}\}$ reads

$$
\{\mathbf{F}\} = \begin{bmatrix} F_1 & F_2 & F_3 & F_4 \end{bmatrix}^{\mathrm{T}}.
\tag{10.9}
$$

Let $\{\boldsymbol{\varphi}\}$ be the solution of the system (10.8), written in the form

$$
\{\boldsymbol{\varphi}\} = \begin{bmatrix} \varphi_2 & s_2 & s_5 & \varphi_6 \end{bmatrix}^{\mathrm{T}}.
\tag{10.10}
$$

Denoting by $\{\boldsymbol{\varphi}\}_0$ the approximate solution

$$
\{\boldsymbol{\varphi}\}_0 = \begin{bmatrix} \varphi_2^0 & s_2^0 & s_5^0 & \varphi_6^0 \end{bmatrix}^{\mathrm{T}}.
\tag{10.11}
$$

and taking into account the (2.28) (see Section 2.3.3), the system (10.8) may be written in the form

$$
\{\boldsymbol{\varphi}\}_1 = \{\boldsymbol{\varphi}\}_0 - [\mathbf{W}]_0^{-1} \{\mathbf{F}\}_0,
\tag{10.12}
$$

where we denoted by $\{\boldsymbol{\varphi}\}_1$ the solution obtained at the first iteration, and by $[\mathbf{W}]$ the Jacobian matrix

$$
[\mathbf{W}] =
\begin{bmatrix}
-s_2 \sin \varphi_2 & \cos \varphi_2 & 0 & 0 \\
s_2 \cos \varphi_2 & \sin \varphi_2 & 0 & 0 \\
-s_2 \sin \varphi_2 & 0 & \cos \varphi_2 & -l_6 \sin \varphi_6 \\
s_2 \cos \varphi_2 & 0 & \sin \varphi_2 & l_6 \cos \varphi_6
\end{bmatrix}.
\tag{10.13}
$$

After $n + 1$ iterations the solution is

$$
\{\boldsymbol{\varphi}\}_{n+1} = \{\boldsymbol{\varphi}\}_n - [\mathbf{W}]_n^{-1} \{\mathbf{F}\}_n,
\tag{10.14}
$$

the iterative procedure continuing until

$$
\left| \{\boldsymbol{\varphi}\}_{n+1} - \{\boldsymbol{\varphi}\}_n \right| \leq \xi,
\tag{10.15}
$$

where ξ is the error in the determination of the solution.

To obtain the linear and angular velocities, we differentiate (10.7) with respect to time and obtain

$$
\begin{aligned}
-s_2 \omega_2 \sin \varphi_2 + v_2 \cos \varphi_2 &= l_1 \omega_1 \sin \varphi_1, \\
s_2 \omega_2 \cos \varphi_2 + v_2 \sin \varphi_2 &= -l_1 \omega_1 \cos \varphi_1, \\
-s_5 \omega_2 \sin \varphi_2 + v_5 \cos \varphi_2 - l_6 \omega_6 \sin \varphi_6 &= 0, \\
s_5 \omega_2 \cos \varphi_2 + v_5 \sin \varphi_2 + l_6 \omega_6 \cos \varphi_6 &= 0.
\end{aligned}
\tag{10.16}
$$

If we denote

$$
\{\dot{\boldsymbol{\varphi}}\} = \begin{bmatrix} \dot{\varphi}_2 & \dot{s}_2 & \dot{s}_5 & \dot{\varphi}_6 \end{bmatrix}^{\mathrm{T}} = \begin{bmatrix} \omega_2 & v_2 & v_5 & \omega_6 \end{bmatrix}^{\mathrm{T}},
\tag{10.17}
$$

system (10.16) can be put in the form

$$
[\mathbf{W}] \{\dot{\boldsymbol{\varphi}}\} = \{\mathbf{V}\},
\tag{10.18}
$$

where $\{\mathbf{V}\}$ stands for the right-hand terms of system (10.16)

$$\{\mathbf{V}\} = \omega_1 \begin{bmatrix} l_1 \sin\varphi_1 & -l_1 \cos\varphi_1 & 0 & 0 \end{bmatrix}^{\mathrm{T}}. \tag{10.19}$$

Multiplying (10.18) on the left by $[\mathbf{W}]^{-1}$ gives

$$\{\dot{\boldsymbol{\varphi}}\} = [\mathbf{W}]^{-1}\{\mathbf{V}\}. \tag{10.20}$$

In a numerical example, we consider the following dimensions for the mechanism in Figure 10.16: $l_1 = OA = 0.01\,\text{m}, l_6 = EF = 0.02\,\text{m}, l_7 = OF = 0.18\,\text{m},$ and $d \in [0.06 \ldots 0.16]\,\text{m},$ and a constant angular velocity of the driving element of 1450 rev/min. To obtain the results we used the program given in the companion website of the book. The program is valid for any number of elementary mechanisms (*nr_mec*) and outputs files that can be used to draw the diagrams of the variation of angular velocity $\omega_6 = \omega_6(\varphi_1, d)$, the average transmission ratio given by (10.3), and the degree of irregularity given by (10.4).

The AutoCAD commands for the drawing of the diagrams $\omega_6 = \omega_6(\varphi_1)$ when the length d varies are in the script file *omega6.scr*. In the text file *Gusa_1* are written the values of the degree of irregularity, while in the file *Gusa_2* are written the values of the average transmission ratio; both sequences of values are functions of the distance d and of the number of elementary mechanisms.

For that purpose, in one *For* loop, described by *For Length:=1 To 11 ...*, the length d is varied in the interval $[0.06 \ldots 0.16]$ with step $\Delta d = 0.01$; in another *For* loop, defined by *For Number_Mechanisms:=2 To 6 ...* the number of elementary mechanisms is varied from 2 to 6 (*nr_max_mec=6*).

Next the kinematic analysis is performed by initializing the iterative procedure with the approximate solution. In a *For* loop described by *For Index:=1 To Intervals ...*, values are given to angle φ_1 in the interval $[0, 2\pi]$ with an angular step of $1°$ (*Intervals=360*). The following are determined:

- the values of the functions given by (10.7), by calling the procedure *Function*
- the values of the components of the Jacobian matrix given by (10.13), by calling the procedure *Jacobian*
- the solution of the system using the procedure *MatInv*, which has as input parameters the matrices $[\mathbf{W}]$ and $[\mathbf{B}]$, in which the first column contains the values of matrix $\{\mathbf{F}\}$; after the resolving of the system, the procedure returns the values of matrix $[\mathbf{W}]^{-1}$ in the matrix $[\mathbf{W}]$, and the solution $\{\boldsymbol{\varphi}\}$ at the first iteration in the first column of the matrix $[\mathbf{B}]$
- the error of determination of the solution using (10.15).

If (10.15) is not fulfilled (in our case, we considered $\xi = Err = 10^{-6}$), then the process moves to the next iteration, where the exact solution from the previous iteration becomes the approximate one for the new iteration. If (10.15) holds true, then the following values are calculated:

- the vector $\{\mathbf{V}\}$ given by (10.19)
- the vector velocity $\{\dot{\boldsymbol{\varphi}}\}$ given by (10.20).

The process moves to a new value for angle φ_1. The approximate solution for this new position is the exact one from the previous step.

The graph in Figure 10.17 was obtained by running the file *omega6.scr* in AutoCAD, giving the variation of angular velocity $\omega_6 = \omega_6(\varphi_1)$ as a function of the distance d. Using the procedures *Separation*, *Forming*, *OmegaVariator*, and *Maximum_ Minimum_Average*, which were all presented in Section 10.3, the kinematic parameters of the variator can also be determined; these parameters are written in the files *Gusa_1*, and *Gusa_2*. The values obtained in our example are presented in Table 10.2.

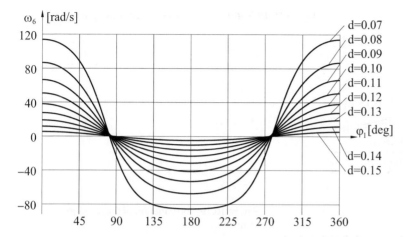

Figure 10.17 Graph of the functions $\omega_6 = \omega_6(\varphi_1)$.

Table 10.2 The values obtained for the kinematic analysis.

nr. mech.	2		3		4		5		6	
	δ	$i_{V_{av}}$	δ	$i_{V_{av}}$	δ	$i_{V_{av}}$	δ	$i_{V_{av}}$	δ	$i_{V_{av}}$
$d = 0.06$	0.7778	1.2857	1.0000	0.0000	1.0000	0.0000	1.0000	0.0000	1.0000	0.0000
$d = 0.07$	0.6742	1.4601	0.6499	0.5339	0.6995	0.2378	0.7194	0.1359	0.7294	0.0887
$d = 0.08$	0.4850	1.5266	0.4735	0.6440	0.5180	0.3191	0.5379	0.1914	0.5484	0.1281
$d = 0.09$	0.3607	1.5701	0.3553	0.6874	0.3912	0.3559	0.4079	0.2183	0.4170	0.1480
$d = 0.10$	0.2706	1.6113	0.2681	0.7064	0.2959	0.3739	0.3092	0.2322	0.3165	0.1586
$d = 0.11$	0.2031	1.6249	0.2001	0.7141	0.2211	0.3828	0.2313	0.2394	0.2369	0.1642
$d = 0.12$	0.1477	1.6887	0.1452	0.7161	0.1605	0.3867	0.1680	0.2429	0.1721	0.1670
$d = 0.13$	0.1033	1.7480	0.0998	0.7149	0.1103	0.3878	0.1154	0.2442	0.1183	0.1681
$d = 0.14$	0.0660	1.8762	0.0614	0.7119	0.0679	0.3872	0.0710	0.2442	0.0728	0.1683
$d = 0.15$	0.0337	2.2764	0.0285	0.7079	0.0315	0.3855	0.0330	0.2434	0.0338	0.1678
$d = 0.16$	0.0338	0.1678	0.0338	0.1678	0.0338	0.1678	0.0338	0.1678	0.0338	0.1678

Similar to variators with mono-loop mechanisms with bars, we can draw graph of the variation of the degree of irregularity (Figure 10.18) as a function of the number of elementary mechanisms and as a function of the distance d, and graphs of the variation of the average transmission ratio (Figure 10.19) as a function of the number of elementary mechanisms and as a function of the distance d.

From the graphs it can be seen that for this type of variator the value of the degree of irregularity is very little influenced by how much distance d varies, while the value of the average transmission ratio is very little influenced by the number of elementary mechanisms.

The Zero-Max variator

For the Zero-Max variator having the kinematic schema in Figure 10.20, we proceed in a similar wa to the Gusa variator. The vector equations are

$$\begin{cases} \mathbf{OA} + \mathbf{AB} + \mathbf{BC} = \mathbf{OC}, \\ \mathbf{OA} + \mathbf{AB} + \mathbf{BE} + \mathbf{EF} = \mathbf{OF}. \end{cases} \tag{10.21}$$

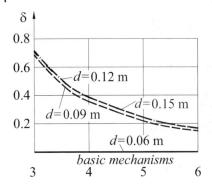

Figure 10.18 The variation of the degree of irregularity δ.

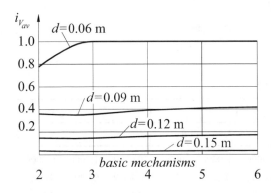

Figure 10.19 The variation of the average transmission ratio $i_{V_{av}}$.

With the notation in Figure 10.20 we determine the functions of the positions

$$\{\mathbf{F}\} = \begin{bmatrix} l_1 \cos \varphi_1 + l_2 \cos \varphi_2 + l_3 \cos \varphi_3 - x_C \\ l_1 \sin \varphi_1 + l_2 \sin \varphi_2 + l_3 \sin \varphi_3 - y_C \\ l_1 \cos \varphi_1 + l_2 \cos \varphi_2 + l_5 \cos \varphi_5 + l_6 \cos \varphi_6 - l_7 \\ l_1 \sin \varphi_1 + l_2 \sin \varphi_2 + l_5 \sin \varphi_5 + l_6 \sin \varphi_6 \end{bmatrix} = \begin{bmatrix} 0 \\ 0 \\ 0 \\ 0 \end{bmatrix}, \tag{10.22}$$

where

$$x_C = l_4 \cos \alpha_R, \; y_C = y_D + l_4 \sin \alpha_R. \tag{10.23}$$

The solution of system (10.22) is the column vector

$$\{\boldsymbol{\varphi}\} = \begin{bmatrix} \varphi_2 & \varphi_3 & \varphi_5 & \varphi_6 \end{bmatrix}^{\mathrm{T}}, \tag{10.24}$$

while the Jacobian matrix reads

$$[\mathbf{W}] = \begin{bmatrix} -l_2 \sin \varphi_2 & -l_3 \sin \varphi_3 & 0 & 0 \\ l_2 \cos \varphi_2 & l_3 \cos \varphi_3 & 0 & 0 \\ -l_2 \sin \varphi_2 & 0 & -l_5 \sin \varphi_5 & -l_6 \sin \varphi_6 \\ l_2 \cos \varphi_2 & 0 & l_5 \cos \varphi_5 & l_6 \cos \varphi_6 \end{bmatrix} \tag{10.25}$$

As a numerical example, we consider the following dimensions for the elements of the mechanism in Figure 10.20: $l_1 = 0.01$ m, $l_2 = 0.03$ m, $l_3 = 0.05$ m, $l_4 = 0.045$ m, $l_5 = 0.067$ m, $l_6 = 0.025$ m, $\alpha_R = 0 \dots 51°$, $l_7 = 0.072$ m, $x_D = 0$, and $y_D = -0.025$, with a constant angular velocity of the driving element of 1800 rev/min.

We use the same program as for the Gusa variator, in which only the procedures for the determination of the functions of position, and the Jacobian matrix are different; the expressions used

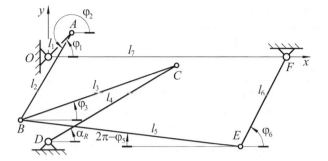

Figure 10.20 The kinematic schema of the Zero-Max variator.

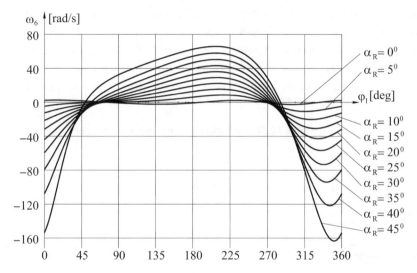

Figure 10.21 The graph of the functions $\omega_6 = \omega_6(\varphi_1)$.

are shown in (10.22) and (10.25), respectively. The approximate solution $\{\boldsymbol{\varphi}\}_0$ is

$$\{\boldsymbol{\varphi}\}_0 = \begin{bmatrix} \varphi_2^0 & \varphi_3^0 & \varphi_5^0 & \varphi_6^0 \end{bmatrix}^{\mathrm{T}} = \begin{bmatrix} \dfrac{3\pi}{2} & \dfrac{\pi}{4} & 0 & \dfrac{\pi}{4} \end{bmatrix}^{\mathrm{T}}, \tag{10.26}$$

while the vector $\{\mathbf{V}\}$ has the expression

$$\{\mathbf{V}\} = \begin{bmatrix} l_1 \sin\varphi_1 & -l_1 \cos\varphi_1 & l_1 \sin\varphi_1 & -l_1 \cos\varphi_1 \end{bmatrix}^{\mathrm{T}}. \tag{10.27}$$

After running the program, we obtain the graph of angular velocity $\omega_6 = \omega_6(\varphi_1)$ as a function of the value of the adjustment angle α_R, in Figure 10.21. Table 10.3 shows the degree of irregularity and the average transmission ratio as functions of the angle α_R and the number of elementary mechanisms. These values were used to draw Figures 10.22 and 10.23.

The Morse variator

The kinematic schema of the Morse variator is shown in Figure 10.24. The similarity of form and notation to the Zero-Max variator in Figure 10.20 can be seen. This similarity means that only minor modifications are required in the kinematic analysis.

Using the notation in Figure 10.24, the vector equations

$$\begin{cases} \mathbf{OA} + \mathbf{AB} + \mathbf{BC} = \mathbf{OC}, \\ \mathbf{OA} + \mathbf{AB} + \mathbf{BE} + \mathbf{EF} = \mathbf{OF}, \end{cases} \tag{10.28}$$

Table 10.3 The values obtained for the kinematic analysis.

Nr of mechanisms	2		3		4		5		6	
	δ	$i_{V_{av}}$	δ	$i_{V_{av}}$	δ	$i_{V_{av}}$	δ	$i_{V_{av}}$	δ	$i_{V_{av}}$
$\alpha_R = 0^0$	0.0091	1.4915	0.0088	1.5357	0.0105	1.2836	0.0110	0.6727	0.0122	0.2536
$\alpha_R = 5^0$	0.0333	1.3407	0.0231	1.0699	0.0410	0.3385	0.0404	0.4005	0.0420	0.2774
$\alpha_R = 10^0$	0.0647	1.2409	0.0505	0.6486	0.0767	0.2163	0.0744	0.2867	0.0769	0.1996
$\alpha_R = 15^0$	0.0959	1.2115	0.0789	0.5662	0.1121	0.2131	0.1087	0.2372	0.1120	0.1688
$\alpha_R = 20^0$	0.1273	1.1955	0.1083	0.5154	0.1477	0.2016	0.1435	0.2117	0.1475	0.1485
$\alpha_R = 25^0$	0.1591	1.1811	0.1394	0.4622	0.1837	0.1860	0.1790	0.1872	0.1835	0.1357
$\alpha_R = 30^0$	0.1918	1.1640	0.1726	0.4115	0.2207	0.1686	0.2155	0.1690	0.2204	0.1198
$\alpha_R = 35^0$	0.2246	1.1851	0.2087	0.3587	0.2589	0.1503	0.2535	0.1504	0.2586	0.1049
$\alpha_R = 40^0$	0.2603	1.1533	0.2488	0.3034	0.2988	0.1318	0.2935	0.1290	0.2986	0.0894
$\alpha_R = 45^0$	0.2971	1.1540	0.2945	0.2439	0.3411	0.1137	0.3361	0.1054	0.3409	0.0751
$\alpha_R = 51^0$	0.3442	1.1540	0.3613	0.1438	0.3958	0.0933	0.3918	0.0731	0.3959	0.0524

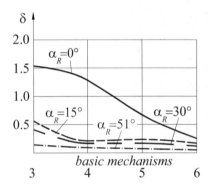

Figure 10.22 The variation of the degree of irregularity δ.

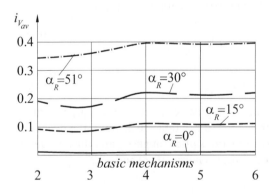

Figure 10.23 The variation of the average transmission ratio $i_{V_{av}}$.

projected onto the system of axes Oxy, determine the system of equations of projections

$$\{F\} = \begin{bmatrix} l_1 \cos\varphi_1 + l_2 \cos\varphi_2 + l_3 \cos\varphi_3 - x_C \\ l_1 \sin\varphi_1 + l_2 \sin\varphi_2 + l_3 \sin\varphi_3 - y_C \\ l_1 \cos\varphi_1 + l_2 \cos\varphi_2 + l_5 \cos\varphi_5 + l_6 \cos\varphi_6 \\ l_1 \sin\varphi_1 + l_2 \sin\varphi_2 + l_5 \sin\varphi_5 + l_6 \sin\varphi_6 - l_7 \end{bmatrix} = \begin{bmatrix} 0 \\ 0 \\ 0 \\ 0 \end{bmatrix}, \tag{10.29}$$

Figure 10.24 The kinematic schema of the Morse variator.

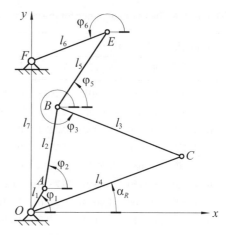

where

$$x_C = l_4 \cos \alpha_R, \quad y_C = l_4 \sin \alpha_R. \tag{10.30}$$

Proceeding as for the previous two variators, we determine the components of the Jacobian matrix $[\mathbf{W}]$

$$[\mathbf{W}] = \begin{bmatrix} -l_2 \sin \varphi_2 & -l_3 \sin \varphi_3 & 0 & 0 \\ l_2 \cos \varphi_2 & l_3 \cos \varphi_3 & 0 & 0 \\ -l_2 \sin \varphi_2 & 0 & -l_5 \sin \varphi_5 & -l_6 \sin \varphi_6 \\ l_2 \cos \varphi_2 & 0 & l_5 \cos \varphi_5 & l_6 \cos \varphi_6 \end{bmatrix}. \tag{10.31}$$

The solution of system (10.29) is, as in the case of the Zero-Max variator,

$$\{\boldsymbol{\varphi}\} = \begin{bmatrix} \varphi_2 & \varphi_3 & \varphi_5 & \varphi_6 \end{bmatrix}^{\mathrm{T}} \tag{10.32}$$

It can be seen that the Jacobian matrix is identical to that obtained for the Zero-Max variator. The vector velocity $\{\mathbf{V}\}$ has the expression in (10.27), which implies that only minor modifications are required in the calculation program that gives the numerical results.

In practice, only the approximate solution $\{\boldsymbol{\varphi}\}_0$ and the expressions of the functions of position given by (10.29), found in the procedure *Functions*, will be different. The other expressions

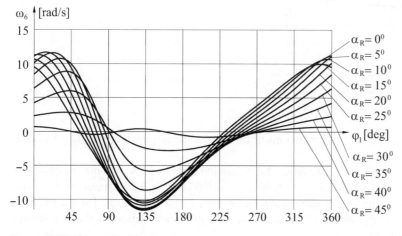

Figure 10.25 The graph of the functions $\omega_6 = \omega_6(\varphi_1)$.

Table 10.4 The values obtained for the kinematic analysis.

nr. mech.	2		3		4		5		6	
	δ	$i_{V_{av}}$	δ	$i_{V_{av}}$	δ	$i_{V_{av}}$	δ	$i_{V_{av}}$	δ	$i_{V_{av}}$
$\alpha_R = 0^0$	0.2268	1.6201	0.2959	0.7321	0.3278	0.4341	0.3449	0.2914	0.3551	0.2021
$\alpha_R = 5^0$	0.2399	1.6435	0.3140	0.7588	0.3489	0.4560	0.3680	0.3017	0.3794	0.2190
$\alpha_R = 10^0$	0.2526	1.7000	0.3272	0.7879	0.3652	0.4824	0.3862	0.3251	0.3989	0.2302
$\alpha_R = 15^0$	0.2646	1.6758	0.3325	0.8278	0.3727	0.5118	0.3955	0.3411	0.4093	0.2432
$\alpha_R = 20^0$	0.2644	1.6740	0.3234	0.9091	0.3656	0.5397	0.3892	0.3667	0.4037	0.2662
$\alpha_R = 25^0$	0.2460	1.7640	0.2898	1.0186	0.3341	0.6111	0.3583	0.3816	0.3723	0.2800
$\alpha_R = 30^0$	0.2068	1.9088	0.2306	1.1054	0.2688	0.6871	0.2917	0.4574	0.3065	0.2783
$\alpha_R = 35^0$	0.1515	2.1162	0.1553	1.1164	0.1810	0.6872	0.1966	0.4532	0.2089	0.2510
$\alpha_R = 40^0$	0.0909	2.4076	0.0801	0.9510	0.0917	0.5137	0.0976	0.3104	0.1009	0.1743
$\alpha_R = 45^0$	0.0409	2.6097	0.0188	1.2618	0.0223	0.5774	0.0239	0.3888	0.0249	0.2633

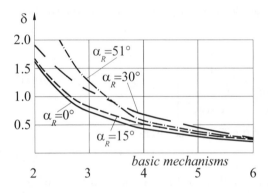

Figure 10.26 The variation of the degree of irregularity δ.

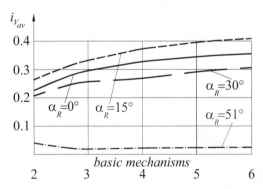

Figure 10.27 The variation of the average transmission ratio $i_{V_{av}}$.

remain unchanged, except for the dimensions of the mechanisms, which are declared as constants in the program.

As a numerical example, we consider the following dimensions for the mechanism: $l_1 = 0.01$ m, $l_2 = 0.03$ m, $l_3 = 0.05$ m, $l_4 = 0.06$ m, $l_5 = 0.033$ m, $l_6 = 0.03$ m, $\alpha_R = 0 \dots 45^0$, and $l_7 = 0.055$ m, with an input angular velocity at element 1 of 250 rev/min. The graph of the angular velocity $\omega_6 = \omega_6(\varphi_1)$ for an elementary mechanism as a function of the adjustment angle α_R is shown in Figure 10.25.

Varying the position of element 4 by a modification of angle α_R and the number of elementary mechanisms, we obtain the data in Table 10.4, from which we can draw Figures 10.26 and 10.27.

The degree of irregularity for the Gusa variator (Figure 10.18) is distinctly influenced by the adjustment angle α_R, a statement that does not hold for the Zero-Max variator (Figure 10.22) and the Morse variator (Figure 10.26). However, there is a similarity in the diagrams of variation of the average transmission ratio $i_{V_{av}}$ (Figures 10.19, 10.23, and 10.27), where the value of the average transmission ratio is little influenced by the number of elementary mechanisms; the adjustment angle has a strong influence.

Optimizing these variators to meet design requirements in made much easier using the calculation programs described in this section; the diagrams and tables are stepping stones on the way to new structural dimensions for the mechanisms involved.

Further Reading

1 Ambekar AG (2007). *Mechanism and Machine Theory*. Boston: Prentice-Hall.
2 Angeles J (Editor), Zakhariev E (Editor) (1998). *Computational Methods in Mechanical Systems: Mechanism Analysis, Synthesis, and Optimization*, 1st Edition. Berlin: Springer-Verlag.
3 Antonescu P (1980). *Mechanisms*. Bucharest: Polytechnic Institute (in Romanian).
4 Arakelian V, Briot S (2015). *Balancing of Linkages and Robot Manipulators: Advanced Methods with Illustrative Examples*. New York: Springer.
5 Artobolevski I (1977). *Théorie des Mécanismes et des Machines*. Moscow: MIR.
6 Artobolevski I (1978). *Les Mécanismes dans la Technique Moderne*: Tome 1–5, Moscow: MIR.
7 Asada H, Slotine JJE (1986). *Robot Analysis and Control*, 1st Edition. New York: Wiley-Interscience.
8 Aziz K (1995). Synthesis of RSSR spatial function generation mechanism by the condition of crank existence. *IFToMM Ninth World Congress on the Theory of Machines and Mechanisms*, Milan, Italy.
9 Barbu Gh, Vaduva I, Boloşteanu M (1997). *Bases of Informatics*. Bucharest: Technical Publishing House (in Romanian).
10 Barton LO (1993). *Mechanism Analysis: Simplified and Graphical Techniques*, Second Edition. Boca Raton: CRC Press.
11 Bickford JH (1972). *Mechanisms for Intermittent Motion*. Malabar: Krieger Publishing.
12 Bolton WC (2006). *Mechanical Science*, 3rd Edition. New York: Wiley-Blackwell.
13 Buculei M, Bagnaru D, Nanu Gh, Marghitu D (1986). *Calculation Methods in the Analysis of the Mechanisms with Bars*. Craiova: Scrisul Românesc Publishing House (in Romanian).
14 Buculei M *et al* (1994). *Elements of Technical Mechanics: Theory and Applications*. Craiova: Universitaria Publishing House (in Romanian).
15 Chaillet N (Editor), Régnier S (Editor) (2010). *Microrobotics for Micromanipulation*, 1st Edition. New York: Wiley.
16 Chang KH (2015). *Mechanism Design and Analysis Using Creo Mechanism 3.0*. Mission: SDC Publications.
17 Chiang CH (2000). *Kinematics and Design of Planar Mechanisms*. Malabar: Krieger Publishing.
18 Chiang CH (2000). *Kinematics of Spherical Mechanisms*, 2nd Edition. Malabar: Krieger Publishing.
19 Childs PRN (2015). *Mechanical Design Engineering Handbook*, 1st Edition. Oxford: Butterworth-Heinemann.
20 Choogin VV, Bandara P, Chepelyuk EV (2013). *Mechanisms of Flat Weaving Technology*, 1st Edition. Oxford: Woodhead Publishing.

Classical and Modern Approaches in the Theory of Mechanisms, First Edition.
Nicolae Pandrea, Dinel Popa and Nicolae-Doru Stănescu.
© 2017 John Wiley & Sons Ltd. Published 2017 by John Wiley & Sons Ltd.
Companion Website: www.wiley.com/go/pandmech17

21 Gans RF (2015). *Mechanical Systems: A Unified Approach to Vibrations and Controls.* New York: Springer.

22 Collins JA, Busby HR, Staab GH (2003). *Mechanical Design of Machine Elements and Machines: A Failure Prevention Perspective*, 2nd Edition. New York: Wiley.

23 Conley PL (Editor) (1988). *Space Vehicle Mechanisms: Elements of Successful Design*, 1st Edition. New York: Wiley.

24 Constantinescu G (1985). *Theory of Sonics.* Bucharest: The Publishing House of the Romanian Academy (in Romanian).

25 Davis WO (1993). *Gears for Small Mechanisms.* Warwickshire: TEE Publishing.

26 Demian T, Tudor D, Grecu E (1982). *Mechanisms of Fine Mechanics.* Bucharest: Didactical and Pedagogical Publishing House (in Romanian).

27 Dijksman EA (1976). *Motion Geometry of Mechanisms.* Cambridge: Cambridge University Press.

28 Dimarogonas AD (2000). *Machine Design: A CAD Approach*, 1st Edition. New York: Wiley-Interscience.

29 Dooner DB (2012). *Kinematic Geometry of Gearing*, 2nd Edition. Chichester: Wiley.

30 Dresig H, Holzweißig (2010). *Dynamics of Machinery: Theory and Applications.* New York: Springer.

31 Drimer D, Oprean A, Dorin A, Alexandrescu P, Paris A, Panaitopol N, Udrea C, Crişan I (1985). *Industrial Robots and Manipulators.* Bucharest: Technical Publishing House (in Romanian).

32 Dudiţă F, Diaconescu D, Gogu G (1989). *Jointed Mechanisms.* Bucharest: Technical Publishing House (in Romanian).

33 Dudiţă F (1977). *Mechanisms.* Braşov: The Publishing House of the University of (in Romanian).

34 Dudiţă F (1974) *Homokinetic Mobile Joints.* Bucharest: Technical Publishing House (in Romanian).

35 Dudiţă F, Diaconescu D (1987). *Structural Optimization of the Mechanisms.* Bucharest: Technical Publishing House (in Romanian).

36 Duffy J (1996). *Statics and Kinematics with Applications to Robotics*, 1st Edition. Cambridge: Cambridge University Press.

37 Dukkipati RV (2001). *Spatial Mechanisms: Analysis and Systems*, 1st Edition. Boca Raton: CRC Press.

38 Eckhardt HD (1988). *Kinematic Design of Machines and Mechanisms*, 1st Edition. New York: McGraw-Hill.

39 Erdman AG, Sandor GN, Kota S (2000). *Mechanism Design: Analysis and Synthesis.* Boston: Prentice Hall.

40 Fahimi F (2008). *Autonomous Robots: Modeling, Path Planning, and Control*, 1st Edition. New York: Springer.

41 Featherstone R (1987). *Robot Dynamics Algorithms.* 1st Edition. Dordrecht: Springer.

42 Goldfarb V (Editor), Barmina N (Editor) (2016). *Theory and Practice of Gearing and Transmissions: In Honor of Professor Faydor L. Litvin*, 1st Edition. New York: Springer.

43 Hahn H (2002). *Rigid Body Dynamics of Mechanisms I: Theoretical Basis.* Berlin: Springer.

44 Hahn H (2003). *Rigid Body Dynamics of Mechanisms 2: Applications Softcover.* Berlin: Springer

45 Hall AS Jr (1987). *Notes on Mechanism Analysis.* Prospect Heights: Waveland Press.

46 Handra Luca V, Stoica IA (1983). *Introduction in the Theory of Mechanisms.* Cluj-Napoca: Dacia Publishing House (in Romanian).

47 Handra Luca V, Stoica IA (1980). *Mechanisms.* Cluj-Napoca: The Publishing House of the Polytechnic Institute of Cluj-Napoca (in Romanian).

48 Hartenberg RS, Denavit J (1964). *Kinematic Synthesis of Linkages.* New York: McGraw-Hill.
49 Howell LL (2001). *Compliant Mechanisms,* 1st Edition. New York: Wiley.
50 Huang Z, Li Q, Ding H (2012). *Theory of Parallel Mechanisms.* New York: Springer.
51 Hunt KH (1979). *Kinematic Geometry of Mechanisms.* Oxford: Oxford University Press.
52 IFToMM Commission on Standards for Terminology (1983). *Terminology for the Theory of Machines and Mechanisms.* Oxford: Pergamon Press.
53 Jones FD (2016). *Mechanisms and Mechanical Movements.* Chicago: Leopold Classic Library.
54 Kelly R, Santibánez Davila V, Loría Perez JL (2005). *Control of Robot Manipulators in Joint Space.* London: Springer-Verlag.
55 Klebanov BM, Groper M (2015). *Power Mechanisms of Rotational and Cyclic Motions.* Boca Raton: CRC Press.
56 Kong X, Gosselin CM (2007). *Type Synthesis of Parallel Mechanisms.* Berlin: Springer.
57 Kovacs F, Cojocaru G (1982). *Manipulators, Robots and their Industrial Applications.* Timişoara: Facla Publishing House (in Romanian).
58 Kumar V, Schmiedeler J, Sreenivasan SV, Su HJ (2013). *Advances in Mechanisms, Robotics and Design Education and Research.* Berlin: Springer-Verlag.
59 Lamb F (2013). *Industrial Automation: Hands On,* 1st Edition. New York: McGraw-Hill.
60 Lenarcic J, Bajd T, Stanisic MM (2013). *Robot Mechanisms.* Dordrecht: Springer.
61 Lewis FL, Dawson DM, Abdallah CT (2003). *Robot Manipulator Control: Theory and Practice,* 2nd Edition. Boca Raton: CRC Press.
62 Liu XJ, Wang J (2014). *Parallel Kinematics: Type, Kinematics, and Optimal Design.* Berlin: Springer-Verlag.
63 Lumelsky VJ (2005). *Sensing, Intelligence, Motion: How Robots and Humans Move in an Unstructured World,* 1st Edition. New York: Wiley-Interscience.
64 Mabie HH, Ocvirk FW (1963). *Mechanisms and Dynamics of Machinery.* New York: Wiley.
65 Mabie HH, Reinholtz CF (1987). *Mechanisms and Dynamics of Machinery,* 4th Edition. New York: Wiley.
66 Mallik AK, Ghosh A, Dittrich G (1994). *Kinematic Analysis and Synthesis of Mechanisms,* 1st Edition. Boca Raton: CRC Press.
67 Manolea D (1996). *Programming in AutoLisp under AutoCAD.* Cluj-Napoca: Albastră Publishing House (in Romanian).
68 Manolescu N, Kovacs F, Orănescu A (1972). *The Theory of Mechanisms and Machines.* Bucharest: Didactical and Pedagogical Publishing House (in Romanian).
69 Marghitu DB (2009). *Mechanisms and Robots Analysis with MATLAB.* New York: Springer.
70 Marghitu DB, Crocker MJ (2001). *Analytical Elements of Mechanisms,* 1st Edition. Cambridge: Cambridge University Press.
71 Mason MT (2001). *Mechanics of Robotic Manipulation.* Boston: Massachusetts Institute of Technology.
72 Miloiu G, Dudiţă F, Diaconescu DV (1980). *Modern Mechanical Transmissions.* Bucharest: Technical Publishing House (in Romanian).
73 Molian S (1997). *Mechanism Design. The Practical Kinematics and Dynamics of Machinery,* 2nd Edition. Oxford: Pergamon Press.
74 Mostafa MA (2012). *Mechanics of Machinery,* 1st Edition. Boca Raton: CRC Press.
75 Myszka DH (2011). *Machines and Mechanisms: Applied Kinematic Analysis,* 4th Edition. Boston: Prentice Hall.
76 Nango J, Watanabe K (1995). *Kinematic analysis of general spatial 7-link 7R mechanisms.* Milan: IFToMM Ninth World Congress on the Theory of Machines and Mechanisms, Milan-Italy.

77 Norton RL (2004). *Design of Machinery: An Introduction to the Synthesis and Analysis of Mechanisms and Machine*s. New York: McGraw-Hill.

78 Orănescu A, Bocioacă R, Bumbaru S (1979). *Mechanisms*. The Publishing House of the University of Galați (in Romanian).

79 Pandrea N, Bărăscu E (1990). *Mechanisms*. Pitești: The Publishing House of the University of Pitești (in Romanian).

80 Pandrea N, Popa D (2000). *Mechanisms. Theory and CAD Applications*. Bucharest: Tehnical Publishing House (in Romanian).

81 Pandrea N, Stănescu ND (2015). *Dynamics of the Rigid Solid with General Constraints by a Multibody Approach*. Chichester: Wiley.

82 Pandrea N, Stănescu ND (2002). *Mechanics*. Bucharest: Didactical and Pedagogical Publishing House (in Romanian).

83 Pelecudi C (1973). *The Bases of the Analysis of the Mechanisms*. Bucharest: The Publishing House of the Romanian Academy (in Romanian).

84 Pelecudi C, Comănescu A, Grecu B, State D, Moise V, Ene M (1986). *Structural and Kinematic Analysis of the Mechanisms*. Bucharest: The Publishing House of the Polytechnic Institute of Bucharest (in Romanian).

85 Pelecudi C, Maroş D, Merticaru V, Pandrea N, Simionescu I (1985). *Mechanisms*. Bucharest: Didactical and Pedagogical Publishing House (in Romanian).

86 Pelecudi C, Simionescu I, Ene M, Moise V, Candrea A (1982). *Problems of Mechanisms*. Bucharest: Didactical and Pedagogical Publishing House (in Romanian).

87 Popa D (2010). *Elastodynamics of the Spatial Mechanisms*. Pitești: The Publishing House of the University of Pitești (in Romanian).

88 Popa D, Chiroiu V, Munteanu L (2007). *Research Trends in Mechanics*. Bucharest: The Publishing House of the Romanian Academy1.

89 Popa D, Popa CM (2003). *Assisted Design in Mechanical Engineering*. Bucharest: Technical Publishing House (in Romanian).

90 Popa D, Stănescu, ND (2013). CAD-Project medium and machine tool. *Proceedings of the Advances in Production, Automation and Transportation Systems, MEQAPS' 13*, Brașov, 2013, 277–282.

91 Popescu I (1977). *Design of the Planar Mechanisms*. Craiova: Scrisul Românesc Publishing House (in Romanian).

92 Reuleaux F (1963). *The Kinematics of Machinery: Outlines of a Theory of Machines*. New York: Dover

93 Rider MJ (2015). *Design and Analysis of Mechanisms: A Planar Approach*, 1st Edition. Chichester: Wiley.

94 Russell K, Shen Q, Sodhi RS (2013). *Mechanism Design: Visual and Programmable Approaches*. Boca Raton: CRC Press.

95 Russell K, Shen Q, Sodhi RS (2015). *Kinematics and Dynamics of Mechanical Systems: Implementation in MATLAB and SimMechanics*. Boca Raton: CRC Press.

96 Sacks E, Joskowicz L (2010). *The Configuration Space Method for Kinematic Design of Mechanisms*. Boston: Massachusetts Institute of Technology Press.

97 Sandin P (2003). *Robot Mechanisms and Mechanical Devices Illustrated*, 1st Edition. New York: McGraw-Hill.

98 Sandler BZ (1991). *Robotics: Designing the Mechanisms for Automated Machinery*. Boston: Prentice Hall.

99 Sarafin TP (Editor), Larson W (Managing Editor) (1995). *Spacecraft Structures and Mechanisms: From Concept to Launch*. Dordrecht: Kluwer.

100 Sciavicco L, Siciliano B (2001). *Modelling and Control of Robot Manipulators*. London: Springer-Verlag.

101 Sclater N (2011). *Mechanisms and Mechanical Devices.* New York: McGraw-Hill.

102 Shigley JE (1969). *Kinematic Analysis of Mechanisms,* 2nd Edition. New York: McGraw-Hill.

103 Shigley JE (Editor), Mischke CR (Editor), Brown TH Jr (Editor) (2004). *Standard Handbook of Machine Design,* 3rd Edition. New York: McGraw-Hill.

104 Siciliano B, Sciavicco L, Villani L, Oriolo G (2009). *Robotics: Modelling, Planning and Control,* 1st Edition. London: Springer-Verlag.

105 Siegwart R, Nourbakhsh IR, Scaramuzza D (2011). *Introduction to Autonomous Mobile Robots,* 2nd Edition. Boston: Massachusetts Institute of Technology.

106 Smith ST (2003). *Flexures: Elements of Elastic Mechanisms,* 1st Edition. London: Taylor and Francis.

107 Soni AH (1974). *Mechanism Synthesis and Analysis Hardcover.* New York: McGraw-Hill.

108 Stănescu ND (2005). *Mechanics.* Piteşti: The Publishing House of the University of Piteşti (in Romanian).

109 Stănescu ND (2013). *Mechanics of Systems.* Bucharest: Didactical and Pedagogical Publishing House (in Romanian).

110 Stănescu ND (2011). *Mechanical systems with neo-Hookean elements. Stability and Behavior.* Saarbrücken: Lambert Academic Publishing.

111 Stănescu ND (2007). *Numerical Methods.* Bucharest: Didactical and Pedagogical Publishing House (in Romanian).

112 Stănescu ND, Munteanu L, Chiroiu V, Pandrea N (2007). *Dynamical Systems. Theory and Applications,* Vol. 1. Bucharest: The Publishing House of the Romanian Academy (in Romanian).

113 Stănescu ND, Munteanu L, Chiroiu V, Pandrea N (2011). *Dynamical Systems. Theory and Applications,* Vol. 2. Bucharest: The Publishing House of the Romanian Academy (in Romanian).

114 Stanisic MM (2014). *Mechanisms and Machines: Kinematics, Dynamics, and Synthesis,* 1st Edition. Stanford: CL Engineering.

115 Talabă D (2001). *Jointed Mechanisms. Computer Assisted Design.* Braşov: The Publishing House of the University of Transylvania of Braşov (in Romanian).

116 Teodorescu P, Stănescu ND, Pandrea N (2013). *Numerical Analysis with Applications in Mechanics and Engineering.* Hoboken: Wiley.

117 Tsai LW (2000). *Mechanism Design: Enumeration of Kinematic Structures According to Function,* 1st Edition. Boca Raton: CRC Press.

118 Vinogradov O (2000). *Fundamentals of Kinematics and Dynamics of Machines and Mechanisms,* 1st Edition. Boca Raton: CRC Press.

119 Uicker J, Pennock G, Shigley J (2010). *Theory of Machines and Mechanisms,* 4th Edition. Oxford: Oxford University Press.

120 Uicker JJ, Ravani B, Sheth PN (2013). *Matrix Methods in the Design Analysis of Mechanisms and Multibody Systems,* 1st Edition. Cambridge: Cambridge University Press.

121 Vepa R (2009). *Biomimetic Robotics: Mechanisms and Control,* 1st Edition. Cambridge: Cambridge University Press.

122 Wittenburg J (2016), *Kinematics: Theory and Applications,* 1st Edition. New York: Springer.

123 Wu Y, Li S, Liu S, Qian Z (2013). *Vibration of Hydraulic Machinery.* Dordrecht: Springer.

124 You Z, Chen Y (2011). *Motion Structures: Deployable Structural Assemblies of Mechanisms,* 1st Edition. Boca Raton: CRC Press.

125 Zhao Z, Feng Z, Chu F, Ma N (2013). *Advanced Theory of Constraint and Motion Analysis for Robot Mechanisms,* 1st Edition. Oxford: Academic Press.

Index

Classical and Modern Approaches in the Theory of Mechanisms, First Edition.
Nicolae Pandrea, Dinel Popa and Nicolae-Doru Stănescu.
© 2017 John Wiley & Sons Ltd. Published 2017 by John Wiley & Sons Ltd.
Companion Website: www.wiley.com/go/pandmech17